Introductory Statistics

FOR BUSINESS AND ECONOMICS

SECOND EDITION

WILEY SERIES IN PROBABILITY AND MATHEMATICAL STATISTICS

ESTABLISHED BY WALTER A. SHEWHART AND SAMUEL S. WILKS

Editors

Ralph A. Bradley   David G. Kendall
J. Stuart Hunter   Geoffrey S. Watson

PROBABILITY AND MATHEMATICAL STATISTICS

APPLIED PROBABILITY AND STATISTICS

BENNETT and FRANKLIN • Statistical Analysis in Chemistry and the Chemical Industry

BHAT • Elements of Applied Stochastic Processes

BLOOMFIELD • Fourier Analysis of Time Series: An Introduction

BOX and DRAPER • Evolutionary Operation: A Statistical Method for Process Improvement

BROWN and HOLLANDER • Statistics: A Biomedical Introduction

BROWNLEE • Statistical Theory and Methodology in Science and Engineering, *Second Edition*

BURY • Statistical Models in Applied Science

CHATTERJEE and PRICE • Regression Analysis by Example

CHERNOFF and MOSES • Elementary Decision Theory

CHOW • Analysis and Control of Dynamic Economic Systems

CLELLAND, deCANI, BROWN, BURSK, and MURRAY • Basic Statistics with Business Applications, *Second Edition*

COCHRAN • Sampling Techniques, *Third Edition*

COCHRAN and COX • Experimental Designs, *Second Edition*

COX • Planning of Experiments

COX and MILLER • The Theory of Stochastic Processes, *Second Edition*

DANIEL • Application of Statistics to Industrial Experimentation

DANIEL and WOOD • Fitting Equations to Data

DAVID • Order Statistics

DEMING • Sample Design in Business Research

DODGE and ROMIG • Sampling Inspection Tables. *Second Edition*

DRAPER and SMITH • Applied Regression Analysis

DUNN and CLARK • Applied Statistics: Analysis of Variance and Regression

ELANDT-JOHNSON • Probability Models and Statistical Methods in Genetics

FLEISS • Statistical Methods for Rates and Proportions

GNANADESIKAN • Methods for Statistical Data Analysis of Multivariate Observations

GOLDBERGER • Econometric Theory

GROSS and CLARK • Survival Distributions

GROSS and HARRIS • Fundamentals of Queueing Theory

GUTTMAN, WILKS, and Hunter • Introductory Engineering Statistics, *Second Edition*

HAHN and SHAPIRO • Statistical Models in Engineering

HALD • Statistical Tables and Formulas

HALD • Statistical Theory with Engineering Applications

HARTIGAN • Clustering Algorithms

HILDEBRAND, LAING and ROSENTHAL • Prediction Analysis of Cross Classifications

HOEL • Elementary Statistics, *Fourth Edition*

HOLLANDER and WOLFE • Nonparametric Statistical Methods

HUANG • Regression and Econometric Methods

JAGERS • Branching Processes with Biological Applications

JOHNSON and KOTZ • Distributions in Statistics
       Discrete Distributions
       Continuous Univariate Distributions-1
       Continuous Univariate Distributions-2
       Continuous Multivariate Distributions

JOHNSON and KOTZ • Urn Models and Their Application

JOHNSON and LEONE • Statistics and Experimental Design: In Engineering and the Physical Sciences, Volumes I and II, *Second Edition*

KEENEY and RAIFFA • Decisions with Multiple Objectives

LANCASTER • The Chi Squared Distribution

LANCASTER • An Introduction to Medical Statistics

FACTS ON PROBABILITY AND STATISTICS

Introductory Statistics
FOR BUSINESS AND ECONOMICS

SECOND EDITION

Thomas H. Wonnacott
University of Western Ontario

and Ronald J. Wonnacott
University of Western Ontario

JOHN WILEY AND SONS
Santa Barbara · New York · London · Sydney · Toronto
A Wiley/Hamilton Publication

Library of Congress Cataloging in Publication Data:
Wonnacott, Thomas H.   1935–
   Introductory statistics for business and economics.

   (Wiley series in probability and mathematical statistics)
   "A Wiley/Hamilton publication."
   1. Statistics. I. Wonnacott, Ronald J., joint author. II. Title.
HA29.W622   1977   519.5   76-55773
ISBN 0–471–95980–4

Printed in the United States of America.

10 9 8 7 6 5 4

This book was copyedited by Naomi
Steinfeld and set in 11 point
Baskerville by Typothetae. The
cover was designed by Tri-Arts
and printing and binding was by
Halliday Lithograph Corporation.
Chuck Pendergast and Jean Varven
supervised production.

To our parents

ABOUT THE AUTHORS

Thomas H. Wonnacott
Ronald J. Wonnacott

The authors studied mathematics, statistics, and economics together as undergraduates at the University of Western Ontario. Ron then went on to a Ph.D. in economics at Harvard, and Tom a Ph.D. in mathematical statistics at Princeton. They have taught at Wesleyan University, the University of Minnesota, Duke University, and the University of California at Berkeley, and now both teach at the University of Western Ontario.

Ron has written books on Canadian-U.S. trade, published by North-Holland and Harvard University Press (with another brother, Paul). Tom has written *Calculus*. Together they have written *Econometrics,* also published by Wiley.

The hobbies they share include skiing, tennis, and music. In addition, Ron plays 7 handicap golf while Tom enjoys touch football.

PREFACE

This is a two-semester introduction to statistics for students in business, economics, and social science. It also is designed so that the first eight to twelve chapters can be used in a one-semester course. Our objective is to introduce students to most of the underlying logic that normally is available only in much more advanced mathematical statistics texts; for example, how sampling and inference are based on the theory of probability and random variables. But we use the simplest possible mathematics consistent with a sound presentation. Fortunately, this is only slightly more difficult than the mathematics in most elementary books.

TO THE STUDENT

Statistics is the fascinating study of how you can describe an unknown world by opening only a few windows on it. You will discover the excitement of thinking in a way you have never thought before.

This book is not a novel, and it cannot be read that way. Although each student has his or her own way of mastering a textbook, we present a method that is recommended by many successful students. Read a section carefully, underlining the most important sections—but don't get held up by difficult details. Whenever you come to a numbered example in the text, try first to answer it by yourself. Only after you have given it hard thought (and hopefully solved it) should you consult the solution that we provide. Then do the problems that follow. It is this problem solving that will draw you back into the text and highlight the most important concepts. Since the solution to one problem often clarifies the next, remember the important student's rule: "Don't get behind in your problem-solving courses." We have kept the problems as computationally simple as possible, so that you can concentrate on the concepts. At the same time, we have tried to make them realistic by frequently using real data.

Now for the traffic signs: problems are starred (*) if they are more difficult, hence optional. They are set with an arrow ($\Rightarrow$) if they introduce important ideas taken up later in the text. They are in parentheses ( ) if they duplicate previous problems and thus provide optional exercise only.

An outline of how various concepts are related is provided inside the front cover; in our experience, this overview is very useful both for review and for future reference. A glossary of symbols is given following the Table of Contents.

TO THE INSTRUCTOR

Since a basic objective of this book is to explain statistical concepts as simply as possible, the only prerequisite is high-school mathematics. Calculus appears only in the few sections where the argument is difficult to develop without it; but even those sections are designed so that a student without calculus also can follow. Similarly, problems and examples, rather than abstract theory, are used to introduce new material; the necessary theory is presented only after the student has gained a clear, intuitive idea of what is happening.

Another major objective is to show the logical relation between topics that often have appeared in texts as separate and isolated chapters. A few examples are the equivalence of: interval estimation and hypothesis testing; the t test and F test; and analysis of variance and regression using dummy variables. In every case, our motivation has been twofold: to help the student appreciate the underlying logic, and to help him arrive at answers to practical problems.

We have placed high priority on the regression model, not only because regression is widely regarded as the most powerful tool of the practicing statistician but also because it provides a good focal point for understanding such related techniques as correlation, analysis of variance, and chi square. We also have placed considerable emphasis on extending simple regression to include multiple regression and nonlinear regression.

A natural question is: how does this book compare with our briefer *Introductory Statistics* (third edition)? Since this present book is designed for students with more concentrated interests, it contains the same material plus five special-interest chapters at the end.

While we have worked within the constraint of keeping the treatment concise and simple enough for beginning students, it has still been possible to provide modern material that often is omitted in other elementary texts. For example, the economists' theory of revealed preference is very simply introduced in order to provide a much more profound understanding of what index numbers are all about. Similarly, the time-series chapter includes not only the standard topics (such as seasonal and trend adjustment) but also a simplified discussion of autocorrelated error and what to do about it, and an introduction to spectral analysis. There is also a chapter on estimating simultaneous equations—a very important topic for both business and economics, and like spectral analysis, one that previously has been available only in much more advanced books and articles. Originally, we felt that these topics could not possibly be introduced at this elementary level; however, by teaching them in an increasingly simplified way over several years, we hope we have been able to remove much of the academic mystery that has kept them beyond the reach of undergraduates in the past.

This book is designed for maximum flexibility. Basic classical statistics are presented in the first fifteen chapters, while the last ten chapters are self-contained topics, including the ones mentioned above plus non-parametric statistics, chi square, sampling designs, Bayesian estimation, and game theory. The instructor can choose any combination of these last ten chapters that he judges suitable to round out his course. While the text is kept simple, the more sophisticated interpretations and developments are reserved for footnotes and starred sections. In all instances these are optional; a special effort has been made to allow the more elementary student to skip these completely without losing continuity. Moreover, some of the finer points (along with solutions to the problems) are included in the student's manual and instructor's manual. Thus, the level as well as the content of the course can be tailored to the student's background.

THIS NEW EDITION

This edition includes a substantial reworking of the previous edition, with new examples and problems that frequently use real data. The material on Bayesian statistics has been expanded into two chapters as well as reworked, including Bayesian estimation in a 0–1 population and in regression. Hypothesis testing has been completely rewritten in order to make it simpler, and to introduce topics in the order of their importance. Thus, there are now several points at which this chapter can be left in order to move on to other topics. This leaves the instructor free to deal quickly with hypothesis testing as a corollary procedure to interval estimation, or to give it a more extended and traditional "center-stage" treatment.

The final major change has been to illustrate sampling theory by Monte Carlo simulation, so that students can watch sampling distributions actually building up; in teaching this, we have found it provides an exciting and enlightening supplement to the analytical derivation.

ACKNOWLEDGEMENTS

So many people have contributed that it is impossible to thank all of them. Special thanks go, without implication, to all those cited in the previous edition, and to the following for their generous help: Jon Baskerville, David Brillinger, Robin Carter, Peter Chinloy, Howard d'Abrera, Franklin Fisher, Joseph Gastwirth, Cornelius Groot, Winston Klass, John Powelson, D. W. G. Treloar, and Roger Trudel. Finally, we warmly thank the many instructors and students who have forwarded suggestions to us based on their experience with the first edition.

Thomas H. Wonnacott
London, Ontario, Canada, 1976   *Ronald J. Wonnacott*

CONTENTS

Glossary of Symbols

SYMBOL	MEANING	DEFINITION OR OTHER IMPORTANT REFERENCE
(a) ENGLISH LETTERS		
a	sample regression intercept, $= \hat{\alpha}$	Figure 11-7, (11-13)
ANOCOVA	analysis of covariance	Figure 13-13
ANOVA	analysis of variance	Table 10-5
b	sample regression slope, $= \hat{\beta}$	Figure 11-7, (11-16)
BES	best easy systematic estimator	(7-14), (7-15)
BLUE	best linear unbiased estimator	Figure 12-3
c	number of columns in two-way table	Table 10-9(a), (17-12)
	or center of a unimodal, symmetric distribution	Figure 7-5(a), Figure 16-5
C	constant coefficient in a contrast	(10-24)
C^2	modified chi-square variable	(8-50), (21-16)
d	statistic used to test randomness	(16-29)
d.f.	degrees of freedom	(8-17)
D	difference in two matched observations	(8-37) Table 16-1
	or Durbin-Watson statistic	(21-38)
e	regression error	(12-3)
E	(also F, G, etc.) $=$ event	(3-8)
$\overline{E}$	not E	(3-20)
$E(\ )$	expected value, $= \mu$	(4-17c), (5-32)

SYMBOL	MEANING	REFERENCE
E_i	expected value in ith category	Table 17-1
F	variance ratio	(10-11), Table 10-5, (14-21), (14-22)
H_0	null hypothesis	(9-9), (9-13)
H_1	alternative hypothesis	(9-10), Table 9-1
iff	if and only if	(3-29)
IV	instrumental variable(s)	(22-17)
L	average loss	(20-4), Table 25-3
$L(\ )$	likelihood function	Table 18-1
MAD	mean absolute deviation	(2-4a)
MLE	maximum likelihood estimate(tion)	Table 18-1
MSD	mean squared deviation	(2-5a), (7-3)
MSS	mean sum of squares	Table 10-5
MSSD	mean squared successive difference	(16-26)
n	sample size	(6-7), (6-10)
N	population size	(6-27)
$N(\ ,\ )$	normal distribution, with specified mean and variance	(19-27)
O_i	observed value in ith category	Table 17-1
OLS	ordinary least squares	(11-8), (22-14)
P	sample proportion	(1-2), (6-17), (6-18), (6-19)
$\Pr(E)$	probability of event E	(3-12)
$\Pr(E/F)$	conditional probability of E, given F	(3-26)
$p(x)$	probability distribution of X	(5-5)
$p(x,y)$	joint probability distribution of X and Y	(5-2)
$p(x/y)$	conditional probability distribution of X, given $Y = y$	(5-10)
r_0, r_1	regrets in Bayesian testing	(20-23), (20-24)
r	simple correlation or number of rows in two-way table	(14-4), (14-14) Table 10-9(a) (17-12)

SYMBOL	MEANING	REFERENCE
r^2	coefficient of determination	(14-29)
$r_{XY/Z}$	partial correlation of X and Y, if Z were held constant	(14-32)
R	multiple correlation	(14-33), (14-34),
	or	(14-39)
	statistic in runs test	(16-24)
s^2	variance of sample, or residual variance	(2-6a), Table 10-9(a) (12-23)
s_p^2	pooled variance of samples	(8-34), (10-6), Table 10-5(a)
$s_{\hat{\beta}}$	standard error of $\hat{\beta}$	(12-25), (12-26), (13-9), (13-10)
s_{XY}	sample covariance of X and Y	(14-5), (22-5)
S	number of successes	(4-7), (6-23), Example 6-10
SS	sum of squares (variation)	Table 10-5
t	student's t variable	(8-16), (9-15), (12-35), (14-22)
var	variance, $= \sigma^2$	(4-4), (5-40)
V	variance spectrum	(21-12), (21-18)
w_i	weights of strata	Table 24-2
W	weighted sum	(5-30)
	or Wilcoxon-Mann-Whitney test statistic	(16-18)
X	(also Y, V, W, etc.) = random variable	(4-1), Figure 4-1
	or regressor in original form	(11-5)
x	(also y, v, etc.) = (realized) value of X	Figure 4-1
	or regressor in terms of deviations from the mean	(11-5)
$\overline{X}$	sample mean of X (note this is a different usage than $\overline{E}$)	(2-1a), (6-10)
Y	independent variable (response) in regression	Table 11-1, (12-1)

SYMBOL	MEANING	REFERENCE
$\hat{Y}$	fitted value of Y	Figure 11-4, Table 11-2
Z	standard normal variable or a second regressor	(4-16), (8-15) Table 13-1

(b) GREEK LETTERS are generally reserved for population *parameters*. A hat, for example $\hat{\alpha}$, means a sample *estimator* of α.

α	probability of type I error or population regression intercept	Table 9-1 (12-1)
β	probability of type II error or population regression slope	Table 9-1 (12-1)
γ	population regression coefficient	(13-1)
Δ	population mean difference	(8-37), (16-38)
θ	any population parameter, or state of nature	(7-1) Table 20-1
μ	population mean	(4-3), (4-17a), (5-32)
μ_0	regression mean at X_0 or mean of prior distribution or population mean, assuming null hypothesis (H_0)	Figure 12-6, (12-40) (19-27) (9-9), (20-31)
μ_1	population mean, assuming alternative hypothesis (H_1)	(9-10), (20-31)
ν	population median	(16-12)
π	population proportion	(1-2), (8-40), (19-17)
Π	product of	(18-15)
Φ	population variance spectrum	(21-18)
ρ	population correlation coefficient, or serial correlation of time series error	(14-3), Figure 14-9, (21-30)
ρ_{XY}	population correlation of X and Y, $= \rho$	(5-25), (5-27), (14-3)
σ	population standard deviation	(4-4)
σ^2	population variance	(4-4), (4-5), (5-40)
σ_{XY}	population covariance of X and Y	(5-23), (5-24)
Σ	sum of	(2-1a)
χ^2	chi-square variable	(17-2), (17-10)

SYMBOL	MEANING	REFERENCE

(c) OTHER MATHEMATICAL SYMBOLS

SYMBOL	MEANING	REFERENCE
$G \cup H$	G or H, or both	(3-13), (3-17)
$G \cap H$	G and H	(3-14), (3-27)
$\overset{\Delta}{=}$	equals, by definition	(2-1a)
$\simeq$	approximately equals	(1-3)
$\sim$	is distributed as	(19-27)
*	optional, difficult section or problem	
( )	easy problem, for drill	
$\Rightarrow$	important problem	

PART I

Basic Probability and Statistics

CHAPTER 1

Introduction

He uses statistics as a drunken man uses lampposts—for support rather than for illumination.

Andrew Lang

The word "statistics" originally meant the collection of population and economic information vital to the state. From that modest beginning, statistics has grown into a scientific method of analysis that now is applied to all the social and natural sciences. The present aims and methods of statistics are best illustrated by a familiar example.

1-1   EXAMPLE

Before every presidential election, the pollsters try to pick the winner; specifically, they try to guess the proportion of the population that will vote for each candidate. Clearly, canvassing all voters would be an impossible task. As the only alternative, pollsters survey a sample of a few thousand in the hope that the sample proportion will constitute a good estimate of the total population proportion. This is a typical example of *statistical inference* or *statistical induction:* the (voting) characteristics of an unknown population are inferred from the (voting) characteristics of an observed sample.

As any pollster will admit, it is an uncertain business. To be *sure* of the population, we have to wait until election day, when all votes are counted. Yet if the sampling is done fairly and adequately, we can have high hopes that the sample proportion will be close to the population proportion. This will allow us to estimate the unknown population proportion π from the observed sample proportion P, as follows:

$$\pi = P \pm \text{ a small error} \tag{1-1}$$

with crucial questions being, "How small is this error?" and "How sure are we that we are right?" Since this typifies the very core of the book, we state the precise formula (taken from Chapter 8, where it is explained in detail):

If the sampling is random, we can state with 95% confidence that:

$$\boxed{\pi = P \pm 1.96\sqrt{\frac{P(1 - P)}{n}}} \tag{1-2}$$

where π and P are the population and sample proportion, and n is the sample size.

Before we illustrate this formula with an example, we repeat the warning that we gave in the preface: Every numbered example in this text is an exercise that you should actively work out yourself, rather than passively read. We therefore put each example in the form of a question for you to answer; if you get stuck, then you may read the solution. But in all cases remember that *statistics is not a spectator sport.* You cannot learn it by watching, any more than you can learn to ride a bike by watching. You have to jump on and take a few spills.

Example 1-1. Just before the 1972 presidential election, a Gallup poll[1] of 2,000 voters showed 760 for McGovern and 1,240 for Nixon. Calculate the 95% confidence interval for the population proportion π that voted for McGovern.

Solution. The sample size n is 2000, and the sample proportion is:

$$P = \frac{760}{2000} = .38$$

Substitute these into (1-2):

$$\pi = .38 \pm 1.96\sqrt{\frac{.38(1 - .38)}{2000}}$$

$$\pi \simeq .38 \pm .02 \tag{1-3}$$

That is,[2] with 95% confidence, the proportion for McGovern among the whole population of voters was between 36% and 40%.

Remarks. In the actual election, the proportion of the voting population who chose McGovern was 38.2% (neglecting third-party candidates). This is well within the confidence interval.

---

[1]Reconstructed from the *Gallup Opinion Index,* August 1974.

[2]In equation (1-3), $\simeq$ means "approximately equals." A glossary of such symbols may be found in front.

Constructing confidence intervals like (1-3) will be our major objective in this book. Another related objective is to test hypotheses. To use the same example, suppose that an ardent Democrat claimed that McGovern would win the election. In mathematical terms, this hypothesis may be written: $\pi > .50$. On the basis of the information in equation (1-3) we would reject this hypothesis, of course. In general, there is a very close association of this kind between confidence intervals and hypothesis tests.

We can make several other crucial observations about equation (1-2).

1. The estimate is *not* made *with certainty;* we are only 95% confident. We must concede the possibility that we are wrong—simply because we were unlucky enough to draw a misleading sample. For example, if less than half the population is in fact Democratic, it is still possible, although unlikely, for us to run into a string of Democrats in our sample. In such circumstances, our confidence interval (1-2) would be dead wrong. Since this sort of bad luck is possible but not likely, we are just 95% confident.

2. As sample size n increases, we note that the error allowance in (1-2) decreases. In Example 1-1, if Gallup increased the sample to 10,000 voters and continued to observe a Democratic proportion of .38, the 95% confidence interval would become more precise:

$$\pi \simeq .38 \pm .01 \tag{1-4}$$

This also is intuitively correct: a larger sample contains more information, and hence allows a more precise conclusion.

3. Suppose that we feel that 95% confidence is not good enough, and that instead we want to be 99% sure of our conclusion. If the additional resources for further sampling are not available, then we can increase our confidence only by making a less precise statement. As we will show in Chapter 8, for 99% confidence the formula (1-2) must have the coefficient 1.96 enlarged to 2.58; this yields the 99% confidence interval:

$$\pi \simeq .38 \pm .03$$

This is broader and less precise than the 95% confidence interval (1-3); we must be less precise because we wish to be more certain of being right. In any case, we note that any statistical statement must be prefaced by *some* degree of uncertainty.

1-2   DEDUCTION AND INDUCTION

Figure 1-1 illustrates the difference between deductive and inductive reasoning. Deduction in panel (a) involves arguing from the general to the specific—i.e., from the population to the sample. Induction in panel (b) is the reverse—arguing from the specific to the general, i.e., from the sample to the population.[3] Equation (1-1) represents inductive reasoning; we are arguing from a sample proportion to a population proportion. But this is possible only if we study the simpler problem of deduction first. Specifically, in Equation (1-1) the *inductive* statement (that the population proportion can be inferred from the sample proportion) is based on a *prior deduction* (that the sample proportion is likely to be close to the population proportion).

Chapters 3 through 7 are devoted to deduction. This involves probability theory, leading up to such questions as, "With a given population, how will a sample behave? Will the sample be on target?" Only when this deductive issue is resolved can we move to questions of statistical inference in later chapters. There, we turn the argument around and ask, "From a

---

[3]To keep these terms straight, remember that the population is the point of reference. The prefix *de* means "away from." Thus *de*duction is arguing away from the population. The prefix *in* means "into" or "towards." Thus *in*duction is arguing towards the population. Finally, statistical *in*ference is based on *in*duction.

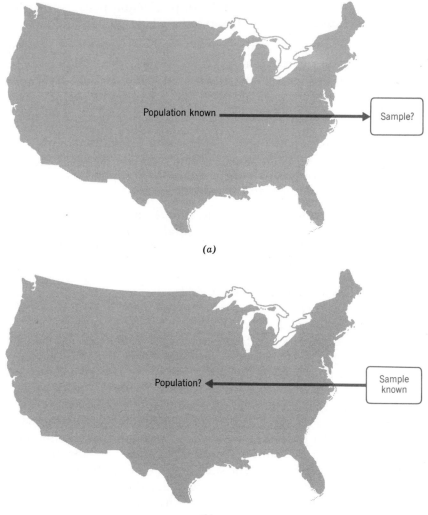

FIGURE 1–1 Deduction and Induction contrasted. (a) Deduction (probability). (b) Induction (statistical inference).

given observed sample, what can we conclude about the unknown population?"

1-3   SAMPLING—WHY AND HOW?

We draw a sample, rather than examine the whole population, for several reasons:

1. *Limited resources.* For example, in preelection polls, neither funds nor time are available to observe the whole population.

2. *Scarcity.* Sometimes only a small sample is available. For example, in heredity versus environment controversies, identical twins provide ideal data because they have identical heredity. Yet very few such twins are available.

There are many examples in business. An allegedly more efficient machine may be introduced for testing, with a view to the purchase of additional similar units. The manager of quality control simply cannot wait around to observe the entire population that this machine will produce. Instead, a sample run is observed, and the decision on efficiency is based on an inference from this sample.

3. *Destructive testing.* For example, suppose that we wish to know the average life of all the light bulbs produced by a certain factory. It would be absurd to insist on observing the whole population of bulbs until they burn out.

If sampling is required, how should it be done? In statistics, as in business or any other profession, it is essential to distinguish between bad luck and bad management. For example, suppose that someone bets you $100 at even odds that a fair die will turn up 6. You accept the challenge, but the die happens to turn up 6 after all, and he wins. He has merely overcome his bad management with good luck. Your only defense against this combination is to get him to keep playing the game—with your dice.

If we now return to our original example of preelection polls, we note that the sample proportion of Democrats may misrepresent the population proportion for either of these reasons. No matter how well-managed and designed our sampling procedure may be, we may be unlucky enough to turn up a Democratic sample from a Republican population. Equation (1-2) relates to this case; it is assumed that the only complication is the luck of the draw, and not mismanagement. From that equation we confirm that the best defense against bad luck is to "keep playing"; by increasing our sample size, we improve the reliability of our estimate.

The other problem is that sampling can be badly mismanaged or biased. For example, in sampling a population of voters, it is a mistake to take their names from a phone book, since poor voters who often cannot afford telephones are badly underrepresented.

Other examples of biased samples are easy to find. Informal polls of people on the street often are biased because the interviewer tends to select people who seem civil and well-dressed; a surly worker or harassed mother is overlooked. A congressman cannot rely on his mail as an unbiased sample of his constituency, since this is a sample of people with strong opinions and includes an inordinate number of cranks and members of pressure groups.

The simplest way to ensure an unbiased sample is to give each member of the population an equal chance of being included in the sample. This, in fact, is essentially the definition of a "random" sample.[4] For a sample to be random, it cannot be chosen in a sloppy or haphazard way; it must be designed carefully. One possibility is to number all the individuals in the population, and draw the sample by using a chance device such as a bowlful of numbered chips, a roulette wheel, or the random digits in Appendix Table II(a). If a sample is random, not only will it be free of bias, but it also will satisfy the assumptions of probability theory, and allow us to make scientific inferences of the form (1-2).

In some circumstances, the only available sample will be a non-random one. While probability theory often cannot be applied strictly to such a sample, it still may provide the basis for a good educated guess— or what we might term the *art* of inference. Although this art is very important, it cannot be taught in an elementary text; we therefore consider only scientific inference based on the assumption that samples are random. The techniques for ensuring this are discussed further in Chapter 6.

PROBLEMS

1-1 Project yourself back in time to four recent U.S. presidential elections (with the Gallup preelection poll of 1500 voters shown in each case in brackets):[5]

Year	Democrat	Republican
1960	Kennedy (51%)	Nixon (49%)
1964	Johnson (64%)	Goldwater (36%)
1968	Humphrey (50%)	Nixon (50%)
1972	McGovern (38%)	Nixon (62%)

(a) In each case, construct a 95% confidence interval for the proportion of Democratic supporters in the population.

(b) In which case would you be prepared to predict a Democratic majority? predict a Republican majority? unprepared to predict either way? Compare your prediction with the actual voting results: Kennedy got 50.1% in 1960, Johnson got 61.3% in 1964, Humphrey got 49.7% in 1968, and McGovern got 38.2% in 1972.

---

[4] Strictly speaking, this is called "simple random sampling," to distinguish it from more complex types of random sampling. A more complete and mathematical definition is given in Section 6-1.

[5] From the *Gallup Opinion Index,* February 1976, or any other recent month, last page.

1-2 Criticize each of the following sampling plans, pointing out the bias and suggesting how to reduce it.

(a) Ten million ballots were mailed to a random sample of American voters found in telephone directories. On the basis of the two million ballots that were returned, a Republican victory was forecast.[6]

(b) A house-to-house survey of a city selected every corner house and asked the housewife, if she was in, which way she intended to vote on a municipal finance question.

(c) To estimate the average income of its alumni five years after graduation, a university polled all the alumni who returned to their fifth reunion.

---

[6]A poll similar to this was actually carried out in 1936 by a prominent American magazine called the *Literary Digest,* which accordingly predicted a Republican victory. The Democratic candidate, Roosevelt, won 60% of the votes in the election. The next year the *Literary Digest* folded.

CHAPTER 2

Descriptive Statistics for Samples

Figures won't lie, but liars will figure.

General Charles H. Grosvenor

We already have discussed the primary purpose of statistics—to make an inference from a sample to the whole population. As a preliminary step, the sample must be simplified and reduced to a few descriptive numbers, called sample *statistics*.

For instance, in Example 1-1, the pollster would record the answers of the 2000 people in the sample, obtaining a sequence such as $D\ D\ R\ D\ R$ . . . , where D and R represent Democrat and Republican. The best way of describing this sample by a single number is the statistic P, the sample proportion of Democrats; this can be used to make an inference about π, the population proportion. Admittedly, this statistic P was trivial to compute. It merely required counting the number of Democrats (760), then dividing by the sample size (2000).

We next consider the more substantial computations of statistics for summarizing two other samples:

(a) The number of children in a sample of 50 American families.

(b) Heights in a sample of 200 American men.

2-1   FREQUENCY TABLES AND GRAPHS

(a) Discrete Example

In a sample of 50 U.S. families, let us record the number of children, X, which takes on the values 0, 1, 2, 3, ... We call X a "discrete" random variable because it can take on only a finite number of values.[1] Suppose that the 50 values of X turn out to be:

$$0,2,2,3,5,1,2,0, \ldots\ldots\ldots\ldots\ldots\ldots\ldots 4,2.$$

To simplify, we keep a running tally of each of the possible outcomes in Table 2-1. In column (3) we record, for example, that 13 is the frequency (f) that we observed for a two-child family. That is, we obtained this outcome on 13/50 of our sample observations; this proportion (.26 or 26%)[2] is called relative frequency (f/n), and is recorded in the last column.

TABLE **2-1**

Calculation of the Frequency and Relative Frequency of the
Number of Children in a Sample of 50 American Families

(1) Number of Children	(2) Tally	(3) Frequency (f)	(4) Relative Frequency $\left(\dfrac{f}{n}\right)$
0	卌 卌 卌	15	.30
1	卌 卌	10	.20
2	卌 卌 |||	13	.26
3	卌		6	.12
4	|||	3	.06
5	|||	3	.06

$$\sum f = 50 = n \qquad \sum\left(\frac{f}{n}\right) = 1.00$$

where Σ means "the sum of." Thus, for example, Σf means "the sum of the frequencies."

---

[1] X still is called discrete if it can take on a countably infinite number of values. It then still is possible to sum over all the possible values (rather than integrate, as is required for continuous random variables).

[2] Throughout this book, relative frequencies (and probabilities) will be expressed as either decimals or percentages, whichever seems more convenient at the time.

Usually, mathematicians prefer decimals, while applied statisticians prefer percentages. Therefore, we usually do our calculations in decimal form and usually give the verbal interpretations in percentage form.

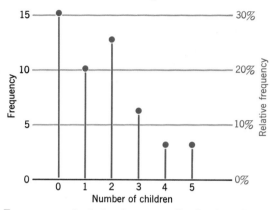

FIGURE 2–1 Frequency and relative frequency distribution of the number of children in a sample of 50 American families

The information in column (3) is called a "frequency distribution," and is graphed in Figure 2-1. The "relative frequency distribution" in the last column could be graphed similarly; note that the two graphs are identical except for the vertical scale. Hence, a simple change of vertical scale transforms Figure 2-1 into a relative frequency distribution.

(b) Continuous Example

Suppose that we take a sample of 200 men, each of whose height is recorded in inches. We call height X a "continuous" random variable, since an individual's height might be any value, such as 64.328 inches.[3] It no longer makes sense to talk about the frequency of this specific value of X, since never again will we observe anyone who is exactly 64.328 inches tall. Instead we can tally the frequency of heights within a class or cell (e.g., 58.5″ to 61.5″), as in Table 2-2. Then the frequency and relative frequency are tabulated, as before.

We have chosen the cells somewhat arbitrarily, but with the following conveniences in mind:

---

[3]We shall overlook the fact that although height is conceptually continuous, in practice the measured height is rounded to a few decimal places at most, and is therefore discrete. In this case we sometimes find observations occurring right at the cell boundary. Where should they be placed—in the cell above or the cell below?

One of the best solutions is systematically to move the first such borderline observation up into the cell above, move the second such observation into the cell below, the third up, the fourth down, etc. This procedure avoids the bias that would occur if, for example, all borderline observations were moved to the cell above.

TABLE 2-2

Frequency, and Relative Frequency of the Heights of 200 Men

Cell No.	(1) Cell Boundaries	(2) Cell Midpoint	(3) Tally	(4) Frequency, f	(5) Relative Frequency $\dfrac{f}{n}$
1	58.5–61.5	60	||	2	.01
2	61.5–64.5	63	⊞ ⊞	10	.05
.	.	66	.	48	.24
.	.	69	.	64	.32
.	.	72	.	56	.28
.	.	75	.	16	.08
7	76.5–79.5	78	||||	4	.02

$$\sum f = 200 = n \qquad \sum \frac{f}{n} = 1.00$$

1. The number of cells is a reasonable compromise between too much detail and too little. Usually, 5 to 15 cells is appropriate.

2. Each cell midpoint, which hereafter will represent all observations in the cell, is a convenient whole number.

The grouping of the 200 observations into cells is illustrated in Figure 2-2(a), where each observation is represented by a dot. The grouped data then are graphed in Figure 2-2(b). We use bars instead of lines to represent frequencies as a reminder that the observations occurred throughout the cell, and not just at the midpoint. Such a graph is called a *bar diagram* or *histogram.*

We next turn to the question of how we may characterize a sample frequency distribution with a single descriptive number (statistic). There are two very useful concepts: the first is the center of the distribution, and the second is the spread. These concepts will be illustrated with the continuous distribution of men's heights; but their application to discrete distributions (such as family size) is even more straightforward, and will be dealt with in the exercises.

2-2   CENTER OF A DISTRIBUTION

There are many different ways to measure the center of distribution. Three of these—the mode, the median, and the mean—are discussed below, starting with the simplest.

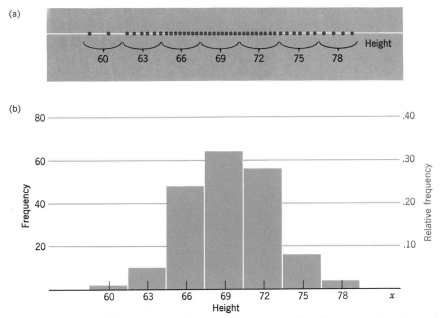

FIGURE 2–2 (a) The grouping of observations into cells, illustrating the first two columns of Table 2-2. (b) The bar graph for the grouped data.

(a) The Mode

Since mode is the French word for fashion, the mode of a distribution is defined as the most frequent (fashionable) value. In the example of men's heights, the mode is 69 inches, since this cell has the greatest frequency, or highest bar in Figure 2-2. Generally, the mode is *not* a good measure of central tendency, since often it depends on the arbitrary grouping of the data. (In Problem 2-5, we note that by redefining cell boundaries, we can shift the mode up or down considerably.) We also can draw a sample in which the largest frequency (highest bar in the group) occurs twice; this unfortunate ambiguity is left unresolved, and the distribution is called "bimodal."

(b) The Median

The median is just the 50th percentile, i.e., the value below which 50% of the values in the sample fall. Since it splits the observations into two halves, it sometimes is called the middle value. In the sample of 200 detailed heights shown in Figure 2-2(a), the median (say, 69.3) easily is found by reading off the 100th value[4] from the left. But if the only information available is the grouped frequency distribution in Figure 2-2(b), the median can only be approximated, by choosing an appropriate value within the median cell.[5]

(c) The Mean, $\overline{X}$

This sometimes is called the arithmetic mean, or simply the average, and is the most common central measure. The original observations (X_1, $X_2, \ldots, X_n$) simply are summed, then divided by n. Thus we define:

$$\overline{X} \triangleq \frac{1}{n}(X_1 + X_2 + \cdots + X_n)$$

$$\boxed{\overline{X} \triangleq \frac{1}{n} \sum_{i=1}^{n} X_i} \qquad (2\text{-}1a)$$

where $\triangleq$ means "equals, by definition," and Σ (capital Greek sigma) is the customary mathematical notation for summation.

The average for the sample of heights could be computed by summing all 200 observations and dividing by 200. However, this tedious calculation can be greatly simplified by using the grouped data in Table 2-3. Let f_1 represent the number of observations in cell 1, where each observation may be approximated[6] by the cell midpoint, x_1. Similar approximations hold for all the other cells too, so that:

---

[4]Or 101st value. This ambiguity is best resolved by defining the median as the average of the 100th and 101st values. In a sample with an odd number of observations, this ambiguity does not arise.

[5]The median cell is clearly the fourth, since this leaves 30% of the observations below and 38% above. To get a specific value of the median that will leave 50% of the observations below, we must progress through this median cell far enough to pick up another 20% of the observations. Since this cell includes 32% of the observations, we move 20/32 of the way through it. Starting at the cell boundary 67.5, and remembering that the cell width is 3, we therefore approximate the median as:

$$67.5 + (\tfrac{20}{32})3 = 69.4$$

[6]In approximating each observed value by the midpoint of its cell, we sometimes err positively, sometimes negatively; but unless we are very unlucky, these errors will tend to cancel. Even in the unluckiest case, however, the overall error in the sample mean will be less than half the cell width. Note that cell midpoints are designated by small x_i, to distinguish them from the observed values X_i.

For a discrete distribution such as family size, however, there is no approximation necessary; then (2-1b) will be exactly true.

$$\overline{X} \simeq \frac{1}{n}\left\{ \underbrace{(x_1 + x_1 + \cdots + x_1)}_{f_1 \text{ times}} + \underbrace{(x_2 + x_2 + \cdots x_2)}_{f_2 \text{ times}} + \cdots \right.$$

$$\left. + \underbrace{(x_7 + \cdots x_7)}_{f_7 \text{ times}} \right\}$$

$$\overline{X} \simeq \frac{1}{n}\{f_1 x_1 + f_2 x_2 + \cdots f_7 x_7\}$$

$$\simeq \frac{f_1}{n} x_1 + \frac{f_2}{n} x_2 + \cdots \frac{f_7}{n} x_7$$

$$\simeq \sum_{i=1}^{7} \left(\frac{f_i}{n}\right) x_i$$

In general, for grouped data:

$$\boxed{\overline{X} \simeq \sum_{i=1}^{m} x_i \left(\frac{f_i}{n}\right)} \tag{2-1b}$$

where (f_i/n) = relative frequency in the ith cell, and m = number of cells. We number this equation (2-1b) to emphasize that it is the equivalent of (2-1a), appropriate for grouped data. Formula (2-1b) is used to calculate the mean height in column (3) of Table 2-3. We can think of this as a "weighted" average, with each x value weighted appropriately by its relative frequency.

(d) Comparison of Mean, Median, and Mode

These three measures of center are compared in Figure 2-3. In panel (a) we show a distribution that has a single peak and is symmetric (i.e., one half is the mirror image of the other); in this case, all three central measures coincide. But when the distribution is skewed to the right, as in (b), the median falls to the right of the mode; with the long scatter of observations strung out in the right-hand tail, we have to move from the mode to the right to pick up half the observations. Moreover, the mean generally will lie even further to the right, as we explain below.

(e) The Mean Interpreted as the Balancing Point

The 200 heights appear in Figure 2-2 as points along the X-axis. If we think of each observation as a one-pound mass, and the X-axis as a weight-

TABLE 2-3

Calculation of Mean and Variance of a Sample of 200 Men's Heights[a]

Given		Calculation of $\overline{X}$ Using (2-1b)	Calculation of MSD Using (2-5b) ($\overline{X}$ Rounded to 69)		
(1)	(2)	(3)	(4)	(5)	(6)
x_i	$\dfrac{f_i}{n}$	$x_i\left(\dfrac{f_i}{n}\right)$	$(x_i - \overline{X})$	$(x_i - \overline{X})^2$	$(x_i - \overline{X})^2\left(\dfrac{f_i}{n}\right)$
60	.01	.60	−9	81	.81
63	.05	3.15	−6	36	1.80
66	.24	15.84	−3	9	2.16
69	.32	22.08	0	0	0
72	.28	20.16	3	9	2.52
75	.08	6.00	6	36	2.88
78	.02	1.56	9	81	1.62

$$\overline{X} \simeq \sum x_i\left(\frac{f_i}{n}\right)$$

$$= 69.39$$

$$\text{MSD} \simeq \sum (x_i - \overline{X})^2\left(\frac{f_i}{n}\right)$$

$$= 11.79$$

Comparing (2-5b) and (2-6b):

$$s^2 = \text{MSD}\left(\frac{n}{n-1}\right)$$

$$= 11.79\left(\frac{200}{199}\right) = 11.85$$

$$s = \sqrt{11.85} = 3.44$$

[a]An easier coded method is shown later in Table 2-4.

less supporting rod, we might ask where this rod balances. Our intuition suggests "the center."

The precise balancing point, also called the center of gravity, is given by the physics formula:[7]

$$\frac{1}{n}\sum X_i$$

which is exactly the formula for the mean. Thus, we may think of the sample mean as the "balancing point" of the data, symbolized by ▲ in our graphs.

---

[7] Here and wherever else it will cause no ambiguity, we shall abbreviate by using plain Σ instead of $\overset{n}{\underset{i=1}{\Sigma}}$.

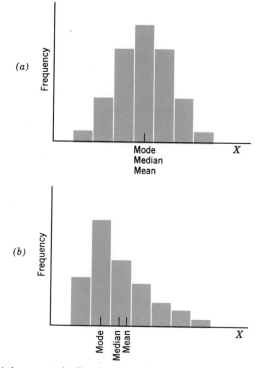

FIGURE 2–3 (a) A symmetric distribution with a single peak. The mode, median, and mean coincide at the point of symmetry. (b) A right-skewed distribution, showing mode < median < mean.

Now it easily can be seen why the mean lies to the right of the median in a right-skewed distribution, as shown in Figure 2-3(b). We can experiment by trying to balance at the median. Fifty percent of the observed values then lie on either side, but the observations on the right are further out and exert more leverage. Balance can be achieved only by placing the mean to the right of the median.

PROBLEMS

2-1 In a large American university in 1969, a random sample of 5 women professors gave the following annual salaries (in thousands of dollars, rounded):[8]

---

[8]From D. A. Katz, "Faculty Salaries, Promotions, and Productivity at a Large University," *American Economic Review,* June 1973. In many of the examples based on real data such as this, we have tried to preserve the reality yet keep the arithmetic manageable by using the following twist: the available data usually is a whole population, or at least a very large sample. We construct by simulation a small sample that has a matching mean and variance (where possible).

9, 12, 8, 10, 16

Without sorting into cells,
(a) Graph the salaries as dots on the x-axis.
(b) Calculate the average (mean), and mark it on the graph.

2-2 At the same university, a random sample of 25 men professors gave the following annual salaries (in thousands of dollars, rounded):

$$12, \quad 11, \quad 19, \quad 16, \quad 22, \qquad 8, \quad 13, \quad 16, \quad 17, \quad 15,$$
$$20, \quad 14, \quad 17, \quad 14, \quad 15, \qquad 26, \quad 9, \quad 20, \quad 16, \quad 18,$$
$$21, \quad 21, \quad 16, \quad 9, \quad 15.$$

Repeat Problem 2-1. Then, as you compare the graphs of the men and women, do you think there is good evidence that, over the whole university, on the average women earn less than men? (This issue will be answered more precisely in Chapter 8.)

2-3 In Problem 2-2, sort the data into cells with midpoints of 10, 15, 20, and 25. Then:
(a) Graph the relative frequency distribution.
(b) Calculate the approximate mean and mark it on the graph.

2-4 Sort the following daily profit figures of 25 newsstands into eight cells, whose midpoints are 55, 60, . . . , 90.

55.31   81.47   64.90   70.88   86.02   77.25   76.73   84.21   56.02

84.92   90.23   78.01   88.05   73.37   87.09   57.41   85.43

74.76   86.51   86.37   76.15   88.64   84.71   66.05   83.91

(a) Approximately what are the mean and mode?
(b) Graph the relative frequency distribution.

2-5 Sort the data of Problem 2-4 into four cells, whose midpoints are 60, 70, 80, 90. Then answer the same questions.

2-6 Summarize the answers to the previous two problems by completing the following table.

2-2 CENTER OF A DISTRIBUTION

	Mean	Mode
Original data (exact values)	77.78	Not defined
—Fine grouping (Problem 2-4)		
—Coarse grouping (Problem 2-5)		

(a) Why is the mode not a good measure?

(b) Which gives a closer approximation to the true ungrouped mean: the coarse or the fine grouping?

2-7 Explain in your own words the relative advantages of the mean, the median, and the mode.

2-8 In selling a house, a real estate agent stated that average income in the subdivision was $20,000; yet at a protest meeting of taxpayers from the subdivision (where he also is a resident), he stated that the typical income is only $12,000. Can this apparent contradiction be explained? Is he necessarily being dishonest?

2-9 The annual tractor production of a multinational corporation in seven different countries totals 105 thousand. Broken down by country, the figures are (in thousands):

$$6, \quad 8, \quad 6, \quad 9, \quad 11, \quad 5, \quad 60$$

(a) Graph the distribution, representing each figure as a dot on the x-axis.

(b) What is the mean production per country? the median? the mode? Mark these on the graph.

(c) For another corporation operating in ten countries, production (in thousands) has a mean of 7.8 per country, a median of 6.5, and a mode of 5.0. What is the total production of this corporation?

2-10 (a) Calculate the mean of the following five numbers:

$$3, \quad 7, \quad 8, \quad 12, \quad 15$$

(b) Calculate the deviations from the mean, $X_i - \overline{X}$. Then calculate the average of these deviations.

(c) Write down any four or five numbers. Calculate their mean, and then the average deviation from the mean.

(d) Prove that, for every possible sample, the average deviation from the mean is exactly zero.

2-11 (a) Two samples had means of $\overline{X}_1 = 100$ and $\overline{X}_2 = 110$. Then the samples were pooled into one large sample, for which the mean $\overline{X}$ was calculated. What is $\overline{X}$ if the sample sizes are:

    (i) $n_1 = 30,\quad n_2 = 70$?
    (ii) $n_1 = 50,\quad n_2 = 50$?
    (iii) $n_1 = 80,\quad n_2 = 20$?
    (iv) $n_1 = 15,\quad n_2 = 15$?

(b) Answer true or false; if false, correct it:

To generalize part (a),

$$\overline{X} = \left(\frac{n_1}{n_1 + n_2}\right)\overline{X}_1 + \left(\frac{n_2}{n_1 + n_2}\right)\overline{X}_2$$

Incidentally, $\overline{X}$ is called a *weighted* average of $\overline{X}_1$ and $\overline{X}_2$, where the weights (coefficients) sum to 1.

2-3   SPREAD OF A DISTRIBUTION

Although average height may be the most important single statistic, it also is important to know how spread out or varied the observations are. As with measures of center, we find that there are several measures of spread. We will start with the simplest.

(a) The Range

The range is simply the distance between the largest and smallest value:

$$\text{Range} \overset{\Delta}{=} \text{largest} - \text{smallest observation}$$

For men's heights, in Figure 2-2(b), the range is 21 (i.e., 79.5–58.5). It may be criticized fairly on the grounds that it tells us nothing about the distribution except where it ends. And using only these two observations may be very unreliable. We therefore turn to measures of spread that take account of all observations.

(b) Mean Absolute Deviation (MAD)

The average deviation, as its name implies, is found by calculating the deviation of each observation from the mean; these deviations $(X_i - \overline{X})$

then are averaged by summing and dividing by n. Although this sounds like a promising measure, in fact it is worthless; positive deviations always cancel negative deviations, leaving an average of zero.[9] This sign problem can be avoided by ignoring all negative signs and taking the average of the *absolute* values of the deviations:

$$\text{Mean Absolute Deviation, MAD} \triangleq \frac{1}{n} \sum_{i=1}^{n} |X_i - \overline{X}| \qquad (2\text{-}4a)$$

For grouped data, of course, each cell midpoint x_i must be counted according to the frequency in the cell (f_i times) so that:
For grouped data:

$$\text{MAD} \simeq \frac{1}{n} \sum_{i=1}^{m} |x_i - \overline{X}| f_i \qquad (2\text{-}4b)$$

(c) Mean Squared Deviation (MSD)

Although MAD intuitively is a good measure of spread, it is mathematically intractable.[10] We therefore turn to an alternative means of avoiding the sign problem—namely, squaring each deviation:

$$\boxed{\text{Mean Squared Deviation, MSD} \triangleq \frac{1}{n} \sum_{i=1}^{n} (X_i - \overline{X})^2 \qquad (2\text{-}5a)}$$

$$\text{For grouped data, MSD} \simeq \sum_{i=1}^{m} (x_i - \overline{X})^2 \left(\frac{f_i}{n}\right) \qquad (2\text{-}5b)$$

---

[9]Although this was proved in Problem 2-10, its proof is so important that we repeat it here:

$$\text{Average deviation} \triangleq \frac{1}{n} \sum (X_i - \overline{X}) \qquad (2\text{-}2)$$

$$= \frac{1}{n}\left(\sum X_i - n\overline{X}\right)$$

$$= \frac{1}{n}\sum X_i - \overline{X}$$

$$= \overline{X} - \overline{X} = 0$$

That is,

$$\text{average deviation} = 0 \qquad (2\text{-}3)$$

[10]One difficulty is the problem of differentiating the absolute value function.

(d) Variance and Standard Deviation

MSD is a good measure, provided that we only wish to describe the sample. But typically we shall want to go one step further and use this to make a statistical inference about the population. For this purpose it is better to use the divisor $n - 1$ rather than[11] n:

$$\text{Variance, } s^2 \triangleq \frac{1}{n - 1} \sum_{i=1}^{n} (X_i - \overline{X})^2 \qquad (2\text{-}6a)$$

$$\text{For grouped data, } s^2 \simeq \frac{1}{n - 1} \sum_{i=1}^{m} (x_i - \overline{X})^2 f_i \qquad (2\text{-}6b)$$

Finally, we define:

$$\text{Standard Deviation, } s \triangleq \sqrt{\text{variance}} \qquad (2\text{-}7)$$

Note that by taking the square root, we compensate for having squared terms in defining the variance in (2-6a), so that s is reduced to the same units as the X observations.

The values of MSD, s^2, and s are calculated in Table 2-3, where $\overline{X}$ has been rounded[12] in order to avoid fractions. (*Warning:* In calculating $\overline{X}$ or s^2 from grouped data, the divisor n is the sample size, not the number of rows in the table.)

We note that the various deviations $|X_i - \overline{X}|$ in Table 2-3 go all the way up to 9. The standard deviation s is 3.44, which is smaller than some of the deviations and larger than others. Thus s is, in a sense, a typical deviation. This provides a rough check on our arithmetic, which is

---

[11]Technically, this makes the sample variance an unbiased estimator of the population variance, as we will show in Chapter 7.

[12]This rounding makes very good sense for hand calculations. If we denote the rounding error by e, we shall prove in (4-23) that, as a result:

$$\text{the variance is overestimated by approximately } e^2 \qquad (2\text{-}8)$$

which usually is negligible. But if full accuracy is required, the rounding error can easily be compensated for by subtracting e^2 from the rounded s^2. For example, in Table 2-3, since $e = 69.39 - 69 = .39$, if we subtract $e^2 = .15$ from the rounded $s^2 = 11.85$, we obtain the true $s^2 = 11.70$. Since this rounding error usually is not worth considering, however, henceforth we shall ignore it.

worthwhile stating in general:

> The standard deviation s has a value[13] that lies somewhere among the various deviations $|X_i - \overline{X}|$. (2-9)

In conclusion, the sample mean $\overline{X}$ is the most common measure of center, and the sample standard deviation s is the most common measure of spread. We refer to $\overline{X}$ and s^2 as the first and second *moments* of the sample.

PROBLEMS

2-12 Compute the variance and standard deviation of the data of Problem 2-1. Check that s satisfies the rule (2-9).

2-13 For the grouped data of Problem 2-3, compute the range, MAD, MSD, variance, and standard deviation.

(2-14)[14] For the grouped data of Problem 2-5, compute the standard deviation.

2-4   LINEAR TRANSFORMATIONS

(a) Change of Origin

Table 2-4(a) gives the elevations X_i of four midwestern states,[15] in feet above sea level. Their mean $\overline{X}$ and standard deviation s_X also are calculated.

It often happens that a new reference level is more convenient—for example, Lake Superior, which is 600 feet above sea level. In Table 2-4(b), we therefore give the elevations in feet above Lake Superior, X_i'. Of course, each new elevation will be 600 feet less than the old (600 feet being the elevation of Lake Superior).

$$X_i' = X_i - 600$$

---

[13]This rule of thumb would be invariably true (a theorem) if s^2 had a divisor of n. (Even with s^2 having a divisor of $n - 1$, however, the rule still is valid for every distribution we shall encounter.)

[14]As we noted in the preface, problems in parentheses closely resemble previous problems, and so are recommended only if further drill is desired.

[15]Taken, with a slight adjustment, from the 1974 *American Almanac*, Table 284, p. 174.

Incidentally, these four states all have about the same area (56,000 square miles), so that it is meaningful to calculate their simple unweighted mean $\overline{X}$.

TABLE 2-4
Altitudes of Four Midwestern States

State	(a) Altitude in Feet above Sea Level X_i	$(X_i - \bar{X})$	$(X_i - \bar{X})^2$	(b) Altitude in Feet above Lake Superior $X_i' = X_i - 600$	$(X_i' - \bar{X}')$	$(X_i' - \bar{X}')^2$	(c) Altitude in Yards above Sea Level $X_i^* = X_i/3$	$(X_i^* - \bar{X}^*)$	$(X_i^* - \bar{X}^*)^2$
Illinois	600	−330	108,900	0	−330	108,900	200	−110	12,100
Iowa	1,170	240	57,600	570	240	57,600	390	80	6,400
Michigan	900	−30	900	300	−30	900	300	−10	100
Wisconsin	1,050	120	14,400	450	120	14,400	350	40	1,600

(a) $\bar{X} = \dfrac{3,720}{4}$ $0\checkmark$ $s_{\bar{X}}^2 = \dfrac{181,800}{3}$

$= 930$ $= 60,600$

$s_X = 246$

(b) $\bar{X}' = \dfrac{1,320}{4}$ $0\checkmark$ $s_{\bar{X}}^2 = \dfrac{181,800}{3}$

$= 330$ $= 60,600$

$s_{X'} = 246$

(c) $\bar{X}^* = \dfrac{1,240}{4}$ $0\checkmark$ $s_{\bar{X}^*}^2 = \dfrac{20,200}{3}$

$= 310$ $= 6,733$

$s_{X^*} = 82$

Now what will the new mean elevation $\overline{X}'$ be? It will, of course, be 600 feet less than the old mean $\overline{X}$:

$$\overline{X}' = \overline{X} - 600$$

On the other hand, the spread of the new elevations will be exactly the same as the old:

$$s_{X'} = s_X$$

These two equations are easy to verify by actual calculation: in Table 2-4(b), we find that $\overline{X}' = 330$, which indeed is 600 feet less than $\overline{X} = 930$; we also find that $s_{X'} = 246 = s_X$.

These issues are illustrated in Figure 2-4, and generalized in the following theorem.[16]

$$\left.\begin{array}{ll}\text{If} & X_i' = X_i - a \\ \text{then} & \overline{X}' = \overline{X} - a \\ \text{and} & s_{X'} = s_X\end{array}\right\} \qquad (2\text{-}11)$$

(b) Change of Scale

Sometimes a new scale of measurement is more convenient—for example, yards instead of feet. In Table 2-4(c), we therefore give the elevations in

---

[16]**proof**   To prove (2-11), consider first the mean:

$$\overline{X}' \triangleq \frac{1}{n}\sum X_i'$$

$$= \frac{1}{n}\sum (X_i - a)$$

$$= \frac{1}{n}\left[\sum X_i - na\right]$$

$$= \overline{X} - a$$

Consider next the variance:

$$s_{\tilde{X}}^2 \triangleq \frac{1}{n-1}\sum (X_i' - \overline{X}')^2$$

$$= \frac{1}{n-1}\sum [(X_i - a) - (\overline{X} - a)]^2$$

$$= \frac{1}{n-1}\sum (X_i - \overline{X})^2 = s_{\tilde{X}}^2 \qquad (2\text{-}10)$$

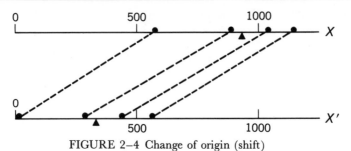

FIGURE 2–4 Change of origin (shift)

yards above sea level, X_i^*. Of course, each new elevation will be 1/3 of the old (since there are three feet in every yard):

$$X_i^* = \tfrac{1}{3} X_i$$

What will the new mean elevation $\overline{X}^*$ be? It will, of course, be 1/3 of the old mean $\overline{X}$:

$$\overline{X}^* = \tfrac{1}{3}\overline{X}$$

Similarly, the spread of the new elevations will be 1/3 of the old:

$$s_{X^*} = \tfrac{1}{3}s_X$$

These two equations again are easy to verify by actual calculation: in Table 2-4(c), we find that $\overline{X}^* = 310$, which indeed is 1/3 of $\overline{X} = 930$. We also find that $s_{X^*} = 82$, which is 1/3 of $s_X = 246$.

These issues are illustrated in Figure 2-5, and generalized in the following theorem.[17]

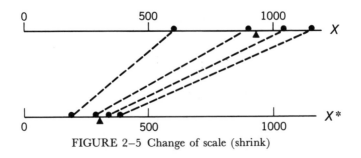

FIGURE 2–5 Change of scale (shrink)

---

[17]The proof resembles that of (2-11) and is left as an exercise.

$$\left.\begin{array}{ll} \text{If} & X_i^* = bX_i \\ \text{then} & \overline{X}^* = b\overline{X} \\ \text{and} & s_{X*} = |b|\,s_X \end{array}\right\} \qquad (2\text{-}12)$$

(c) General Linear Transformation

It now is appropriate to combine the above two theorems into one. For the general linear[18] transformation, we have the following theorem:

$$\left.\begin{array}{ll} \text{If} & Y_i = a + bX_i \\ \text{then} & \overline{Y} = a + b\overline{X} \\ \text{and} & s_Y = |b|\,s_X \end{array}\right\} \qquad (2\text{-}13)$$

This theorem may be interpreted very simply: if the *individual* observations (X_i) are linearly transformed (into corresponding Y_i values), then the *mean* observation is transformed in exactly the same way, and the *standard deviation* is changed by the factor $|b|$, with no effect from a.

(d) Application to Coding

In future chapters, we shall find linear transformations an indispensible theoretical tool. But they also provide immediate practical benefits. For example, linear transformations give a simpler computation of $\overline{X}$ and s_X than that shown in Table 2-3. This involves three steps:

1. Code all the X_i values into a new set of Y_i values. The computations will be simplest if we use the formula:

$$Y_i = \frac{X_i - \text{one of the cell midpoints}}{\text{cell width}} \qquad (2\text{-}14)$$

Which cell midpoint is the most convenient? Usually, one is chosen near the center of the data, since it makes all the coded numbers small. Therefore, in the height example, the cell midpoint of 69 is chosen, so the coding becomes:

---

[18]The transformation (2-13) is called linear because, for fixed values of a and b, the graph of $Y = a + bX$ is a straight line (with slope b and Y-intercept a).

The proof of (2-13) resembles that of (2-11), and is left as an exercise. Of course, proving all of theorems (2-13), (2-12), and (2-11) would be redundant. Only theorem (2-13) needs proof; then (2-12) and (2-11) follow as special cases.

$$Y_i = \frac{X_i - 69}{3} \qquad (2\text{-}15)$$

It is evident that when $X_i = 69$, $Y_i = 0$. Furthermore, as X_i progresses by steps of 3, Y_i progresses conveniently by steps of 1. With these guidelines we can fill in the appropriate Y values in column (2) of Table 2-5.

2. Compute the mean and standard deviation of the Y values. We note in the successive columns of Table 2-5 how easily this can be done.

3. With $\overline{Y}$ and s_Y now in hand, we are in a position to decode back into $\overline{X}$ and s_X. The theory of linear transformations (2-13) applied to (2-15) yields:

TABLE 2-5

Coded Computation of Mean and Standard Deviation
of a Sample of 200 Men's Heights[a]

Coding			For $\overline{Y}$	For s_Y^2, using $\overline{Y}$ rounded to 0	
(1)	(2)	(3)	(4)	(5)	(6)
x_i	$y_i = \dfrac{x_i - 69}{3}$	f_i	$f_i y_i$	$(y_i - \overline{Y})^2$	$(y_i - \overline{Y})^2 f_i$
60	−3	2	−6	9	18
63	−2	10	−20	4	40
66	−1	48	−48	1	48
69	0	64	0	0	0
72	1	56	56	1	56
75	2	16	32	4	64
78	3	4	12	9	36

$$\sum f_i y_i = 26 \qquad \sum (y_i - \overline{Y})^2 f_i = 262$$

$$\overline{Y} = \frac{\sum f_i y_i}{n} = \frac{26}{200} \qquad s_Y^2 = \frac{262}{199}$$

$$= .13 \qquad = 1.317$$

$$\text{Decoding:} \quad \overline{X} = 3\overline{Y} + 69 \qquad s_X = 3s_Y$$

$$\overline{X} = 69.39 \qquad = 3\sqrt{1.317}$$

$$s_X = 3.44$$

[a]Compare with Table 2-3.

$$\overline{Y} = \frac{\overline{X} - 69}{3} \tag{2-16}$$

and
$$s_Y = \tfrac{1}{3} s_X \tag{2-17}$$

From (2-16):

$$\overline{X} = 3\,\overline{Y} + 69$$
$$= 3(.13) + 69 = 69.39$$

From (2-17):

$$s_X = 3s_Y$$
$$= 3\sqrt{1.317} = 3.44$$

Thus our simplified computations of $\overline{X}$ and s_X are complete—and agree with the more complex computations of Table 2-3.

PROBLEMS

2-15 Use coding to find the mean and standard deviation of the data in Problem 2-4.

(2-16) Use coding to find the mean and standard deviation of the data in Problem 2-5.

(2-17) Find the mean and standard deviation of the following sample of 50 executive ages. Graph the relative frequency distribution.

35	46	63	69	54	50	62	68	38	40
55	43	42	59	45	44	57	47	48	46
43	64	49	36	59	60	42	60	42	38
51	50	66	63	57	56	51	38	61	54
50	44	48	69	64	37	56	53	62	52

2-18 Find the mean of the following eleven measurements:

239510	239250	239860	239360
239480	239430	239230	239680
239370	239290	239850	

(*Hint:* It is natural to simply drop the first three digits of every number, and just work with the numbers 510, 250, ... This is mathematically justified—it is just the linear transformation

$$Y = X - 239{,}000)$$

REVIEW PROBLEMS

The review problems at the end of each chapter are the most important problems. Because they may not fit neatly into a simple pigeonhole, they require more thought. Consequently, they provide the best preparation for meeting real problems—and exams.

2-19 The number of incoming calls at a telephone exchange in each of 50 successive minutes were recorded as follows:

1, 0, 1, 1, 0, 0, 2, 2, 0, 1, 1, 0, 1, 4, 0, 3, 0, 1, 0, 2, 0, 0, 1, 1, 0, 0, 1, 2, 1, 0, 0, 1, 3, 1, 0, 1, 1, 4, 0, 1, 2, 1, 1, 0, 2, 0, 0, 1, 2, 1.

(a) Tabulate and graph the relative frequency distribution.
(b) On the average, how many calls are there per minute?
(c) Calculate the variance and standard deviation.

2-20 The 1971 populations and growth rates for various regions are given below.[19] Find the growth rate for the world as a whole.

Region	Population (millions)	Annual Growth Rate (%)
Europe	470	0.8
USSR	240	1.1
N. America	230	1.3
Oceana	20	2.1
Asia	2,100	2.3
Africa	350	2.6
S. America	290	2.9
World	3,700	?

2-21 The following table gives a 1959 regional breakdown of U.S. farmland that was harvested (as opposed to pastured, left fallow, etc.). Compute the percentage that was harvested in the U.S. as a whole.

---

[19]From the 1974 *American Almanac*, Table 1322.

Region	Amount of Farmland (millions of acres)	Percentage Harvested
North	421	46.7%
South	357	21.0%
Mountain	264	8.7%
Pacific	80	18.8%
U.S.A.	1,122	?

2-22 Suppose that the disposable annual incomes of the five million residents of a certain country had a mean of $4,800 and a median of $3,400.

(a) What is the disposable income of the whole country?

(b) What would you say is the disposable income of the "typical resident?"

(c) Is disposable income distributed symmetrically—i.e., what is the shape of the distribution?

⇒2-23[20] Throw a die 50 times. (Or simulate this by consulting the random numbers in Appendix Table II(a).[21] Disregard the digits 0, 7, 8, and 9. The remaining digits (1, 2 . . . 6) will, of course, still be equally likely, and hence will provide an accurate simulation of a die.)

Graph the relative frequency distribution, and calculate the sample mean:

(a) After 10 throws.

(b) After 20 throws.

(c) After 50 throws.

(d) After millions of throws (guess).

---

[20]Problems with arrows are especially important; they introduce later sections of the text.

[21]Reference tables in the Appendix at the *back* of the book are consistently given Roman numerals to distinguish them from tables within the text

CHAPTER 3

Probability

The urge to gamble is so universal and its practice so pleasurable that I assume it must be evil.

Heywood Broun

3-1 INTRODUCTION

In the next four chapters, we make deductions about a sample from a known population. For example, if the population of American voters is 55% Democrat, we can hardly hope to draw exactly that same percentage of Democrats in a random sample. Nevertheless, it is "likely" that "close to" this percentage will turn up in our sample. Our objective is to define "likely" and "close to" more precisely; in this way we shall be able to make useful predictions. First, however, we must lay a good deal of groundwork. Predicting in the face of uncertainty requires a knowledge of the laws of *probability,* and this chapter is devoted exclusively to their development. We shall begin with the simplest example—rolling dice—which was also the historical beginning of probability theory, several hundred years ago.

(a) Concept of Probability

Suppose that a gambler has a die he suspects is loaded, and asks us the probability that it will come up an ace (one dot). One solution would be to roll it over and over again, observing whether or not the relative frequency of aces is 1/6. Of course, rolling it five or ten times would not be enough to average out chance fluctuations. But over the long run, the relative frequency of aces would settle down to a limiting value,[1] which is probability. That is:

$$\boxed{\text{probability} = \text{proportion, in the long run.}} \qquad (3\text{-}1)$$

Or, more formally:

$$\Pr(e_1) = \lim \frac{n_1}{n} \qquad (3\text{-}2)$$

where e_1 is the outcome ("ace")

   n is the total number of times that the trial is repeated (die is thrown)

   n_1 is the number of times that the outcome e_1 occurs (also called $n(e_1)$ or the frequency f)

   $\dfrac{n_1}{n}$ is therefore the relative frequency of e_1

   lim is "the limit of . . . , as n approaches infinity."

Throughout this book, we shall continue to think of probabilities as proportions, because this is such a clear and intuitive concept. Strictly speaking, however, (3-2) should be taken as a way to *empirically determine* or *interpret* probability. As a *definition* of probability, it is flawed by circular reasoning that requires a sophisticated, axiomatic approach to untangle, as we discuss in Section 3-6.

(b) Elementary Properties of Probability

We generalize by considering an experiment with N outcomes (e_1, e_2, . . . , e_i, . . . , e_N). The relative frequency n_i/n of any outcome e_i must be positive, since both the numerator and denominator are positive; moreover, since the numerator cannot exceed the denominator, relative frequency cannot exceed 1. Thus:

$$0 \le \frac{n_i}{n} \le 1$$

---

[1]This tendency to settle down to a limit has already been observed in Problem 2-23, in the case of a fair die. It has also been observed innumerable times by gamblers, prisoners-of-war, and other bored or curious people.

This is also true in the limit; that is:

$$0 \leq Pr(e_i) \tag{3-3}$$

and

$$Pr(e_i) \leq 1 \tag{3-4}$$

Next, we note that the frequencies of all possible outcomes sum to n:

$$n_1 + n_2 + \cdots + n_N = n$$

Dividing this equation by n, we find that all the *relative* frequencies sum[2] to 1:

$$\frac{n_1}{n} + \frac{n_2}{n} + \cdots + \frac{n_N}{n} = 1$$

This also is true in the limit; that is:

$$Pr(e_1) + Pr(e_2) + \cdots + Pr(e_N) = 1 \tag{3-5}$$

Example 3-1. If a fair die is tossed, what is the probability of an ace?

Solution. A "fair die" means that no face is favored; that is, all 6 faces must have the same probability. According to (3-5), all 6 probabilities must add to 1. Thus each probability must be 1/6.

PROBLEMS

3-1 (a) Throw a thumbtack 50 times. Define tossing the point up as e_1. Record your results as in the following table:

----

[2]This first was noted in the last column of Table 2-1.

Trial Number (n)	Point Up?	Frequency of "Ups" (n_1) Accumulated	Relative Frequency (n_1/n)
1	No	0	.00
2	Yes	1	.50
3	Yes	2	.67
4	No	2	.50
5	Yes	3	.60
⋮	⋮	⋮	⋮
10			
20			
30			
40			
50			

(b) Show your results on the following graph:

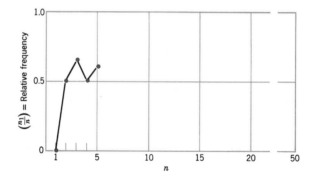

(c) What is your best guess of the probability of tossing the point up?

3-2 Toss a coin 25 times. Define a head as e_1, and proceed as in Problem 3-1.

3-3 Roll a pair of dice, and define "a total of 7 or 11" as the event E. (You may simulate this by drawing a pair of random numbers—properly restricted to the digits 1 through 6, of course.) Repeat 50 times and proceed as in Problem 3-1.

3-2 OUTCOMES AND THEIR PROBABILITIES

(a) The Outcome Set—An Example

In the previous section, the die example was an experiment in which the outcomes $e_1, e_2, \ldots, e_6$ were extremely simple. Usually, an experiment has a more complex set of outcomes.

For example, suppose that the "experiment" consists of planning a family of three children: boys (B) or girls (G). A typical outcome occurs when a boy is followed by two girls, denoted by the sequence $(B\ G\ G)$. The list of all possible outcomes, or *outcome set*, is shown in Figure 3-1. (Since we shall often refer to Figure 3-1, we suggest that you make a copy of it so that you do not have to turn the page too often.) The outcome set is also called the *sample space*, since most experiments that interest the practical statistician are sampling experiments.

We note several features. The order in which the set of eight outcomes is listed is irrelevant. Whenever this is the case, the appropriate mathematical convention is to use curly brackets. Thus, the two outcome sets $\{e_1, e_2, \ldots, e_8\}$ and $\{e_2, e_8, \ldots, e_1\}$ are the *same* set.

However, $(B\ B\ G)$ and $(B\ G\ B)$ are separate and distinct outcomes; in this case, when the order in which B and G appear is an essential feature, we use *round* brackets and call the result an *ordered* triple.

$$(B\ B\ B) = e_1$$
$$(B\ B\ G) = e_2$$
$$(B\ G\ B) = e_3$$
$$(B\ G\ G) = e_4$$
$$(G\ B\ B) = e_5$$
$$(G\ B\ G) = e_6$$
$$(G\ G\ B) = e_7$$
$$(G\ G\ G) = e_8$$

FIGURE 3–1 Outcome set in planning a family of 3 children

To simplify calculations without restricting our concepts in any way, let us suppose that boys and girls are equally likely, and that births are independent[3] (that is, having a boy on the first birth does not change the

---

[3]Actually, in the U.S., the probability of a boy is about .52 rather than $1/2$. There also is some evidence that births are not quite independent. However, these inevitable discrepancies between reality and our mathematical model are small enough to be ignored in this problem.

If the sex of a baby ever becomes widely controllable, then the probability of a boy might change considerably. In Example 3-2, for instance, we consider a model where the probability of a boy is .60.

probability of having a boy on the second).[4] Then all eight outcomes are equally probable. Since all eight probabilities must sum to 1 according to (3-5), we have:

$$\Pr(e_1) = \Pr(e_2) = \cdots = \Pr(e_8) = \frac{1}{8} \qquad (3\text{-}6)$$

(b) Outcome Trees

Often, the probability of an outcome is much more difficult to calculate. For example, consider a biased coin that has a 2/3 probability of heads and a 1/3 probability of tails. If we flip it three times and list the outcomes toss by toss, we obtain Figure 3-2.

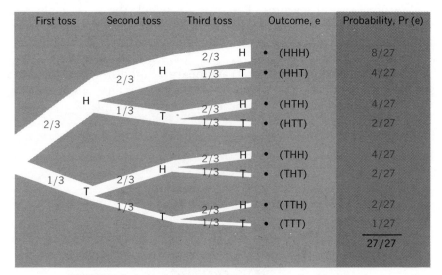

FIGURE 3–2 An outcome tree for 3 tosses of a biased coin

Now consider a typical outcome—for example, the third one listed (*HTH*). Its probability can be calculated using the concept of relative frequency. Imagine millions of people gathered together, each performing this experiment of flipping a biased coin three times. If we summarize the result after the first toss, 2/3 of the people would report *H*, and 1/3 would report *T*. This is the very definition of probability, and it is indicated by the first branching of the tree in Figure 3-2. Of the people who

---

[4]A more thorough discussion of independence is given in Section 3-5. Meanwhile, we deal with it intuitively.

initially had H, only 1/3 would report T on the second toss; and of these, only 2/3 would finally report H on the third toss. Thus:

$$\frac{2}{3} \quad \text{of} \quad \frac{1}{3} \quad \text{of} \quad \frac{2}{3} = \frac{4}{27}$$

of the people would report the complete outcome (HTH). This relative frequency, or probability, is recorded in the last column of Figure 3-2.

As another example, suppose that a *fair* coin is flipped three times. Then the outcome tree is the same as in Figure 3-2, except that the probabilities on each branching are now 1/2. Thus, every outcome has probability:

$$\frac{1}{2} \quad \text{of} \quad \frac{1}{2} \quad \text{of} \quad \frac{1}{2} = \frac{1}{8} \tag{3-7}$$

We note that flipping a fair coin three times is the mathematical equivalent (*simulation*) of having three children—just replacing H and T with B and G. Thus (3-7) confirms (3-6).

Example 3-2. Suppose that the probability of a boy were 60%, and of a girl were 40%. Calculate the probability of each possible outcome for a couple having three children.

Solution. For example,
$\Pr(B\ B\ B) = 60\%$ of 60% of 60% $= .216$, or
$\Pr(G\ B\ G) = 40\%$ of 60% of 40% $= .096$.
Continuing in this way, we obtain the outcome set in Table 3-1.

TABLE 3-1
Outcome set, when
$\Pr(B) = 60\%$

e	$\Pr(e)$
$e_1 = (B\ B\ B)$	.216
$e_2 = (B\ B\ G)$	.144
$e_3 = (B\ G\ B)$	.144
$e_4 = (B\ G\ G)$	.096
$e_5 = (G\ B\ B)$	.144
$e_6 = (G\ B\ G)$	.096
$e_7 = (G\ G\ B)$	.096
$e_8 = (G\ G\ G)$	.064

Sum $= 1.000$

3-3   EVENTS AND THEIR PROBABILITIES

(a) Definition of an Event and its Probability

In planning three children, suppose that a couple is hoping for the event

$$E: \text{ at least 2 girls.}$$

This event includes outcomes e_4, e_6, e_7, and e_8 in Figure 3-1. We find it convenient to move to a cleaner, more abstract representation of each outcome as just a *point*. The result is Figure 3-3, called a *Venn diagram*. Here, we can clearly mark off the event E as a collection of points:

$$E = \{e_4, \quad e_6, \quad e_7, \quad e_8\}$$

In fact, this method provides a convenient way to define an event in general.

$$\boxed{\text{An event } E \text{ is a subset of the outcome set } S} \qquad (3\text{-}8)$$

Now, the interesting question is: "What is the probability of E?" If we imagine many families carrying out this experiment, 1/8 of the time e_4 will occur; 1/8 of the time e_6 will occur; etc. Thus, in one way or another, E will occur $1/8 + 1/8 + \cdots = 4/8$ of the time; that is:[5]

---

[5]Here is a more rigorous argument that we can use to arrive at the conclusion (3-11). Using the definition of limiting relative frequency, we may write:

$$\Pr(E) = \lim \frac{n_E}{n} \qquad (3\text{-}9)$$

where n_E = frequency of E. But E (at least two girls) occurs, of course, whenever the outcomes e_4, e_6, e_7, or e_8 occur. Thus:

$$n_E = n_4 + n_6 + n_7 + n_8$$

and from (3-9):

$$\Pr(E) = \lim \frac{n_4 + n_6 + n_7 + n_8}{n} \qquad (3\text{-}10)$$

$$= \lim \left(\frac{n_4}{n} + \frac{n_6}{n} + \frac{n_7}{n} + \frac{n_8}{n} \right)$$

$$= \Pr(e_4) + \Pr(e_6) + \Pr(e_7) + \Pr(e_8)$$

$$= \frac{1}{8} + \frac{1}{8} + \frac{1}{8} + \frac{1}{8} = .50 \qquad (3\text{-}11) \quad \text{proved}$$

$$\Pr(E) = \frac{1}{8} + \frac{1}{8} + \frac{1}{8} + \frac{1}{8} = \frac{4}{8} = .50 \qquad (3\text{-}11)$$

The obvious generalization of (3-11) is that the probability of an event is the sum of the probabilities of all the points (or outcomes) included in that event. That is:

$$\boxed{\Pr(E) = \sum \Pr(e_i)} \qquad (3\text{-}12)$$

where we sum over just those outcomes e_i that are in E. Again, note an analogy between mass and probability: the mass of an object is the sum of the masses of all the atoms in that object; the probability of an event is the sum of the probabilities of all the outcomes included in that event.

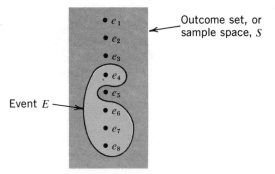

FIGURE 3–3 An event as a subset of points within an outcome set

Example 3-3. Refer again to the planning of three children in Figure 3-1.

(a) Make a Venn diagram to show the event:

$F =$ second child a girl, followed by a boy.

(b) List the outcomes in event F, and so calculate its probability. Then repeat for the following events:

$G =$ fewer than 2 girls $\qquad H =$ all the same sex

$I =$ no girls $\qquad I_1 =$ exactly one girl

$I_2 =$ exactly 2 girls $\qquad I_3 =$ exactly 3 girls

$J =$ less than 2 boys

Solution. (a) Using the sample space of Figure 3-1, we look at the points one by one. We find that there are two in the event F.

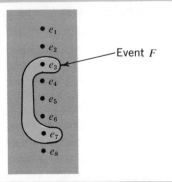

Event F

(b) The event F is just $\{e_3, e_7\}$. We list it in Table 3-2, along with all the other events. The probability of each event simply requires counting up the outcomes in it (because every outcome has the same probability: 1/8).

TABLE **3-2**
Several Events in Planning 3 Children[a]

Three Alternative Ways of Naming an Event			
(1) Arbitrary Symbol for Event	(2) Verbal Description	(3) Outcome List	(4) Probability
F	Second child a girl followed by a boy	$\{e_3, \quad e_7\}$	2/8
G	Fewer than 2 girls	$\{e_1, \quad e_2, \quad e_3, \quad e_5\}$	4/8
H	All the same sex	$\{e_1, \quad e_8\}$	2/8
I	No girls	$\{e_1\}$	1/8
I_1	Exactly 1 girl	$\{e_2, \quad e_3, \quad e_5\}$	3/8
I_2	Exactly 2 girls	$\{e_4, \quad e_6, \quad e_7\}$	3/8
I_3	Exactly 3 girls	$\{e_8\}$	1/8
J	Less than 2 boys	$\{e_4, \quad e_6, \quad e_7, \quad e_8\}$	4/8
E	At least 2 girls	$\{e_4, \quad e_6, \quad e_7, \quad e_8\}$	4/8

[a]See Figure 3-1.

Remarks. In Table 3-2, we see that there are three ways to name or specify an event. The value of specifying an event by its outcome list is evident from the second last event J (less than 2 boys). This list is the same event as E (at least 2 girls) shown in Figure 3-3. That is, $J = E$, an equality that is not so evident from the verbal description.

(b) Combining Events

In planning their three children, suppose that the couple would be disappointed if there were fewer than two girls, or if all were the same sex. Referring to Table 3-2, you can see that this is the event "G or H," also denoted by $G \cup H$, and read "G union H." From the lists of Table 3-2, we can pick out the points that are in G or in H, and so obtain:

$$G \cup H = \{e_1, \quad e_2, \quad e_3, \quad e_5, \quad e_8\}$$

And in general, we define:

$$G \cup H \overset{\Delta}{=} \text{set of points that are in } G, \text{ or in } H, \text{ or in both.} \quad (3\text{-}13)$$

Figure 3-4(a) illustrates this definition. Since five outcomes are included in $G \cup H$, its probability is 5/8.

The couple would be doubly disappointed if there were fewer than two girls *and* if all children were the same sex. This clearly is a much more restricted combined event, consisting only of those outcomes that satisfy both G and H. This is denoted by $G \cap H$[6], and is read "G intersect H" as well as "G and H." From the lists of Table 3-2, we see there is only one point in both G and H:

$$G \cap H = \{e_1\}$$

And in general, we define:

$$G \cap H \overset{\Delta}{=} \text{set of points that are in both } G \text{ and } H \quad (3\text{-}14)$$

---

[6]To remember when $\cup$ or $\cap$ is used, it may help to recall that $\cup$ stands for "Union," and that $\cap$ resembles the letter A in the word "And." These technical symbols are used to avoid the ambiguity that might occur if we used ordinary English. For example, the sentence "$E \cup F$ has 5 points" has a precise meaning, but the informal "E or F has 5 points" is ambiguous.

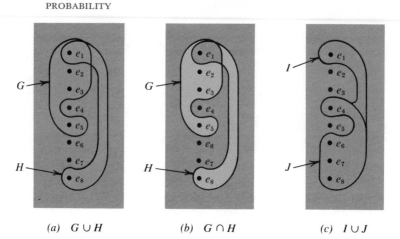

(a) $G \cup H$ (b) $G \cap H$ (c) $I \cup J$

FIGURE 3–4 Venn diagrams, illustrating combined events. (a) $G \cup H$ shaded, "G or H." (b) $G \cap H$ shaded, "G and H." (c) $I \cup J$ shaded.

Figure 3-4(b) illustrates this definition. Since only one outcome is included in $G \cap H$, its probability is 1/8.

Example 3-4.   From the lists in Table 3-2, construct the lists for the following events and hence find their probability.

$$F \cup G, \quad F \cap G, \quad I \cup J, \quad I \cap J$$

Solution.

$F \cup G = \{e_1, \ e_2, \ e_3, \ e_5, \ e_7\}$ Hence $\Pr(F \cup G) = \frac{5}{8}$

$F \cap G = \{e_3\}$ Hence $\Pr(F \cap G) = \frac{1}{8}$

$I \cup J = \{e_1, \ e_4, \ e_6, \ e_7, \ e_8\}$ Hence $\Pr(I \cup J) = \frac{5}{8}$

$I \cap J = \{ \ \ \}$ Hence $\Pr(I \cap J) = 0$

(c) Probability of $G \cup H$

We already have shown how $\Pr(G \cup H)$ may be found from the Venn diagram in Figure 3-4. Now we will develop a formula. First, consider a pair of events that do not have any points in common, such as I and J from Table 3-2. (We also say that they are mutually exclusive, or do not overlap). From Figure 3-4(c), it is obvious that:

$$\Pr(I \cup J) = \Pr(I) + \Pr(J) \qquad \text{(3-15)}$$

$$\frac{5}{8} = \frac{1}{8} + \frac{4}{8}$$

But this simple addition does not always work. For example:

$$\Pr(G \cup H) \neq \Pr(G) + \Pr(H) \qquad \text{(3-16)}$$

$$\frac{5}{8} \neq \frac{2}{8} + \frac{4}{8}$$

What has gone wrong in this case? Since G and H overlap, in summing $\Pr(G)$ and $\Pr(H)$ we count the intersection $G \cap H$ *twice;* this is why (3-16) overestimates. This is easily corrected; subtracting $\Pr(G \cap H)$ eliminates this double counting. Accordingly, it is generally true that:

$$\boxed{\Pr(G \cup H) = \Pr(G) + \Pr(H) - \Pr(G \cap H)} \qquad \text{(3-17)}$$

In our example:

$$\frac{5}{8} \overset{\checkmark}{=} \frac{4}{8} + \frac{2}{8} - \frac{1}{8}$$

Formula (3-17) applies not only to those cases where G and H overlap. It also applies in cases like (3-15), where I and J do not overlap, where $\Pr(I \cap J) = 0$, and this last term in (3-17) disappears. Then we obtain the special case:

$$\boxed{\begin{aligned} &\Pr(I \cup J) = \Pr(I) + \Pr(J) \\ &\text{if } I \text{ and } J \text{ are mutually exclusive.} \end{aligned}} \qquad \text{(3-18)}$$

(d) Partitions and Complements

A collection of several events is defined as mutually exclusive if there is no overlap whatsoever, i.e., if no outcome belongs to more than one event. For example, in Table 3-2, the collection of events $\{I_1, I_2, I_3\}$ is mutually exclusive; but $\{E, F, I\}$ is not, because E and F overlap at e_7.

As another example, in Table 3-2 the collection of events $\{I, I_1, I_2, I_3\}$

is mutually exclusive and also covers the whole sample space S. We therefor call it a *partition* of S. In general, we define:

A partition of a sample space S is a collection of mutually exclusive events whose union is the whole of S.   (3-19)

Thus, a partition completely divides the sample space into nonoverlapping events, as illustrated in Figure 3-5(b).

In Table 3-2, note that G consists of exactly those points that are not in E. We therefore call G the *"complement of E,"* or *"not E,"* and denote it by $\overline{E}$. In general, for any event E:

$$\overline{E} \triangleq \text{set of points that are } not \text{ in } E. \qquad (3\text{-}20)$$

An event and its complement $\{E, \overline{E}\}$ form a very simple partition. Because these events are mutually exclusive, by (3-18):

$$\Pr(E \cup \overline{E}) = \Pr(E) + \Pr(\overline{E}) \qquad (3\text{-}21)$$

and since $E \cup \overline{E}$ constitutes all of the sample space:

$$\Pr(E \cup \overline{E}) = 1 \qquad (3\text{-}22)$$

Substituting (3-22) into (3-21):

$$1 = \Pr(E) + \Pr(\overline{E})$$

$$\Pr(\overline{E}) = 1 - \Pr(E) \qquad (3\text{-}23)$$

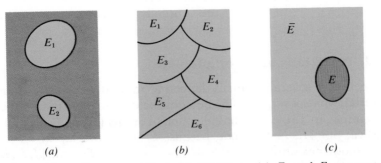

FIGURE 3–5 Venn diagrams to illustrate definitions. (a) E_1 and E_2 are mutually exclusive. (b) $E_1, E_2, \ldots, E_6$ form a partition. (c) $\overline{E}$ is the complement of E.

Example 3-5. To make sure that they get at least one boy, a couple plans on having 5 children. What are their chances of success?

Solution. It would be tedious to list the sample space and pick out the event of at least one boy. Let us see if we can avoid this mess by working with the complement.

Let E = at least one boy

then $\overline{E}$ = no boys; that is, all girls.

We see that $\text{Pr}(\overline{E})$ is much easier to calculate: the probability of getting girl after girl is:

$$\text{Pr}(\overline{E}) = \frac{1}{2} \times \frac{1}{2} \times \frac{1}{2} \times \frac{1}{2} \times \frac{1}{2} = \frac{1}{32}$$

Finally, we can obtain the required probability as the complement:

$$\text{Pr}(E) = 1 - \text{Pr}(\overline{E})$$
$$= 1 - \frac{1}{32} = \frac{31}{32} = .97$$

In other words, it is 97% certain that they will get at least one boy in five births.

Remarks. This was not the only way to answer this question, but is was by far the simplest since Pr (no boys) was so easy to evaluate. You should be on the alert for similar problems: the key words to watch for are "at least," "more than," "less than," "no more than," etc.

Example 3-6. In a certain college, the men engage in various sports in the following proportions:

football (F), 60% of all men

basketball (B), 50%

both football and basketball, 30%

If a man is selected at random for an interview, what is the chance that he will:
(a) Play football or basketball?
(b) Play neither sport?

Solution.   (a) From (3-17):

$$\Pr(F \cup B) = \Pr(F) + \Pr(B) - \Pr(F \cap B)$$

$$= .60 + .50 - .30 = .80$$

(b) $1 - P(F \cup B) = 1 - .80 = .20$

PROBLEMS

3-4 (a) Use a tree diagram to derive the outcome list if a blindfolded man is to draw two chips from an urn containing 5 red chips, 4 white, and 1 black. Assume that he replaces the first chip before drawing the second. What is the probability that both chips will be the same color?
(b) Repeat (a), except that the first chip, once drawn, is not replaced.

3-5 When a penny and a nickel are tossed, the outcome set could be written as a tree or as a rectangular array:

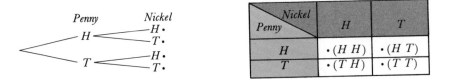

In the same two ways, list the outcome set when a pair of dice are thrown—one red, one white. Then calculate the probability of:
(a) A total of 4 dots.
(b) A total of 7 dots.
(c) A total of 7 or 11 dots (as in Problem 3-3).
(d) A double.
(e) A total of at least 8 dots.
(f) A 1 on one die, 5 on the other.
(g) A 1 on one die, 1 on the other ("snake eyes").
(h) Would you get the same answers if both dice were painted white? In particular, would you get the same answers as before, for (f) and (g)?

3-6 Suppose that a coin was unfairly tossed 3 times in such a way that over the long run, the following relative frequencies were observed:

e	$Pr(e)$
$\cdot (H\ H\ H)$	.15
$\cdot (H\ H\ T)$	.10
$\cdot (H\ T\ H)$	.10
$\cdot (H\ T\ T)$	.15
$\cdot (T\ H\ H)$	.15
$\cdot (T\ H\ T)$	.10
$\cdot (T\ T\ H)$	.10
$\cdot (T\ T\ T)$	.15

Suppose that we are interested in the following events:

G: fewer than 2 heads

H: all coins the same

K: fewer than 2 tails

L: some coins different

Find the following probabilities.
(a) $Pr(G)$, $Pr(H)$, $Pr(G \cup H)$, $Pr(G \cap H)$.
(b) Verify that (3-17) holds true.
(c) $Pr(K)$, $Pr(L)$, $Pr(K \cup L)$, $Pr(K \cap L)$.
(d) Verify that (3-17) holds true.

3-7 In planning a family of 4 children, a couple is interested in the following events:

A: all the same sex

B: precisely 1 boy

C: at least 2 boys

(a) Evaluate $Pr(A) + Pr(B) + Pr(C)$. Do these events form a partition?
(b) Define A' as "no boys." Do A', B, C now form a partition? What is $Pr(A') + Pr(B) + Pr(C)$?

3-8 Continuing Problem 3-7, let Y denote the number of changes in sex sequence. For example, the outcome $(BGBB)$ may be written $(B/G/BB)$,

where the two changes are indicated by slashes; similarly, the outcome
(B/GGG) has only one change. Find:
(a) $\Pr(Y = 1)$.
(b) $\Pr(Y = 2)$.

3-9 (a) What is the probability of at least 1 boy occurring in 10 children?
(b) What is the probability of boys and girls both occurring in 10
children?

⇒3-10 Suppose that a class of 100 students consists of several groups, in the
following proportions:

	Men	Women
Taking math	17%	38%
Not taking math	23%	22%

If a student is chosen by lot to be class president, what is the chance that
the student will be:
(a) A man?
(b) A woman?
(c) Taking math?
(d) A man, or taking math?
(e) A man, and taking math?
(f) If the class president in fact turns out to be a man, what is the
chance that he is taking math? not taking math?

3-11 Of the Canadian population:

30% are from the province of Quebec

28% are French-speaking

24% are from Quebec and French-speaking.

If a Canadian is selected at random, what is the chance that he or she
will be:
(a) From Quebec or French-speaking?
(b) Neither from Quebec nor French-speaking?
(c) French-speaking, but not from Quebec?

3-12 The men of a certain college engage in various sports in the following
proportions:

Football, 30% of all men.

Basketball, 20%.

Baseball, 20%.

Both football and basketball, 5%.

Both football and baseball, 10%.

Both basketball and baseball, 5%.

All three sports, 2%.

If a man is chosen by lot for an interview, use a Venn diagram to calculate the chance that he will be:
(a) An athlete (playing at least one sport)?
(b) A football player only?
(c) A football player or a baseball player?
    If an *athlete* is chosen by lot, what is the chance that he will be:
(d) A football player only?
(e) A football player or a baseball player?

*3-13[7] Generalize the result in Problem 3-12(a). That is, write out a formula like (3-17) that expresses $\Pr(E_1 \cup E_2 \cup E_3)$ in terms of $\Pr(E_1)$ . . . , $\Pr(E_1 \cap E_2)$ . . . , and $\Pr(E_1 \cap E_2 \cap E_3)$.

*3-14 Use Venn diagrams to determine which of the following statements are true:
(a) $\overline{E \cup F} = \overline{E} \cup \overline{F}$
(b) $\overline{E \cup F} = \overline{E} \cap \overline{F}$
(c) $\overline{E \cap F} = \overline{E} \cap \overline{F}$
(d) $\overline{E \cap F} = \overline{E} \cup \overline{F}$
Incidentally, the true statements are known as *De Morgan's Laws.*

3-4   CONDITIONAL PROBABILITY

(a) Definition

Conditional probability is just the familiar concept of limiting relative frequency, but with a slight twist—the set of relevant outcomes is restricted by a condition. An example will illustrate.

---

[7] As we mentioned in the preface, a star next to a problem (or section) indicates that it is more difficult. Although it may be an interesting challenge for the good student, it can be omitted by the student with a weaker background.

Example 3-7.   In a family of 3 children, suppose it is known that G (fewer than two girls) has occurred. What is the probability that H (all the same sex) has occurred? That is, if we imagine many repetitions of this experiment and consider just those cases in which G has occurred, how often will H occur? This is called the *conditional probability* of H, given G, and is denoted $\Pr(H/G)$. Answer in two different cases, where the probability of a boy is:

(a) 50% (as in Example 3-3).

(b) 60% (as in Example 3-2).

Solution.   (a) As shown in Figure 3-6, there are 4 outcomes in G, and only one of them is in H. Thus, when all outcomes are equally likely:

$$\Pr(H/G) = \frac{1}{4} = .25 \qquad (3\text{-}24)$$

(b) When the outcomes are not equally likely, we must be more subtle. Suppose, for example, that the experiment is carried out 100 million times. Then how often will G occur? The answer is about 64 million times; (from our calculations in Table 3-1, $\Pr(G) = .22 + .14 + .14 + .14 = .64$). Of these times, how often will H occur? The answer is about 22 million times. Thus, from our fundamental notion of probability as relative frequency:

$$\Pr(H/G) = \frac{22 \text{ million}}{64 \text{ million}} \simeq .34 \qquad (3\text{-}25)$$

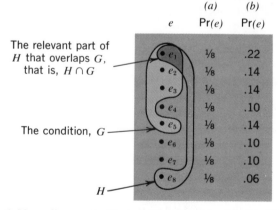

	(a)	(b)
e	$\Pr(e)$	$\Pr(e)$
e_1	⅛	.22
e_2	⅛	.14
e_3	⅛	.14
e_4	⅛	.10
e_5	⅛	.14
e_6	⅛	.10
e_7	⅛	.10
e_8	⅛	.06

The relevant part of H that overlaps G, that is, $H \cap G$

The condition, G

H

FIGURE 3–6 Venn diagram to illustrate the conditional probability $\Pr(H/G)$

Now let us express our answer in general terms. The ratio that appears in (3-25) is essentially $\Pr(H \cap G)$ divided by $\Pr(G)$. Thus:[8]

$$\Pr(H/G) = \frac{\Pr(H \cap G)}{\Pr(G)} \tag{3-26}$$

(b) An Application of Conditional Probability

Formula (3-26) often is very useful when it is reexpressed. Let us cross-multiply by $\Pr(G)$, and note of course that $\Pr(H \cap G) = \Pr(G \cap H)$. Then:

$$\Pr(G \cap H) = \Pr(G)\Pr(H/G) \tag{3-27}$$

This formula breaks down $\Pr(G \cap H)$ into two easy steps: $\Pr(G)$ and then $\Pr(H/G)$.

Example 3-8. Three defective light bulbs inadvertently got mixed with 6 good ones. If 2 bulbs are chosen at random for a ceiling lamp, what is the probability that they both are good?

Solution. We can break this problem down very naturally if we imagine the bulbs being picked up one after the other. Then let us denote:

$$G_1 = \text{first bulb is good}$$

$$G_2 = \text{second bulb is good}$$

Thus:

$$\Pr(\text{both good}) = \Pr(G_1 \cap G_2)$$

by (3-27):

---

[8]Note that the general formula (3-26) also covers the special case of equally likely outcomes. For instance, in Example 3-7(a), it would yield the same answer as the one we derived previously:

$$\Pr(H/G) = \frac{1/8}{4/8} = .25 \qquad \text{(3-24) confirmed}$$

$$= \Pr(G_1)\Pr(G_2/G_1)$$

Now on the first draw, there are 6 good bulbs among 9 altogether, so that the probability of drawing a good bulb is $\Pr(G_1) = 6/9$. After that, however, there are only 5 good bulbs among the 8 left, so that $\Pr(G_2/G_1) = 5/8$. Thus:

$$\Pr(G_1 \cap G_2) = \frac{6}{9} \times \frac{5}{8} = \frac{5}{12} = .42 \qquad (3\text{-}28)$$

Remarks. This problem could have been solved just as well using a tree. If this experiment were repeated many times, 6/9 of the time there would be a good bulb drawn first; of these times, 5/8 of the time there would be another good bulb drawn second. Thus, the probability of drawing two good bulbs altogether would be:

$$\frac{6}{9} \quad \text{of} \quad \frac{5}{8} = \frac{5}{12} \qquad (3\text{-}28) \text{ confirmed}$$

Thus, the product formula (3-27) has a strong intuitive basis.

(c) Bayesian Analysis

Our next example introduces an important branch of statistics called Bayesian analysis, which is based on an imaginative use of formulas (3-27) and (3-26).

Example 3-9. In a population of workers, suppose that 40% are grade-school graduates (class C_1), 50% are high-school graduates (C_2), and 10% are college graduates (C_3). Among the grade-school graduates, 10% are unemployed (E); among the high-school graduates, 5% are unemployed; and among the college graduates, 2% are unemployed.

If a worker is chosen at random and found to be unemployed, (E), what is the probability that he or she is a college graduate (C_3)?

Solution. In Figure 3-7, we represent each worker with a dot, so that probability is proportional to area. Our problem is to calculate $\Pr(C_3/E)$; according to (3-26), this is:

$$Pr(C_3/E) = \frac{Pr(C_3 \cap E)}{Pr(E)}$$

First, then, it is necessary to calculate $Pr(E)$, the proportion of workers who are unemployed. Each class produces some unemployed, shaded in gray and evaluated in the right-hand column of Figure 3-7. Since they total 6.7% of the population, $Pr(E) = .067$.

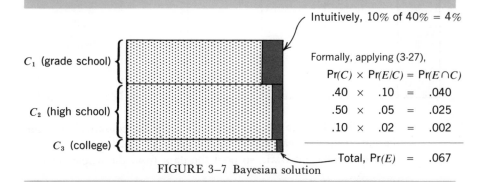

C_1 (grade school)

C_2 (high school)

C_3 (college)

Intuitively, 10% of 40% = 4%

Formally, applying (3-27),

Pr(C)	×	Pr(E/C)	=	Pr(E∩C)
.40	×	.10	=	.040
.50	×	.05	=	.025
.10	×	.02	=	.002

Total, $Pr(E)$ = .067

FIGURE 3–7 Bayesian solution

Now we can evaluate the required conditional probability:

$$Pr(C_3/E) = \frac{.002}{.067} = .03$$

where .002 is also recognized as one of the calculations in Figure 3-7.

This specific example illustrates the general form of Bayesian analysis. Certain *causes* (such as class of education) $C_1, C_2, \ldots$, have *prior probabilities* $Pr(C)$. These causes produce an *effect* E (such as unemployment), not with certainty but with conditional probabilities $Pr(E/C)$. We eventually calculate $Pr(C/E)$, the *posterior probability* of a cause, once the effect has been observed. In other words:

Given	Deduced
$\left.\begin{array}{c} Pr(C) \\ Pr(E/C) \end{array}\right\}$ →	$Pr(C/E)$

PROBLEMS

3-15 In this problem, we shall again simulate the family planning problem. Flip three coins over and over again (or use the random digits three at a time, letting an even number be H, and an odd number be T). Record your results in the following table:

Trial Number n	Does G Occur? (Less than 2 Heads)	Accumulated Frequency $n(G)$	If G Occurs, Does H Occur? (All Coins the Same)	Accumulated Frequency $n(H \cap G)$	Conditional Relative Frequency $n(H \cap G)/n(G)$
1	No				
2	Yes	1	Yes	1	1.00
3	No				
4	Yes	2	No	1	.50
5	Yes	3	Yes	2	.67
⋮					

After 50 trials, is the relative frequency in the last column close to the probability that was calculated theoretically in (3-24)? (If not, it is because of insufficient trials, so pool the data from the whole class.)

3-16 In Middletown USA in 1974,[9] the population was classified as male or female, and as being in favor of or opposed to abortion. The proportions in each category were as follows (note that all the proportions add up to $1.00 = 100\%$ of the population):

	Favor	Opposed
Male	.27	.21
Female	.24	.28

What is the probability that an individual drawn at random will be:
(a) In favor of abortion?
(b) In favor of abortion, if male?
(c) In favor of abortion, if female?

---

[9]The given figures are approximately the same as those in the U.S. as a whole, as reported in the *Gallup Opinion Index*, April 1974, p. 24. The exact question was, "The U.S. Supreme Court has ruled that a woman may go to a doctor to end pregnancy at any time during the first 3 months of pregnancy. Do you favor or oppose this ruling?" The 10% who had no opinion are not included.

3-17 Suppose that 4 defective light bulbs inadvertently have been mixed up with 6 good ones.

(a) If 2 bulbs are chosen at random, what is the chance that they both are good?

(b) If the first 2 are good, what is the chance that the next 3 are good?

(c) If we started all over again and chose 5 bulbs, what is the chance they all would be good?

3-18 Two cards are drawn from an ordinary deck. What is the probability of drawing:

(a) 2 aces?

(b) The 2 black aces?

(c) 2 honor cards (ace, king, queen, jack or ten)?

*3-19 A poker hand (5 cards) is drawn from an ordinary deck of cards. What is the chance that:

(a) The first 4 cards are the aces?

(b) The first and last 2 cards are the aces?

(c) The 4 aces are somewhere among the 5 cards ("4 aces")?

(d) "4 of a kind" (4 aces, or 4 kings, or 4 queens, . . . , or 4 deuces)?

3-20 A company employs 100 persons—75 men and 25 women. The accounting department provides jobs for 12% of the men and 20% of the women. If a name is chosen at random from the accounting department, what is the probability that it is a man? That it is a woman?

3-21 Two baskets are taken on a cookout. The first contains 2 ham and 5 cheese sandwiches; the second contains 6 ham and 3 cheese. In the confusion of darkness, a basket is picked at random.

(a) If a sandwich is drawn from the basket, what is the chance that it is a ham sandwich?

*(b) If a second sandwich is drawn from the same basket, what is the conditional probability that this second sandwich also will be ham, if the first is ham?

3-22 Suppose that a test has been discovered for a disease that afflicts 2% of the population. In clinical trials, the test was found to have the following error rates:

(1) 8% of the people free of the disease had a positive reaction ("false alarm").

(2) 1% of the people having the disease had a negative reaction ("missed alarm").

A mass testing program is being proposed for the whole population; anyone with a positive reaction will be suspected of having the disease, and will be brought to the hospital for further observation.

(a) Of all the people with positive reactions, what proportion actually will have the disease?

(b) Of all the people with negative reactions, what proportion actually will be free of the disease?

(c) Of the whole population, what proportion will be misdiagnosed one way or another?

⇒3-23 In a certain country, it rains 40% of the days and shines 60% of the days. A barometer manufacturer, in testing his instrument in the lab, has found that it sometimes errs: on rainy days it erroneously predicts "shine" 10% of the time, and on shiny days it erroneously predicts "rain" 30% of the time.

In predicting tomorrow's weather before looking at the barometer, the (prior) chance of rain is 40%. *After* looking at the barometer and seeing it predict "rain," what is the (posterior) chance that it will:

(a) Rain?

(b) Shine?

3-24 Is there anything wrong with the following arguments?

(a) In the three-year period that followed the murder of President Kennedy and Lee Harvey Oswald, fifteen material witnesses died—six by gunfire, three in motor accidents, two by suicide, one from a cut throat, one from a karate chop to the neck, and two from natural causes. An actuary concluded that on the day of the assassination, the odds against these witnesses being dead within three years were one hundred thousand trillion to one. Since all these things couldn't have just happened, they must reflect an organized coverup of the assassination.

(b) A man who heard that there was one chance in a million of a bomb being on an aircraft calculated that there would be only one chance in a million million of there being two bombs on an aircraft. Therefore, in order to enjoy these longer odds, he always carried a bomb on with him.

⇒3-25 Two dice are thrown, and we are interested in the following events:

$$E: \text{first die is 5}$$

$$F: \text{total is 7}$$

$$G: \text{total is 10}$$

By calculating the probabilities using Venn diagrams, show that:
(a) $\Pr(F/E) = \Pr(F)$.
(b) $\Pr(G/E) \neq \Pr(G)$.
(c) $\Pr(E/F) = \Pr(E)$.
(d) Is the following a correct verbal conclusion? If not, correct it.

If I'm going to bet on whether the dice show 10, it will help (change the odds) to peek at the first die to see whether it is a 5. But if I'm going to bet on whether the dice show 7, a peek won't help.

3-5   INDEPENDENCE

(a) Definition

Independence is a very precise concept that we define in terms of certain probabilities. An example will illustrate.

Example 3-10. In Middletown U.S.A. in 1974,[10] the population was classified as White or Nonwhite and in favor of or opposed to abortion. The proportions in each category were as follows (note that all the proportions add up to 1.00 = 100% of the population):

	Favor (F)	Opposed (O)
White (W)	.468	.432
Nonwhite (B)	.052	.048

If a person is drawn at random, what is:
(a) $\Pr(F)$?
(b) $\Pr(F/W)$?

Solution.   (a) $\Pr(F) = .468 + .052 = .52$ like (3-12)

$$(b) \ \Pr(F/W) = \frac{\Pr(F \cap W)}{\Pr(W)} \qquad \text{like (3-26)}$$

$$= \frac{.468}{.468 + .432} = .52$$

---

[10]*Gallup Opinion Index*, April 1974, p. 24. See footnote 9.

In this example, the probability of being F is not in any way affected by being W. This kind of independence, defined in terms of probability, is called *statistical independence*. In general, we may state the exact definition as

> An event F is called *statistically independent* of an event E iff[11] $\Pr(F/E) = \Pr(F)$

(3-29)

Of course, if $\Pr(F/E)$ is different from $\Pr(F)$, we call F statistically *dependent* on E. Statistical dependence is the usual case, since it is much easier for two probabilities to be somewhat unequal than to be exactly equal. For example, in Problem 3-16 we found that being in favor of abortion was statistically dependent on being male.[12]

So far we have insisted on the phrase "*statistical* independence," in order to distinguish it from other forms of independence—philosophical, logical, or whatever. For example, we might be tempted to say that for the dice of Problem 3-25, F was "somehow" dependent on E because the total of the two tosses depends on the first die. Although this vague notion of dependence is of no use in statistics and will be considered no further, we mention it as a warning that *statistical* independence is a very precise concept, defined by probabilities (as in (3-29)).

Now that we clearly understand statistical independence and have stated that this is the only kind of independence that we shall consider, we shall run no risk of confusion if we are lazy and drop the word "statistical."

(b) Implications

If an event F is independent of another event E, we can develop some interesting logical consequences. According to (3-27), it is always true that

$$\Pr(E \cap F) = \Pr(E)\Pr(F/E)$$

When we substitute (3-29) we obtain,

> For independent events,
> $$\Pr(E \cap F) = \Pr(E)\Pr(F)$$

(3-30)

---

[11]iff is an abbreviation for "if, and only if." Although the "only if" part is customarily understood implicitly in definitions, it helps to write it out explicitly this way.

[12]And if, in Example 3-10, the exact probabilities for the U.S.A. were quoted instead of the convenient "Middletown" model, $\Pr(F/W)$ would turn out to be slightly different from $\Pr(F)$; that is, F would be slightly dependent on W.

In other words, for independent events, we may apply the simplest kind of multiplication rule—just as we did for the simplest kind of tree in Figure 3-2 (which we now recognize was an early illustration of independence).

Furthermore, by dividing (3-30) by $Pr(F)$ we obtain,

$$\frac{Pr(E \cap F)}{Pr(F)} = Pr(E)$$

i.e.,

$$Pr(E/F) = Pr(E) \tag{3-31}$$

That is, E is independent of F. In other words,

| whenever[1] F is independent of E, then E must be independent of F | (3-32) |

In view of this symmetry, we may henceforth simply state that E and F are statistically independent of each other, whenever any of the three logically equivalent statements (3-29), (3-30), or (3-31), is true. Usually, statement (3-30) is the preferred form, in view of its symmetry. Sometimes, in fact, this "multiplication formula" is taken as the definition of statistical independence. But this is just a matter of taste.

(c) Conclusion

Now we have completed our development of the most important formulas of probability. To review them, Table 3-3 sets out our basic conclusions for $Pr(E \cup F)$ and $Pr(E \cap F)$.

TABLE **3-3**
Review of Probability Formulas

	$Pr(E \cup F)$	$Pr(E \cap F)$
general theorem	$= Pr(E) + Pr(F) - Pr(E \cap F)$	$= Pr(E)Pr(F/E)$
special case	$= Pr(E) + Pr(F)$ if E and F are mutually exclusive; i.e., if $Pr(E \cap F) = 0$	$= Pr(E)Pr(F)$ if E and F are independent; i.e. if $Pr(F/E) = Pr(F)$

---

[1] This theorem was illustrated already in Problem 3-25, and will be illustrated again in Problem 3-26.

PROBLEMS

3-26 In Example 3-10, we found that F was independent of W.
  (a) Can you guess, or better still, state for certain on the basis of theoretical reasoning:
  (i) Whether W will be independent of F?
  (ii) Whether F will be independent of B?
  (b) Calculate the appropriate probabilities to verify your answers in (a).

3-27 Three coins are fairly tossed, and we define:

$$E_1: \text{ first two coins are heads.}$$

$$E_2: \text{ last coin is a head.}$$

$$E_3: \text{ all three coins are heads.}$$

Try to answer the following questions intuitively (does knowledge of the condition affect your betting odds?). Then verify by drawing the sample space and calculating the relevant probabilities for (3-29).
  (a) Are E_1 and E_2 independent?
  (b) Are E_1 and E_3 independent?

3-28 Repeat Problem 3-27, using the unfairly tossed coins whose sample space is as follows (as in Problem 3-6):

e	$\Pr(e)$
•$(H H H)$	.15
•$(H H T)$	.10
•$(H T H)$	.10
•$(H T T)$	.15
•$(T H H)$	.15
•$(T H T)$	.10
•$(T T H)$	.10
•$(T T T)$	.15

3-29 A single card is drawn from a standard deck, and we define:

$$E: \text{ it is an ace}$$

$$F: \text{ it is a heart.}$$

Are E and F independent, when we use:
(a) An ordinary 52-card deck?
(b) An ordinary deck, with all the spades deleted?
(c) An ordinary deck, with all the spades from 2 to 9 deleted?

3-30 If E and F are two mutually exclusive events, what can be said about their independence? (*Hint:* What is $\Pr(E \cap F)$? Then, using (3-26), what is $\Pr(E/F)$? Can it equal $\Pr(E)$?)

3-6   OTHER VIEWS OF PROBABILITY

In Section 3-1, we defined probability as the limit of relative frequency. There are several other possible approaches, including *symmetric* probability, *axiomatic* probability, and *subjective* probability, which we shall treat in historical order.

(a) Symmetric Probability

Symmetric probability was first developed for fair games of chance such as dice, where the outcomes were equally likely. This permitted probabilities to be calculated even before the dice were thrown; the empirical determination (3-1) was not necessary, although it did provide a reassuring confirmation.

> *Example 3-11.*   In throwing a single die, what is the probability of getting an even number?
>
> *Solution.*   We already showed in Example 3-1 that since the die is symmetric, each outcome must have the same probability, 1/6. Thus:
>
> $$\Pr(\text{even number}) = \Pr(2 \text{ or } 4 \text{ or } 6)$$
> $$= \frac{1}{6} + \frac{1}{6} + \frac{1}{6} = \frac{3}{6} = .50$$

It is easy to generalize. Suppose that there is an experiment with N equally probable outcomes altogether, and N_E of them are in an event E. Then:

$$\Pr(E) = \frac{N_E}{N} \tag{3-33}$$

Symmetric probability theory begins with (3-33) as the very definition of probability, and it is a little simpler than the relative frequency approach. However, it is severely limited because it lacks generality—it cannot even handle crooked dice.

Symmetric probability theory also has a major philosophical weakness. Note how the preamble to the definition (3-33) involved the phrase "equally probable." In using the word "probable" in defining probability, we are guilty of circular reasoning.

Our own relative frequency approach to probability suffers from the same philosophical weakness, incidentally. What sort of limit is meant in equation (3-2)? It is *logically* possible that the relative frequency n_1/n behaves badly, even in the limit; for example, no matter how often we toss a die, it is just conceivable that the ace will keep turning up every time, making $\lim n_1/n = 1$. Therefore, we should qualify equation (3-2) by stating that the limit occurs with high *probability*, not logical certainty. If we tried to use (3-2) as the *definition* of probability, we would be using the concept of probability to define probability—circular reasoning again. To break such a circle, we shall turn next to an axiomatic approach.

(b) Axiomatic Probability

All attempts so far to define probability have failed because they require using probability itself within the definition of probability. The only philosophically satisfactory way to break this circular reasoning is to let probability be a basic undefined term. In such a mathematical model, we make no attempt to say what probability *really* is. We simply state the rules (axioms) that it follows.[13]

In a simplified version, the following properties are taken as axioms:[14]

---

[13]Perhaps an analogy from chess will help. In chess, no attempt is made to define what a queen, for example, *really* is. Instead, a queen is merely characterized by the set of rules (axioms) she must obey: she can move in any direction any number of spaces, in a straight line. She is no more and no less than this. Similar rules (axioms) are made for the other pieces, to complete the definition of the game. It is then possible to draw some conclusions (prove some theorems) such as: a king and a rook can win against a king alone; a king and a knight cannot.

In the spirit in which it is set up, probability theory is just like the game of chess: it starts with basic undefined terms and axioms, and from them draws conclusions (theorems). But in its intellectual content, of course, probability theory is much richer than chess: probability theory has many more interesting theorems; even more important, probability theory provides models for many practical problems. By a good model, we mean an abstraction of the real world that captures sufficient detail to be realistic, but omits the inessential detail that would complicate it uselessly. For example, in Chapter 4 we shall study the binomial model that provides excellent predictions for things as diverse as genetics, polls, and radioactive decay.

[14]Historically, probability theory was set up on a proper axiomatic basis by the Russian mathematician Kolmogoroff in the 1930s. Of course, Kolmogoroff's axioms and theorems were much richer than the simplified version that we are setting out.

axioms

$$Pr(e_i) \geq 0 \qquad \text{(3-3) repeated}$$
$$Pr(e_1) + Pr(e_2) + \cdots + Pr(e_N) = 1 \qquad \text{(3-5) repeated}$$
$$Pr(E) = \sum Pr(e_i) \qquad \text{(3-12) repeated}$$

Then the other properties, such as (3-2) and (3-4), are theorems derived from these axioms—with axioms and theorems together comprising a mathematical model that allows us to analyze and predict physical situations such as tossing dice, sampling, etc.

Equation (3-2) is a particularly important theorem, and is known as the *law of large numbers*.[15] Unfortunately, its proof is too complicated to introduce at this point. However, we can prove some simpler theorems, to illustrate how this axiomatic theory can be developed. For any event E:

Theorem

$$0 \leq Pr(E) \qquad \text{(3-34)}$$
$$Pr(E) \leq 1 \qquad \text{(3-35)}$$
$$Pr(\overline{E}) = 1 - Pr(E) \qquad \text{(3-36),}$$
$$\text{like (3-23)}$$

proof According to axioms (3-12) and (3-3), $Pr(E)$ is the sum of terms that are positive or zero, and is therefore itself positive or zero; thus (3-34) is proved.

To prove (3-36), we write out axiom (3-5):

$$\underbrace{Pr(e_1) + Pr(e_2) +}_{\text{Terms for } E} \cdots \underbrace{+ Pr(e_N)}_{\text{Terms for } \overline{E}} = 1$$

According to (3-12), this is just:

$$Pr(E) + Pr(\overline{E}) = 1 \qquad \text{(3-37)}$$

from which (3-36) follows.

---

[15] Equation (3-2), you recall, states that relative frequency settles down to probability. More precisely, it should be called the *first (or weak) law of large numbers,* and should be written out in equation form, like (7-11).

In (3-34), we proved that every probability is positive or zero. In particular, $\Pr(\overline{E})$ is positive or zero; substituting this into (3-37) finally ensures that:

$$\Pr(E) \leq 1 \qquad\qquad \text{(3-35) proved}$$

(c) Subjective Probability

Subjective or personal probability is an attempt to deal with unique historical events that cannot be repeated, and that hence cannot be given any frequency interpretation. For example, consider events such as a doubling in the stock market average within the next decade, or the overthrow of a certain government within the next month. These events are described by the layman as "likely" or "unlikely," even though there is no hope of estimating this by observing their relative frequency. Nevertheless, their likelihood vitally influences policy decisions, and as a consequence must be estimated in some way. Only then can decisions be made on what risks are worth taking.

Roughly speaking, personal probability may be interpreted as the odds one would give in betting on an event; detail on how it may be estimated is given in Problems 3-32 through 3-36.

PROBLEMS

3-31 A gambler has obtained the following long-run relative frequencies for a loaded tetrahedral (4-sided) die. He wants you to deduce $\Pr(e_4)$. Therefore, calculate $\Pr(e_4)$, or state that it is undetermined, or show that he has given you misinformation, in each of the following cases:

(a) $\Pr(e_1) = .2$; $\Pr(e_2) = .4$; $\Pr(e_3) = .1$.
(b) $\Pr(e_1) = .7$; $\Pr(e_2) = .5$.
(c) $\Pr(e_1) = .8$; $\Pr(e_2) = .2$.
(d) $\Pr(e_1) = .6$; $\Pr(e_3) = .2$.

3-32 (a) If you had your choice today between the following two bets, which would you take?
Bet 1 (Election Bet)
If the Democratic candidate wins the next presidential election, you will then win a $100 prize (and win nothing otherwise).
Bet 2 (Jar Bet)
A chip will be drawn at random from a jar containing 1 black chip and 999 white chips. If the chip turns out black, you will then win the $100 prize (and win nothing otherwise).

(b) Repeat choice (a), with a slight change—make the composition of the urn the opposite extreme: 999 black chips and 1 white chip.

(c) Obviously, your answers in parts (a) and (b) depended upon your subjective estimate of American politics; there were no objective "right answers." However, the odds in the urn were so lopsided that there undoubtedly is widespread agreement in part (a) to prefer the election bet, and in part (b) to prefer the jar bet. The question is: as you gradually increase the black chips from 1 to 999, at what point do you become indifferent between the two bets? Is it reasonable to call this your personal probability of a Democrat winning?

3-33 Using the "calibrating jar" of Problem 3-32, roughly evaluate your personal probability that:

(a) The Dow Jones average (of certain stock-market prices) will advance at least 10% in the next twelve months.

(b) U.S. population will increase by at least 1.5 million in the next twelve months. (In percentage terms, this is an increase of 0.7%.)

(c) U.S. population will be double or more at the end of 100 years. (That is, grow at an average rate of at least 0.7% per year.)

(d) The next vice president of the U.S. will be a female.

(e) The next president of your student council will be a female.

(f) The next president of your student council will be a female, if the president is a senior student chosen at random.

3-34 In Problem 3-33, for which answers do you think there will be least agreement among the students in your class? most agreement? Which questions are amenable to a brief investigation that would result in everyone agreeing? (If you have time in class, check out your answers.)

3-35 Do you think the following conclusions are valid? If not, correct them. Certain probabilities (as in part (f) of Problem 3-33) can be agreed upon by all reasonable people; we may call them "objective probabilities." Other probabilities (as in part (a) or (c)) are disagreed upon vigorously, even by experts; we may call them "subjective probabilities." But there is a continuous range in between of probabilities that are subjective to a greater or lesser degree.

REVIEW PROBLEMS

3-36 To reduce pilfering, suppose that a company screens its workers with a lie-detector test—a test that has been proven correct 90% of the time (for guilty subjects, and also for innocent subjects).[16] The company fires all the workers who fail the test. If, in fact, 5% of the workers are pilferers before the firings:

(a) Of the workers who are fired, what proportion actually will be pilferers?

(b) Of the workers who remain (who are not fired), what proportion will be pilferers?

(c) Do you think this kind of screening should be:

  (i) Illegal;

  (ii) Ignored by the government; or

  (iii) Subsidized by the government as a means of raising the general level of honesty?

3-37 Suppose that A and B are independent events, with $Pr(A) = .6$ and $Pr(B) = .2$. What is

(a) $Pr(A/B)$?

(b) $Pr(A \cap B)$?

(c) $Pr(A \cup B)$?

3-38 Repeat Problem 3-37 if A and B are mutually exclusive instead of independent.

3-39 True or False? If false, correct it:

(a) When two events are independent, the occurrence of one event will not change the probability of the second event.

(b) Two events are mutually exclusive if they do not contain any outcomes in common.

(c) A and B are mutually exclusive if $P(A \cap B) = P(A)P(B)$.

3-40 For various forms of transportation, the 1968–1970 U.S. death rates[17] were as follows (deaths per billion passenger miles):

auto	22.0
auto on turnpike	11.7
bus	2.0
plane	1.3
train	0.9

(a) Graph these rates.

In questions (b) and (c) below, first guess the answer, then calculate it. Then compare the two.

---

[16]This example, and some interesting conclusions based on it, are taken from D. T. Lykken, "The Right Way to Use a Lie Detector," *Psychology Today*, March 1975, p. 60.

[17]From the 1972 *World Almanac*, p. 96.

(b) Suppose that you travel by auto about 10,000 miles per year, with an average safety record. Over a lifetime of 75 years, what is your (approximate) chance of being killed in an auto accident?

(c) How would your chances in (b) be changed:

    (i) If you travelled by bus instead of auto?

    (ii) If you travelled by bus 2/3 of the time, and by auto 1/3 of the time?

3-41 Suppose that the last 3 customers out of a restaurant all lose their hat-checks, so that the girl has to hand back their 3 hats in random order. What is the probability

(a) That no man will get the right hat?

(b) That exactly 1 man will?

(c) That exactly 2 men will?

(d) That all 3 men will?

3-42 Find (without bothering to multiply out the final answer) the probability that:

(a) A group of 3 people (picked at random) all have different birthdays?

(b) A group of 30 people all have different birthdays?

(c) In a group of 30 people there are at least two people with the same birthday?

(d) What assumptions did you make above?

3-43 (a) A fair coin is flipped 10 times, and happens to come up heads every time. What is the probability of this?

(b) What is the conditional probability, then, that the eleventh toss also will be a head?

*3-44 A bag contains a thousand coins. One of the coins is badly loaded, so that it comes up heads 3/4 of the time. A coin is drawn at random. What is the probability that it is the loaded coin, if it is flipped and turns up heads without fail:

(a) 3 times in a row?

(b) 10 times in a row?

(c) 20 times in a row?

*3-45 Repeat Problem 3-44—but this time the loaded coin in the bag has heads on both sides.

CHAPTER 4

Probability Distributions

We must believe in luck. For how else can we explain the success of those we don't like?

Jean Cocteau

4-1 DISCRETE RANDOM VARIABLES

Suppose that a couple, in planning three children, is primarily interested in the number of boys. This is an example of a *random variable* and is customarily denoted by a capital letter:

$$X = \text{the number of boys}$$

The possible values of X are 0, 1, 2, 3; however, they are not equally likely. To find what the probabilities are, we must examine the original sample space, given in Figure 4-1(a). Thus, for example, the event "one boy" ($X = 1$) consists of three of the eight equiprobable outcomes; hence its probability is 3/8. Similarly, the probability of each of the other events is computed. Thus, in Figure 4-1(a), we obtain the probability distribution of X, shown in color. In general:

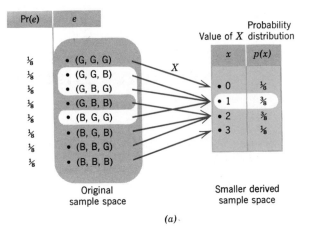

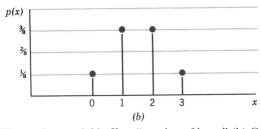

FIGURE 4–1 (a) The random variable X = "number of boys." (b) Graph of its probability distribution.

> A discrete random variable takes on various values with probabilities specified by its probability distribution.[1]   (4-1)

As shown in Figure 4-2, we begin in the original sample space by considering events such as $(X = 0)$, $(X = 1)$, . . . , in general $(X = x)$; note that capital X represents the random variable while small x represents a specific value that it may take. For these events we calculate the prob-

---

[1]Although the intuitive definition (4-1) will serve our purposes well enough, it is not always as satisfactory as the more rigorous mathematical definition: "A discrete random variable is a numerical-valued function defined over a sample space, that takes on a finite (or countably infinite) number of values."

This definition stresses the random variable's relation to the original sample space. Thus, for example, the random variable Y = the number of girls, is seen to be a different random variable from X = the number of boys. Yet X and Y have the *same probability distribution,* and anyone who used the loose definition (4-1) might be deceived into thinking that they were the *same random variable.* In conclusion, there is more to a random variable than its probability distribution.

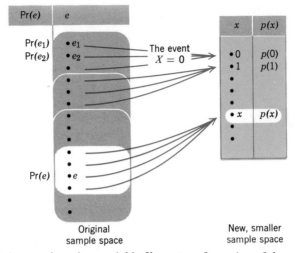

FIGURE 4–2 A general random variable X as a transformation of the original sample space to a new, condensed sample space shown as 0, 1, 2, . . . , the set of positive integers. (To be more general, however, we should allow negative, fractional, and even irrational values as well. Thus our notation should be $x_1, x_2, . . . , x_i, . . .$, rather than 0, 1, 2, . . . , x, ...)

abilities and denote[2] them $p(0)$, $p(1)$, . . . , $p(x)$, ... This probability distribution $p(x)$ may be presented equally well in any of the three customary forms for a function:

1. Table form, as in the right-hand side of Figure 4-1(a).

2. Graph form, as in Figure 4-1(b).

3. By formula, as in Equation (4-7), given later on.

In Figure 4-1 and 4-2, the original sample space (outcome set) is reduced to a much smaller and more convenient numerical sample space. The original sample space was introduced to enable us to calculate the probability distribution $p(x)$ for the new space; having served its purpose, the old unwieldy space is then forgotten. The interesting questions can be answered very easily in the new space. For example, referring to Figure 4-3, what is the probability of one boy or fewer? We simply add up the relevant probabilities in the new sample space:

$$Pr(X \leq 1) = p(0) + p(1) = \tfrac{1}{8} + \tfrac{3}{8} = \tfrac{1}{2} \qquad (4\text{-}2)$$

---

[2]This notation, like any other, may be regarded simply as an abbreviation for convenience. Thus, for example, $p(1)$ is short for $Pr(X = 1)$, which in turn is short for "the probability that the number of boys is one."

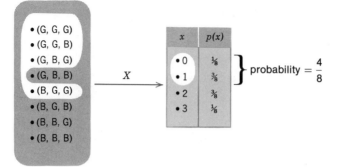

FIGURE 4–3 The event $X \leq 1$ in both sample spaces, illustrating the easier calculation in the new sample space

The answer could have been found, but with more trouble, in the original sample space.

PROBLEMS

4-1 In planning a family of 4 children, find the probability distribution of:
(a) $X =$ the number of boys.
(b) $Y =$ the number of changes in sex sequence.

4-2 When 2 fair dice are thrown, find the probability distribution of:
(a) The total number of dots, $S = X_1 + X_2$.
(b) The difference (absolute value) between the numbers, $D = |X_1 - X_2|$.

⇒4-3 To review Chapter 2, consider the planning of three children, and again let $X =$ the number of boys (0, 1, 2, 3). Simulate this "experiment" either by tossing 3 coins, or by using the random numbers in Appendix Table II(a).
Repeat this experiment 50 times, recording the frequency table of X. Then calculate:
(a) The relative frequency distribution.
(b) The mean $\overline{X}$.
(c) The mean squared deviation MSD.
(d) The variance s^2.

⇒4-4 If the experiment in Problem 4-3 were repeated millions of times (rather than 50 times), to what value would the calculated quantities tend?

4-2 MEAN AND VARIANCE

Notice the close relation between the relative frequency distribution observed in Problems 4-3 and 4-4 and the probability distribution calculated in Figure 4-1 for planning 3 children: if the sample size were increased without limit, the relative frequency distribution would settle down to the probability distribution. This is an old story: relative frequency becomes probability in the limit.

From the relative frequency distribution, we calculated the mean $\overline{X}$ and variance s^2 of the *sample*.[3] It is natural to calculate analogous values from the probability distribution and call them the mean μ and variance σ^2 of the probability distribution $p(x)$, or of the random variable X itself (or even call them μ and σ^2 of the conceptual *population* of all possible repetitions of this experiment, of which we chose a sample of fifty). Thus, we define:

$$\text{Population mean, } \mu \triangleq \sum_x x p(x)$$

(4-3)
like (2-1b)

$$\text{Population variance, } \sigma^2 \triangleq \sum_x (x - \mu)^2 p(x)$$

(4-4)
like (2-5b)

Here we are following the usual custom of reserving *Greek* letters[4] for *population* values. The computation of σ^2 often can be simplified by using the formula:[5]

$$\sigma^2 = \sum x^2 p(x) - \mu^2$$

(4-5)

---

[3] We also calculated MSD. But as sample size $n \to \infty$, it becomes indistinguishable from s^2.

[4] μ is the Greek equivalent of m for mean, and σ is the Greek equivalent of s for standard deviation.

[5] *Proof that* (4-5) *is equivalent to* (4-4). Reexpress (4-4) as:

$$\sigma^2 = \sum (x^2 - 2\mu x + \mu^2) p(x)$$

and, noting that μ is a constant:

$$\sigma^2 = \sum x^2 p(x) - 2\mu \sum x p(x) + \mu^2 \sum p(x)$$

Since $\Sigma xp(x) = \mu$ and $\Sigma p(x) = 1$,

$$\sigma^2 = \sum x^2 p(x) - 2\mu(\mu) + \mu^2(1)$$
$$= \sum x^2 p(x) - \mu^2$$

(4-5) proved

Example 4-1. (a) Calculate the mean and variance of X = the number of boys in a family of 3 children.

(b) Calculate the standard deviation σ, and make sure that it satisfies the same interpretation as s in (2-9).

Solution. (a) The computations are similar to Table 2-3, and are set out in Table 4-1.

TABLE **4-1**

Calculation of the Mean and Variance of X = number of boys

Given Probability Distribution		Calculation of μ Using (4-3)	Calculation of σ^2 Using (4-4)			Easier Calculation of σ^2 Using (4-5); Multiply Columns (1) and (3)
(1) x	(2) $p(x)$	(3) $xp(x)$	(4) $(x-\mu)$	(5) $(x-\mu)^2$	(6) $(x-\mu)^2p(x)$	(7) $x^2p(x)$
0	1/8	0	$-3/2$	9/4	9/32	0
1	3/8	3/8	$-1/2$	1/4	3/32	3/8
2	3/8	6/8	$+1/2$	1/4	3/32	12/8
3	1/8	3/8	$+3/2$	9/4	9/32	9/8
		$\mu = 12/8$ $= 1.5$			$\sigma^2 = 24/32$ $= .75$	$\sum x^2p(x) = 24/8$ from column (3), $\mu^2 = 18/8$ difference $\sigma^2 = 6/8$ $= .75 \checkmark$

(b) $\sigma = \sqrt{.75} = .87$

Since the deviations $|x - \mu|$ range from $1/2$ to $3/2$, we note that σ lies among them.

Thus σ is in a sense a typical deviation, and (2-9) is satisfied.

Remarks. To calculate σ^2, the definition (4-4) was more work because it involved several values of $(x - \mu)$, which are fractional because μ is fractional. It is easier to use the alternate form (4-5), where the fractional μ appears only once.

A clear distinction must be made between sample and population values: μ is called the population mean since it is based on the population of all possible repetitions of the experiment; on the other hand, we call $\overline{X}$ the sample mean since it is based on a mere sample drawn from the parent population. Similarly, σ^2 and s^2 represent population and sample variance, respectively.

Since the definitions of μ and σ are similar to those of $\overline{X}$ and s, we find similar interpretations. We continue to think of the mean μ as a weighted average, using probability weights rather than relative frequency weights. The mean is also a fulcrum and center. The standard deviation σ is a measure of spread—in a sense, a typical deviation.

When a random variable is linearly transformed, the new mean and standard deviation behave exactly as they did when sample observations were transformed in (2-13) (the proof is quite analogous and is left to Problem 4-26). For future reference, we state these results in Table 4-2. We could write out verbally all the information in this table, working across the rows, as follows: consider any random variable X, with mean μ_X and variance σ_X^2; if we define a new random variable Y as a linear function of X (specifically $Y = a + bX$), then the mean of Y will be $a + b\mu_X$, and its variance will be $b^2\sigma_X^2$.

TABLE **4-2**
Linear Transformation of a Random Variable

Random Variable	Mean	Variance	Standard Deviation
X $Y \triangleq a + bX$	μ_X $\mu_Y = a + b\mu_X$	σ_X^2 $\sigma_Y^2 = b^2\sigma_X^2$	σ_X $\sigma_Y = \lvert b \rvert \sigma_X$

Example 4-2. An instructor gave a short test, consisting of 5 true—false questions, to a large population of students. The distribution of grades is given in the first 2 columns on p. 80 (with $p(x)$ being the proportion of the population giving x correct answers).

(a) Calculate the mean grade μ_X and the standard deviation σ_X.

(b) Suppose that the instructor rescaled the marks to go from 10 to 20, according to the transformation:

$$Y = 10 + 2X$$

What are μ_Y and σ_Y?

(c) Graph the distribution of X and of Y.

Solution. (a) We calculate μ_X in the third column of the table. To calculate σ_X^2, we again note that the definition (4-4) would involve working with several values of $(x - \mu)$, which are fractional because μ is fractional. It is therefore much easier to use the alternate form (4-5); these calculations for σ_X^2 are given in the last column of the table.

x	$p(x)$	$xp(x)$	$x^2p(x)$
0	.001	0	0
1	.015	.015	.015
2	.087	.174	.348
3	.264	.792	2.376
4	.396	1.584	6.336
5	.237	1.185	5.925

$$\mu_X = 3.750 \qquad \sum x^2p(x) = 15.000$$
$$\mu^2 = 14.062$$

$$\text{difference } \sigma^2 = \quad .938$$
$$\sigma = \sqrt{.938} = .968$$

(b) Since $Y = 10 + 2X$ is a linear transformation, we can use Table 4-2:

$$\mu_Y = 10 + 2\mu_X$$
$$= 10 + 2(3.75) = 17.5$$
$$\sigma_Y = |2|\sigma_X$$
$$= 2(.968) = 1.936$$

(c) The graphs are given in Figure 4-4.

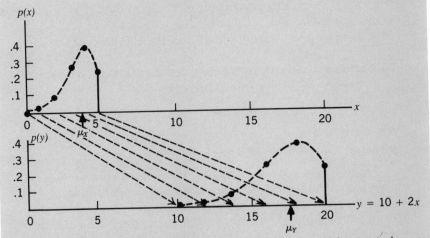

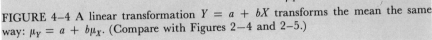

FIGURE 4-4 A linear transformation $Y = a + bX$ transforms the mean the same way: $\mu_Y = a + b\mu_X$. (Compare with Figures 2—4 and 2—5.)

PROBLEMS

4-5 Compute μ and σ^2 for the probability distributions in Problem 4-1. As a check, compute σ^2 in 2 ways—from the definition (4-4), and from the easy formula (4-5).

(4-6) Compute μ and σ^2 for the random variables of Problem 4-2.

4-7 Letting $X =$ the number of dots rolled on a fair die, find μ_X and σ_X. If $Y = 2X + 4$, calculate μ_Y and σ_Y in two ways:
(a) By tabulating the probability distribution of Y, then using (4-3) and (4-4).
(b) By Table 4-2.

4-8 An exam consists of 4 multiple-choice questions, each with a choice of 3 answers. Let X be the number of correct answers when the student has to resort to pure guessing for each answer.
(a) Tabulate the distribution of X.
(b) Calculate μ_X and σ_X.
(c) If the instructor calculates a rescaled mark $Y = 22.5X + 10$, what are μ_Y and σ_Y?

⇒4-9 Let X be a random variable with mean μ and standard deviation σ. What are the mean and standard deviation of Z, where:
(a) $Z = 2X - 8$
(b) $Z = 3/(5X - 2)$
(c) $Z = \dfrac{X - 5}{10}$
(d) $Z = \dfrac{X - \mu}{\sigma}$

*4-10 Suppose that in 1972, the whole population of American families yielded the following table for family size.[6]

Children	0	1	2	3	4
Proportion of families	.44	.19	.18	.10	.09

---

[6] For families consisting of husband, wife, and children under eighteen. Since this table includes young couples who have not yet had a chance to have children, it understates the fertility of the U.S. population. Abridged, with slight modification, from the 1974, *American Almanac*, p. 42, Table 55.

(a) Let X be the number of children in a family selected at random. (This selection may be done by lot: imagine each family being recorded on a chip, the chips well mixed, and then one drawn.) Find μ_X and σ_X.

(b) Now let a child be selected at random (rather than a family), and let Y be the number of children in his or her family. (This selection may be done by a teacher, for example, who picks a child by lot from the register of children.) What are the possible values of Y? Find the probability distribution, and compute μ_Y and σ_Y.

(c) Is μ_X or μ_Y more properly called the "average family size?"

4-3   THE BINOMIAL DISTRIBUTION

There are many types of discrete random variables. We shall study the commonest type—the binomial—as an example of how a general formula can be developed for a probability distribution.

The classical example of a binomial variable is:

$$S = \text{number of heads in } n \text{ tosses of a coin}$$

TABLE **4-3**
Examples of Binomial Variables

Trial	Success	Failure	π	n	S
Tossing a fair coin	Head	Tail	1/2	n tosses	Number of heads
Birth of a child	Girl	Boy	Practically 1/2	Family size	Number of girls in family
Throwing 2 dice	7 dots	Anything else	6/36	n throws	Number of sevens
Drawing a voter in a poll	Democrat	Republican	Proportion of Democrats in the population	Sample size	Number of Democrats in the sample
The history of one atom which may radioactively decay during a certain time period	Decay	No change	Very small	Very large, the number of atoms in the sample	Number of radioactive decays

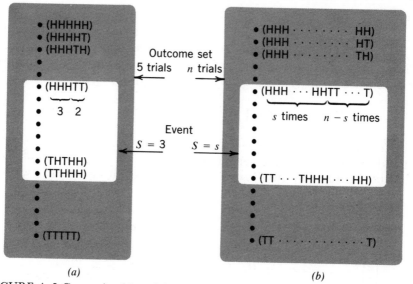

FIGURE 4–5 Computing binomial probability. (a) Special case: 3 heads in 5 tosses of a coin. (b) General case: s successes in n trials.

In order to generalize, we shall speak of n independent "trials," each resulting in either "success" or "failure," with respective probabilities π and $(1 - \pi)$. Then the total number of successes S is called a binomial random variable.

There are many, many random variables of this type, some of which are listed in Table 4-3. We shall now derive a simple formula for the probability distribution $p(s)$. As an example, consider the special case in which we compute the probability of getting three heads in tossing a coin five times, as shown in Figure 4-5(a). Each point in the outcome set is represented as a sequence of five of the letters H (success) or T (failure). We concentrate for example on the event "three heads" ($S = 3$), and show all outcomes that comprise this event. In each of these outcomes, H appears three times, and T twice. Since the probability of H is π and T is $(1 - \pi)$:

Then, for example, the sequence

$$\underbrace{HHH}_{3}\ \underbrace{TT}_{2}$$

has probability

$$\pi \cdot \pi \cdot \pi \cdot (1 - \pi) \cdot (1 - \pi)$$
$$= \pi^3 (1 - \pi)^2$$

Then, in general, the sequence

$$\underbrace{HH \cdots H}_{s \text{ times}}\ \underbrace{TT \cdots T}_{(n - s) \text{ times}}$$

has probability

$$\pi \cdot \pi \cdots (1 - \pi) \cdot (1 - \pi) \cdots$$
$$= \pi^s (1 - \pi)^{n-s}$$

where the simple multiplication is justified by the independence of the trials. We further note that all other outcomes in this event have the same probability. For example, another outcome that yields three heads is $(HTHHT)$; its probability is again the product of five individual probabilities:

$$\pi \cdot (1 - \pi) \cdot \pi \cdot \pi \cdot (1 - \pi) = \pi^3(1 - \pi)^2 \qquad \text{as before}$$

Now we only have to determine how many such outcomes are included in the event "three heads." This is precisely the number of ways that the three Hs and two Ts can be rearranged. This number of ways is denoted by $\binom{5}{3}$ or C_3^5,

and is:[7]

$$\binom{5}{3} = \frac{5!}{3!(5 - 3)!} = 10$$

or, in general:

$$\binom{n}{s} = \frac{n!}{s!(n - s)!}$$

To summarize:
The event

$$(S = 3)$$

includes

$$\binom{5}{3} = 10$$

outcomes, each with probability

$$\pi^3(1 - \pi)^2 = (\tfrac{1}{2})^3(\tfrac{1}{2})^2 = \tfrac{1}{32}$$

Hence its probability is:

$$p(3) = \binom{5}{3}\pi^3(1 - \pi)^2$$

$$= \frac{5!}{3!2!}(\tfrac{1}{2})^3(\tfrac{1}{2})^2 = \tfrac{10}{32}$$

In general, the event

$$(S = 3)$$

includes

$$\binom{n}{s}$$

outcomes, each with probability

$$\pi^s(1 - \pi)^{n-s}$$

Hence its probability is:

$$p(s) = \binom{n}{s}\pi^s(1 - \pi)^{n-s} \qquad (4\text{-}7)$$

---

[7] This formula is developed as follows. In how many ways can we fill five spots with five distinct objects, designated H_1, H_2, H_3, T_1, T_2? We have a choice of 5 objects to fill the first spot, 4 the second, and so on; thus the number of options we have is:

$$5 \cdot 4 \cdot 3 \cdot 2 \cdot 1 = 5! \qquad (4\text{-}6)$$

(cont'd)

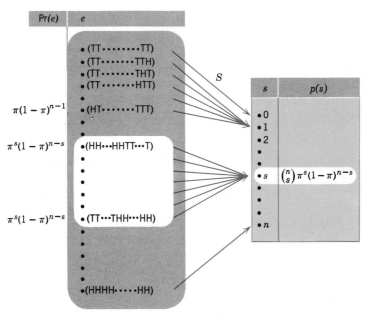

FIGURE 4–6 Computing the binomial probability of s successes in n trials (generalization of Figure 4–1(a))

The derivation of (4-7) is illustrated in Figure 4-6. A special case already has been derived in Figure 4-1(a), where we considered the probability of one boy in three births. Note that it nicely fits the general formula (4-7), with $n = 3$, $s = 1$, and $\pi = 1/2$:

$$p(1) = \binom{3}{1}(\tfrac{1}{2})^1(\tfrac{1}{2})^2 = \frac{3!}{1!2!}(\tfrac{1}{2})^3 = \tfrac{3}{8}$$

But this is not quite the problem at hand; in fact we cannot distinguish between H_1, H_2 and H_3—all of which appear as H. Thus many of the separate and distinct arrangements counted in (4-6) cannot be distinguished, and appear as a single arrangement (for example, $H_1 H_2 H_3 T_1 T_2$ and $H_2 H_3 H_1 T_1 T_2$ and many others appear as the single arrangement $HHHTT$). Thus (4-6) involves serious overcounting. How much?

We overcounted $3 \cdot 2 \cdot 1 = 3!$ times because we assumed in (4-6) that we could distinguish between H_1, H_2, and H_3, when in fact we could not. (3! is simply the number of distinct ways of rearranging H_1, H_2, and H_3). Similarly, we overcounted $2 \cdot 1 = 2!$ times because we assumed in (4-6) that we could distinguish between T_1 and T_2, when in fact we could not. When (4-6) is deflated for overcounting in both these ways, we have:

$$\frac{5!}{3!2!} = \frac{5!}{3!(5-3)!}$$

which is the given formula for $\binom{5}{3}$.

From (4-7), a table of the binomial distribution has been computed for selected n and π, and appears in Appendix Table III(b). It should be used whenever possible, to avoid repeating the tedious calculations of (4-7). For example, when $n = 3$ and $\pi = .50$, we find in Table III(b) that $p(1) = .375$, which agrees with the value $\frac{3}{8}$ we just calculated.

Some problems require the accumulated probability in the right-hand tail that is computed in Appendix Table III(c).

Example 4-3.   A poll of 5 voters is to be drawn from a large population, 60% of whom are Democrats (as in the 1964 Presidential election, for example). What is the probability that:

(a) Exactly 3 of the voters will be Democrats?
(b) At least 3 of the voters (a majority) will be Democrats?

Solution.   (a) Each voter who is drawn constitutes "a trial." There are $n = 5$ trials, and the probability of "success" on each trial is $\pi = .60$. We want the probability of exactly $s = 3$ successes altogether, which is found from Table III(b) to be .3456 $\simeq$ 35%.
(b) In Table III(c), we find the probability of *at least* three successes to be .6826 $\simeq$ 68%.

Remarks.   Alternatively, we could have solved using the formula (4-7); for part (a):

$$p(3) = \binom{5}{3}.60^3.40^2 = 10 \times .6^3 \times .4^2 = .3456$$

For part (b), we would have calculated and summed three such numbers.

Clearly, using Table III was much easier. However, there are many cases that are not covered by the table (for example, $n = 12$, $\pi = .64$), where formula (4-7) must be used.

PROBLEMS

4-11 (a) One hundred coins are spilled at random on the table, and the total number of heads X is counted. The distribution of X is binomial, with $n = $ _____ and $\pi = $ _____. Although this distribution would be tedious to tabulate, the *average number* of heads is easily guessed to be _____.
(b) Repeat (a) for $X = $ the number of aces when 100 dice are spilled at random.

(c) Repeat (a) for $X =$ the number of correct answers when 25 true—false questions are answered by pure guessing.

4-12 (a) On the basis of Problem 4-11, guess the formula for the mean of a general binomial variable, in terms of n and π.

(b) It can be proved that the formula for the variance is $n\pi(1 - \pi)$. Use this to calculate the variance in Problem 4-11.

4-13 (a) Construct a diagram similar to Figure 4-1 to obtain the probability distribution for the number of heads S when 4 fair coins are tossed.

(b) Check that your answer agrees with (4-7),[8] and with Table III(b).

(c) Calculate μ and σ^2.

(d) Graph the distribution, showing μ.

4-14 A chip is drawn from a bowl containing 2 red, 1 blue, and 7 black chips. The chip is replaced, and a second chip is drawn, and so on until 5 chips have been drawn (sampling with replacement).

(a) Let $S =$ the total number of red chips drawn. Tabulate and graph its probability distribution. Find μ and σ^2.

(b) Repeat (a) for $Y =$ the total number of blue chips.

4-15 Check the answers to Problem 4-8(a) and (b), using the binomial formulas.

4-16 Suppose that the probability of a warship hitting a target on any shot is $\tfrac{1}{5}$. What is the probability that in 6 shots it will hit the target:

(a) Exactly 2 times?

(b) At least 3 times?

(c) At most 2 times?

(d) What crucial assumption are you implicitly making? Why may it be questionable? (To appreciate this point, put yourself in the position of the captain of the British battleship *Prince of Wales* in 1941. The gunners on the German *Bismark* have just homed in on the British *Hood,* sinking it. They now turn their fire on you, and after several misses, they make a direct hit.)

---

[8]In evaluating the binomial formula (4-7) for $s = 0$, you will get stuck at $\binom{4}{0} = 4!/4!0!$, because 0! has not been defined. We therefore take this opportunity to define 0! so that it will give the right answer in this case. Since there is just one way to arrange zero Hs and four Ts, we would like $\binom{4}{0}$ to be 1. This requires 0! to be 1, and so we define:

$$0! \stackrel{\Delta}{=} 1$$

4-17 (a) A city council consists of the mayor and six councilmen, with a majority vote among these seven people deciding any given issue. Suppose that the mayor wants to pass a certain motion, but is not sure of its support. Suppose that the six councilmen vote independently, each with probability 40% of voting for the motion. What is the chance that it will pass?

(b) If the mayor had two firm allies who, he was certain, would vote for the motion, how would that improve the chances of its passing? (Assume that the other four members are the same as in part (a).)

(c) Assume that all six councilmen vote as in part (a), except that two are friends of the mayor and so caucus with him before the meeting. All three members agree to vote within the caucus to determine their majority position, and then go into the council meeting with a solid block of three votes to support that position.

(d) Does this or does this not illustrate the motto, "united we stand, divided we fall"?

*4-18 (Requires calculus). Graph the function $f(z) = e^{-(1/2)z^2}$, showing its:

(a) Symmetry.

(b) Asymptotes.

(c) Maximum.

(d) Points of inflection.

4-4   CONTINUOUS DISTRIBUTIONS

In Figure 2-2, we saw how a continuous quantity such as height could be nicely represented by a bar graph showing relative frequencies. This graph is reproduced in Figure 4-7(a), below (with height now measured in feet, rather than inches; furthermore, the y-axis has been shrunk to the same scale as the x-axis.) The sum of all the relative frequencies (i.e., the sum of all the heights of the bars) in Figure 4-7(a) is of course 1, as we first noted in Table 2-2. We find it convenient to change the vertical scale to relative frequency *density* as in Figure 4-7(b). This rescaling is designed specifically to make the total *area* equal to 1. We accomplish this by defining:

$$\boxed{\text{relative frequency density} \triangleq \frac{\text{relative frequency}}{\text{cell width}}} \qquad (4\text{-}8)$$

$$= \frac{\text{relative frequency}}{1/4}$$

$$= 4 \,(\text{relative frequency}). \qquad (4\text{-}9)$$

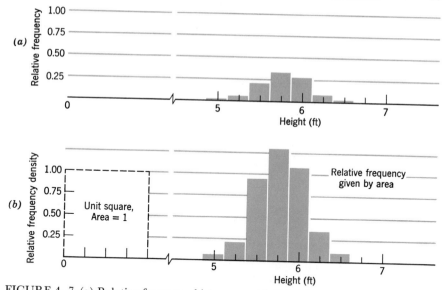

FIGURE 4–7 (a) Relative frequency histogram with Σ heights $= 1$. (b) Rescaled into relative frequency density, making total area $= 1$.

Thus, in Figure 4-7, panel (b) is four times as high as panel (a); we also confirm that panel (b) has area equal to 1.[9]

In Figure 4-8, we show what happens to the relative frequency density of a continuous random variable as:

1. Sample size increases.

2. Cell size decreases.

---

[9]To prove in general that the total area of the relative frequency density is always 1, we first obtain from (4-8) the following statement for each cell:

relative frequency $=$ (relative frequency density) $\times$ (cell width)

$=$ (height of relative frequency density bar) $\times$ (cell width)

$=$ area of relative frequency density bar.

Summing over all cells:

$\sum$ relative frequency $=$ total area of relative frequency density

Since the relative frequencies sum to 1, we have proved:

$$\boxed{\text{total area of relative frequency density} = 1}$$ (4-10)

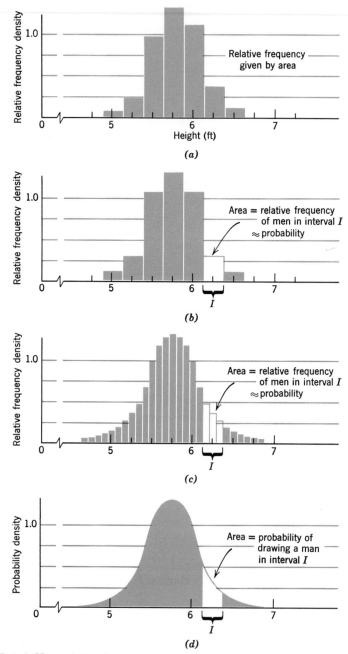

FIGURE 4–8 How relative frequency density may be approximated by a probability density as sample size increases, and cell size decreases. (a) Small n, as in Figure 4–7(b). (b) Large enough n to stabilize relative frequencies. (c) Even larger n, to permit finer cells while keeping relative frequencies stable. (d) For very large n, this becomes (approximately) a smooth probability density curve.

With a small sample, chance fluctuations influence the picture. But as sample size increases, chance is averaged out, and relative frequencies settle down to probabilities. At the same time, the increase in sample size allows a finer definition of cells. While the area remains fixed at 1, the relative frequency density becomes approximately a curve, the so-called probability density function, which we shall refer to simply as the probability distribution, denoted by $p(x)$.

If we wish to compute the mean and variance from Figure 4-8(c), the discrete formulas (4-3) and (4-4) can be applied. But if we are working with the probability density function in Figure 4-8(d), then integration (which, as calculus students will recognize, is the limiting case of summation) must be used; if a and b are the limits of X, then (4-3) and (4-4) become:

$$\text{Mean, } \mu = \int_a^b xp(x)\,dx \tag{4-11}$$

$$\text{Variance, } \sigma^2 = \int_a^b (x - \mu)^2 p(x)\,dx \tag{4-12}$$

All the theorems that we state about discrete random variables are equally valid for continuous random variables, with summations replaced by integrals. Proofs also are very similar. Therefore, to avoid tedious duplication, we give theorems for discrete random variables only.

4-5 THE NORMAL DISTRIBUTION

For many random variables, the probability distribution is a specific bell-shaped curve, called the *normal* curve, or *Gaussian* curve. It is the most useful probability distribution in statistics. For example, errors made in measuring physical and economic phenomena often are distributed normally. In addition, many other probability distributions (such as the binomial) often can be approximated by the normal curve.

(a) Standard Normal Distribution

A random variable Z is called *standard normal* if its probability distribution is:

$$p(z) = \frac{1}{\sqrt{2\pi}} e^{-(1/2)z^2} \tag{4-13}$$

The constant $1/\sqrt{2\pi}$ is a scale factor required to make the total area 1. The symbols π and e denote important mathematical constants, approxi-

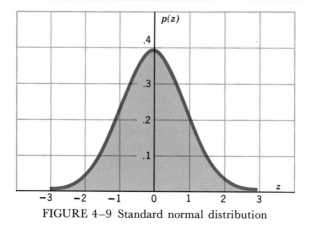

FIGURE 4–9 Standard normal distribution

mately 3.14 and 2.72 respectively. We draw the normal curve[10] in Figure 4-9 to reach a maximum at $z = 0$; we confirm in (4-13) that this is so. As we move to the left or right of 0, z^2 increases in the negative exponent; therefore $p(z)$ decreases, approaching zero in both tails. This curve also is symmetric: since z appears only in squared form, $-z$ generates the same probability in (4-13) as $+z$.

The mean and variance of Z can be calculated by integration using (4-11) and (4-12); since this requires calculus, we quote the results without proof:

$$\mu_Z = 0$$

$$\sigma_Z = 1$$

It is for this very reason, in fact, that Z is called a *standard* normal variable. Later when we speak of "standardizing" any variable, this is precisely what we mean: shifting it so that its mean is zero and rescaling it so that its standard deviation (or variance) is one.

The probability (area) enclosed by the normal curve above any specified value z_0 also requires calculus to evaluate precisely, but may be easily pictured, as in Figure 4-10. This evaluation of probability, done once and for all, has been recorded in Table IV of the Appendix. Students who do not know calculus can think of this as accumulating the area of the approximating rectangles, as in Figure 4-8(c). Table IV is very similar to the binomial probabilities in Table III(c).

---

[10]Problem 4-18 confirmed that the graph of (4-13) is the one shown in Figure 4-9.

The mathematical constant $\pi = 3.14$ must not be confused with the π used in Section 4-3 to designate probability of success.

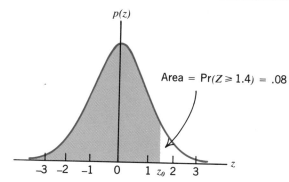

FIGURE 4–10 Probability enclosed by the standard normal curve above a given point, say $z_0 = 1.4$

There is one difference, however, between discrete and continuous random variables. Since a continuous random variable has zero probability at any specific point[11]—a feature already noted in Section 2-1—it makes no difference whether we include or exclude a single point in calculating a probability. Therefore, if Z is a *continuous* random variable, and c is any point:

$$\boxed{\Pr(Z \geq c) = \Pr(Z > c)}$$
(4-14)

In other words, $\geq$ and $>$ can be used interchangeably for continuous random variables.

Example 4-4. If Z has a standard normal distribution, find:
(a) $\Pr(Z > 1.0)$.
(b) $\Pr(Z < -1.0)$.
(c) $\Pr(1.0 < Z < 1.5)$.
(d) $\Pr(-1 < Z < 2)$.
(e) $\Pr(|Z| < 2)$.

Solution. Each is illustrated in the corresponding panel of Figure 4-11. Calculations are as follows:
(a) By Table IV:

---

[11]How can any specific point have zero probability, yet an interval consisting of points have positive probability? The resolution of this paradox rests on the fact that an interval consists of an *infinite* number of points. A rigorous discussion of this issue belongs to a branch of pure mathematics called *measure theory*.

$$\Pr(Z > 1.0) = .1587$$

(b) By symmetry:

$$\Pr(Z < -1) = \Pr(Z > +1)$$

by Table IV:

$$= .1587$$

(c) Take the probability above 1.0, and subtract from it the probability above 1.5:

$$\Pr(1.0 < Z < 1.5) = \Pr(Z > 1.0) - \Pr(Z > 1.5)$$

by Table IV:

$$= .1587 - .0668 = .0919$$

(d) Subtract the two tail areas from the total area of 1:

$$\Pr(-1 < Z < 2) = 1 - \Pr(Z < -1) - \Pr(Z > 2)$$

by Table IV and symmetry:

$$= 1 - .1587 - .0228 = .8185$$

(e) $\Pr(|Z| < 2) = \Pr(-2 < Z < 2)$

$$= 1 - \Pr(Z < -2) - \Pr(Z > 2)$$

$$= 1 - 2(.0228) = .9544$$

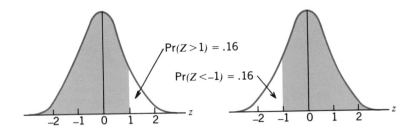

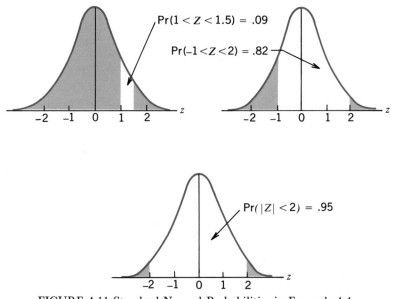

FIGURE 4-11 Standard Normal Probabilities in Example 4-4

Because of its importance as a frequent reference, Table IV is reproduced inside the back cover of this book.

(b) General Normal Distribution

In part (a), we considered only a very special normal distribution—the standard normal Z with mean 0 and standard deviation 1. Now consider the general form of the normal distribution, centered on *any* mean μ with *any* standard deviation σ. Its probability distribution has the formula:[12]

$$p(x) = \frac{1}{\sqrt{2\pi}\sigma}e^{-\left(\frac{1}{2}\right)\left(\frac{x-\mu}{\sigma}\right)^2} \qquad (4\text{-}15)$$

By using the transformation:

---

[12]To prove that (4-15) is centered at μ, we note that the peak of the curve occurs when the negative exponent attains its smallest value 0, i.e., when $x = \mu$. It also may be shown that (4-15) is scaled by the factor σ. Finally, it is bell-shaped for the same reasons given in part (a).

   We notice that in the very special case in which $\mu = 0$ and $\sigma = 1$, (4-15) reduces to the standard normal distribution (4-13).

$$Z = \frac{X - \mu}{\sigma} \qquad\qquad (4\text{-}16)$$

we transform X into a variable Z that has mean 0 and standard deviation 1 (as proved in Problem 4-9). That is, Z is the *standard* normal variable, as illustrated in Figure 4-12. In summary, to evaluate any normal variable X, we first transform X into Z, and then evaluate Z in the standard normal table (Appendix Table IV).

In Figure 4-12, we also note a feature that is very helpful in sketching normal curves: at a distance of 1σ from the mean μ, the normal distribution falls to 60% of its peak value; at a distance of 2σ from the mean μ, it falls to 14% of its peak value.

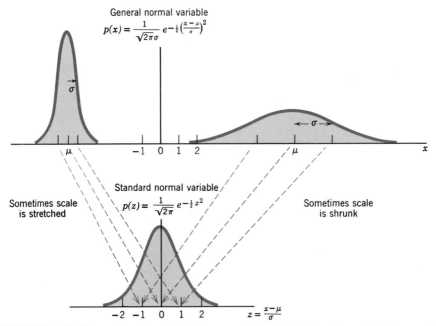

FIGURE 4–12 Linear transformation of any normal variable into the standard normal variable

Example 4-5. If X is normal with $\mu = 100$ and $\sigma = 5$, what is $\Pr(X > 110)$?

Solution. Any inequality (imbalance) is preserved if both sides are diminished by the same amount ($\mu = 100$) and divided by the same positive amount ($\sigma = 5$). We therefore rewrite the problem equivalently as:

$$\Pr(X > 110) = \Pr\left(\frac{X - \mu}{\sigma} > \frac{110 - 100}{5}\right)$$

by (4-16):

$$= \Pr(Z > 2.0)$$

by Table IV:

$$= .0228$$

Alternative Intuitive Solution. To get the probability that X exceeds 110, how many standard deviations from the mean is that? We may picture using a measuring stick that is $\sigma = 5$ units long, as in Figure 4-13. Clearly, the standardized Z value is $+2$. Thus the answer is, from Table IV,

$$\Pr(Z > 2) = .0228$$

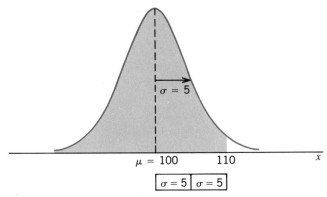

FIGURE 4–13 Intuitive view of normal probability

Example 4-6.   A bolt picked at random from a production line has a length X that is distributed normally with $\mu = 78.3$ mm. and $\sigma = 1.4$ mm. If bolts longer than 80 mm. have to be discarded, what proportion of output is wasted?

Solution.

$$\Pr(X > 80) = \Pr\left(\frac{X - \mu}{\sigma} > \frac{80 - 78.3}{1.4}\right)$$

$$= \Pr(Z > 1.21)$$

$$= .1131 \simeq 11\%$$

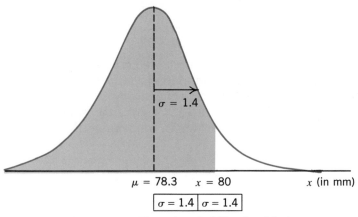

$\mu = 78.3$ $x = 80$ x (in mm)

$\sigma = 1.4$ $\sigma = 1.4$

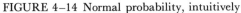

FIGURE 4-14 Normal probability, intuitively

Alternative Intuitive Solution.   Since the exact value of the standardized Z is not immediately clear from Figure 4-14, we need an algebraic approach:

(a) The critical value of 80 differs from the mean by $80 - 78.3 = 1.7$

(b) How many standard deviations is this? Since there are 1.4 mm. in each standard deviation, 1.7 mm. is:

$$\frac{1.7}{1.4} = 1.21 \text{ standard deviations}$$

Thus the answer, from Table IV, is:

$$Pr(Z > 1.21) = .1131 \simeq 11\%$$

PROBLEMS

4-19 If Z is a standard normal variable, calculate:
(a) $Pr(Z > 1.6)$.
(b) $Pr(1.6 < Z < 2.3)$.
(c) $Pr(Z < 1.64)$.
(d) $Pr(-1.64 < Z < -1.02)$.
(e) $Pr(|Z| > 1)$.
(f) $Pr(|Z| < 3)$.

⇒4-20 (a) How far above the mean of the Z distribution must we go so that only 1% of the probability remains in the right-hand tail? That is:

$$\text{if } Pr(Z \geq z_0) = .01, \text{ what is } z_0?$$

(b) How far on either side of the mean of the Z distribution must we go to include 95% of the probability? That is:

$$\text{if } Pr(-z_0 < z < z_0) = .95, \text{ what is } z_0?$$

4-21 If X is normal, calculate:
(a) $Pr(4.5 < X < 6.5)$ where $\mu_X = 5$ and $\sigma_X = 3$.
(b) $Pr(X < 800)$ where $\mu_X = 500$ and $\sigma_X = 200$.
(c) $Pr(9.9 < X < 10.1)$ where $\mu_X = 10.0$ and $\sigma_X = 0.2$.

4-22 Suppose that a population of men's heights is normally distributed with a mean of 68 inches, and standard deviation of 3 inches. Find the proportion of the men who are:
(a) Under 66″.
(b) Over 72″.
(c) Between 66″ and 72″.
To check your three answers, see whether they sum to 1.

4-6 A FUNCTION OF A RANDOM VARIABLE

Considering again the planning of three children, suppose for example, that the cost of sports equipment R is a function of the number of boys X in the family, that is:

$$R = g(X)$$

Specifically, suppose that:[15]

$$R = -X^2 + 4X + 5.$$

which is equally well given by Table 4-4.

TABLE **4-4**
Tabled Form of
$R = g(X) = -X^2 + 4X + 5$

Value of X	Value of R = g(X)
0	5
1	8
2	9
3	8

The values of R customarily are rearranged in order, as shown in the third column of Table 4-5. Furthermore, the values of R have certain probabilities that may be deduced from the previous probabilities[16] of X.

TABLE **4-5**
Mean of R Calculated[a] by First Deriving the Probability Distribution of R

(1) x	(2) p(x)	(3) r = g(x)	(4) p(r)	(5) rp(r)
• 0	1/8	• 5	1/8	5/8
• 1	3/8	• 8	4/8	32/8
• 2	3/8	• 9	3/8	27/8
• 3	1/8			$\mu_R = 8$

[a]In this table, recall our notation: whereas capital X represents a random variable, small x represents a value that it may take.

For example, the two X values of 1 and 3 give rise to an R value of 8, as shown with arrows in Table 4-5.

---

[15]A brief justification of this relationship might be that boys tend to use more sports equipment than girls, so that R tends to increase with X. Yet certain savings are possible if all the children are the same sex, so that R finally drops slightly when $X = 3$.

[16]Deducing the probabilities of R from earlier probabilities is similar to deducing the probabilities of X from the earlier probabilities in the original sample space, in Figure 4-1.

What will be the average cost μ_R? From the distribution of R in Table 4-5, this is calculated to be $\mu_R = 8$. However, often it is more convenient to omit the distribution of R and calculate the mean of R directly from the distribution of X, as shown in Table 4-6.

TABLE **4-6**
Mean of R Calculated
Directly from $p(x)$, as an
Easier Alternative to
Table 4-5

x	$g(x)$	$p(x)$	$g(x)p(x)$
0	5	1/8	5/8
1	8	3/8	24/8
2	9	3/8	27/8
3	8	1/8	8/8

$$\mu_R = 8 \, \checkmark$$

It is easy to see why this works: in a disguised way, we are calculating μ_R in the same way as in Table 4-5. For example, the rows for $X = 1$ and $X = 3$ in Table 4-6 appear condensed together as the single row for $R = 8$ in Table 4-5. Similarly, the other rows in Table 4-6 correspond to rows in Table 4-5, so that both tables yield the same value for μ_R. The only difference, really, is that Table 4-6 is ordered according to X values, while Table 4-5 is ordered (and condensed) according to R values.

This example can easily be generalized. If X is a random variable and g is any function, then $R = g(X)$ is a random variable. μ_R may be calculated either from the probability function of R, or alternatively from the probability function of X according to:

$$\mu_R = \sum_x g(x)p(x) \qquad (4\text{-}17a)$$

4-7 NOTATION

Some new notation will help us better understand the various viewpoints of the mean. For any random variable X, all the following terms have exactly the same meaning:[17]

---

[17] The reason for so many names is historical. For example, gamblers and economists use the term "*expected gain,*" meteorologists use the term "*mean* annual rainfall," and teachers use the term "*average* grade."

$$\mu_X = \text{mean of } X$$

$$= \text{average of } X$$

$$= \text{expectation of } X$$

$$= E(X), \text{ the expected value of } X$$

The term $E(X)$ is introduced because it is useful as a reminder that it represents a weighted sum (E looks like Σ). With this new notation, the result (4-17a) can be written:

$$E(R) = \sum_x g(x)p(x) \tag{4-17b}$$

Finally, since we recall that R was just an abbreviation for $g(X)$, we may equally well write (4-17b) as:

$$\boxed{E[g(X)] = \sum_x g(x)p(x)} \tag{4-17c}$$

As an example of this notation, we may write:

$$E(X - \mu)^2 = \sum_x (x - \mu)^2 p(x) \tag{4-18}$$

By (4-4):

$$\boxed{E(X - \mu)^2 = \sigma^2} \tag{4-19}$$

Thus we see that σ^2 may be regarded as just a kind of expectation— namely, the expectation of the random variable $(X - \mu)^2$. In the new notation, we also may rewrite (4-5) as:[18]

$$\sigma^2 = E(X^2) - \mu^2 \tag{4-20}$$

---

[18] An interesting corollary to (4-20) is obtained by applying the theory of linear transformations. From Table 4-2:

$$\text{if we let } Y = X - a, \text{ then } \sigma_Y^2 = \sigma_X^2 \text{ and } \mu_Y = \mu_X - a \tag{4-21}$$

Now (4-20) is true for any random variable, Y in particular. Thus:

$$\sigma_Y^2 = E(Y^2) - \mu_Y^2 \tag{4-22}$$

(cont'd)

PROBLEMS

4-23 Let X be the number of boys in a 4-child family. If $R = -X^2 + 5X + 3$, find $E(R)$:
(a) By first calculating $p(r)$ as in Table 4-5.
(b) Directly from $p(x)$, as in Table 4-6.

4-24 Again let X be the number of boys in a 4-child family. If $R = |X - 2|$:
(a) Find the mean and variance of R.
(b) In Table 4-2, we stated that for a linear function, the mean follows the same formula as the individual values. Does this hold true for non-linear functions? In particular, does $\mu_R = |\mu_X - 2|$?

4-25 The time T, in seconds, required for a rat to run a maze, is a random variable with the following probability distribution.

t	2	3	4	5	6	7
$p(t)$	.1	.1	.3	.2	.2	.1

(a) Find the average time.
(b) Suppose that the rat is rewarded with 1 biscuit for each second faster than 6. (For example, if he takes just 4 seconds, he gets a reward of 2 biscuits. Of course, if he takes 6 seconds or longer, he gets no reward.) What is the rat's average reward?
(c) Suppose that the rat is punished by getting a shock that increases sharply as his time increases—specifically, a shock of T^2 volts for a time of T seconds. What is the rat's average punishment?
(d) What is the variance of T?

Substituting (4-21), we finally obtain:

$$\sigma_X^2 = E(X - a)^2 - (\mu_X - a)^2 \qquad (4\text{-}23)$$

The practical application of (4-23) comes from letting a be the rounded value of μ_X, so that the rounding error $e = \mu_X - a$ is small. Thus, $e^2 = (\mu_X - a)^2$ is negligibly small, and:

$$\sigma_X^2 \simeq E(X - a)^2 \qquad (4\text{-}24)$$

We conclude that calculating σ^2 around a good approximation of μ (rather than the true μ) introduces very little error; in fact, the error in σ^2 will be merely the squared error in approximating μ (i.e., the last term in (4-23)).

The analogous formula for samples was used extensively in Chapter 2, especially (2-8).

4-26 Consider the simple linear function:

$$R = a + bX$$

We already have stated (in Table 4-2) a formula for its mean:

$$\mu_R = a + b\mu_X$$

Prove this, using (4-17a).

4-27 (a) A certain game has a reward that depends on the number of dots X when a die is rolled. Tabulate the distribution of X, and calculate the average $E(X)$.
(b) The reward is a linear function $Y = 2X + 8$. Tabulate the distribution of Y, and from it calculate the average reward $E(Y)$.
(c) Is it true that the means satisfy the same relation as the individual values; that is:

$$\text{if } Y = 2X + 8$$

$$\text{then } E(Y) = 2E(X) + 8?$$

Or, to put it more directly:

$$E(2X + 8) = 2E(X) + 8?$$

(d) Graph the distributions of X and of Y as in Figure 4-4, illustrating what you found in part (c).

4-28 Repeat Problem 4-27 for the nonlinear function $Y = X^2$.

REVIEW PROBLEMS

4-29 Suppose that X is a very simple discrete random variable, having $p(2) = p(4) = 1/2$. For each of the following, state true or false, and back up your answer either by calculating both sides of the equation, or by appealing to a general theorem.
(a) $E(X + 10) = E(X) + 10$.
(b) $E(X/10) = E(X)/10$.
(c) $E(10/X) = 10/E(X)$.
(d) $E(5X^2 + 10) = 5E(X^2) + 10$.

4-30 When is it true that $E(X^2) = [E(X)]^2$? Select the one correct answer.
   (a) Always.
   (b) Never.
   (c) Iff X has zero mean.
   (d) Iff X has zero variance.
   (e) Iff X has positive mean.

4-31 IQ scores are distributed approximately normally, with a mean of 100 and standard deviation of 15. What proportion of IQs are:
   (a) Over 140?
   (b) Between 100 and 140?

4-32 Referring ahead to Problem 6-29, suppose that a student answers it by sheer guessing. Let $X =$ the number of correct matches. Find the probability distribution of X, and its mean and variance.

4-33 In the 1964 presidential election, 60% voted Democratic and 40% voted Republican. Calculate the probability that a random sample would correctly forecast the election winner; i.e., that a majority of the sample would be Democrats, if the sample size were:
   (a) $n = 1$.
   (b) $n = 3$.
   (c) $n = 9$.
   Note how the larger sample increases the probability of a correct forecast.

⇒4-34 (The "Sign Test.") Eight volunteers are to have their breathing capacity measured before and after a certain treatment, and recorded in a layout like the following:

Person / Breathing Capacity	Before	After	Improvement
A	2750	2850	+100
B	2360	2380	+20
C	2950	2800	−150
D	.	.	.
E	.	.	.
:	:	:	:

   (a) Suppose that the treatment has no effect whatever, so that the "improvements" represent random fluctuations (resulting from measurement error or minor variation in a person's performance). Also assume that measurement is so precise that a zero improvement is never observed.

What is the probability that seven or more signs will be $+$?

(b) If it actually turned out that seven of the eight signs were $+$, would you question the hypothesis in (a) that the treatment has no effect whatever?

4-35 In a learning experiment, a subject attempts a certain task twice in a row. Each trial can result in either success S or failure F.

On his first trial, the chance of F is 1/3.

If his first trial was F, he has a 1/2 chance of F on the second trial. If his first trial was S, however, he has only a 1/8 chance of F on the second trial. That is, he seems to be "encouraged by success."

(a) Find the probability table and mean of the total number of failures X.

(b) Find the (unconditional) probability of failure on the second trial.

4-36 Another subject runs through the learning experiment in Problem 4-35, with the following difference: on the second trial, his probability of failure is 1/4, independent of the first trial's outcome.

(a) Find the probability table and mean of the total number of failures X.

*(b) Suppose that in continuing the experiment for two more trials, he consistently learned, so that his probability of failure was 1/5 on the third trial, and 1/6 on the fourth trial. Without taking the trouble to tabulate the distribution of X, can you guess what the mean of X is now?

4-37 The following scheme, called a guilty-knowledge test, has been proposed to help the police narrow down the list of suspects to a crime. Suppose, for example, that a loan company has been robbed, and that a suspect is being interrogated. The police officer reads him the following statement:

Before showing his gun, the robber pretended to take out a loan for a certain purpose. If you're the guilty man, you will know whether that purpose was to buy a car, to pay doctor bills, to pay for a vacation trip, to buy a color TV, or to get a present for his wife. I'm going to name each of these five possibilities in order, and I want you to sit quietly and just repeat what I say.

The suspect is hooked up to a polygraph (lie detector) that records which of the 5 possibilities he reacts most strongly to. If the suspect is guilty, it is likely that his strongest reaction will be to the correct answer. If he is not guilty, it is equally likely that his strongest reaction will be to any one of the 5 possibilities.

Several similar questions are given the suspect (for example, a question about the name of the loan company, or a question about the color of the thief's shirt, etc.). Now answer true or false; if false, correct it.
(a) With 10 guilty-knowledge questions of this type, there is only about one chance in 10 million that an innocent suspect will react strongly to the "correct" alternative in all 10 questions.
(b) And if a suspect shows guilty knowledge in as many as 6 out of 10 questions, the chances that he is innocent are only 1 in 1,000.[19]

*4-38 (Requires calculus.) Suppose that a continuous random variable X has the probability distribution:

$$p(x) = 3x^2 \qquad 0 \le x \le 1$$

$$= 0 \qquad \text{otherwise}$$

(a) Graph $p(x)$.
(b) Find the mean, median, and mode. Graph them. Are they in the order you expect?
(c) Find σ^2.

*4-39 Repeat Problem 4-38 for the following alternatives:
   (i) $p(x) = 6x(1 - x)$ $0 \le x \le 1$.
   (ii) $p(x) = 12x^2(1 - x)$ $0 \le x \le 1$.

---

[19]Quoted from D. T. Lykken, "The Right Way to Use a Lie Detector," *Psychology Today*, March 1975, p. 58.

CHAPTER 5

Two Random Variables

If you bet on a horse, that's gambling. If you bet you can make three spades, that's entertainment. If you bet cotton will go up three points, that's business. See the difference?

Blackie Sherrod

5-1   DISTRIBUTIONS

This first section is a simple extension of the last two chapters. The main problem will be to recognize the old ideas behind the new names. We therefore give an outline in Table 5-1, both as an introduction and a review.

(a) Joint Distributions

In the planning of three children, let us define two random variables:

$$X = \text{number of boys}$$

$$Y = \text{number of changes in sequence}$$

Suppose that we are interested in the probability of two boys and one change of sequence occurring together. As usual, we refer to the sample space of the experiment, shown in Table 5-2. The intersection of these two events has probability:

$$\Pr(X = 2 \cap Y = 1) = \frac{2}{8} \tag{5-1}$$

which we denote $p(2, 1)$ for simplicity.

TABLE **5-1**
Outline of Section 5-1

Old Idea		New Terminology	
$\Pr(G \cap H)$ $\qquad\qquad$ (3-14) applied to: $\Pr(X = 2 \cap Y = 1)$ $\Pr(X = x \cap Y = y)$ in general		*Joint* distribution: $p(2, 1)$ $p(x, y)$ in general $\qquad\qquad$ (5-2)	
$\Pr(H/G) = \dfrac{\Pr(H \cap G)}{\Pr(G)}$ $\quad$ (3-26) applied to: $\Pr(X = 2/Y = 1)$ $\Pr(X = x/Y = y)$ in general		*Conditional* distribution: $p(2/Y = 1)$ $p(x/y)$ in general $\qquad\qquad$ (5-10)	
Event F is independent of E iff: $\qquad \Pr(F/E) = \Pr(F)$ $\quad$ (3-29) or: $\qquad \Pr(E \cap F) = \Pr(E)\Pr(F)$ $\quad$ (3-30)		Variable X is independent of Y iff: $p(x/y) = p(x)$ $\qquad\qquad$ (5-12) or: $p(x, y) = p(x)\,p(y)$ $\qquad\qquad$ (5-13) for all x and y.	

TABLE **5-2**
Two Random Variables Defined on the Same Sample Space

(1) Outcome e	(2) Corresponding X value	(3) Corresponding Y value
• BBB	3	0
• BBG	2	1
• BGB	2	2
• BGG	1	1
• GBB	2	1
• GBG	1	2
• GGB	1	1
• GGG	0	0

$X = 2$ $\qquad$ $Y = 1$

TABLE **5-3**

$p(x,y)$, The Joint Distribution of X and Y,
Summarizing Table 5-2

$p(x)\longrightarrow$	1/8	3/8	3/8	1/8	1 ✓
2	0	1/8	1/8	0	2/8
1	0	2/8	2/8	0	4/8
0	1/8	0	0	1/8	2/8
↑y = value of Y x = value of X →	0	1	2	3	↑ $p(y)$

Similarly, we could compute $p(0, 0)$, $p(0, 1)$, $p(0, 2)$, $p(1, 2)$, . . . , obtaining in Table 5-3 what is called the *joint* (or *bivariate*) *probability distribution of X and Y.* (Actually, the easiest way to derive the joint distribution is to run down the last two columns of Table 5-2, tabulating all this information line by line into the appropriate cell of Table 5-3.[1] Note that x and y in Table 5-3 are designated in the lower and left-hand margins; the information in the upper and right-hand margins can be temporarily ignored.)

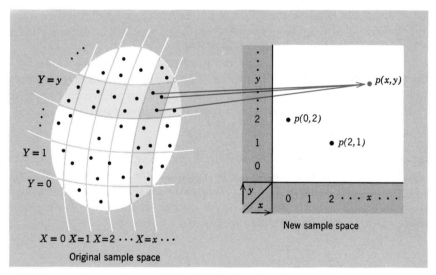

FIGURE 5–1 Two random variables (X, Y), showing their joint probability distribution derived from the original sample space (compare with Figure 4–2)

---

[1]For example, the first line in Table 5-2 provides information on the southeast element of Table 5-3 (where $X = 3$ and $Y = 0$).

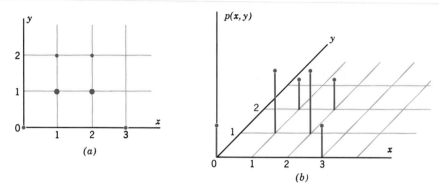

FIGURE 5–2 Two graphic presentations of the joint distribution in Table 5–3. (a) $p(x, y)$ is represented by the size of the dot. (b) $p(x, y)$ is represented by height.

The formal definition of the joint probability distribution is:

$$p(x, y) \stackrel{\Delta}{=} \Pr(X = x \cap Y = y) \tag{5-2}$$

The general case is illustrated in Figure 5-1. The events $X = 0$, $X = 1$, $X = 2$, . . . , form a partition of the sample space, shown schematically as a vertical slicing. Similarly, the events $Y = 0$, $Y = 1$, . . . , form a partition shown as a horizontal slicing. The intersection of the slice $X = x$ and the slice $Y = y$ is the event $(X = x \cap Y = y)$. Its probability is collected into the table, and denoted $p(x, y)$.

The distribution $p(x, y)$ may be graphed. For example, Table 5-3 is represented graphically in Figure 5-2. The easiest method is to represent each probability by an appropriately sized dot, as shown in panel (a); or represent each probability by a height, as shown in panel (b).

(b) Marginal Distributions

Suppose that we are interested only in X, yet have to work with the joint distribution of X and Y. How can we compute the distribution of X; for example, $p(2)$? The event $X = 2$ is a vertical slice in the schematic sample space of Figure 5-1. Its probability, of course, is the sum of the probabilities of all those chunks comprising it:

$$p(2) = p(2, 0) + p(2, 1) + p(2, 2) + p(2, 3) + \cdots + p(2, y) + \cdots \tag{5-3}$$

$$= \sum_y p(2, y) \tag{5-4}$$

and in general, for any given x:

$$p(x) = \sum_y p(x,y)$$ (5-5)

This idea may be applied to Table 5-3. We thus find, for example:

$$p(2) = 0 + \frac{2}{8} + \frac{1}{8} = \frac{3}{8}$$ (5-6)

and place this sum in the upper margin. Similarly, $p(x)$ is computed for every x, thus providing the whole row in the upper margin. This is sometimes called the *marginal* distribution of X, to describe how it was obtained. But, of course, it is just the ordinary distribution of X (which could have been found without any reference to Y, as indeed it was in Figure 4-1).

In conclusion, the word "marginal" merely describes how the distribution of X may be calculated when another variable Y is in play; a column sum is calculated and placed "in the margin."

In the same way, we calculate $p(y)$, the marginal distribution[2] of Y, set out in the right-hand margin of Table 5-3; each value of $p(y)$ is a row sum.

(c) Conditional Distributions

From Table 5-3, suppose that we wish to know the probabilities of various numbers of boys, given one change in sex sequence. In other words, how should we evaluate the conditional distribution of X, given $Y = 1$? First, the appropriate row when $Y = 1$ is reproduced in Table 5-4. The problem is that the probabilities in this row sum to only $1/2$; hence they cannot represent a true probability distribution. They do, however, give us the *relative* probabilities of various X values. (Thus, if we know that $Y = 1$, we know that X cannot be 0 or 3, and that X values of 1 or 2 are equally probable.) Therefore, to finally obtain a true probability distribution, we double these probabilities so that they will now sum to 1; the resulting distribution is denoted $p(x/Y = 1)$ in Table 5-4.

---

[2]Strictly speaking, $p(y)$ is not an adequate notation, and may cause ambiguity on occasion. For example, from Table 5-3 we can find any particular value for the marginal distribution of Y; when $y = 2$, for instance, we find $p(2) = 2/8$. This seems to contradict (5-6). We could resolve this contradiction by more careful notation:

$$p_Y(2) = \frac{2}{8} \quad \text{whereas} \quad p_X(2) = \frac{3}{8}$$

Such subscripts are so awkward, however, that normally, we will avoid them. For example, we shall avoid writing $p_X(x)$ for the general value of the probability function, even though it is, strictly speaking, the most correct form.

TABLE 5-4
Derivation of the Conditional Distribution of X,
Given $Y = 1$

x	0	1	2	3	
$p(x, 1)$	0	2/8	2/8	0	Sum = Pr$(Y = 1)$ = $p(1) = 1/2$
$p(x/Y = 1)$	0	1/2	1/2	0	Sum = 1 ✓

Formally, this doubling can be justified rigorously by the theory in Chapter 3, where conditional probability was:

$$\Pr(H/G) = \frac{\Pr(H \cap G)}{\Pr(G)} \qquad \text{(3-26) repeated}$$

For H and G, we simply substitute events defined in terms of random variables, as follows:

$$\text{For } H, \text{ substitute } (X = x)$$
$$\text{For } G, \text{ substitute } (Y = 1) \qquad (5\text{-}7)$$

Thus:

$$\Pr(X = x/Y = 1) = \frac{\Pr(X = x \cap Y = 1)}{\Pr(Y = 1)}$$

In new notation:

$$p(x/Y = 1) = \frac{p(x, 1)}{p(1)} \qquad (5\text{-}8)$$

In our example, $p(1) = \frac{1}{2}$, so that (5-8) becomes:

$$p(x/Y = 1) = 2p(x, 1)$$

which justifies the doubling in Table 5-4.
   The generalization of (5-8) is clearly:

$$p(x/Y = y) = \frac{p(x, y)}{p(y)} \qquad (5\text{-}9)$$

This conditional distribution may be further abbreviated to $p(x/y)$, giving:

$$p(x/y) = \frac{p(x,y)}{p(y)} \tag{5-10}$$

Note how similar this is to equation (3-26).

(d) Independence

We define the independence of two random variables in terms of the independence of two events, developed in Chapter 3:

The random variables X and Y are called independent iff for every x and y, the events $(X = x)$ and $(Y = y)$ are independent. (5-11)

The consequences are derived easily. From (3-29) we know that the independence of events $(X = x)$ and $(Y = y)$ means that:

$$\Pr(X = x/Y = y) = \Pr(X = x)$$

In other words:

X and Y are independent iff $p(x/y) = p(x)$ for all x and y

(5-12)
like (3-29)

By substituting (5-10) and then solving for $p(x,y)$, we can obtain the equivalent form:

X and Y are independent iff $p(x,y) = p(x)\,p(y)$ for all x and y

(5-13)
like (3-30)

For the distribution in Table 5-3, we can easily show that X and Y are not independent. For independence, (5-13) must hold for every (x, y) combination. We ask whether it holds, for example, when $x = 0$ and $y = 0$? The answer is no:

$$\frac{1}{8} \neq \frac{1}{8} \cdot \frac{2}{8}$$

Thus X and Y are dependent.

Example 5-1. Suppose that X and Y have the following joint distribution:

$$p(x,y)$$

40	.12	.24	.24	.15
20	.04	.08	.08	.05

y				
x	20	40	60	80

(a) Graph $p(x, y)$.
(b) Find $p(y)$, $E(Y)$, var(Y).
(c) For $X = 40$, find the conditional distribution of Y, and the conditional mean and variance.
(d) Are X and Y independent?

Solution. (a)

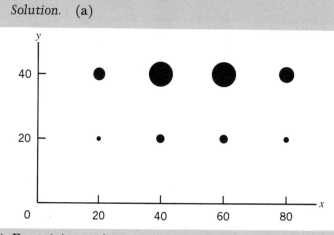

(b) For $p(y)$, we just sum each row in the table above, obtaining:

y	$p(y)$	$yp(y)$	$(y - \mu)$	$(y - \mu)^2 p(y)$
40	.75	30	5	18.75
20	.25	5	−15	56.25

$$E(Y) = 35 \qquad\qquad \sigma^2 = 75$$

(c) For $p(y/X = 40)$, we start with the column where $x = 40$. We "norm" it by dividing by its total probability .32, obtaining:

y	$p(y/X = 40)$	$yp(y/X = 40)$	$(y - \mu)$	$(y - \mu)^2 p(y/X = 40)$
40	.24/.32 = .75	30	5	18.75
20	.08/.32 = .25	5	−15	56.25

$$1.00 \checkmark \quad\quad E(Y/X = 40) \quad\quad\quad\quad \text{var}(Y/X = 40)$$
$$= 35 \quad\quad\quad\quad\quad\quad\quad = 75$$

Incidentally, if we had noticed, in comparing (b) and (c), that $p(y/X = 40) = p(y)$, we *immediately* could have concluded that the conditional mean and variance are the same as the unconditional mean and variance, and our further calculations in (c) would have been unnecessary.

(d) Since $p(y/X = 40) = p(y)$, we already have partly shown that X and Y are independent. To finish the job, we could show that $p(y/X = 20) = p(y)$. This would establish that $p(y/x) = p(y)$ for all x and y, which proves independence (according to equation (5-12)).

However, to prove independence, it is generally more systematic to use (5-13). First calculate the marginal distributions $p(y)$ and $p(x)$ (by the row and columnwise summation of $p(x,y)$ in its table above, just as we did in Table 5-3). Then multiply them cell by cell (as illustrated in the first column below) to get the following table of $p(x)p(y)$:

$\overrightarrow{p(x)}$		.16	.32	.32	.20	
40	(.16)(.75) = .12	.24	.24	.15	.75
20	(.16)(.25) = .04	.08	.08	.05	.25
y		20	40	60	80	$p(y)\uparrow$
$\quad\quad x$						

Since this table agrees everywhere with the original table of $p(x, y)$, we have proved that $p(x, y) = p(x) p(y)$ for all x, y. Thus, according to (5-13), X and Y are indeed independent.

PROBLEMS

5-1 In a family of 4 children, again let:

$$X = \text{number of boys}$$
$$Y = \text{number of changes of sequence}$$

List the sample space, and then:

(a) Find the bivariate distribution and its graph.

(b) Are X and Y independent?

5-2 Suppose that X and Y have the following joint distribution:

6	.10	.15
4	.20	.30
2	.10	.15
y / x	5	10

Answer the same questions as you did in Problem 5-1.

(5-3) Suppose that X and Y have the following bivariate distribution:

3	0	.1	.1
2	.1	.4	.1
1	.1	.1	0
y / x	0	1	2

Answer the same questions as in Problem 5-1.

5-2   A FUNCTION OF TWO RANDOM VARIABLES

In Section 4-6, we analyzed a derived random variable R, which was some function of a random variable X:

$$R = g(X) \qquad (5-14)$$

In this chapter, we similarly shall analyze a derived variable R, which is some function of a *pair* of random variables X, Y:

$$R = g(X, Y) \qquad (5-15)$$

The concepts and proofs of this section therefore will be similar to those of the previous chapter, the main difference being that the joint

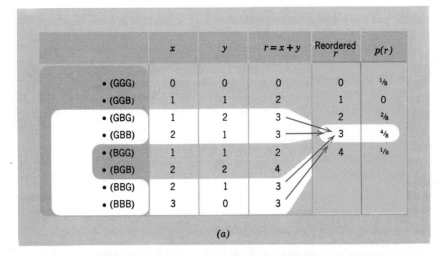

	x	y	$r = x + y$	Reordered r	$p(r)$
• (GGG)	0	0	0	0	1/8
• (GGB)	1	1	2	1	0
• (GBG)	1	2	3	2	2/8
• (GBB)	2	1	3	3	4/8
• (BGG)	1	1	2	4	1/8
• (BGB)	2	2	4		
• (BBG)	2	1	3		
• (BBB)	3	0	3		

(a)

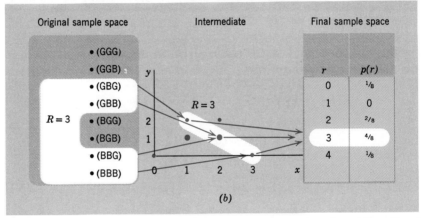

Original sample space   Intermediate   Final sample space

r	$p(r)$
0	1/8
1	0
2	2/8
3	4/8
4	1/8

(b)

FIGURE 5–3 Two views of the derivation of the probability function of $R = X + Y$. (a) Directly. (b) Using $p(x, y)$ as an intermediate condensation.

probability distribution $p(x, y)$ will replace the probability distribution $p(x)$. We shall be particularly interested in the distribution and mean of R.

Following our usual procedure, we develop the argument first in terms of a simple example, the sample space for a family of three children, as shown in Figure 5-3. Suppose, for example, that excess clothing costs are simply the sum[3] of X and Y:

---

[3] The sum $X + Y$ would reflect excess clothing costs if, for example, boys were more expensive to clothe than girls, and if changes in sex sequence interfered with the convenient passing-on of clothing.

$$R = g(X, Y) = X + Y \tag{5-16}$$

In Figure 5-3, we show how $p(r)$ may be derived directly from the original sample space, or indirectly by means of $p(x, y)$. In either case, the result is the same.

On the one hand, consider the direct derivation in panel (a). We note that the event $R = 3$, for example, consists of four of the eight equiprobable outcomes. Thus $p(3)$ is $4/8$. On the other hand, in panel (b) we derive $p(3)$ indirectly, using $p(x, y)$ in the intermediate sample space.

The expectation $E(R)$ may similarly be derived in two ways. On the one hand, applying the definition (4-3) to the probability distribution of R, we have:

$$
\begin{aligned}
E(R) &= \sum rp(r) \\
&= 0(\tfrac{1}{8}) + 1(0) + 2(\tfrac{2}{8}) + 3(\tfrac{4}{8}) + 4(\tfrac{1}{8}) \\
&= 2\tfrac{1}{2}
\end{aligned}
\tag{5-17}
$$

On the other hand, let us try to arrive at the same result[4] by using $p(x, y)$ in an extension of (4-17c):

$$
\begin{aligned}
E(R) = E(X + Y) &= \sum_x \sum_y (x + y)p(x, y) \\
&= (0 + 0)\tfrac{1}{8} + (0 + 1)0 \\
&\quad + (1 + 0)0 + (1 + 1)\tfrac{2}{8} \cdots \\
&\quad + (1 + 2)\tfrac{1}{8} + (2 + 1)\tfrac{2}{8} + (3 + 0)\tfrac{1}{8} \\
&= 2\tfrac{1}{2} \quad \text{again}
\end{aligned}
\tag{5-18}
$$

So (5-18) does in fact work, and it is easy to see why. That last line of (5-18), for example, amounts to:

$$3(\tfrac{1}{8} + \tfrac{2}{8} + \tfrac{1}{8}) = 3(\tfrac{4}{8})$$

which is the same as the second last term of (5-17). Continuing in this fashion, we see that (5-18) is just a disguised form of the more condensed form (5-17).

---

[4]Although the calculations in (5-17) and (5-18) (and indeed throughout this chapter) are displayed in a long line for typographical reasons, in practice it is better to carry out the computations in tabular form, as in Table 4-1, for example.

This example can be generalized easily. If $R = g(X, Y)$ is a function of two random variables, then:

$$\boxed{E(R) = E[g(X, Y)] = \sum_x \sum_y g(x,y)p(x,y)} \qquad \begin{array}{l}(5\text{-}19)\\ \text{like (4-17c)}\end{array}$$

PROBLEMS

5-4 Suppose that:

$$R = XY$$

$$T = (X - 1)(Y - 2)$$

and that X and Y have the following joint distribution:

4	0	.1	.1
2	.1	.4	.1
0	.1	.1	0
y			
x	0	1	2

(a) Find the mean of R using (5-19).
(b) Find the mean of R, by first finding its distribution and then proceeding as we did in (5-17).
(c) Find $E(T)$ any way you like.

(5-5) Repeat Problem 5-4, where:

$$R = (2X - Y)^2$$

$$T = 4X + 2Y$$

⇒5-6 In a certain gambling game, a pair of honest, three-sided dice are thrown. Let:

$$X = \text{number on first die}$$

$$Y = \text{number on the second die}$$

The joint probability distribution of X and Y is, of course,

3	1/9	1/9	1/9
2	1/9	1/9	1/9
1	1/9	1/9	1/9
y / x	1	2	3

and the total number of dots S is:

$$S = X + Y$$

(a) Find the distribution of S, and its mean and variance.
(b) Find the mean and variance of X and of Y.
(c) Do you see the relation between (a) and (b)?

⇒5-7 Suppose that the gambling game of Problem 5-6 is complicated by using loaded dice, as follows:

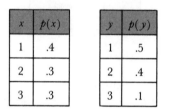

x	$p(x)$
1	.4
2	.3
3	.3

y	$p(y)$
1	.5
2	.4
3	.1

Assuming that the dice are tossed independently, tabulate the joint distribution of X and Y, and then answer the same questions as before.

5-3   COVARIANCE AND CORRELATION

(a) Covariance

In this section, we shall develop a measure of how well two variables are linearly related. As an example, consider the joint distribution tabled in Figure 5-4(a). We notice some tendency for these two variables to move together: a large x tends to be associated with a large y, and a small x with a small y.

Our measure of how the variables move together should be independent of where the variables happen to be centered. We therefore consider the deviations from the mean:

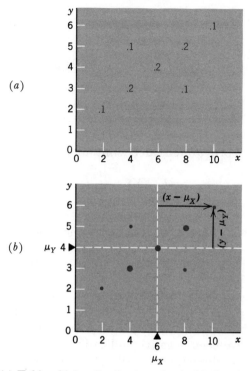

FIGURE 5–4 (a) Table of joint distribution $p(x, y)$. (b) Graph of $p(x, y)$, showing the definition of new variables $(X - \mu_X)$ and $(Y - \mu_Y)$ that translate axes into new dotted position.

$$(x - \mu_X) \quad \text{and} \quad (y - \mu_Y) \tag{5-20}$$

which are shown in Figure 5-4(b) for a typical point; the geometric interpretation of (5-20) is the dotted translation of the axes. Now let us multiply the deviations, obtaining the product:

$$(x - \mu_X)(y - \mu_Y) \tag{5-21}$$

For any point in the NE (North-East) quadrant of Figure 5-4(b), both deviations are positive, so their product (5-21) is positive. This also holds for any point in the SW quadrant, since both deviations are negative. However, for points in the other two quadrants, the product is negative. We can obtain a good measure of how X and Y vary together if we sum all these products, attaching the appropriate probability weights to each. This is called the *covariance:*

$$\text{covariance, } \sigma_{XY} \overset{\Delta}{=} \sum_x \sum_y (x - \mu_X)(y - \mu_Y) \, p(x,y) \qquad (5\text{-}22)$$

We may rewrite this with our E notation (5-19) as:

$$\boxed{\sigma_{XY} = E(X - \mu_X)(Y - \mu_Y)} \qquad \begin{array}{l}(5\text{-}23)\\ \text{like } (4\text{-}19)\end{array}$$

The computation of σ_{XY} often can be simplified by using the formula:[5]

$$\sigma_{XY} = E(XY) - \mu_X \mu_Y \qquad \begin{array}{l}(5\text{-}24)\\ \text{like } (4\text{-}20)\end{array}$$

For the distribution in Figure 5-4, the covariance (5-22) is:

$$\sigma_{XY} = (4)(2)(.1) + (2)(1)(.2) + \cdots + (-4)(-2)(.1) = +2.0$$

σ_{XY} was positive in this case because the variables moved together; that is, the heavier probabilities occurred in the NE and SW quadrants. If the heavy probabilities had occurred in the other two quadrants, the covariance would have been negative, indicating the tendency for X and Y to move in opposite directions. Finally, if the probabilities had been evenly distributed in all four quadrants, the covariance would have been zero, indicating no tendency for X and Y to move together.

(b) Correlation

The covariance still can be improved. As it now stands, it depends upon the units in which X and Y are measured. If X were measured, for example, in feet instead of inches, each x-deviation in (5-21), and hence σ_{XY} itself, would unfortunately change by a factor of 12. To eliminate this difficulty, consider a modified concept called the correlation[6] ρ:

$$\boxed{\text{correlation,}^7 \; \rho_{XY} \overset{\Delta}{=} \frac{\sigma_{XY}}{\sigma_X \, \sigma_Y}} \qquad (5\text{-}25)$$

---

[5] The proof of (5-24) is very similar to the proof of (4-5), and is left to Problem 5-13. Incidentally, if $Y = X$, then (5-24) reduces to (4-20), and (5-23) reduces to (4-19).

[6] ρ is the Greek letter rho, which corresponds to the Roman r (for relation).

[7] The correlation may be reexpressed as an expectation by substituting (5-23) into (5-25), obtaining:

(cont'd)

This does, in fact, work: measuring X in terms of feet rather than inches still changes σ_{XY} in the numerator by a factor of 12, but this is exactly cancelled by a change in σ_X in the denominator by the same factor 12. Since (5-25) similarly neutralizes any change in the scale of Y, the correlation coefficient ρ is, as desired, completely independent of the units of measurement of either variable.

Another reason that ρ is a very useful measure of the relation between X and Y is that it is always bounded by:[8]

$$-1 \leq \rho \leq +1 \tag{5-27}$$

Whenever X and Y have a perfect positive linear relation (as would occur if the entire distribution in Figure 5-4 were located on a straight line with positive slope), then ρ takes on the limiting value of $+1$; similarly, if there is a perfect negative linear relation, then ρ would be -1. To illustrate these bounds, we calculate ρ for the data in Figure 5-4. From (5-25):

$$\rho_{XY} = \frac{2}{\sqrt{5.6}\,\sqrt{1.4}} = .71$$

which indeed is less than 1.

Finally, we must ask how correlation and independence are related. An important theorem[9] states:

$$\rho_{XY} = E\left(\frac{X - \mu_X}{\sigma_X} \cdot \frac{Y - \mu_Y}{\sigma_Y}\right) \tag{5-26}$$

In this form, we see that ρ_{XY} simply uses the standardized values of X and of Y.

[8]Proofs of (5-27) and later statements, as well as more interpretations of ρ, are given in Chapter 14.

[9]**proof**   If X and Y are independent, then from (5-13):

$$p(x,y) = p(x)p(y)$$

Thus (5-22) becomes:

$$\sigma_{XY} = \sum \sum (x - \mu_X)(y - \mu_Y)\, p(x)\, p(y)$$

$$= \left[\sum (x - \mu_X)\, p(x)\right]\left[\sum (y - \mu_Y)\, p(y)\right]$$

Both of these quantities are zero, however, since the average deviation from the mean always equals zero. Thus:

$$\sigma_{XY} = 0 \cdot 0 = 0 \tag{5-28} \text{ proved}$$

From (5-25), ρ_{XY} also must be zero.

> If X and Y are independent, then they are uncorrelated, i.e., $\sigma_{XY} = \rho_{XY} = 0$.   (5-28)

PROBLEMS

5-8 For the following joint probability table:

2	0	.4
1	.2	.2
0	.2	0
y╲x	0	1

(a) Calculate σ_{XY} from the definition (5-22).
(b) Calculate σ_{XY} from the easier formula (5-24).
(c) Calculate ρ_{XY}.

(5-9) Repeat Problem 5-8 for the following joint probability distribution:

2	.4	0	0
1	0	.2	.2
0	0	0	.2
y╲x	0	1	2

5-10 Suppose that X and Y have the following joint distribution:

2	.10	.05	.05
1	.40	.20	.20
y╲x	1	2	3

(a) Are X and Y independent?

(b) What is σ_{XY}?

(c) What is ρ_{XY}?

5-11 Referring to Problem 5-4, is it true that $E(T) = \sigma_{XY}$?

5-12 Referring to Problem 5-1:

(a) What is σ_{XY}?

(b) Are X and Y independent?

(c) Looking beyond this particular example, which of the following statements are true for all X and Y?

(1) If X and Y are independent, then they must be uncorrelated.

(2) If X and Y are uncorrelated, then they must be independent.

*5-13 Prove (5-24).

5-4   LINEAR COMBINATION OF TWO RANDOM VARIABLES

In this section, we shall find easy ways to analyze the sum of two random variables:

$$S = X + Y \tag{5-29}$$

and, more generally, the *weighted sum:*

$$W = aX + bY$$

where the coefficients a and b are called weights. This is also called a linear combination (or linear function) of X and Y. For example, the average $(X + Y)/2 = \frac{1}{2}X + \frac{1}{2}Y$ is just a weighted sum with weights $1/2$. Similarly, any weighted sum whose weights add up to one is called a *weighted average.*

    To illustrate, suppose that a couple is drawn at random from a certain population of working couples. In thousand dollar units, let:

$$X = \text{man's income}$$

$$Y = \text{woman's income}$$

Then:

$$S = X + Y = \text{couple's total income}$$

Suppose that the tax rates on X and Y are fixed at 20% and 30%, respectively; then the couple's income after taxes is the weighted sum:

$$W = .8X + .7Y \qquad (5\text{-}30)$$

Of course, since X and Y are random variables, S and W are also. How can we find their moments easily?

(a) Mean

Continuing our example, if we know that:

$$E(X) = 12 \quad \text{and} \quad E(Y) = 9$$

it is natural to guess that:

$$E(S) = 12 + 9 = 21$$

This guess is correct, because it is always true[10] that:

$$\boxed{E(X + Y) = E(X) + E(Y)} \qquad (5\text{-}31)$$

(This is the same result that we guessed and confirmed in Problems 5-6 and 5-7.) Similarly, it is natural to guess that:

---

[10]**proof**   For $g(X, Y) = X + Y$, (5-19) becomes:

$$E(X + Y) = \sum_x \sum_y (x + y)\, p(x,y)$$

$$= \sum_x \sum_y xp(x,y) + \sum_x \sum_y yp(x,y)$$

Considering the first term, we may write it as:

$$\sum_x \sum_y xp(x,y) = \sum_x x \left[\sum_y p(x,y) \right]$$

by (5-5):

$$= \sum_x xp(x)$$

$$= E(X)$$

Similarly, the second term reduces to $E(Y)$, so that:

$$E(X + Y) = E(X) + E(Y) \qquad (5\text{-}31) \text{ proved}$$

$$E(W) = .8(12) + .7(9) = 15.9$$

This guess also is correct, because it is always true[11] that:

$$\boxed{E(aX + bY) = aE(X) + bE(Y)} \tag{5-32}$$

Statisticians often refer to this important property as the "additivity" or "linearity" of the expectation operator.

To appreciate the simplicity of (5-32), we should compare it to (5-19). Both formulas provide a means of calculating the expected value of a function of X and Y. However, (5-19) applies to *any* function of X and Y, whereas (5-32) is restricted to *linear* functions only. But when we are dealing with this restricted class of linear functions, (5-32) generally is preferred to (5-19) because it is much simpler. Whereas evaluation of (5-19) involves working through the whole joint probability distribution of X and Y, (5-32) requires only the marginal distributions of X and Y.

(b) Variance

The variance of a sum is a little more complicated[12] than its mean:

$$\boxed{\boxed{\operatorname{var}(X + Y) = \operatorname{var} X + \operatorname{var} Y + 2 \operatorname{cov}(X, Y)}} \tag{5-33}$$

where var and cov are abbreviations for variance and covariance, of course. Similarly, for a weighted sum:[13]

---

[11] Since the proof resembles the proof of (5-31), it is left as an exercise.

[12] **proof of (5-33)**   It is time to simplify our proofs by using brief notation such as $E(W)$ rather than the awkward $\Sigma\ wp(w)$, or the even more awkward $\underset{x\ y}{\Sigma\Sigma}\ w(x,y)\,p(x,y)$. First, from (4-19):

$$\operatorname{var} S = E(S - \mu_S)^2$$

Substituting for S and μ_S:

$$\operatorname{var} S = E[(X + Y) - (\mu_X + \mu_Y)]^2$$
$$= E[(X - \mu_X) + (Y - \mu_Y)]^2$$
$$= E[\ \underbrace{(X - \mu_X)^2}\ + 2\ \underbrace{(X - \mu_X)(Y - \mu_Y)}\ +\ \underbrace{(Y - \mu_Y)^2}\ ]$$
$$\text{each of these is a random variable}$$

Realizing that (5-32) holds for *any* random variables (in particular, each of the above):

$$\operatorname{var} S = E(X - \mu_X)^2 + 2E(X - \mu_X)(Y - \mu_Y) + E(Y - \mu_Y)^2$$
$$= \operatorname{var} X + 2 \operatorname{cov}(X, Y) + \operatorname{var} Y \qquad \text{(5-33) proved}$$

[13] Since the proof resembles the proof of (5-33), it is left as an exercise.

$$\boxed{\text{var}\,(aX + bY) = a^2\text{var}\,X + b^2\text{var}\,Y + 2ab\,\text{cov}\,(X, Y)} \qquad (5\text{-}34)$$

TABLE 5-5

The Mean and Variance of Various Functions of Two Random Variables

Function of X and Y	Mean and Variance Derived by:	Mean	Variance
1. Any function $g(X, Y)$		$E[g(X, Y)] =$ $\displaystyle\sum_x \sum_y g(x,y)\,p(x,y)$	
2. Linear combination $aX + bY$		$E(aX + bY)$ $= aE(X) + bE(Y)$	$\text{var}(aX + bY)$ $= a^2\text{var}\,X + b^2\text{var}\,Y$ $+ 2ab\,\text{cov}(X, Y)$
3. Simple sum $X + Y$	Setting $a = b = 1$ in row 2	$E(X + Y)$ $= E(X) + E(Y)$	$\text{var}(X + Y)$ $= \text{var}\,X + \text{var}\,Y$ $+ 2\,\text{cov}(X, Y)$
4. Function of one variable, aX	Setting $b = 0$ in row 2	$E(aX) = aE(X)$	$\text{var}(aX) = a^2\text{var}\,X$

For example, suppose that for our randomly selected couple:

$$\text{var}\,X = 16, \quad \text{var}\,Y = 10, \quad \text{and} \quad \text{cov}\,(X, Y) = 8 \qquad (5\text{-}35)$$

Then:

$$\text{var}\,S = 16 + 10 + 2(8) = 42 \qquad (5\text{-}36)$$

$$\text{var}\,W = (.8)^2(16) + (.7)^2(10) + 2(.8)(.7)(8) = 24.1 \qquad (5\text{-}37)$$

There is a simple intuitive reason for the covariance term in (5-33). A positive covariance (as in (5-35), for example) means that when X is high, Y tends to be high as well. Then the sum $X + Y$ tends to be very high. (Similarly, when X is low, Y tends to be low, making the sum tend to be very low.) These extreme values of S make its variance high (as borne out in the formula (5-36), for example).

A very simple and common case occurs when X and Y are uncorrelated, i.e., $\sigma_{XY} = 0$. Then (5-33) reduces to:

$$\text{var}\,(X + Y) = \text{var}\,X + \text{var}\,Y \qquad (5\text{-}38)$$

Since independence assures us that X and Y are uncorrelated according to (5-28), we may finally conclude that:

If X and Y are independent:

$$\text{var}\,(X + Y) = \text{var}\,X + \text{var}\,Y \qquad (5\text{-}39)$$

Similarly:

$$\text{var}\,(aX + bY) = a^2\text{var}\,X + b^2\text{var}\,Y \qquad (5\text{-}40)$$

For example, the independent dice thrown in Problems 5-6 and 5-7 obeyed the simple rule (5-39)

The theorems of this chapter are summarized in Table 5-5 for future reference. The general function $g(X, Y)$ is dealt with in the first row, while the succeeding rows represent increasingly simpler special cases.

PROBLEMS

5-14 According to (5-33), the variance of $(X + Y)$ is reduced if cov (X, Y) is negative. Give an intuitive explanation.

5-15 In a small community of ten working couples, yearly income (in thousands of dollars) has the following distribution:

Couple	Husband's Income X	Wife's Income Y
1	10	5
2	15	15
3	15	10
4	10	10
5	10	10
6	15	5
7	20	10
8	15	10
9	20	15
10	20	10

A couple is drawn by lot to represent the community at a convention. Let X and Y be the income of the husband and wife, respectively. Find:
(a) The bivariate probability distribution, and its graph.
(b) The distribution of X, and its mean and variance.

(c) The distribution of Y, and its mean and variance.

(d) The covariance σ_{XY}.

(e) $E(Y/X = 10)$, $E(Y/X = 15)$, and $E(Y/X = 20)$. Note that as X increases, the conditional mean of Y increases too. This is another expression of the positive relation between X and Y.

(f) If S is the total combined income of the couple, what is its mean and variance?

(g) Suppose that $W = .6X + .8Y$ is the couple's income after taxes; what is its mean and variance?

(h) To measure the degree of sexual discrimination against wives, a certain organization is interested in the difference $D = X - Y$. What is the mean and variance of D?

(i) Incidentally, do you think that $E(D)$ is a good measure of sexual discrimination?

5-16 Continuing Problem 5-15, we shall consider some alternative schemes for collecting the tax T on the couple's income S. Find the mean and variance of T:

(a) If S is taxed at a straight 20%, i.e.,

$$T = .20 \, S$$

(b) If S is taxed according to the formula:

$$T = .5(S - 12)$$

(c) If S is taxed according to the following progressive tax table:

Combined Income S	Tax T
10	1
15	2
20	3
25	5
30	7
35	10
40	13

*(d) Remembering that T is the tax on the representative couple (the couple drawn at random), compare the three tax schemes above in terms of the following criteria:

(i) Which scheme yields the most revenue to the government?

(ii) Which scheme is most egalitarian—i.e., which scheme results in the smallest variance in net income left after taxes?

(iii) Which of these schemes has the smallest marginal tax rate at high-income levels—i.e., which takes the smallest tax bite out of the last dollar earned by a rich couple ($S = 35$), and hence provides the strongest incentive for such a couple to continue working?

5-17 The students of a certain large class wrote two exams, obtaining a distribution of grades with the following characteristics:

	Class Mean μ	Standard Deviation σ	Variance σ^2	
1st exam, X_1	50	20	?	covariance
2nd exam, X_2	80	20	?	$\sigma_{12} = 50$
(a) Average, $\overline{X}$	?	?	?	
(b) Weighted average W	?	?	?	

Fill in the blanks in the table, assuming that:

(a) The instructor calculated a simple average of the two grades, $\overline{X} = (X_1 + X_2)/2$.

(b) The instructor thought the second exam was twice as important, so she took a weighted average:

$$W = \frac{1}{3}X_1 + \frac{2}{3}X_2$$

5-18 Repeat Problem 5-17:

(i) If the covariance $\sigma_{12} = -200$. How might you interpret such a negative covariance? What has it done to the variance of the average grade?

(ii) If the covariance $\sigma_{12} = 0$.

5-19 Continuing Problems 5-6 and 5-7, suppose that the pair of three-sided dice are not only loaded but dependent, so that the joint probability function of the pair of numbers is:

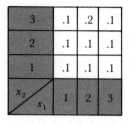

(a) Find the distribution of S (the total number of dots), and its mean and variance.

(b) Find the mean and variance of X_1 and of X_2.

(c) Find the covariance of X_1 and X_2, and then verify that (5-31) and (5-33) hold true.

REVIEW PROBLEMS

5-20 When is it true that:

$$E(XY) = E(X)E(Y)?$$

Select the one correct answer:

(a) Always.

(b) Never.

(c) Iff X and Y are independent.

(d) Iff X and Y are uncorrelated.

(e) Iff X and Y each has zero mean.

5-21 Suppose that X and Y are independent random variables with the following distributions:

x	$p(x)$
1	.4
2	.2
3	0
4	.4

y	$p(y)$
0	.4
1	.2
2	0
3	0
4	.4

As a preliminary, tabulate the distribution of $Z \triangleq 2X + 5$, and the distribution of $S \triangleq X + Y$.

(a) Find the medians $M(X)$, $M(Y)$, $M(Z)$, $M(S)$.

(b) Is it true that $M(Z) = 2M(X) + 5$?

i.e., $M(2X + 5) = 2M(X) + 5$?

(c) Is it true that $M(S) = M(X) + M(Y)$?
i.e., $M(X + Y) = M(X) + M(Y)$?

5-22 Repeat Problem 5-21 for the means, instead of the medians.

5-23 When a coin is fairly tossed 3 times, let:

$$X = \text{number of heads on the first two coins}$$

$$Y = \text{number of heads on the last coin}$$

$$S = \text{total number of heads}$$

(a) Are X and Y independent? What is their covariance?
(b) For each of X, Y, and S, find the distribution, mean, and variance.
(c) Verify that (5-31) and (5-33) hold true.

(5-24) Repeat Problem 5-23 for a coin that is not fairly tossed but that, in fact, has the sample space given in Problem 3-6.

⇒5-25 (Random sampling with replacement.) A bowl contains three chips numbered 2, 6 and 7. One chip is selected at random, replaced, and then a second chip is selected at random. Let X_1 and X_2 be the first and second numbers drawn.
(a) Tabulate the joint distribution of X_1 and X_2.
(b) Tabulate the (marginal) distribution of X_1 and of X_2.
(c) Are X_1 and X_2 independent? What is their covariance?
(d) Find the mean and variance of X_1 and of X_2.
(e) Find the mean and variance of $\overline{X} = \frac{1}{2}(X_1 + X_2)$. Then find the distribution of $\overline{X}$ and verify directly.
(f) How would your answers above be different if the bowl contained 1,000 chips of each kind?
(g) Is this a correct conclusion from part (f):
        In sampling with replacement, the important issue is the *relative* frequency of the various kinds of chips. The *absolute* frequency is immaterial.

⇒5-26 Continuing Problem 5-25, if 10 chips are sampled with replacement:
(a) What are the mean and variance of $\overline{X}$?
(b) What is the range of possible values of $\overline{X}$?

*5-27 (Random sampling without replacement.) Repeat Problem 5-25 with the following change. The first chip is kept out when the second is drawn (or the two chips could be drawn simultaneously).

REVIEWS PROBLEMS (Chapters 1-5)

5-28 The following table gives, for a thousand newborn American baby boys, the approximate number dying, in successive decades.[14]

Age, x	$n(x) =$ Number Dying within the Decade	$L(x) =$ Number Living to the Beginning of the Decade	Mortality Rate, $m(x) = \dfrac{n(x)}{L(x)}$
0 to 10	20	1,000	.020
10 to 20	14	980	·
20 to 30	18	966	·
30 to 40	24	948	.025
40 to 50	48	·	·
50 to 60	106	·	·
60 to 70	211	·	·
70 to 80	296		
80 to 90	216		
90 to 100	47		

Total $= 1,000$

(a) The third column shows the number surviving at the beginning of each decade. Complete this tabulation of $L(x)$.

(b) The mortality rate is just the relative number who die in a decade (i.e., relative to the number living at the beginning of the decade); for example, for men in their thirties, the mortality rate is

$$m(x) = \frac{n(x)}{L(x)} = \frac{24}{948} = .025 = 2.5\% \text{ per decade.}$$

Complete the tabulation of the mortality rate $m(x)$. Then answer true or false. If false, correct it.

    (i) The mortality rate is lowest during the first decade of life.

    (ii) Roughly speaking, from age twenty until age ninety, the mortality rate nearly doubles every decade.

(c) Ten-year term insurance is an agreement whereby a man pays the insurance company $x (the "premium," which usually is spread throughout the decade in 120 monthly payments) in return for a payment of

---

[14]From the 1958 Mortality Table, *Chemical Rubber Co. Standard Mathematical Table*, 14th ed., 1964, p. 583.

$1,000 to the man's estate if he dies. In order for this to be a "fair bet" (and ignoring interest), what should the premium x be for a man:

    (i) In his 20s?

    (ii) In his 40s?

(d) The work that we have done so far could be just as well expressed in probability terms. For example, find:

    (i) the probability of a man dying in his 40s.

    (ii) the probability of a man surviving to age 40.

    (iii) the conditional probability of a person dying in his 40s, given that he survived to 40.

5-29 Referring to the mortality table in Problem 5-28:

(a) Find the mean age at death (i.e., mean length of life, also called life expectancy).

(b) If the mortality rate were 50% higher after age 40, how much would this reduce life expectancy? (Incidentally, heavy cigarette smoking— two or more packs per day—seems to be associated with roughly this much increase in mortality).[15]

5-30 The proportion of people living in various regions of Canada, and the proportion in each region whose mother tongue is French, are as follows (1961 Census, roughly):

Region, R	$P(R)$	$P(F/R)$
Atlantic Provinces	.10	.14
Quebec	.29	.81
Ontario	.34	.07
Prairie Provinces	.18	.04
British Columbia and North	.09	.02

(a) What proportion of Canadians have French as their mother tongue?

(b) If a person selected at random is known to have French as his mother tongue, what is the conditional probability that he comes from Quebec?

5-31 An apartment manager in Cincinatti orders three new refrigerators, which the seller guarantees. Each refrigerator has a 20% probability of being defective.

---

[15] Reconstructed from "*Smoking and Health: Report of the Advisory Committee to the Surgeon General of the Public Health Service,*" U.S. Department of Health, Education and Welfare, 1964, p. 143.

(a) Tabulate the probability distribution for the total number of defective refrigerators, X.

(b) What are the mean and variance of X?

(c) Suppose that the cost of repair, in order to honor the guarantee, consists of a fixed fee ($10) plus a variable component ($15 per defective refrigerator). That is:

$$c(x) = 0 \qquad \text{if } x = 0$$

$$= 10 + 15x \qquad \text{if } x > 0$$

Find the average cost of repair.

5-32 To test whether a drug has an effect, a group of 10 matched pairs of volunteers are used. In each pair, one man is treated with the drug while the other man is treated with a placebo as a control. At the end of the experiment, a doctor examines each pair and declares which of the two is healthier. (He is forced to declare a decision—no ties allowed). Then he counts up X, the number of pairs in which the treated patient was declared healthier than the control. Of course, the random variable X could be any integer from 0 to 10.

(a) If the drug were absolutely ineffective (neither good nor bad), what would be $\Pr(X \geq 8)$?

(b) Suppose now that the drug were so effective that for each matched pair, the probability is 90% that the treated patient will be declared healthier. What would be $\Pr(X \geq 8)$ now?

(c) Should the doctor who makes the diagnosis know who has been treated and who has not, or should he be kept "blind" about this?

5-33 In putting up a new building, suppose that there are two consecutive stages, surveying and construction. The times, in years, which are required to complete the surveying (S) and construction (C) are two independent random variables, with the following probability distributions:

$$p(s) = (.8)(.2)^{s-1} \qquad s = 1, 2, 3, \ldots$$

$$p(c) = (.5)(.5)^{c-2} \qquad c = 2, 3, 4, \ldots$$

The total building time T is just $S + C$, of course. A penalty is incurred if completion takes more than $T = 4$ years. What is the probability of incurring this penalty?

5-34 (a) Answer Problem 5-33, if S and C are now normally distributed with means of 1.0 and 2.0 years, and standard deviations of .3 and .4 years, respectively. (*Hint:* look ahead to (6-9).)

(b) Repeat (a), if S and C are no longer independent, but instead have a correlation $\rho = .60$.

5-35 (a) Suppose that a secretary spills 3 different form letters and their envelopes; she so hopelessly scrambles them that, in despair, she stuffs each letter in an envelope at random and then mails them. Let S = number of people who receive the right letter; thus S = 0, 1, 2, 3. Find the mean of S.

(b) Repeat (a) if the 3 letters are not all different but instead, if 2 of the letters are identical.

*(c) Repeat (a) if there are n different letters instead of 3.

*(d) Repeat (a) if there are n letters, k of which are identical, while the remaining $(n - k)$ all are different.

5-36 A certain millionaire devised the following "sure-fire" sequential scheme for making $1,000 by selecting the right color (black or red) at roulette.

The first time, bet $1,000. If you win, stop. If you lose, double your bet.

If you win this second time, stop. If you lose, double your bet, and continue in this way until you win—at which point your net winning will be $1,000. What do you think of this idea?

PART II

Basic Inference: Estimating and Testing Means

CHAPTER 6

Sampling

*The
normal
law of error
stands out in the
experience of mankind
as one of the broadest
generalizations of natural
philosophy. It serves as the
guiding instrument in researches
in the physical and social sciences and
in medicine agriculture and engineering.
It is an indispensable tool for the analysis and the
interpretation of the basic data obtained by observation and experiment.*

W. J. Youden

*I know of scarcely anything so apt to impress the imagination as the
wonderful form of cosmic order expressed by the "Law of Frequency of
Error." The law would have been personified by the Greeks and deified, if
they had known it.*

Sir Francis Galton

In Part I, we have studied probability and random variables so that we can now answer the basic deductive question in statistics: What can we expect of a random sample drawn from a known population?

6-1   RANDOM SAMPLING

We already have considered several examples of sampling: the poll of voters sampled from the population of all voters; the sample of light bulbs drawn from the whole production of bulbs; a sample of men's heights drawn from the whole population; and a sample of two chips

143

drawn from a bowl of chips. In cases such as these, the sample is called *random* if each individual in the population is equally likely to be sampled. For example, suppose that a random sample is to be drawn from the population of students in the classroom. There are several ways to actually carry out the physical process of random sampling.

1. The most graphic method is to record each person on a cardboard chip, mix all these chips in a large bowl, and then draw the sample.

2. A more practical method is to assign each person a number, and then draw a random sample of numbers. For example, suppose that a random sample of 12 students is to be drawn from a class (population) of 100 students. By counting off, each student can be assigned a different 2-digit number. Then 12 such numbers can be read out of a table of random digits, such as Appendix Table II(a).[1]

These two sampling methods are mathematically equivalent. Since the random number method is simpler to employ, it is common in practical sampling. However, the bowlful of chips is easier conceptually; consequently, in our theoretical development of random sampling, we shall often visualize drawing chips from a bowl.

(a) Illustration

Suppose that a population of a million men's heights has the distribution shown in Table 6-1 and Figure 6-1. For future reference, we also compute μ and σ^2, the population mean and variance. Random sampling from this population is equivalent mathematically to placing the million chips of column (2) in a bowl. Then the first chip selected at random can take on any of the x values shown in column (1), with probabilities shown in column (3). This first random observation (chip) is denoted by X_1; the second observation is denoted by X_2, and so on. Each observation is a random variable with the same probability distribution—the distribution of the population:[2]

$$p(x_1) = p(x_2) = \cdots = p(x_n) = \text{population distribution, } p(x) \quad (6\text{-}1)$$

---

[1] Reference tables in the Appendix at the back of the book will consistently be given *Roman* numerals.

[2] Strictly speaking, (6-1) is not precise enough. It would be more accurate to let p_1 denote the probability function of X_1, etc., and then write:

$$p_1(x) \equiv p_2(x) \equiv p_3(x) \equiv \cdots \equiv p_n(x) \equiv p(x)$$

where $\equiv$ means "identically equal for all x." This is the same mathematical subtlety we first encountered in footnote 2 in Chapter 5.

TABLE **6-1**

A Population of Men's Heights[a], and the Calculation of μ and σ^2

(1) Height x	(2) Frequency	(3) Relative Frequency, also $p(x)$	(4) $xp(x)$	(5) $(x - \mu)^2 p(x)$
60	10,000	.01	.60	.81
63	60,000	.06	3.78	2.16
66	240,000	.24	15.84	2.16
69	380,000	.38	26.22	0
72	240,000	.24	17.28	2.16
75	60,000	.06	4.50	2.16
78	10,000	.01	.78	.81
$N = 1,000,000 \checkmark$		$1.00 \checkmark$	$\mu = 69.00$	$\sigma^2 = 10.26$ $\sigma = 3.20$

[a]We approximate each man's height by the cell midpoint to keep concepts simple. If we had chosen to be more precise, we would have used a very fine subdivision of height into so many cells that we would closely approximate the continuous distribution shown in Figure 6-1.

Of course, the calculation of μ and σ^2 could have been simplified by coding, as in Table 2-5.

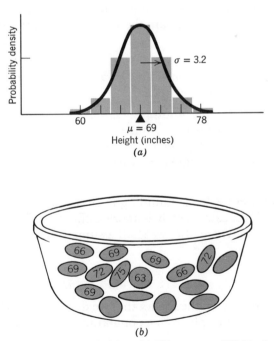

FIGURE 6–1 Population of men's heights. (a) Histogram of Table 6–1, and its continuous approximation. (b) Bowl-of-chips equivalent.

To summarize, a random sample consists of n observations X_1, X_2, . . . X_n; each observation may range over the whole population, with probabilities $p(x)$ given by the relative frequencies in the population.

(b) Sampling With or Without Replacement

In large populations (such as the million heights in Table 6-1), it makes practically no difference whether or not we replace each chip before we draw the next. After all, what is one chip in a million? It cannot substantially change the relative frequencies, $p(x)$.

However, in small populations, replacement of each sampled chip is an important issue. In Problems 5-25 and 5-27, this difference was made clear. When the first chip X_1 was replaced, it restored the population to exactly its original state, so that X_2 was completely independent of whatever X_1 happened to be. On the other hand, when the first chip X_1 was not replaced, the population was changed significantly. That is, the distribution of X_2 was dependent on what X_1 happened to be.

In general, then, we see that the random observations $X_1, X_2, \ldots,$ X_n will be independent if we sample with replacement. And, in large populations, even if we sample without replacement, it is practically the same as with replacement, so that we still essentially have independence. All these cases in which observations can be assumed to be independent are very easy to analyze, and are therefore called *simple random sampling*.

(c) Conclusion

We may restate the definition of simple random sampling in more mathematical terms for future reference:

> A *simple random sample* is a sample whose n observations X_1, $X_2, \ldots, X_n$ are independent. The distribution of each X_i is the population distribution $p(x)$ (with mean μ and variance σ^2).   (6-2)

The exception to this is sampling from a small population, without replacement. This case, which is more difficult, will be dealt with in Section 6-6. Everywhere else, we shall assume simple random sampling.

6-2   MOMENTS OF THE SAMPLE MEAN

We now are ready to use the theory that we developed in earlier chapters. In taking a random sample from a population, consider the sample mean:

$$\overline{X} \triangleq \frac{1}{n}[X_1 + X_2 + \cdots + X_n] \tag{6-3}$$

Being a linear combination of random variables, $\overline{X}$ itself will also be a random variable. How does it fluctuate? In particular, what is its expectation and variance?

By applying (5-32) to (6-3),[3] we can easily calculate the expectation:

$$E(\overline{X}) = \frac{1}{n}[E(X_1) + E(X_2) + \cdots + E(X_n)]$$

In (6-1), we noted that each observation X_i has the population distribution $p(x)$, with expectation μ, so that:

$$E(\overline{X}) = \frac{1}{n}[\mu + \mu + \cdots + \mu] \tag{6-4}$$

$$\boxed{E(\overline{X}) = \mu} \tag{6-5}$$

Similarly, we obtain the variance of $\overline{X}$. Because all the components of $\overline{X}$ are independent, the simple formula (5-40) applies:

$$\text{var } \overline{X} = \frac{1}{n^2}[\text{var } (X_1) + \text{var } (X_2) + \cdots + \text{var } (X_n)] \tag{6-6}$$

Again, we note that each observation X_i has the population distribution $p(x)$, with variance denoted by σ^2, so that:

$$\text{var } \overline{X} = \frac{1}{n^2}[\sigma^2 + \sigma^2 + \cdots + \sigma^2] = \frac{1}{n^2}[n\sigma^2]$$

$$\text{var } \overline{X} = \frac{\sigma^2}{n} \tag{6-7}$$

$$\boxed{\sigma_{\overline{X}} = \frac{\sigma}{\sqrt{n}}} \tag{6-8}$$

---

[3]Extending (5-32) and (5-40) to cover more than two variables is left as an exercise.

This is our first important deduction: we have deduced the behavior of a *sample* mean from knowledge of the *population*.

For example, suppose that a sample of $n = 4$ observations is drawn from the population of heights in Figure 6-1. Then $\overline{X}$ would fluctuate around:

$$E(\overline{X}) = \mu \qquad \text{(6-5) applied}$$

with standard deviation:

$$\sigma_{\overline{X}} = \frac{\sigma}{\sqrt{n}} = \frac{\sigma}{2} \qquad \text{(6-8) applied}$$

The distribution of $\overline{X}$ is shown in Figure 6-2. It is also intuitively clear: $\overline{X}$ fluctuates around the same central value as an individual observation, but with less deviation because of "averaging out."

A concrete view of "averaging out" may be helpful: we might get a seven-foot man as an *individual* observation from the population, but we would be far less likely to get a seven-foot *average* in a sample of four men. This is because any seven-foot man that appears in the sample will likely be partially cancelled by a short man; or at least his effect will be diluted, because he is averaged in with other (more typical) men.

Since it is very important to distinguish between the distribution of the sample mean $\overline{X}$ and the population distribution, we introduce two conventions:

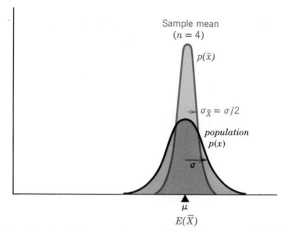

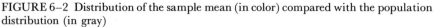

FIGURE 6–2 Distribution of the sample mean (in color) compared with the population distribution (in gray)

1. Since the population "gives birth" to the sample, we shall speak of the population distribution as the *parent* distribution. The distribution of $\overline{X}$ is then called a derived distribution or a *sampling* distribution.

2. In the rest of the diagrams in this text, *color is reserved for samples and sampling distributions*. In contrast, parent populations are shown in gray. This convention first appears in Figure 6-2, where the distribution of the sample mean $\overline{X}$ is shown in color, while the distribution of the parent population is shown in gray.

Example 6-1.   The workers in a certain industry earned a weekly wage that averaged \$280, with a standard deviation of \$40. Each student at the local university was given a project: to sample 25 workers and report the mean weekly wage $\overline{X}$. Of course, by the luck of the draw, each student would get a different sample and hence a different sample mean.

Around what value would the $\overline{X}$s fluctuate, and with what standard deviation? (Assume a huge number of students, so that the sampling experiment is carried out enough times to build up the (limiting) probability distribution of $\overline{X}$).

Solution.

$$E(\overline{X}) = \mu = \$280$$

$$\sigma_{\overline{X}} = \frac{\sigma}{\sqrt{n}} = \frac{\$40}{\sqrt{25}} = \$8$$

Remarks.   Note that the sample mean fluctuates very little, because of averaging out: the typical sample will contain both high- and low-wage workers who tend to offset one another.

PROBLEMS

6-1 Suppose that 10 men were sampled randomly from the population of Table 6-1 and that their average $\overline{X}$ was calculated. Then imagine that the experiment was repeated many, many times. Answer true or false; if false, correct it.

(a) The expectation of $\overline{X}$ would be $\mu = 69$ inches, and its standard deviation would be $\sigma/n = .32$ inches.

(b) The long and short men in the sample tend to "average out," making $\overline{X}$ fluctuate less than a single observation.

6-2 The population of employees in a certain large office building has weights distributed around a mean of 150 pounds, with a standard deviation of 20 pounds. A random group of 25 employees enters the elevator each morning. Find the expectation, variance, and standard deviation of the average weight $\overline{X}$.

6-3 A bowl contains six chips, numbered 1 to 6. A sample of 2 chips is drawn, with replacement.
(a) By tabulating the joint distribution of X_1 and X_2, calculate the distribution of $\overline{X} = (X_1 + X_2)/2$.
(b) From the distribution of $\overline{X}$, calculate its mean and variance.
(c) Find the population mean and variance, and hence verify (b).
(d) Graph the parent distribution (for a single observation X), and the sampling distribution of $\overline{X}$.

Answer true or false; if false, correct it.
(e) $\overline{X}$ fluctuates around $3\frac{1}{2}$, as does an individual observation X. This illustrates $E(\overline{X}) = \mu$.
(f) $\overline{X}$ ranges from 1 to 6, as does an individual observation X. However, the extreme values of $\overline{X}$ are rare (probability $\frac{1}{36}$, compared to probability $\frac{1}{6}$ for an extreme value of an individual observation X.) Thus, $\overline{X}$ has a smaller standard deviation than X. This illustrates $\sigma_{\overline{X}} = \sigma/\sqrt{n}$.

   Incidentally, this also illustrates why the standard deviation is a better measure of spread than the range.

6-4 A bowl is full of many chips, one-third marked 2, one-third marked 4, and one-third marked 6.
(a) When one chip is drawn, let X be its number. Find μ and σ (the population mean and standard deviation).
(b) When a sample of 2 chips is drawn, let $\overline{X}$ be the sample mean.
    (i) Find the probability table of $\overline{X}$. From this, calculate $\mu_{\overline{X}}$ and $\sigma_{\overline{X}}$.
    (ii) Check your answers, using (6-5) and (6-8).
(c) Repeat (b) for a sample of 3 chips.
(d) Graph $p(\overline{x})$ for each case above; i.e., for sample size $n = 1, 2, 3$. Comparison is facilitated by using probability density; i.e., by using a bar graph with probability = area = (height) $\times$ (width).
    As n increases, notice that $p(\overline{x})$ becomes more concentrated around μ. What else is happening to the *shape* of $p(\overline{x})$?

6-3   THE CENTRAL LIMIT THEOREM

In the preceding section, we found the mean and standard deviation of $\overline{X}$. Now we shall investigate the *shape* of its distribution.

(a) The Distribution of $\overline{X}$ from a Normal Population

There is a very important theorem about linear combinations of normal variables, which we quote within proof:

> If X and Y are normal,[4] then any linear combination $Z = aX + bY$ is also a normal random variable.   (6-9)

To see how this theorem will answer questions about sampling, suppose we have a parent population that is normal. Then each observation in the sample $X_1, X_2, \ldots, X_n$ has this same normal distribution, according to (6-1). Since the sample mean $\overline{X}$ in (6-3) is a linear combination of these normal variables, (6-9) establishes that $\overline{X}$ is normal.

(b) The Distribution of $\overline{X}$ from a Nonnormal Population

Across the top of Figure 6-3, we show three different examples of a nonnormal population; in each case, successive graphs reading down the page show how the distribution of the sample mean changes shape as sample size n increases. These three examples display an astonishing pattern— *the sample mean becomes approximately normally distributed as n grows, no matter what the parent population is.* This is especially remarkable in the right-hand example, where even a skewed population eventually generates the symmetric normal distribution for the sample mean. This pattern is so important that mathematicians have formulated it as:

> **the central limit theorem**   As the sample size n increases, the distribution of the mean $\overline{X}$ of a random sample taken from practically any population approaches a *normal* distribution[5] (with mean μ and standard deviation $\sigma/\sqrt{n}$).   (6-10)

---

[4] Strictly speaking, it is required that X and Y be *jointly* normal, as described in Figure 14-5.

[5] To be precise, the central limit theorem proves the following: the probability that the standardized sample mean $(\overline{X} - \mu)/(\sigma/\sqrt{n})$ falls in a given interval converges to the probability that a standardized normal variable falls in that interval.

The one qualification is that the population have finite variance. The proof of this theorem requires a very heavy mathematical background, and so we omit it.

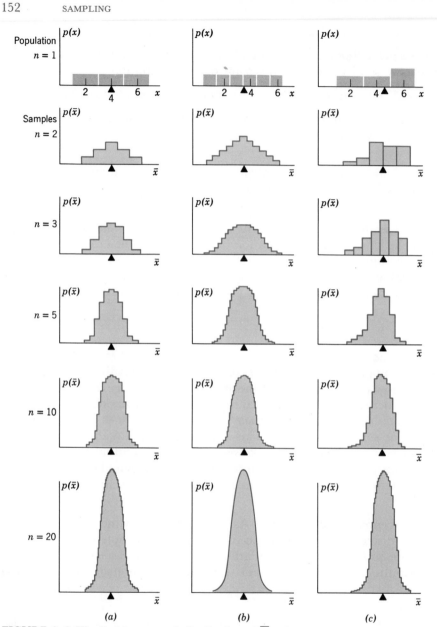

FIGURE 6–3 The limiting normal distribution of $\overline{X}$, where the parent population is (a) Bowl of 3 kinds of chips.[a] (b) Bowl of 6 kinds of chips (or die). (c) Bowl of 3 kinds of chips of different frequency.

[a]The sampling in panel (a) already was discussed in Problem 6-4, and panel (b) in Problem 6-3. Panel (c) could be similarly worked out for $n = 2$ at least, as an exercise.

The central limit theorem is not only remarkable, but very practical as well. For it completely specifies the distribution of $\overline{X}$ in large samples, and is therefore the key to large-sample statistical inference. In fact, in most cases when the sample size n reaches about 10 or 20, the distribution of $\overline{X}$ already is practically normal.[6] This is certainly the case in the three examples of Figure 6-3.

> *Example 6-2.*   Suppose that the marks of a large class of statistics have a mean of 72 and standard deviation of 9.
> (a) Find the probability that a random sample of 10 students will have an average mark over 80.
> (b) If the population is normal, find the probability that an individual student drawn at random will have a mark over 80.
>
> *Solution.*   (a) The central limit theorem assures us that $\overline{X}$ has an approximately normal distribution; its mean is μ and its standard deviation is $\sigma/\sqrt{n}$. Using these moments to standardize, we have:

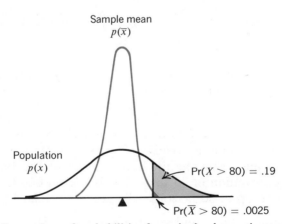

FIGURE 6–4 Comparison of probabilities for a single observation and for a sample mean

---

[6]Since (6-10) assures us that eventually the distribution of $\overline{X}$ is approximately normal, it would be better for applied statisticians to call it the "central *approximation* theorem." However, theoretical statisticians prefer to call it the "central limit theorem," because a limit is a very precise mathematical concept.

$$\Pr(\overline{X} > 80) = \Pr\left(\frac{\overline{X} - \mu}{\sigma/\sqrt{n}} > \frac{80 - 72}{9/\sqrt{10}}\right) \qquad (6\text{-}11)$$

$$= \Pr(Z > 2.81)$$

$$= .0025$$

(b) For an individual observation drawn from a normal population, the standardization is even simpler:[7]

$$\Pr(X > 80) = \Pr\left(\frac{X - \mu}{\sigma} > \frac{80 - 72}{9}\right) \qquad (6\text{-}12)$$

$$= \Pr(Z > .89)$$

$$= .1867 \simeq .19$$

Remarks. Although there is a reasonable chance (about 19%) that a single student will get over 80, there is very little chance (less than 1%) that a sample average of ten students will perform this well. This is shown in Figure 6-4; once again we see how "averaging out" tends to reduce the extremes.

---

[7]We also could get the same answer by using the same formula as in (a), with $n = 1$ instead of 10.

Our next example will show how a problem of load limit can be solved by simply rephrasing it in terms of a sample mean.

Example 6-3. A ski lift is designed with a load limit of 10,000 lbs. It claims a capacity of 50 persons. If the weights of all the people using the lift have a mean of 190 lbs. and a standard deviation of 25 lbs., what is the probability that a random group of 50 persons will exceed the load limit?

Solution. First, rephrase the question: "The probability that 50 persons will total more than 10,000 lbs." is exactly the same as "the probability that 50 persons will *average* more than 10,000/50 = 200 lbs. each." Now we can proceed exactly as before:

$$Pr(\overline{X} > 200) = Pr\left(\frac{\overline{X} - \mu}{\sigma/\sqrt{n}} > \frac{200 - 190}{25/\sqrt{50}}\right)$$

$$= Pr(Z > 2.83)$$

$$= .0023$$

PROBLEMS

6-5 The weights of packages filled by a machine are normally distributed about a mean of 25 ounces, with a standard deviation of 2 ounces. What is the probability that n packages from the machine will have an average weight of less than 24 ounces if:
(a) $n = 1$?
(b) $n = 4$?
(c) $n = 16$?
(d) $n = 64$?

6-6 Suppose that the education level among adults in a certain country has a mean of 11.1 years and a standard deviation of 3 years. What is the probability that in a random survey of 100 adults you will find an average level of schooling between 10 and 12 years?

6-7 A ski lift is designed with a load limit of 18,000 lbs. It claims a capacity of 100 persons. If the weights of all the people using the lift have a mean of 175 lbs and a standard deviation of 30 lbs, what is the probability that a group of 100 persons will exceed the load limit?

6-8 Suppose that bicycle chain links have lengths distributed around a mean $\mu = .50$ cm, with a standard deviation $\sigma = .04$ cm. The manufacturer's standards require the chain to be between 49 and 50 cm long.
(a) If chains are made of 100 links, what proportion meets the standards?
(b) If chains are made of only 99 links, what proportion now meets the standards?
(c) Using 99 links, to what value must σ be reduced (how much must the quality control on the links be improved) in order to have 90% of the chains meet the standards?

6-9 The amount of pocket money that persons in a certain city carry has a nonnormal distribution, with a mean of $9.00 and a standard deviation of $2.50. What is the probability that a group of 225 individuals will be carrying a total of more than $2,100?

6-10 In each of Problems 6-7 to 6-9, what crucial assumption did you implicitly make? Suggest some circumstances where it would be seriously violated. Then how would the correct answer differ?

⇒6-11 The number of completed years of college had the following distribution, in a large class of statistics students:

x = Number of Completed Years	$p(x)$ = Relative Frequency
0	.30
1	.30
2	.40

(a) Find the mean and variance of X.

(b) If a random sample of 10 students were taken, what would be $\Pr(\overline{X} \leq .40)$?

(c) The instructor looked up the background of the 10 best students in the class. He found that 6 were freshmen and 4 were sophomores, so that their average number of completed years of college was $\overline{X} = .40$. This led the instructor to claim that the younger students were doing unusually well; the dean replied that 10 students provided too little evidence for any firm conclusion. How would you settle the dispute?

*6-12 (a) A farmer has 9 wheatfields planted. The distribution of yield from each field has mean 1,000 bushels and variance 20,000. Furthermore, the yields of any 2 fields are correlated, because they share the same weather conditions, weed control, etc.; in fact the correlation is $\rho = .59$. What is the probability that the total yield will be at least 10,000 bushels? (*Hint:* Rephrase in terms of $\overline{X}$. In spite of the correlation, $\overline{X}$ will still be approximately normal, with mean μ. The only change is that its variance requires an adjusted version of (6-6)).

(b) What would be the answer in (a) if the fields were uncorrelated, i.e., $\rho = 0$?

(c) Answer true or false; if false, correct it.

When there is positive correlation among the components of a sample mean $\overline{X}$, the probability of getting an extreme value of $\overline{X}$ is much higher than when there is zero correlation. This is because positive correlation produces a tendency for high yields to occur with other high yields, rather than having highs and lows cancel.

6-4   SAMPLING SIMULATION (MONTE CARLO)

Although this section on Monte Carlo is optional, it is not difficult and it provides a good deal of insight into the sampling process, and into the imaginative ways that statisticians can use computers.

(a) Sampling from any Population

The theory of sampling that we have developed so far can be verified empirically. We simply draw many, many samples, calculate $\overline{X}$ each time, and graph the relative frequency distribution of $\overline{X}$. As we gather more and more $\overline{X}$s, in the limit we should get the probability distribution of $\overline{X}$.

A similar idea occurred in Chapter 3. The probability of obtaining a "7 or 11" with a pair of dice was calculated theoretically to be $8/36 = .22$. This can be verified empirically by throwing the dice many, many times, and observing that the relative frequency settles down to the probability. That is, we can go to the tables of Las Vegas or Monte Carlo for our confirmation.

Example 6-4. In a classroom of 100 men, suppose that heights are distributed as in Table 6-2(a).

TABLE **6-2**

| (a) | | (b) |
Height X (Inches)	Frequency	Serial Number (Address)
60	1	0
63	6	1–6
66	24	7–30
69	38	31–68
72	24	69–92
75	6	93–98
78	1	99

Draw a "Monte Carlo" random sample of $n = 5$ men, and calculate $\overline{X}$.

Solution. As suggested in Section 6-1, the most practical method begins by assigning each man a serial number (address). This could be done alphabetically, or according to where each person sits in class, or in any other arbitrary way we like, since the numbers are to be drawn at random in any case. Therefore number the students from shortest to tallest, so that this enumerating can easily be shown in Table 6-2(b).[8]

Now we consult the random digits in Table IIa. The first random pair is 39, which is the address of an individual 69 inches tall. Continuing to draw the remaining observations in the same way, we obtain the following sample:

Random Pair (Address)	Height
39	$X_1 = 69$
65	$X_2 = 69$
76	$X_3 = 72$
45	$X_4 = 69$
45	$X_5 = 69$

$$\overline{X} = 69.6$$

Note that the random address 45 is repeated, and so we include this student both times in our sample. This is because simple random sampling involves drawing with replacement[9] (as we stated prior to (6-2)). Thus we sample this student, then "replace" him in the population, and by the luck of the draw sample him again.

---

[8]The smallest number appearing in Table II(a) is zero (00), so this is where we begin numbering Table 6-2(b).

[9]Sampling with replacement not only simplifies the formulas, but in this case also simplifies the sampling procedure; we do not need to keep a record of who has been sampled and who has not. In actual human populations, however, most sampling is done without replacement for reasons of efficiency.

Example 6-5. (a) Let everyone in your class perform a fresh repetition of the sampling experiment in Example 6-4, by starting at a different (randomly selected) place in Table II(a). Let the instructor tabulate and graph the frequency distribution of all the values of $\overline{X}$.

(b) Using the theory developed in Section 6-2, derive the mean and standard deviation of the sampling distribution of $\overline{X}$. Graph it. Is it approximated by the Monte Carlo result in (a)?

Solution. (b) We first calculate the moments of the population:[10]

x	$p(x)$	$xp(x)$	$(x - \mu)^2 p(x)$
60	.01	.60	.81
63	.06	3.78	2.16
66	.24	15.84	2.16
69	.38	26.22	0
72	.24	17.28	2.16
75	.06	4.50	2.16
78	.01	.78	.81

$$\mu = 69.00 \qquad \sigma^2 = 10.26$$

Now we can find the sampling moments of $\overline{X}$:

$$E(\overline{X}) = \mu = 69 \qquad \text{(6-5) repeated}$$

$$\sigma_{\overline{X}} = \frac{\sigma}{\sqrt{n}} = \frac{\sqrt{10.26}}{\sqrt{5}} = 1.43 \qquad \text{(6-8) repeated}$$

Using these moments and the central limit theorem, we graph the normal distribution of $\overline{X}$ in Figure 6-5. The Monte Carlo distribution of $\overline{X}$ in part (a) will be approximately like this.

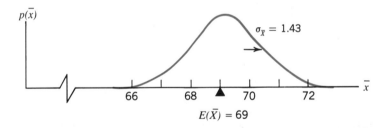

$$E(\overline{X}) = 69$$

FIGURE 6-5 The approximate sampling distribution of $\overline{X}$, for a sample of $n = 5$ observations drawn from the population in Table 6-2 (or Figure 6-1)

---

[10]Incidentally, the population distribution $p(x)$ is the same as in Table 6-1 (although the population size is much smaller—100 instead of 1,000,000). Thus the moments are the same as found in Table 6-1.

If the population size were 1000 (or equivalently, if the population distribution $p(x)$ were given to 3 decimal places), then it would require 3 digits to number the population, and so the digits in Table II(a) would be read in blocks of 3. In this way, sampling could be carried out quite generally.

There remains only one major problem: how do we proceed if the population size is not a power of 10 (such as 100, or 1000, or 10000, etc.)? The answer comes from the following example.

Example 6-6.   In a class of 80 students,[11] each student was asked how many novels he had read in the past three months (X). The distribution, which we represent graphically for a change, is shown in Figure 6-6.

Draw a Monte Carlo random sample of 8 students, then repeat the questions in Example 6-5.

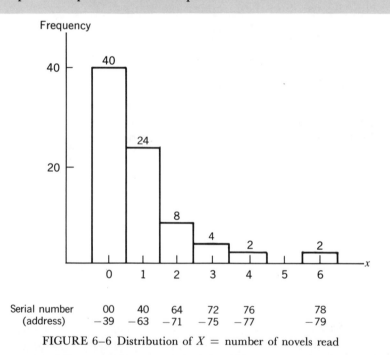

| Serial number | 00 | 40 | 64 | 72 | 76 | 78 |
| (address) | −39 | −63 | −71 | −75 | −77 | −79 |

FIGURE 6–6 Distribution of X = number of novels read

Solution.   (a) It takes two digits to number everybody from 00 to 79, as shown below the graph. We draw our first observation by consulting the first pair of digits in Table II(a); they turn out to be 39, which is the address of a student who read 0 books. This is recorded, and we continue to sample until we draw the address 90, which corresponds to no actual student. What do we do now? We simply ignore such an address, and go on to the next.[12]

We thus obtain the following random sample:

Random Pair (Address)	X_i
39	0
65	2
76	4
45	1
45	1
19	0
ignore 90	
69	2
64	2
	$\overline{X} = 12/8 = 1.50$

(b) To theoretically obtain the sampling distribution of $\overline{X}$, we begin by calculating the moments of the population, by first converting the Figure 6-6 to tabular form:

x	f	$p(x) = \dfrac{f}{N}$	$xp(x)$	$x^2p(x)$
0	40	.50	0	0
1	24	.30	.30	.30
2	8	.10	.20	.40
3	4	.05	.15	.45
4	2	.025	.10	.40
5	0	0	0	0
6	2	.025	.15	.90
	$N = 80$	$1.00 \checkmark$	$\mu = .90$	$E(X^2) = 2.45$ $-\mu^2 = -.81$ $\sigma^2 = 1.64$

Now we can find the sampling moments of $\overline{X}$:

$$E(\overline{X}) = \mu = .90 \qquad \text{(6-5) repeated}$$

$$\sigma_{\overline{X}} = \frac{\sigma}{\sqrt{n}} = \frac{\sqrt{1.64}}{\sqrt{8}} = .45 \qquad\qquad \text{(6-8) repeated}$$

Using these moments and the central limit theorem, we graph the approximate normal distribution of $\overline{X}$ in Figure 6-7. The Monte Carlo distribution of $\overline{X}$ in part (a) will be approximately[13] like this.

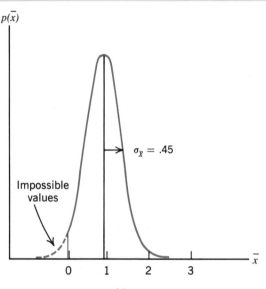

$$\mu_{\overline{x}} = .90$$

FIGURE 6–7 The approximate sampling distribution of $\overline{X}$, for a sample of $n = 8$ drawn from the population in Figure 6–6

---

[11]It might be more interesting to actually use your own class; it simply requires your instructor to gather the population information by a show of hands.

[12]In ignoring addresses 80 to 99, we do not, of course, change the fact that the addresses 00, 01, 02, . . . , 79 all remain equally likely—the key idea in random sampling.

[13]The approximation is twofold: (1) The normal curve in Figure 6-7 cannot be exactly true, because it spills over into negative values for $\overline{X}$, which clearly are impossible. (However, this still is consistent with the central limit theorem, which only claims that $\overline{X}$ is *approximately* normal.) (2) The Monte Carlo distribution of $\overline{X}$ always is just an approximation to the true distribution of $\overline{X}$ (unless the sampling experiment is repeated without end).

(b) Normal Sampling

In Example 6-4, we note that the population is the same as in Figure 6-1(a), which is approximately normal with $\mu = 69$ and $\sigma = 3.2$. In cases such as this, where the population has a standard shape (in this case, normal), the sampling may be speeded up. Instead of numbering all the individuals in the population, we use the standardized normal transformation in reverse. Recall that Z represents a normal variable with mean 0 and standard deviation 1; now let:

$$X = \mu + \sigma Z \qquad (6\text{-}13)$$

Then X has a normal distribution with mean μ and standard deviation[14] σ.

Table II(b) gives a list of random observations Z_i from the standard normal distribution. By multiplying by $\sigma = 3.2$ and adding $\mu = 69$, we generate observations X_i from the normal distribution of men's heights. For example, starting from the first entry in Table II(b), $Z_1 = .5$ and so:

$$X_1 = \mu + \sigma Z_1 = 69 + 3.2(.5) = 70.6$$

Continuing in this way, we simulate a complete sample of five observations of men's heights:

Z_i	X_i
.5	70.6
.1	69.3
2.5	77.0
−.3	68.0
−.1	68.7

$$\overline{X} = 70.7$$

Example 6-7. Let everyone in your class perform a fresh computation of the above sampling experiment (that is, simulating a random sample of $n = 5$ observations from a normal population with $\mu = 69$ and $\sigma = 3.2$). Let the instructor graph the frequency distribution of all the values of $\overline{X}$. Is it

---

[14]The proof, of course, directly follows from the theory of linear transformations in Table 4-2.

approximately the same sampling distribution as in Example 6-5?

Solution.   The sampling distribution of $\overline{X}$ will indeed be similar to the distribution generated in Example 6-5. We have simply done the same problem in two different ways.

(c) Monte Carlo by Computer

The sampling by hand calculation that we have done so far has provided a lot of insight. In practice, however, a computer is much better at these repetitive calculations. A computer can easily draw a sample and calculate its mean, then repeat this procedure over and over, hundreds of times a second. It takes only a few minutes, therefore, to build up a frequency distribution that very closely approximates the probability distribution. This is called *computer Monte Carlo, or sampling simulation.*

Monte Carlo studies are commonly used to check theoretical results, just as we checked the theoretical sampling distribution of $\overline{X}$. But even more importantly, Monte Carlo is used to estimate sampling distributions in cases where theoretical sampling distributions are too difficult to derive (for example, the sampling distribution of $\overline{X}$ when n is very small and the population is not normal).

PROBLEMS

For Problems 6-13 to 6-16, use a random starting point in Table II to generate a random sample, and calculate the indicated statistic. Then have the instructor graph the frequency distribution of all the values obtained by the class. Compare it with the theoretical sampling distribution, where possible.

6-13 The daily demand for fan belts ordered from a large auto supply has the following distribution (rounded to the nearest 5):

x	$p(x)$
15	.01
20	.05
25	.22
30	.33
35	.26
40	.10
45	.03

1.00

Supposing that successive days are independent, simulate the demand for a week (6 days). Then calculate the sample's average daily demand $\overline{X}$.

6-14 (a) In Problem 6-13, graph the given distribution of demands $p(x)$.
(b) Graph the normal distribution with mean $\mu = 31$ and standard deviation $\sigma = 6$; note that this is an excellent approximation to (a).
(c) Using the normal distribution in (b), and the normal Monte Carlo technique of (6-13), redo Problem 6-13.

6-15 A certain small ski tram holding 8 skiers routinely waits until it is filled before going up. Suppose that skiers arrive at random to board it. To be precise, denote the waiting time for the first skier to arrive by T_1, then the additional time for the second skier to arrive by T_2, etc. Suppose that these successive waiting times $T_1, T_2, \ldots T_8$ are independent, and that all have the following common distribution:

t	$p(t)$
0	.33
1	.22
2	.14
3	.09
4	.07
5	.05
6	.04
7	.03
8	.02
9	.01

Simulate one loading of the ski tram. What is the total waiting time required? What, then, is the average waiting time for a skier to arrive?

6-16 In a population of working couples, the man's weekly income X and the woman's weekly income Y are statistically independent and normally distributed. X has $\mu = 250$ and $\sigma = 50$, while Y has $\mu = 200$ and $\sigma = 60$. Five couples are sampled randomly in a sociological survey. Simulate this experiment, and in particular find:
(a) The average male income.
(b) The maximum male income.
(c) The maximum income (male or female).
(d) By how much the husband's income exceeds his wife's, on average.

*6-17 (Requires a computer). Write a computer program to carry out an extensive Monte Carlo study of Problem 6-16. (A random normal generator should be available as a standard subroutine.)

6-5   0–1 VARIABLES

(a) Introduction

In Problem 6-11, we investigated a population of students that we can think of as a bowl of chips, numbered 0, 1, or 2. Now we shall carry this idea to the extreme case where there are only two kinds of chips, numbered 0 or 1.

Example 6-8.   An election may be interpreted as a way of asking every voter in the population: "How many votes do you cast for the Democratic candidate?" If this is an honest election, the voter must reply either 0 or 1. In the 1964 presidential election, the following distribution was recorded for this population:

$x = $ Number of Democratic Votes an Individual Casts	$p(x) = $ Relative Frequency
0	.40
1	.60

(a) What is the population mean? The population variance? The population proportion of Democrats?
(b) When 10 voters were polled randomly, they gave the following answers:

$$1, 0, 1, 1, 0, 1, 1, 0, 1, 1.$$

What is the sample mean? The sample proportion of Democrats?

Solution.   (a) In Table 6-3(a), we calculate the population mean and variance. The proportion of Democrats is .60, the same as the mean.
(b) The sample mean is $7/10 = .70$. The sample proportion is $7/10 = .70$, the same as the mean.

To generalize this example, consider a population of voters in which the proportion voting Democratic is π. For a single voter drawn at random, let:

$$X = \text{the number of Democratic votes cast}$$

If this seems a strange way to define such a simple random variable, we could explicitly define it with the formula:

$$X = 0 \text{ if he is not a Democrat}$$

$$= 1 \text{ if he is a Democrat} \qquad (6\text{-}14)$$

TABLE **6-3**

Population Mean and Variance for a 0–1 Variable, when the Proportion of Democrats is (a) 60%, as in Example 6-8 and (b) π, in General

(a)

$x = $ Number of Democratic Votes an Individual Casts	$p(x)$ = Relative Frequency = Population Proportion	$xp(x)$	$x^2p(x)$
0	.40	0	0
1	.60	.60	.60
		$\mu = .60$	$E(X^2) = \quad .60$ $-\mu^2 = \quad -.60^2$
			$\sigma^2 = \quad .24$

(b)

x	$p(x)$	$xp(x)$	$x^2p(x)$
0	$(1 - \pi)$	0	0
1	π	π	π
		$\mu = \pi$	$E(X^2) = \pi$ $-\mu^2 = -\pi^2$
			$\sigma^2 = \pi - \pi^2 = \pi(1 - \pi)$ $\sigma = \sqrt{\pi(1 - \pi)}$

Thus the population of voters may be thought of as the bowl of chips, marked 0 or 1, shown in Figure 6-8. The moments of this population are worked out in Table 6-3(b), where we find:

$$\boxed{\text{population mean } \mu = \text{population proportion } \pi} \qquad (6\text{-}15)$$

$$\boxed{\text{population standard deviation, } \sigma = \sqrt{\pi(1 - \pi)}} \qquad (6\text{-}16)$$

FIGURE 6–8 A 0–1 population (population of voters)

When we take a sample, the sample mean is calculated by adding up the 1s (counting up the Democrats[15]), and dividing by n. This, of course, yields the sample proportion of Democrats, so that:

$$\boxed{\text{Sample mean } \overline{X} = \text{sample proportion } P} \qquad (6\text{-}17)$$
$$\text{like } (6\text{-}15)$$

Thus we have found an ingenious way to handle proportions: they simply are disguised averages of 0–1 variables. As averages, they can be handled easily with the general theory of sampling that we already developed.

How does the sample proportion P fluctuate around the population proportion π? Since P is just the sample mean $\overline{X}$ in disguise, we may find its expectation by substituting (6-17) and (6-15) into (6-5):

$$\boxed{E(P) = \pi} \qquad (6\text{-}18)$$

The standard deviation is obtained by substituting (6-17) and (6-16) into (6-8):

$$\boxed{\text{standard deviation of } P,\ \sigma_P = \sqrt{\frac{\pi(1-\pi)}{n}}} \qquad (6\text{-}19)$$

Finally, the central limit theorem assures us that, for large samples, the distribution of P is approximately normal.

---

[15]Since X provides such a simple means of counting the number of Democrats, it is called a *counting variable* as well as a *0–1 variable* (or an *off–on variable, dummy variable,* or *Bernoulli variable.*)

Example 6-9. What is the probability that, in a sample of 50 voters, at least 60% will be Democrats? Assume that the population proportion of Democrats is 56%.

Solution. In this large sample of 50, we can use the normal approximation.[16] We standardize, using the moments of P obtained from (6-18) and (6-19):

$$\Pr(P \geq .60) = \Pr\left(\frac{P - \pi}{\sqrt{\pi(1 - \pi)/n}} \geq \frac{.60 - .56}{\sqrt{(.56)(.44)/50}}\right) \tag{6-20}$$

$$= \Pr(Z \geq .57) = .28 \tag{6-21}$$

---

[16]The normal approximation is shown in the figure below. For readers who want an improved approximation, we offer the following subtlety, called the *continuity correction:*

The actual distribution of P is in discrete bars, of course, because there are only $n = 50$ voters in the sample; thus the sample proportion P can only take on the discrete values 0, 1/50, 2/50, . . . , that is, 0, .02, .04, . . . , .58, .60, .62, . . . , .98, 1.00. The exact probability that we desire is the sum of all the bars from .60 up, shown in dark color.

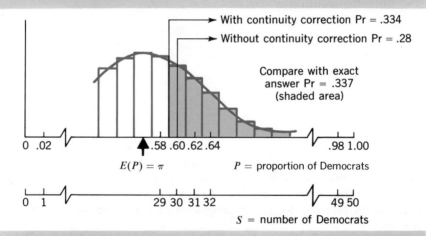

With continuity correction Pr = .334
Without continuity correction Pr = .28
Compare with exact answer Pr = .337 (shaded area)

$E(P) = \pi$ P = proportion of Democrats

S = number of Democrats

To approximate this sum with the normal curve as well as possible, we clearly should include, as closely as we can, the *whole bar* at .60 (as well as all the bars above .60). This means starting at the left-hand side of the bar, at .59 (halfway between .58 and .60). Thus, instead of (6-20), we should calculate:

$$\Pr(P > .59) = \Pr\left(\frac{P - \pi}{\sqrt{\pi(1 - \pi)/n}} \geq \frac{.59 - .56}{\sqrt{(.56)(.44)/50}}\right)$$

$$= \Pr(Z \geq .43) = .334 \tag{6-22}$$

(cont'd, top of p. 170)

This answer is a much better approximation than the value of .28 that is calculated in (6-21). (The absolutely correct answer (.337) involves evaluating all these bars on a computer using (4-7).)

Example 6-10. If the population proportion of Democrats is 40%, what is the probability that a sample of 200 voters will include at least 100 Democrats?

Solution. This is an example of the binomial distribution (as previously recognized in the fourth row of Table 4-3), with $n = 200$ and $\pi = .40$. But it is very difficult to solve this way, because Table III(c) does not cover sample sizes as large as $n = 200$. Moreover, if we tried to evaluate the binomial formula (4-7), we would find the solution very difficult to compute.

A much simpler solution is to recognize this as a problem similar to Example 6-9: in a sample of 200 voters, the probability of getting at least 100 Democrats is exactly the same as the probability of getting at least $100/200 = 50\%$ Democrats. Thus the solution is:

$$\Pr(P \geq .50) = \Pr\left(\frac{P - \pi}{\sqrt{\pi(1 - \pi)/n}} \geq \frac{.50 - .40}{\sqrt{.40(.60)/200}}\right)$$

$$= \Pr(Z \geq 2.89)$$

$$= .0019$$

This result is important enough to review. When we have a binomial problem that is awkward because of large sample size, a simple solution is to first rephrase the problem in terms of a sample proportion, and then proceed on that basis, using the normal approximation. Accordingly, this is often referred to as the large-sample *normal approximation to the binomial*.[17]

PROBLEMS

If you want your answers to be highly accurate, you should do these problems with continuity correction (w.c.c.) as in (6-22).

---

[17]A useful rule of thumb is that n should be large enough to make $n\pi > 5$ and $n(1 - \pi) > 5$. If n is large ($n \geq 100$), yet π is so small that $n\pi < 5$, then there is a better approximation than the normal, called the "Poisson distribution for rare events."

6-18 What is the probability of fewer than 40% heads, if a fair coin is tossed:
(a) $n = 15$ times?
(b) $n = 30$ times?
(c) $n = 100$ times?

6-19 In a certain city, 55% of the eligible jurors are women. If a jury of 12 is picked fairly (at random), what is the probability that there would be 3 or fewer women?

6-20 In the 1964 presidential election, 60% voted Democratic (as in Problem 4-33). If, prior to the election, Gallup took a random sample of 25 voters, what is the probability that the sample would correctly forecast the election winner, i.e., that a majority would be Democratic?

6-21 In the New Year, what is the probability that of the first 50 babies born, more than 60% will be boys?

6-22 What is the chance that, of the first 10 babies born in the New Year, more than 6 will be boys? Answer in three ways:
(a) Exactly, using the binomial distribution.
(b) Approximately, using the normal distribution.
(c) Approximately, using the normal approximation with continuity correction.
(d) Answer true or false; if false, correct it.
Part (c) illustrates that the normal approximation with continuity correction can be an excellent approximation, even for n as small as 10.

6-23 In sampling from a 0–1 population, the sample proportion P is related to the total number of successes S by:

$$P = \frac{S}{n}$$

or $S = nP$

Using the moments of P and the theory of linear transformations, prove that:

$$
\boxed{
\begin{array}{l}
\text{binomial mean, } E(S) = n\pi \\[4pt]
\text{binomial variance, } \mathrm{var}(S) = n\pi(1 - \pi)
\end{array}
}
\qquad (6\text{-}23)
$$

*6-6   SMALL-POPULATION SAMPLING[18]

(a) Introduction

In (6-2), we defined simple random sampling as sampling in which the observations $X_1, X_2, \ldots, X_n$ are independent. In this section, we shall analyze the exception to this—when we sample without replacement from a small population (often called simply "small-population sampling").

To be concrete, suppose that we are drawing a sample of two chips from a bowl containing three chips, marked 2, 6, and 7 (as in Problem 5-27). The joint distribution of X_1 and X_2 is displayed in Table 6-4, from which the marginal distributions are derived. Since the joint distribution is symmetric, the marginal distribution of X_1 and of X_2 must always be exactly the same.

This is also intuitively clear: if we have no knowledge of X_1, then X_2 will have the population distribution, just as X_1 has the population distribution. For example, suppose that the deal of a poker hand is interrupted after two cards have been dealt face down. Which card is likely to be better? Obviously, both cards have the *same* distribution of possible values—the population of fifty-two cards in the deck. Similarly, all successive cards that are dealt (sampled) have the same population distribution, so that we may write:

$$p(x_1) = p(x_2) = \cdots = p(x_n) = \text{population distribution} \qquad (6\text{-}24)$$
$$\text{like } (6\text{-}1)$$

TABLE **6-4**
Joint Distribution of Two Chips
Drawn without Replacement from a Bowl with
3 Chips (Marked 2, 6, and 7)

$p(x_1)$ $\rightarrow$	1/3	1/3	1/3	1 ✓
7	1/6	1/6	0	1/3
6	1/6	0	1/6	1/3
2	0	1/6	1/6	1/3
$\uparrow x_2 \quad x_1 \rightarrow$	2	6	7	$p(x_2)\uparrow$

---

[18]This is a starred section, which, like a starred problem, is optional. You may skip it without loss of continuity.

(b) Expectation

As a consequence of (6-24), all observations have the same mean and variance, so that many of the formulas for simple random sampling still hold true. For example, the formula for expectation has exactly the same proof as before, so that we arrive at the same result:

$$E(\overline{X}) = \mu \tag{6-25}$$
like (6-5)

(c) Variance

We already have noted that the *marginal* distribution for X_1 is the same as for X_2, both being the distribution of the population. However, the *conditional* distribution of X_2, given X_1, is not the same. Since the first chip X_1 is not replaced, the remaining population and hence the conditional distribution of X_2 is different. That is, the distribution of X_2 is dependent on what X_1 happens to be. In fact, X_1 and X_2 have a negative correlation: when X_1 is low, then the remaining chips (possible values for X_2) tend to be high.

This negative correlation can also be seen from the joint distribution of X_1 and X_2 in Table 6-4. The main diagonal is zero, reflecting the impossibility of drawing the same chip twice. This leaves the probability (in the white region) with a negative tilt, so that cov (X_1, X_2) is slightly negative. (Numerical computation bears this out.) When we substitute into the following equation,

$$\text{var } \overline{X} = \text{var } \tfrac{1}{2}[X_1 + X_2]$$

$$= \frac{1}{2^2}[\text{var } X_1 + \text{var } X_2 + 2 \text{ cov } (X_1, X_2)] \qquad \text{like (5-34)}$$

we note that var $\overline{X}$ is slightly reduced by the negative covariance. It similarly could be proved in general, for a sample of n observations drawn from a population of N individuals, that the variance of $\overline{X}$ is reduced by an amount that can be expressed as:

$$\boxed{\text{the reduction factor} = \frac{N - n}{N - 1}} \tag{6-26}$$

Thus, (6-7) becomes:

$$\text{var } \overline{X} = \frac{\sigma^2}{n}\left(\frac{N-n}{N-1}\right) \tag{6-27}$$

Certain special sample sizes shed a great deal of light on the variance of $\overline{X}$:

1. When there is only $n = 1$ chip sampled, it does not matter whether or not it is replaced. This is reflected in the reduction factor (6-26) becoming 1, making var $\overline{X}$ without replacement the same as with replacement. (If you have wondered where the 1 came from in the denominator of (6-26), you can see that it is needed to logically make (6-27) and (6-7) equivalent—as they must be—for a sample size of one.)

2. When $n = N$, the sample coincides with the whole population, every time. Hence every sample mean must be the same—the population mean. The variance of the sample mean, being a measure of its fluctuation, must be zero. This is reflected in (6-27) becoming zero.

3. On the other hand, when n is much smaller than N, (e.g., when 200 men are sampled from one million), then (6-26) is practically 1. That is, as we already noted, it makes very little difference whether or not the observations are replaced before continuing sampling.

A simple example should make all this intuitively clear. Suppose that we sample ten of the heights of the male students on a small college campus; suppose further that the first student we sample is the star of the basketball team at seven feet. Clearly, we now face the problem of a sample average that is too high. If we replace, then in the next nine men who are chosen, the star *could* turn up again, thus distorting our sample mean for the second time. But if we don't replace, then we don't have to worry about this tall individual again. In summary, sampling without replacement yields a less variable sample mean because extreme values, once sampled, cannot return to haunt us again.

(d) Distribution Shape

A modified form of the central limit theorem assures us that, under most circumstances, $\overline{X}$ is still approximately normal, for large n.

PROBLEMS

*6-24 In the game of bridge, cards are allotted points as follows:

Cards	Points
All cards below jack	0
Jack	1
Queen	2
King	3
Ace	4

(a) For the population of 52 cards, find the mean number of points, and the variance.
(b) In a randomly dealt hand of 13 cards, what is the probability that there will be at least 13 points? (Bridge players beware: no points counted for distribution.)

*6-25 Rework Problem 6-7, assuming that the population of people using the ski lift is no longer very large, but rather is:
(a) $N = 500$.
(b) $N = 150$.

*6-26 Rework Problem 6-9, assuming that the population is no longer very large, but rather is:
(a) $N = 30,000$.
(b) $N = 3,000$.
(c) $N = 300$.

*6-27 True or false? If false, correct it.
When sampling from a finite population (without replacement), the variance of $\overline{X}$ contains the reduction factor $(N - n)/(N - 1)$. However, if the population is large relative to the sample, this factor may be ignored just as if the population were infinite. (To be specific, if $N \geq 100n$, then the reduction factor is between .99 and 1.00, and so changes the variance less than 1%.)

*6-28 (a) In a bridge hand (13 out of 52 cards), what is the probability that there will be at least 7 spades? at least 7 of one suit?
(b) In a poker hand (5 out of 52 cards), what is the probability that there will be at least 2 aces?

REVIEW PROBLEMS

6-29 Match the symbol on the left with the phrase on the right:

μ	sample mean
$\overline{X}$	variance of the sample mean
σ^2	population mean
$\sigma_{\overline{X}}^2$	population variance

6-30 (Sample examination questions, American Society of Actuaries.)

(a) A certain industrial process yields a large number of steel cylinders whose lengths are approximately normally distributed with mean 3.25 inches and variance 0.0008 square inch. If two cylinders are chosen at random and placed end to end, what is the probability that their combined length is less than 6.55 inches?

(b) A company annually uses many thousands of electric lamps, which burn continuously day and night. Assume that, under such conditions, the life of a lamp may be regarded as a variable normally distributed about a mean of 50 days with a standard deviation of 19 days. On January 1, 1964, the company put 5,000 new lamps into service. How many would be expected to need replacement by February 1, 1964?

6-31 (a) Suppose that the population of weights of airline passengers has mean 150 lbs. and standard deviation 25 lbs. A certain plane has a capacity of 7,800 lbs. What is the probability that a flight of 50 passengers will overload it?

(b) If we wish to reduce the chance of overload to 1/100, to how much must the capacity be increased?

(c) What assumptions are you making implicitly? In what way do you think they are questionable?

6-32 In Chapter 3, it was implied that in a very large sample, relative frequency is "very likely" to be "close to" probability. To make this statement precise, for the rolling of a die for example, let P denote the proportion of aces in 10,000 throws, and calculate:

$$\Pr(\tfrac{1}{6} - .01 < P < \tfrac{1}{6} + .01)$$

6-33 A man at a carnival pays $1 to play a game (roulette) with the following payoff:

Win	Probability
−1	20/38
+1	18/38

(a) What is the average win in a game?

(b) What is his approximate chance of ending up a loser if he plays the game:

    (i) 5 times?

    (ii) 25 times?

    (iii) 125 times?

(c) Calculate a more exact answer to (b)(i), using the binomial distribution.

(d) How many times should the man plan to play if he wants to be 99% certain of losing?

6-34 Suppose that there are five men in a room, whose heights in inches are 62, 65, 68, 65, 65. One man is drawn at random, and his height is denoted X.

(a) Graph the probability distribution of X, i.e., the population distribution. Find its mean μ, and variance σ^2.

    Suppose that a sample of two men is drawn, with replacement, and that the sample mean $\overline{X}$ is calculated. Then:

(b) Construct a table and graph of the probability distribution of $\overline{X}$.

(c) Find the mean and variance of $\overline{X}$ from its probability distribution.

(d) Check your answers to (c), using the theory of this chapter.

(e) Is the following statement valid for this problem? If not, correct it. $\overline{X}$ fluctuates around μ—sometimes larger, sometimes smaller—but its expectation is exactly μ. Being the average of two observations, $\overline{X}$ does not fluctuate as much as a single observation, however. This is reasonable, because in a sample, a large observation will tend to be cancelled out by a small observation.

6-35 Fill in the blank

    Suppose that in a certain election, the U.S. and California are alike in their proportion of Democrats, π, the only difference being that the U.S. is about 10 times as large a population. In order to get an equally reliable sampling estimate of π, the U.S. sample should be _____ as large as the California sample.

CHAPTER 7

Point Estimation

The only way to save yourself from the pain of lost illusions is to have none.

Charles Marriott

7-1   POPULATIONS AND SAMPLES

In Table 7-1, we review the concepts of population and sample. It is essential to remember that the population mean μ and variance σ^2 are constants (though generally unknown). These are called population *parameters*.

By contrast, the sample mean $\overline{X}$ and sample variance s^2 are random variables, varying from sample to sample, with a certain probability distribution. For example, the distribution of $\overline{X}$ was found to be approximately normal in (6-10). A random variable such as $\overline{X}$ or s^2, which is calculated from the observations in a sample, is given the technical name *sample statistic*. In Table 7-1 and throughout the rest of the text, we shall leave the *population gray* and make the *sample colored* in order to keep the distinction clear, just as we did in Chapter 6.

Now we can address the problem of statistical inference that we posed in Chapter 1: how can the population be estimated by the sample? Suppose, for example, that to estimate the mean family income μ in a certain region, we take a random sample of 100 incomes. Then the sample

TABLE **7-1**

Review of Population versus Sample

A Random Sample is a Random Subset of the Population	
Relative frequencies f_i/n are used to compute	Probabilities $p(x)$ are used to compute
$\overline{X}$ and s^2,	μ and σ^2,
which are examples of random statistics or estimators	which are examples of fixed parameters or targets

mean $\overline{X}$ surely is a reasonable estimator of μ. By the central limit theorem (6-10), we know that $\overline{X}$ fluctuates about μ; sometimes it will be above μ, sometimes below (as shown in Figure 6-2). Even better than estimating μ with the single *point* estimate $\overline{X}$ would be to construct an *interval* estimate around $\overline{X}$ that is likely to bracket μ. We defer this more difficult task to Chapter 8.

For now, we ask whether another statistic, such as the sample median or mode, might do better than $\overline{X}$ as a point estimator of μ. To answer such questions, we require criteria for judging a good estimator.

7-2   DESIRABLE PROPERTIES OF ESTIMATORS

(a) No Bias

We already have noted that the sample mean $\overline{X}$ is, on average, exactly on its target μ. We therefore call $\overline{X}$ an *unbiased* estimator of μ.

To generalize, we consider any population parameter θ, and denote its estimator by $\hat{\theta}$ (read "theta hat" or "theta estimator"). If, on average, $\hat{\theta}$ is exactly on target as shown in Figure 7-1(a), it is called an unbiased estimator. Formally, we state the definition:

$$\boxed{\begin{array}{c} \hat{\theta} \text{ is an unbiased estimator of } \theta \text{ iff[1]} \\ E(\hat{\theta}) = \theta \end{array}}$$

(7-1)

like (6-5)

Of course, an estimator $\hat{\theta}$ is called biased if $E(\hat{\theta})$ is different from θ. In fact, bias is defined as this difference:

---

[1] Recall that "iff" is an abbreviation for "if, and only if."

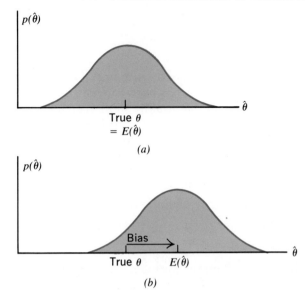

FIGURE 7–1 Comparison of (a) unbiased estimator, and (b) biased estimator

$$\text{Bias} \overset{\Delta}{=} E(\hat{\theta}) - \theta \qquad (7\text{-}2)$$

Bias is illustrated in Figure 7-1(b). The distribution of $\hat{\theta}$ is off-target; since $E(\hat{\theta})$ exceeds θ, there will be a tendency for $\hat{\theta}$ to overestimate θ.

As an example of a biased estimator, consider the sample mean squared deviation:

$$\text{MSD} = \frac{1}{n}\sum (X_i - \overline{X})^2 \qquad (7\text{-}3)$$
$$(2\text{-}5a)\text{ repeated}$$

It will, on the average, underestimate[2] the population variance σ^2. But if we inflate it just a little, by dividing by $n - 1$ instead of n, we obtain the sample variance:

---

[2]This underestimation can be seen very easily in the case of $n = 1$. Then $\overline{X}$ coincides with X_i, so that (7-3) gives MSD $= 0$, which is an obvious underestimate of σ^2.

On the other hand when $n = 1$, (7-4) gives $s^2 = 0/0$, which is undefined. But this is not a drawback. In fact, it gives a good warning that since a sample of just one observation has no "spread," it cannot estimate the population variance σ^2 (assuming that μ is unknown, of course).

$$s^2 = \frac{1}{n-1} \sum (X_i - \overline{X})^2 \qquad\qquad (7\text{-}4)$$
$$(2\text{-}6a) \text{ repeated}$$

which has been proven an unbiased estimator of σ^2. (When we say "has been proven," we mean that it has been proven in advanced texts. If it has been proven in *this* text, we usually shall say "we have proven.") If you were puzzled by the divisor $n-1$ used in defining s^2, you now can see why: we want to use this sample variance as an unbiased estimator of the population variance.

Both the sample mean and sample median are unbiased estimators of μ in a normal population. Thus, in judging which is to be preferred, we must examine their other characteristics.

(b) Minimum Variance: the Efficiency of Unbiased Estimators

As well as being on target on the average, we should also like the distribution of an estimator $\hat{\theta}$ to be highly concentrated, that is, to have a small variance. This is the notion of *efficiency,* shown in Figure 7-2. We describe $\hat{\theta}$ as more efficient because it has smaller variance. Formally, the relative efficiency of two estimators is defined as:

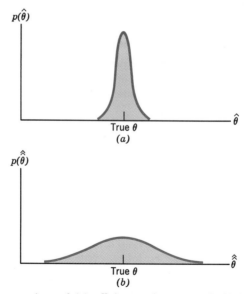

FIGURE 7–2 A comparison of (a) efficient estimator, and (b) inefficient estimator (both are unbiased)

For unbiased estimators,

Relative efficiency of $\hat{\hat{\theta}}$ compared to $\hat{\theta} \triangleq \dfrac{\text{var } \hat{\hat{\theta}}}{\text{var } \hat{\theta}}$ (7-5)

An estimator that is more efficient than any other is called *absolutely efficient,* or simply *efficient.*

Example 7-1.   It has been proven that if the population is normal, the sample median is an unbiased estimator of μ, with a variance of approximately $(\pi/2)(\sigma^2/n)$. On the other hand, we have proven in (6-7) that the variance of the sample mean is always σ^2/n. What therefore is the efficiency of the sample mean relative to the sample median, in normal populations?

Solution.   Since both estimators are unbiased, from (7-5) the relative efficiency is:

$$\frac{(\pi/2)(\sigma^2/n)}{\sigma^2/n} = \frac{\pi}{2} = 157\% \qquad (7\text{-}6)$$

Remarks.   The sample mean is 57% more efficient than the sample median. In fact, it has been proven that, in estimating the mean of a *normal* population, the sample mean is more efficient than *any* other estimator, and so is *absolutely efficient.*

Of course, by increasing sample size n, we can reduce the variance of either the sample mean or median. This provides an alternative way of looking at the greater efficiency of the sample mean (in sampling from *normal* populations). The sample median will yield as accurate an estimate only if we take a larger sample (specifically, 57% larger). Hence the sample mean is more efficient because it costs less to sample. Note how the economic and statistical definitions of efficiency coincide.

(c) Minimum Mean Squared Error: Efficiency of Any Estimator

So far we have concluded that, in comparing unbiased estimators, we choose the one with minimum variance. But suppose that we are com-

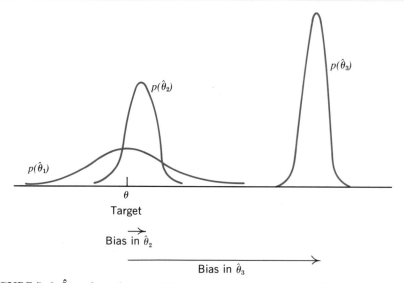

FIGURE 7–3 $\hat{\theta}_2$ as the estimator with the best combination of small bias and variance

paring both biased and unbiased estimators, as in Figure 7-3. Now it is no longer necessarily appropriate to select the estimator with minimum variance; $\hat{\theta}_3$ qualifies on that score, but it is unsatisfactory because it is so badly biased. Nor do we necessarily pick the estimator with least bias; $\hat{\theta}_1$ has zero bias, but it seems unsatisfactory because of its high variance. Instead, the estimator that seems to perform best overall is $\hat{\theta}_2$, because it has the best combination of small bias and small variance.

This intuitive argument suggests using a criterion that appropriately considers both bias and variance. Alternatively, we could say that we are interested not just in the variance of an estimator, since this only measures how it is spread around its (possibly biased) mean. Instead, we are interested in a similar measure of how an estimator is spread around its *true target* θ. The most common measure of this kind is:

$$\text{Mean squared error (MSE)} \triangleq E(\hat{\theta} - \theta)^2 \qquad (7\text{-}7)$$

This is very similar to the variance $E(\hat{\theta} - \mu_{\hat{\theta}})^2$, except that it is measured around the true target θ rather than around the mean of the estimator $\mu_{\hat{\theta}}$. Then, as we hoped, MSE turns out to be a measure of both variance and bias:

$$\text{MSE} = \sigma^2 + \text{bias}^2 \qquad (7\text{-}8)$$

This confirms[3] two earlier conclusions: if two estimators with equal variance are compared (as in Figure 7-1), the one with less bias is preferred; and if two unbiased (or equally biased) estimators are compared (as in Figure 7-2), the one with smaller variance is preferred. In fact, if two estimators are unbiased, it is evident from either (7-7) or (7-8) that the MSE reduces to the variance. Thus, MSE may be regarded as a generalization of the variance concept. Using MSE therefore leads to a generalized definition of the relative efficiency of two estimators:

For any two estimators—whether biased or unbiased—

$$\text{Relative efficiency of } \hat{\hat{\theta}} \text{ compared to } \hat{\theta} \triangleq \frac{\text{MSE}(\hat{\theta})}{\text{MSE}(\hat{\hat{\theta}})} \qquad (7\text{-}9)$$

---

with (7-5) recognized as just a special case of this.

To sum up, because it combines the two attractive properties of small bias and small variance, the concept of efficiency as defined in (7-9) becomes the single most important criterion for judging estimators.

PROBLEMS

7-1 True or false? If false, correct it.

(a) μ is a random variable (varying from sample to sample), and is used to estimate the parameter $\overline{X}$.

(b) The sample proportion P is an unbiased estimator of the population proportion π.

(c) In sampling from a normal population, the sample median and sample mean are both efficient estimators of μ. The difference is that the sample median is biased, whereas the sample mean is unbiased.

7-2 Each of three guns is being tested by firing 12 shots at a target from a clamped position. Gun A was not clamped down hard enough, and wobbled. Gun B was clamped down in a position that pointed slightly to the left, due to a misaligned sight. Gun C was clamped down just right.

(a) Which of the following patterns of shots belongs to gun A? gun B? gun C?

(b) Which guns are biased? Which gun has minimum variance? Which has the largest MSE? Which is most efficient? Which is least efficient?

---

[3] Equation (7-8) can be proved by substituting $\hat{\theta}$ for X, and θ for a, in equation (4-23) and then rearranging it.

7-3 Based on a random sample of 100 observations, consider two estimators of μ:

$$\overline{X}_{100} \triangleq \frac{1}{100}(X_1 + X_2 + \cdots + X_{100})$$

and
$$\overline{X}_{90} \triangleq \frac{1}{90}(X_1 + X_2 + \cdots + X_{90})$$

That is, $\overline{X}_{100}$ is the mean of all the observations, whereas $\overline{X}_{90}$ is just the mean of the first 90.
(a) Are they unbiased?
(b) What is the efficiency of $\overline{X}_{90}$ relative to $\overline{X}_{100}$?

7-4 Based on a random sample of 2 observations, consider two estimators of μ:

$$\overline{X} \triangleq \tfrac{1}{2}X_1 + \tfrac{1}{2}X_2$$

and
$$W \triangleq \tfrac{1}{3}X_1 + \tfrac{2}{3}X_2$$

(a) Are they unbiased?
(b) What is the efficiency of W relative to $\overline{X}$? Which estimator is better?
*(c) (Requires calculus.) Prove that $\overline{X}$ is more efficient than any other unbiased linear combination. (*Hint:* for $cX_1 + (1 - c)X_2$, find the variance in terms of c, and call it $f(c)$. Then set $f'(c) = 0$.)

7-5 Discuss the following statement:
(a) In both Problems 7-3 and 7-4, we have an example of a relatively inefficient estimator (an estimator with relative efficiency of only 90%). In Problem 7-3, this inefficiency was obvious, because 10% of the observations were thrown away in calculating $\overline{X}_{90}$. In Problem 7-4, the inefficiency was more subtle, because it was caused merely by an inefficient analysis, using the wrong weights for W. However, in terms of results (producing an estimator with more variance than necessary), we can say that both

inefficiencies are the same; in other words, the man who uses an inefficient analysis is getting the same results as a man who throws away data.

(b) In view of this, what advice would you give to a researcher who spends $100,000 collecting data, and $100 analyzing it?

7-6 (Monte Carlo.)

(a) Simulate drawing a random sample of 5 observations from a normal population of women's heights, with $\mu = 65''$, $\sigma = 3''$. Calculate the sample mean and median.

(b) Have the instructor graph the relative frequency distributions of the sample mean and median. Which seems more efficient? Are the results compatible with (7-6)?

7-7 When S successes occur in n trials, the sample proportion $P = S/n$ customarily is used as an estimator of the probability of success π. However, sometimes there are good reasons[4] to use the estimator $P* \overset{\Delta}{=} (S + 1)/(n + 2)$. Alternatively, $P*$ can be written as a linear combination of the familiar estimator P:

$$P* = \frac{nP + 1}{n + 2} = \left(\frac{n}{n + 2}\right)P + \left(\frac{1}{n + 2}\right)$$

(a) Find the mean and variance of $P*$. (*Hint:* exploit the familiar moments of P.) Hence, find the MSE of $P*$.

(b) Find the MSE of P.

(c) Using the formulas developed in (a) and (b), compare the MSE of $P*$ and of P, when $n = 10$, and $\pi = 0, .1, .2, \ldots, .9, 1.0$.

(d) State some possible circumstances when you might prefer to use $P*$ instead of P to estimate π.

7-8 Consider a bowl full of many chips—one-third marked 2, one-third marked 4, and one-third marked 6. The population moments were found in Problem 6-4 to be $\mu = 4$ and $\sigma^2 = 8/3$. When a sample of 2 chips is drawn, construct the probability table of $\overline{X}$, and hence answer the following:

(a) Show (once more) that $\overline{X}$ is an unbiased estimator of μ.

(b) Is $(2\overline{X} + 1)$ an unbiased estimator of $(2\mu + 1)$?

(c) Is $(\overline{X})^2$ an unbiased estimator of μ^2?

(d) Is $1/\overline{X}$ an unbiased estimator of $1/\mu$?

(e) How could you have answered parts (a), (b) and (c) theoretically, without going through all the computations?

---

[4] The reasons that lead naturally to the estimator $P*$ are given by Bayesian analysis.

7-9 To illustrate bias very concretely, consider again the sampling in Problem 7-8. We shall study sample estimators in three ways.

(a) *Monte Carlo approach.* Simulate drawing a sample of 2 chips, and calculate $\overline{X}$, s^2, and MSD. Repeat this simulation over and over, until 10 or 20 values of $\overline{X}$, etc., have been obtained. Then tally the distribution of $\overline{X}$, etc., and hence check the following:

(i) Does $\overline{X}$ average close to μ?

(ii) Does s^2 average close to σ^2?

(iii) Does MSD average close to σ^2?

(b) *Analytical approach.* In (a), if the experiment were repeated endlessly, the relative frequencies would settle down to probabilities, which you can calculate easily. Then use them to answer the following questions exactly:

(i) $E(\overline{X}) \overset{?}{=} \mu$

(ii) $E(s^2) \overset{?}{=} \sigma^2$

(iii) $E(\text{MSD}) \overset{?}{=} \sigma^2$

(c) *Theoretical approach.* Find the references in the text that give the same answers as part (b)—only more generally, for any sample size and any population.

*7-10 A random sample of 1,000 students is to be polled from a population of 5,000 students. Judging on the narrow grounds of efficiency, is it better to sample with or without replacement? How much better? (*Hint:* requires Section 6-6.)

*7-3   CONSISTENCY

(a) Introduction

Roughly speaking, a consistent estimator is one that concentrates perfectly on its target as sample size increases indefinitely,[5] as sketched in Figure 7-4.

---

[5] The precise definition of consistency involves a limit statement: $\hat{\theta}$ is defined to be a consistent estimator of θ, iff for any positive δ (no matter how small),

$$\Pr(|\hat{\theta} - \theta| < \delta) \to 1 \qquad (7\text{-}10)$$

$$\text{as} \quad n \to \infty$$

This is seen to be just a formal way of stating that in the limit as $n \to \infty$, it eventually becomes certain that $\hat{\theta}$ will be as close to θ as we please (within δ).

To be concrete, in Example 7-2 we show that the sample proportion P is a consistent estimator of π; for this case, (7-10) therefore becomes:

(*cont'd*)

An estimator $\hat{\theta}$ will be consistent if its MSE approaches zero.[6] In view of (7-8), this may be reexpressed as:

> $\hat{\theta}$ is a consistent estimator if its bias[7] and variance *both* approach zero, as $n \to \infty$.   (7-13)

$$\Pr(|P - \pi| < \delta) \to 1 \qquad (7\text{-}11)$$

$$\text{as} \quad n \to \infty$$

Stated informally, (7-11) means that it is eventually certain that P will get as close to π as we please (within δ). In fact, (7-11) is the precise statement of the law of large numbers (3-2).

[6]The proof may be found in mathematical statistics texts.

Incidentally, we cannot make the implication the other way. That is, consistent estimators exist where the MSE does not approach zero. As an example, consider an estimator $\hat{\theta}$ with the following sampling distribution:

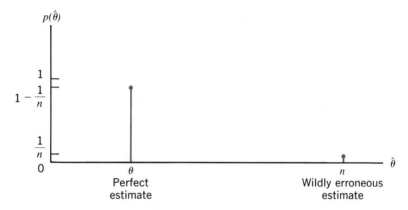

Perfect estimate — Wildly erroneous estimate

As n increases, note how the distribution of $\hat{\theta}$ becomes increasingly concentrated on its target θ while the probability of getting an erroneous estimate becomes smaller and smaller (as $1/n$ decreases) Thus, $\hat{\theta}$ is a consistent estimator according to the definition (7-10).

But now let us consider the MSE of $\hat{\theta}$:

$$E(\hat{\theta} - \theta)^2 = \Sigma(\hat{\theta} - \theta)^2 p(\hat{\theta})$$

$$= 0^2\left(1 - \frac{1}{n}\right) + (n - \theta)^2\left(\frac{1}{n}\right)$$

$$= n - 2\theta + \frac{\theta^2}{n} \qquad (7\text{-}12)$$

As $n \to \infty$, this MSE does not approach zero (in fact, it behaves much worse—it approaches infinity!)

[7]An estimator whose bias approaches zero is often called *unbiased in the limit*, or *asymptotically unbiased*.

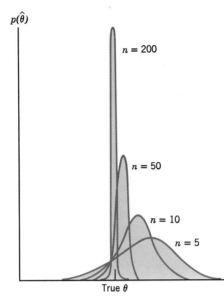

FIGURE 7–4 A consistent estimator, showing how the distribution of $\hat{\theta}$ concentrates on its target θ as n increases

Example 7-2.   Is $\overline{X}$ a consistent estimator of μ?

Solution.[8]   In view of (7-13), it will be enough to prove that the bias and variance of $\overline{X}$ both approach zero.

From (6-5):

$$\text{bias} = 0 \qquad \text{for all } n$$

From (6-7):

$$\text{variance} = \frac{\sigma^2}{n} \to 0 \quad \text{as } n \to \infty$$

Thus $\overline{X}$ is indeed a consistent estimator of μ.

Remarks.   As a special case, we also can conclude that the sample proportion P is a consistent estimator of π.

---

[8]Since we use the population variance σ^2 in this solution, we assume that the population variance does exist. This, incidentally, is the same assumption that we made for the central limit theorem (6-10).

(b) Critique

Just because an estimator is consistent does not mean that it is a good one. For example, as an estimator of μ in a normal population, the sample median has been proven to be consistent. Or, to take a more drastic example, if we throw away half our observations and average the remaining half, we still obtain a consistent estimator. But neither of these consistent estimators is very good; the sample mean is preferred because it is both consistent *and* efficient.

These examples show that there are many consistent estimators, and that the criterion of consistency may not necessarily help us choose the best one, or even a sensible one. Why? Because consistency is a limiting property, defined in very abstract mathematical terms that say nothing about how *fast* the approach to the limit is. To know that an estimator eventually will come close (though it may take a billion billion observations) is cold comfort to an applied statistician who only has a dozen observations in hand.

Why, then, should we put any effort into studying consistency? The answer is that in many difficult situations (especially simultaneous equation estimation), estimators cannot be analyzed easily according to the really important criterion of efficiency. But the weaker property of consistency often can be established, and this is better than nothing. In other words, when a statistician proposes a new estimator, nobody will applaud if he proves it consistent. But everybody will criticize if it turns out to be inconsistent.

7-4   AN INTRODUCTION TO NONPARAMETRIC ESTIMATION

In this section, we shall consider nonnormal populations; however, they will all be perfectly symmetric, with the mean, median, and mode all coinciding exactly at the center of symmetry c, shown in Figure 7-5(a). Suppose that the statistician has taken the sample of 9 observations shown in Figure 7-5(b). How can he use this sample efficiently to estimate c?

Of course, the sample of 9 random observations will not be perfectly symmetric like the population, so that the *sample* mean and median will differ. Which is the better estimator of c?

One statistician, noting that c happens to be the population mean, might recommend using the sample mean, on intuitive grounds. But another statistician, noting that c is the population median, might recommend the sample median, on similar intuitive grounds. In the particular sample shown in Figure 7-5(b), the sample median happens to be closer than the sample mean to the target c; however, in other samples the re-

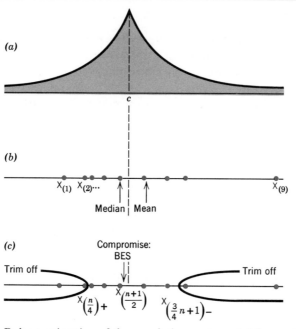

FIGURE 7–5 Robust estimation of the population center c. (a) Symmetrical population distribution. (b) Sample of 9 observations. (c) BES (best easy systematic) estimate.

verse might be true. Since they both are unbiased, a choice between them involves examining their relative efficiency, as defined by (7-5).

If the population is normal, we already know from (7-6) that the sample mean has an efficiency of about 157% relative to the sample median.[9] Unfortunately, this advantage is not maintained for many nonnormal populations, including the Laplace distribution[10] actually shown in Figure 7-5(a). Since the Laplace has much thicker tails than the normal distribution, it is much more likely to produce a very erratic observation that will seriously affect the sample mean. It is thus no surprise that for the Laplace distribution, the sample mean has an efficiency of only 50% relative to the sample median.

We often have emphasized that the population from which we are sampling is a mystery, in the sense that its mean and variance are unknown. But now we see that there is another problem: when its *shape* (normal, Laplace, or whatever) is unknown, it is no longer clear which

---

[9]This efficiency of 157%, like all the other numerical estimates of efficiency of this section, is a very good approximation in large samples, and a reasonable approximation in small samples as well.

[10]Also called the two-tailed exponential distribution, its formula is given following(16-40).

estimator is best. Hence a strong argument can be made for using an estimator that is reasonably efficient for *any* kind of population shape; such an estimator, which is free of the assumption that the population distribution is normal, is called "robust," or "distribution-free," or "nonparametric,"[11]

These remarks on the Laplace distribution suggest that the sample median may be more robust than the sample mean. Yet when we consider all possible population shapes, we find neither really is satisfactory: the sample median uses too few observations (only the middle one), while the sample mean uses too many observations (all of them, including the problem outliers.) Is it not possible to compromise and find something better? One suggestion is to "trim off" about a quarter of the observations from each end of the sample, thus getting rid of possible wild outliers; then, in this trimmed sample, average just the two end and two middle values. The result may be called the Best Easy Systematic (BES) estimate. (Its efficiency is very good, but not quite best in the mathematical sense of surpassing everything else for every population.[12]) The explicit formulas are:

for n even,

$$\text{BES} \triangleq \frac{1}{4}\left\{X_{(\frac{n}{4})+} + X_{(\frac{n}{2})} + X_{(\frac{n}{2}+1)} + X_{(\frac{3}{4}n+1)-}\right\} \qquad (7\text{-}14)$$

for n odd,[13]

$$\text{BES} \triangleq \frac{1}{4}\left\{X_{(\frac{n}{4})+} + X_{(\frac{n+1}{2})} + X_{(\frac{n+1}{2})} + X_{(\frac{3}{4}n+1)-}\right\} \qquad (7\text{-}15)$$

---

[11] To be more precise, when we know the form of the population distribution (e.g., normal) except for specifying parameters (e.g., μ and σ^2), we call it *parametric* estimation. When our assumptions fall short of this, we call it *nonparametric* estimation.

[12] For the normal, Laplace, and many other distributions, the efficiency of the BES estimator has been shown to be almost as good as the very best estimator that you could possibly use (even if you *knew* the shape of the population).

[13] In this case, we may find it instructive to rewrite (7-15) as:

$$\text{BES} = \frac{1}{4}\left\{X_{(\frac{n}{4})+} + 2X_{(\frac{n+1}{2})} + X_{(\frac{3}{4}n+1)-}\right\} \qquad (7\text{-}16)$$

$$= \frac{1}{2}\left\{X_{(\frac{n+1}{2})} + \left[\frac{X_{(\frac{n}{4})+} + X_{(\frac{3}{4}n+1)-}}{2}\right]\right\} \qquad (7\text{-}17)$$

In the form (7-17), we see that the BES estimate is just the average of (1) the median and (2) the average of the two quartiles (in square brackets).

where $X_{(i)} =$ the ith ordered observation; i.e., $X_{(1)}$ has the smallest numerical value, $X_{(2)}$ the next smallest, etc., as in Figure 7-5(b). (This must be distinguished from our previous usage, where X_1, with no brackets on the subscript, was the first value to be sampled, X_2 the second, and so on.) For subscripts involving fractions, $+$ means round up to the nearest integer, while $-$ means round down.

Example 7-3. The increase in sales in 1977 over 1976 for a random sample of 10 hardware stores in a certain chain are as follows (thousands of dollars):

$$3, \quad -8, \quad 56, \quad 38, \quad 19, \quad 2, \quad -23, \quad 94, \quad 15, \quad 24$$

Assume that the complete population (of increases in sales) is symmetrically distributed about an unknown central value c.

(a) Estimate c by calculating:

(i) The sample mean $\overline{X}$.

(ii) The BES estimator.

(b) Is $\overline{X}$ or BES the better (more efficient) estimator of c if, on the basis of past experience, it has been found that:

(i) The population distribution is normal.

(ii) The population distribution has much longer tails than the normal, due to the presence of a few high-risk stores.

Solution.   (a) (i) $\overline{X} = \dfrac{1}{n}\sum X_i = \dfrac{1}{10}(220) = 22.0$

(ii) To calculate BES, we first arrange the observations X_i into ordered values $X_{(i)}$:

$$X_{(1)} = -23 \qquad\qquad X_{(6)} = 19$$
$$X_{(2)} = -8 \qquad\qquad X_{(7)} = 24$$
$$X_{(3)} = 2 \qquad\qquad X_{(8)} = 38$$
$$X_{(4)} = 3 \qquad\qquad X_{(9)} = 56$$
$$X_{(5)} = 15 \qquad\qquad X_{(10)} = 94$$

Since n is even, we shall use (7-14); we first must find:

$$X_{(\frac{n}{4})+} = X_{(\frac{10}{4})+} = X_{(2.5)+} = X_{(3)}$$

where the $+$ sign indicates we round up (to 3). From the data, we find that $X_{(3)} = 2$. Similarly, we find that:

$$X_{\left(\frac{n}{2}\right)} = X_{\left(\frac{10}{2}\right)} = X_{(5)} = 15$$

$$X_{\left(\frac{n}{2}+1\right)} = X_{\left(\frac{10}{2}+1\right)} = X_{(6)} = 19$$

$$X_{\left(\frac{3}{4}n+1\right)-} = X_{\left(\frac{30}{4}+1\right)-} = X_{(8.5)-} = X_{(8)} = 38$$

These are the quantities that we average in (7-14):

$$\text{BES} = \tfrac{1}{4}(2 + 15 + 19 + 38) = \tfrac{74}{4} = 18.5$$

(b) (i) If the population is normal, the remark following Example 7-1 indicates that $\overline{X}$ is more efficient than *any* other estimator, including BES.

(ii) If the population has very long tails, then $\overline{X}$ will be less efficient than BES. This is because $\overline{X}$ gets pulled by out-lying observations. Note, for example, that in this particular sample, $\overline{X}$ is pulled up by the very large observation $X_{(10)}$. (BES, on the other hand, is unaffected because $X_{(10)}$ doesn't enter into the formula for BES at all.)

To summarize, we have considered three estimators of a population center. In increasing order of robustness, they are:

1. Sample mean.

2. Sample median.

3. BES (best easy systematic) estimator given by (7-14) and (7-15).

Thus, if the population shape is unknown, BES may be the preferred estimator. But for most of this book we shall return to the sample mean, not only because it is best if the population is known to be normal, but also because of its popularity and its attractive mathematical properties.

PROBLEMS

7-11 (a) Simulate a random sample of 10 observations from a normal population with $\mu = 0$, $\sigma = 1$. Calculate the mean, median, and BES estimate. Which provides the best point estimate of μ in your specific sample?

(b) Which estimate would outperform the others if this sampling experiment were repeated many times?

7-12 In each of the samples below, assume that the parent population is distributed symmetrically around a central value c. Calculate the most efficient point estimate of c (sample mean, median, or BES), explaining why it is preferred.

(a) A chemist takes the following 12 measurements:

$$8.9 \quad 7.2 \quad 8.5 \quad 8.3 \quad 7.3 \quad 7.8$$
$$7.6 \quad 7.5 \quad 8.6 \quad 7.9 \quad 9.4 \quad 7.9$$

Assume that these observations differ only because of a normally distributed measurement error.

(b) Suppose that a random sample of 5 economists gave the following predictions of the inflation rate in 1975:

$$10\%, \quad 8\%, \quad 8\%, \quad 11\%, \quad 15\%$$

Suppose also that the whole population of economists contains a few who make really extreme predictions, so that the distribution has long tails.

(c) An anthropologist measured the width (in centimeters) of 9 skulls from a certain tribe:

$$13.3, \quad 14.2, \quad 13.5, \quad 16.7, \quad 11.1, \quad 13.1, \quad 13.0, \quad 12.2, \quad 13.0$$

Unfortunately, his measurement technique occasionally erred, being out by several centimeters. Since he was unaware of his blunders, however, he does not know which of the above measurements are invalid.

REVIEW PROBLEMS

7-13 Answer true or false; if false, correct it.

(a) Samples usually are selected with the purpose of making inferences about the population from which the sample is drawn.

(b) The expected value of a random variable is also called its average or mean value.

(c) If we double the sample size, we halve the standard deviation of $\overline{X}$, and consequently double its accuracy in estimating the population mean.

(d) The expected value of $\overline{X}$ usually is equal to the population mean. The only exception is in sampling from a small population.

7-14 Suppose that a surveyor is trying to determine the area of a rectangular field, in which the measured length X and the measured width Y are independent random variables that fluctuate about the true values, according to the following probability distributions:

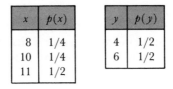

x	$p(x)$
8	1/4
10	1/4
11	1/2

y	$p(y)$
4	1/2
6	1/2

The calculated area $A = XY$, of course, is a random variable, and is used to estimate the true area. If the true length and width are 10 and 5, respectively,

(a) Is X an unbiased estimator of the true length?
(b) Is Y an unbiased estimator of the true width?
(c) Is A an unbiased estimator of the true area?
*(d) What is the variance of A?

7-15 Based on a random sample of 3 observations, consider four possible estimators of μ:

$$\overline{X} = \tfrac{1}{3}X_1 + \tfrac{1}{3}X_2 + \tfrac{1}{3}X_3$$
$$W_1 = \tfrac{1}{4}X_1 + \tfrac{1}{4}X_2 + \tfrac{1}{2}X_3$$
$$W_2 = \tfrac{1}{4}X_1 + \tfrac{1}{4}X_2 + \tfrac{1}{4}X_3$$
$$W_3 = .3X_1 + .3X_2 + .4X_3$$

(a) Which are unbiased?
(b) State a simple rule that the coefficients of an estimator must follow in order to be unbiased.
(c) What is the variance of each estimator?
(d) Of the unbiased estimators, which has least variance (is most efficient)?
(e) Find the efficiency of each unbiased estimator relative to the most efficient estimator found in (d).

7-16 Suppose that two economists estimate μ (the average expenditure of American families on food), with two unbiased (and statistically independent) estimates $\hat{\mu}$ and $\hat{\hat{\mu}}$. The second economist is less careful than the first—the standard deviation of $\hat{\hat{\mu}}$ is five times as large as the standard deviation of $\hat{\mu}$. When asked how to combine $\hat{\mu}$ and $\hat{\hat{\mu}}$ to get a publishable overall estimate, a certain group of statisticians comes up with four proposals:

(1) $W_1 = \tfrac{1}{2}(\hat{\mu} + \hat{\hat{\mu}})$
(2) $W_2 = \tfrac{4}{5}\hat{\mu} + \tfrac{1}{5}\hat{\hat{\mu}}$
(3) $W_3 = \tfrac{5}{6}\hat{\mu} + \tfrac{1}{6}\hat{\hat{\mu}}$
(4) $W_4 = 1\hat{\mu} + 0\hat{\hat{\mu}} = \hat{\mu}$

(a) Which are unbiased?

(b) Rank these estimators in decreasing order of efficiency.

*(c) (Requires calculus.) Find an even more efficient unbiased estimator of μ, or else prove that the best estimator in (b) is the most efficient linear combination possible.

*(d) Find the efficiency of each estimator relative to the most efficient estimator found in (c).

*(e) Answer true or false; if false, correct it.

The inefficiency of the first estimator (only 15% efficient relative to the optimum linear estimator) was very surprising. Here is an example of an estimator that might seem reasonable to some people (the two conflicting parties are simply "splitting their difference"). Yet it is just as damaging to use this as to throw away 85% of the observations in a simple random sample.

*7-17 A farmer has a square field, whose area he wants to estimate. When he measures the length of the field, he makes a random error, so that his *observed length* X_1 is a normal variable centered at μ (the true but unknown value) with standard deviation σ. Aware of his possible error, he decides to take a second independent observation X_2 and average. But he is in a dilemma as to how to proceed:

(1) Should he average X_1 and X_2, and then square? or

(2) Should he square first, and then average?

Mathematically, it's a question whether:

$$\left(\frac{X_1 + X_2}{2}\right)^2 \quad \text{or} \quad \left(\frac{X_1^2 + X_2^2}{2}\right) \text{ is best}$$

(a) Are methods (1) and (2) really different, or are they just two different ways of saying the same thing? (*Hint:* Try a simulation. Use $\mu = 40$ and $\sigma = 10$, for example.)

(b) Which has less bias? (*Hint:* See Equation (4-20), remembering that normality is irrelevant to questions of expectation.)

(c) Generalize answer (b) to a sample of n measurements.

(d) As an alternative estimator of the area, what is the bias of $X_1 X_2$? (*Hint:* See Problem 5-20.)

CHAPTER 8

Interval Estimation

Reconnaissance is as important in the art of politics as it is in the art of war—or the art of love.

Henry Durant

8-1 A SINGLE MEAN

(a) Theory

In Chapter 7, we considered various *point estimators*. For example, we concluded that $\overline{X}$ was a good estimator of μ for populations that are approximately normal. Although on average $\overline{X}$ is on target, however, the specific sample mean $\overline{X}$ that we happen to observe[1] is almost certain to be a bit high or a bit low. Accordingly, if we want to be reasonably confident that our inference is correct, we cannot claim that μ is precisely equal to the observed $\overline{X}$. Instead, we must construct an *interval estimate* or *confidence interval* of the form:

$$\mu = \overline{X} \pm \text{a sampling error} \tag{8-1}$$

---

[1] Strictly speaking, a specific realized value such as this should be denoted by the lower-case letter $\overline{x}$, to distinguish it from the potential value (random variable) denoted by $\overline{X}$. However, from now on we shall not bother to distinguish notationally the realized value from the potential value, and so we shall always use $\overline{X}$ to refer to either. And for certain other variables, such as s, in order to conform to common usage we shall always use a lower-case letter.

199

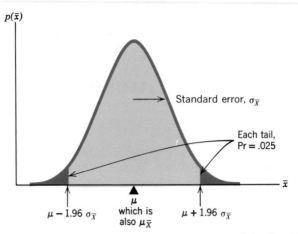

FIGURE 8–1 Normal distribution of the sample mean around the fixed but unknown parameter μ. 95% of the probability is contained within 1.96 standard errors.

The crucial question is: How wide must this allowance for sampling error be? The answer, of course, will depend on how much $\overline{X}$ fluctuates (i.e., on the sampling distribution of $\overline{X}$), which we review in Figure 8-1 (compare with Figure 6-2 and equation (6-10).)

First we must decide how confident we wish to be that our interval estimate is right—that it does indeed bracket μ. It is common to choose 95% confidence; in other words, we will use a technique that will give us, in the long run, a correct interval 19 times out of 20.

To get a confidence level of 95%, we select the smallest range under the normal distribution of $\overline{X}$ that will just enclose a 95% probability. Obviously, this is the middle chunk, leaving $2\frac{1}{2}$% probability excluded in each tail. From Table IV, we find that this requires a z value of 1.96. That is, we must go above and below the mean by 1.96 standard deviations of $\overline{X}$, as shown in Figure 8-1. The standard deviation of $\overline{X}$ (also called the standard error)[2] is denoted by $\sigma_{\overline{X}}$, so that we may write:

$$\Pr(\mu - 1.96\,\sigma_{\overline{X}} < \overline{X} < \mu + 1.96\,\sigma_{\overline{X}}) = 95\% \qquad (8\text{-}2)$$

The bracketed inequalities may be solved for μ—"turned around" so to speak—to obtain the equivalent statement:[3]

---

[2]In the context of estimation, the deviation of $\overline{X}$ from its target μ must be considered an error. So the standard deviation of $\overline{X}$ commonly is called the standard error of the mean.

[3]To prove (8-3) more directly, we could begin by standardizing $\overline{X}$, which then has the standard normal distribution. Thus from the standard normal tables: (*cont'd*)

$$\Pr(\overline{X} - 1.96\,\sigma_{\overline{X}} < \mu < \overline{X} + 1.96\,\sigma_{\overline{X}}) = 95\% \qquad (8\text{-}3)$$

We must be exceeding careful not to misinterpret (8-3). μ has not changed its character in the course of this algebraic manipulation. It has not become a variable but has remained a population constant. Equation (8-3), like (8-2), is a probability statement about the random variable $\overline{X}$, or more precisely, the "random interval" $\overline{X} - 1.96\sigma_{\overline{X}}$ to $\overline{X} + 1.96\sigma_{\overline{X}}$. *It is this interval that varies, not μ.*

(b) Illustration

To appreciate the fundamental point that the confidence interval fluctuates while μ remains constant, consider an example. Suppose that we wish to construct an interval estimate for μ, the mean height of a population of men, on the basis of a random sample of $n = 25$ men. Moreover, to clearly illustrate what is going on, suppose that we have some supernatural knowledge of the population μ and σ. Suppose that we know, for example, that:

$$\mu = 69 \qquad (8\text{-}5)$$

and
$$\sigma = 5.1$$

Then, from (6-8):

$$\boxed{\sigma_{\overline{X}} = \frac{\sigma}{\sqrt{n}}} \qquad (8\text{-}6)$$

$$= \frac{5.1}{\sqrt{25}} = 1.02$$

Now let us observe what happens when the statistician (who, of course, does not have our supernatural knowledge) tries to estimate μ using (8-3). Just for the sake of illustration, let's suppose that he makes

---

$$\Pr\left(-1.96 < \frac{\overline{X} - \mu}{\sigma_{\overline{X}}} < 1.96\right) = 95\% \qquad (8\text{-}4)$$

In (8-4), the bracketed inequalities may be solved for μ, obtaining the equivalent inequalities:

$$\Pr(\overline{X} - 1.96\sigma_{\overline{X}} < \mu < \overline{X} + 1.96\sigma_{\overline{X}}) = 95\% \qquad (8\text{-}3) \text{ proved}$$

twenty such interval estimates, each time from a different random sample of 25 men. Figure 8-2 shows his typical experience.

First, in the top panel we illustrate equation (8-2): $\overline{X}$ is distributed around $\mu = 69$, with a 95% probability that it lies as close as:

$$\pm 1.96\ \sigma_{\overline{X}} = \pm 1.96(1.02) = \pm 2.0 \qquad (8\text{-}7)$$

That is, there is a 95% probability that any $\overline{X}$ will fall in the range 67 to 71 inches.

But the statistician does not know this; he blindly takes his first random sample, from which we suppose that he computes the first mean $\overline{X}$ to be 70. From (8-3), he calculates[4] the appropriate 95% confidence interval for μ:

$$\overline{X} \pm 1.96\ \sigma_{\overline{X}} \qquad (8\text{-}8)$$

$$= 70 \pm 1.96\ (1.02) \qquad (8\text{-}9)$$

$$= 70 \pm 2$$

$$= 68 \quad \text{to} \quad 72 \qquad (8\text{-}10)$$

This interval estimate for μ is the first one shown in Figure 8-2. In his first effort, the statistician is right; μ is enclosed in this interval.

In his second sample, suppose that the statistician happens to draw a shorter group of individuals, and duly computes $\overline{X}$ to be 68.2 inches. From a similar evaluation of (8-3), he comes up with his second interval estimate shown in the diagram, and so on. We observe that, typically, nineteen of these twenty estimates bracket the constant μ. Only one missed the mark.

We easily can see why he is right most of the time. For each interval estimate he is simply adding or subtracting 2 inches from his sample mean; but this is the same ± 2 inches that appears in (8-7) and defines the range ab around μ. Thus, if and only if he observes a sample mean within the range ab will his interval estimate bracket μ. Nineteen of his twenty sample means do fall in the range ab, and in all these instances his interval estimate is right. He is wrong only in the one instance when he observes a sample mean outside ab.

---

[4]We gloss over one difficulty here. In evaluating (8-8), the statistician must have a standard error $\sigma_{\overline{X}}$, which he can calculate from (8-6)—but only if he knows the population standard deviation σ. In this section, we assume that he knows σ, and defer to the next section the more typical case in which he has to estimate σ.

In practice, of course, a statistician would take only one sample, not many. And once this interval estimate was made, he either would be right or wrong; this interval would bracket μ or it would not. But the important point to recognize is that the statistician is using a method with a 95% probability of success; this follows because there is a 95% probability that his observed $\overline{X}$ will fall within the range ab and, as a consequence, that his interval estimate will bracket μ. This is what is meant by a 95% confidence interval: the statistician knows that in the long run, 95% of the intervals that he constructs in this way will bracket μ.

(c) Analogy: Pitching Horseshoes

Constructing 95% confidence intervals is like pitching horseshoes. In each case, there is a fixed target: either the population μ or the stake. We are trying to bracket the target with some chancy device, either the random interval or the horseshoe. This analogy is illustrated in Figure 8-3.

There are several important ways however, in which confidence intervals differ. Customarily, only *one* confidence interval is constructed, for a target μ that is *not* visible. Consequently, the statistician does not know directly whether his confidence interval is correct; he must rely on indirect statistical theory for assurance that, in the long run, 95% of the confidence intervals similarly constructed would be correct.

(d) Review

To review, we briefly emphasize the main points:

1. The population parameter μ is constant, and remains constant. It is the interval estimate that is a random variable, because its center $\overline{X}$ is a random variable. As long as $\overline{X}$ is a random variable that can take on a whole range of values, it is referred to as an "estimator" of μ.

2. But once the sample has been observed and $\overline{X}$ takes on one actual value (e.g., 70 inches), it then is called an "estimate" of μ. Since it is no longer a random variable, probability statements are no longer strictly valid. For this reason, when the actual sample mean is substituted into (8-3), it no longer is called a 95% probability statement, but rather a 95% confidence statement:

$$\overline{X} - 1.96\,\sigma_{\overline{X}} < \mu < \overline{X} + 1.96\,\sigma_{\overline{X}} \qquad (8\text{-}11)$$

Thus, our deduction in (8-2) that $\overline{X}$ is close to μ has been "turned around" into the induction that μ is close to the observed $\overline{X}$.

3. Often (8-11) is abbreviated to:

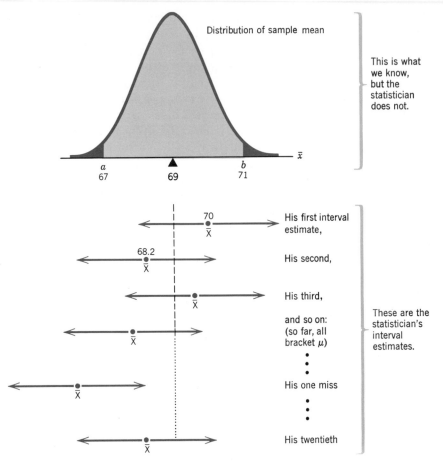

FIGURE 8–2 Constructing 20 interval estimates: typical results

95% confidence interval,

$$\mu = \overline{X} \pm 1.96\,\sigma_{\overline{X}} \qquad (8\text{-}12)$$

i.e., $\mu = \overline{X} \pm z_{.025}$ (standard error)   (8-13)

where $z_{.025}$ is the value 1.96 obtained from Table IV; i.e., the z value that cuts off $2\frac{1}{2}\%$ from the upper tail (and by symmetry, also $2\frac{1}{2}\%$ from the lower tail) of the normal distribution. We can view Equation (8-12) as the prototype for all confidence intervals that we shall study. When we substitute (8-6) into (8-12), we obtain another very useful form:

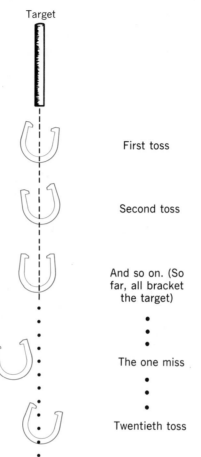

FIGURE 8–3 Pitching 20 horseshoes: typical results

$$\boxed{\begin{array}{c} \text{95\% confidence interval,} \\ \mu = \overline{X} \pm z_{.025}\dfrac{\sigma}{\sqrt{n}} \end{array}} \qquad (8\text{-}14)$$

4. To summarize, once $\overline{X}$ is actually observed, then the "die is cast," and the interval estimate (8-14) is either dead right or dead wrong. Because of our omniscience, we know the one time that the statistician erred. But *he* does not know which confidence intervals, if any, are wrong. All he knows is that he will be right 95% of the time, in the long run.

5. As sample size is increased, the distribution of $\overline{X}$ becomes more concentrated around μ (as n increases, $\sigma/\sqrt{n}$ decreases), and the confidence interval becomes more narrow and precise.

6. Suppose that we wish to be more confident—for example, 99% confident. Then, in Figure 8-2, we require a larger range ab, and so the interval estimate also has a larger range and becomes more vague. Note how this remark, and the one preceding, verify the casual observations made in Chapter 1.

Example 8-1. Sixteen marks were sampled randomly from a very large class that had a standard deviation of 12; these 16 marks had a mean of 58. Find a 95% confidence interval for the mean mark of the whole class.

Solution. Substitute $n = 16$, $\sigma = 12$ and $\overline{X} = 58$ into (8-14):

$$\mu = 58 \pm (1.96)\frac{12}{\sqrt{16}}$$

$$= 58 \pm 6$$

that is, the 95% confidence interval for μ is:

$$52 < \mu < 64$$

PROBLEMS

8-1 An anthropologist measured the heights (in inches) of a random sample of 100 men from a certain population, and found the sample mean to be 71.3. If the population variance $\sigma^2 = 9$,
(a) Find a 95% confidence interval for the mean height μ of the whole population.
(b) Find a 99% confidence interval.

8-2 Answer true or false; if false, correct it.
(a) The standard deviation of $\overline{X}$ also is called the standard error of the mean.
(b) $\overline{X}$ is called a point estimate of the population mean μ, and is almost always incorrect in that it almost always misses somewhat the true value of the population mean. For this reason, an interval estimate (confidence interval) often is more useful, because it deliberately is made sufficiently vague to be usually correct.

(c) The parameter, although unknown, is a fixed quantity. Its confidence interval, however, varies from one sample to another.

(d) The greater the confidence level, the wider the confidence interval must be.

(e) A large confidence interval is like a large horseshoe: it is more likely to bracket the target. But there is a cost: just as the larger horseshoe makes a less interesting game, so the larger confidence interval provides a less interesting (less precise) conclusion.

8-3 (a) Suppose that you took a random sample of 10 accounts in a large department-store chain, and found that the mean balance due was $27.60. If you know that the standard deviation of all balances due is $12.00, find the 95% confidence interval for the mean of all balances due.

(b) Explain to the vice president the meaning of your answer to (a) as simply as you can.

(c) Suppose that the skeptical vice president undertook a complete accounting of the whole population of balances due, and that the mean balance due turned out to be $29.10. What would you say?

8-4 A research study examines the consumption expenditures (in thousands of dollars) of a random sample of 25 American families (all at the same income and asset level). Suppose that the sample mean is 8.2. If $\sigma = .72$, construct a 95% confidence interval for the mean consumption of all American families (at this income and asset level).

8-5 In a certain problem[5] of determining confidence limits for a population parameter θ, suppose that M is a random variable computed from a sample of size n, for which it is known that:

$$\Pr\left(.05^{1/n} < \frac{M}{\theta} < 1\right) = .95$$

If M turns out to be 11.6 in a sample of 25, what is the 95% confidence interval for θ?

8-2   t-DISTRIBUTION WHEN σ^2 UNKNOWN

In the previous section, we assumed that, in constructing a confidence interval, the statistician knows the true population standard deviation σ.

---

[5]The problem happens to be the following: a population is uniformly distributed from 0 up to an unknown value θ. M is the maximum observation in a sample of n observations.

In this section, we consider the more typical case in which he does not.[6]

Since σ is unknown, the statistician who wishes to evaluate the confidence interval (8-14) must use some estimator of σ. The most obvious candidate is the sample standard deviation s (note that s, along with $\overline{X}$, always can be calculated from the sample data). Substituting s into (8-14), he estimates the 95% confidence interval for μ as:

$$\mu = \overline{X} \pm z_{.025}\frac{s}{\sqrt{n}}$$

Provided that his sample is large (50 or 100, depending on the precision required), this will be an accurate enough approximation. But if the sample size is small, this substitution introduces an appreciable source of error. Therefore, if the statistician wishes to remain 95% confident, his interval estimate must be broadened. How much?

Recall that $\overline{X}$ has a normal distribution; when σ was known, we formed the standardized normal variable:

$$Z = \frac{\overline{X} - \mu}{\sigma/\sqrt{n}} \tag{8-15}$$

By analogy, we introduce "Student's t" variable:[7]

$$\boxed{t = \frac{\overline{X} - \mu}{s/\sqrt{n}}} \tag{8-16}$$

The similarity of these two variables immediately is evident. The only difference is that Z involves σ, which usually is unknown; but t involves s, which always can be calculated from the sample. The distribution of t is compared to Z in Figure 8-4, and is tabulated in Appendix Table V.

The t distribution has a wider spread than the normal, of course, since the use of s instead of σ introduces additional uncertainty. Moreover, while there is one standard normal distribution, there is a whole family of t distributions, as shown in Figure 8-4. Note that with small

---

[6]In this section, in order for the results to be perfectly valid, we would have to assume that the population was normal. However, t still may be approximately valid unless the sample is very small and the population is very non-normal.

[7]This t variable was first introduced by W. S. Gosset, writing under the pseudonym, "Student." Later, R. A. Fisher verified its distribution.

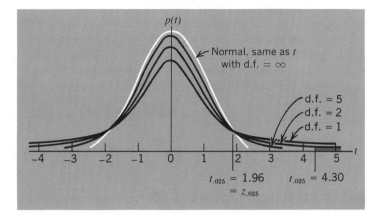

FIGURE 8–4 The standard normal distribution and the t distribution compared

sample size, the t distribution differs substantially from the normal; but as sample size increases, it approaches the normal.

The distribution of t is not tabled according to sample size n, but rather according to the divisor in s^2, which now is called "degrees of freedom."[8] In calculating s^2 in (2-6), we used the divisor:

$$\boxed{\text{degrees of freedom, d.f.} = n - 1} \qquad (8\text{-}17)$$

For example, for a sample with $n = 3$, then d.f. $= 2$, and we find from Table V that the critical t value which leaves $2\frac{1}{2}\%$ probability in the upper tail is:

$$t_{.025} = 4.30$$

This is shown in Figure 8-4. By symmetry, it follows that:

$$\Pr(-4.30 < t < 4.30) = 95\% \qquad (8\text{-}18)$$

---

[8]The phrase "degrees of freedom" is explained in the following intuitive way:

Originally there are n degrees of freedom in a sample of n observations. But one degree of freedom is used up in calculating $\overline{X}$, leaving only $n - 1$ degrees of freedom for the residuals $(X_i - \overline{X})$ to calculate s^2.

For example, consider a sample of two observations, 21 and 15, say. Since $\overline{X} = 18$, the residuals are $+3$ and -3, the second residual necessarily being just the negative of the first. While the first residual is "free," the second is strictly determined; hence there is only 1 degree of freedom in the residuals.

Generally, for a sample of size n, it may be shown that while the first $n - 1$ residuals are free, the last residual is strictly determined by the requirement that the sum of all residuals be zero, i.e., $\Sigma(X_i - \overline{X}) = 0$.

Substitute for t according to (8-16):

$$\Pr\left(-4.30 < \frac{\overline{X} - \mu}{s/\sqrt{n}} < 4.30\right) = 95\% \qquad (8\text{-}19)$$

This deduction now can be turned around, as usual, to produce an inference: for a sample of size 3, the 95% confidence interval for μ is:

$$\mu = \overline{X} \pm 4.30 \frac{s}{\sqrt{n}} \qquad (8\text{-}20)$$

For a sample of size n, we obtain the following generalization of (8-20):

95% confidence interval for the population mean,

$$\mu = \overline{X} \pm t_{.025} \frac{s}{\sqrt{n}} \qquad (8\text{-}21)$$

i.e., $\mu = \overline{X} \pm t_{.025}$ (estimated standard error) $\qquad (8\text{-}22)$

---

where $t_{.025}$ is the critical t value leaving $2\frac{1}{2}\%$ of the probability in the upper tail, with $n - 1$ degrees of freedom.

To sum up, we note the similarity of t estimation in (8-21) and normal estimation in (8-14). The only difference is that the observed sample value s is substituted for σ, and as a consequence the critical t value must be substituted for the critical z value.

An important practical question is: When do we use the t distribution and when do we use the normal? If σ is known, the normal distribution is appropriate; if σ is unknown, then the t distribution is appropriate—regardless of sample size. However, if the sample size is large, the normal is an accurate enough approximation[9] of the t. So in practice, the t distribution is used only for small samples when σ is unknown—and the normal is used otherwise.

---

[9]This may be verified from Table V. For example, a 95% confidence interval with d.f. = 60 should use a critical t value of 2.00; but the use of the normal value of 1.96 as an approximation involves very little error (just 2%).

As we scan down the $t_{.025}$ column in Table V, these critical values approach $z_{.025} = 1.96$, shown in the last row. This verifies Figure 8-4, where the t distributions approach the normal.

Example 8-2.   From a large class, a sample of 4 grades were drawn: 64, 66, 89, and 77. Calculate a 95% confidence interval for the whole class mean μ.

Solution.   Since $n = 4$, d.f. $= 3$; and so from Table V, $t_{.025} = 3.182$. In Table 8-1, we calculate $\overline{X} = 74$ and $s^2 = 132.7$. When all these are substituted into (8-21), we obtain:

$$\mu = 74 \pm 3.182 \frac{\sqrt{132.7}}{\sqrt{4}}$$

$$= 74 \pm 18$$

TABLE **8-1**
Analysis of One Sample

Observed Grade X_i	$(X_i - \overline{X})$	$(X_i - \overline{X})^2$
64	-10	100
66	-8	64
89	15	225
77	3	9
$\overline{X} = \dfrac{296}{4}$ $= 74$	$0\checkmark$	$s^2 = \dfrac{398}{3} = 132.7$

PROBLEMS

8-6 Answer true or false; if false, correct it.

   If σ is unknown, then we must use (8-21) instead of (8-14). This involves replacing σ with its estimator s, an additional source of unreliability; and to allow for this, $z_{.025}$ is replaced by the larger $t_{.025}$ value in order to keep the confidence level at 95%.

8-7 Five cars on a thruway were clocked at an average speed of 54 m.p.h., with a standard deviation of 4 m.p.h. For the mean speed of all cars on this thruway:
   (a) Construct a 95% confidence interval.
   (b) Construct a 99% confidence interval.

8-8 The reaction times of 30 randomly selected drivers were found to have a mean of .83 sec and standard deviation of .20 sec. Find a 95% confidence interval for the mean reaction time of the whole population of drivers:
(a) Using the t distribution.
(b) Using the normal approximation. How much is the normal approximation in error?

8-9 From a very large class in statistics, the following 40 marks were selected randomly:

$$
\begin{array}{cccccccccc}
71 & 74 & 65 & 72 & 64 & 42 & 62 & 62 & 58 & 82 \\
49 & 83 & 58 & 65 & 68 & 60 & 76 & 86 & 74 & 53 \\
78 & 64 & 55 & 87 & 56 & 50 & 71 & 58 & 57 & 75 \\
58 & 86 & 64 & 56 & 45 & 73 & 54 & 86 & 70 & 73
\end{array}
$$

Construct a 95% confidence interval for the average mark of the whole class. (*Hint:* Reduce your work to manageable proportions by grouping into cells of width 5.)

8-10 (a) (Monte Carlo.) Simulate a random sample of 3 observations from a normal population with $\mu = 70$ and $\sigma = 5$. Construct a 95% confidence interval for μ:
(i) Assuming that σ is known.
(ii) Assuming that σ is unknown.
(b) Have the instructor check the proportion of the confidence intervals that are correct in the class. What proportion would be correct in a very large class (in the long run)?
(c) Answer true or false; if false, correct it.
    The confidence intervals in (a) (ii), based on the t distribution, are always wider than those in (a) (i), based on the normal z distribution. This represents the cost of not knowing σ.

8-3   DIFFERENCE IN TWO MEANS, $(\mu_1 - \mu_2)$

Two population means commonly are compared by forming their difference:

$$
\mu_1 - \mu_2 \tag{8-23}
$$

A reasonable point estimate of this is the corresponding difference in *sample* means:

$$\overline{X}_1 - \overline{X}_2 \qquad (8\text{-}24)$$

Our interest, again, is in constructing an interval estimate around this.

(a) Population Variances Known

In this case, we can use an argument similar to the one set out in Section 8-1 and so develop the confidence interval:

$$(\mu_1 - \mu_2) = (\overline{X}_1 - \overline{X}_2) \pm z_{.025} \text{ (standard error)} \qquad \begin{array}{c} (8\text{-}25) \\ \text{like } (8\text{-}12) \end{array}$$

But what is the standard error in this case? To answer this, note that $(\overline{X}_1 - \overline{X}_2)$ is a linear combination; when $\overline{X}_1$ and $\overline{X}_2$ are independent, then (5-40) yields:

$$\text{var }(\overline{X}_1 - \overline{X}_2) = (+1)^2 \text{var } \overline{X}_1 + (-1)^2 \text{var } \overline{X}_2$$

By (6-7):

$$= \frac{\sigma_1^2}{n_1} + \frac{\sigma_2^2}{n_2} \qquad (8\text{-}26)$$

Therefore:

$$\text{standard error} = \sqrt{\frac{\sigma_1^2}{n_1} + \frac{\sigma_2^2}{n_2}} \qquad (8\text{-}27)$$

Substituting (8-27) into (8-25) yields:

> 95% confidence interval, in independent samples,
>
> $$(\mu_1 - \mu_2) = (\overline{X}_1 - \overline{X}_2) \pm z_{.025} \sqrt{\frac{\sigma_1^2}{n_1} + \frac{\sigma_2^2}{n_2}} \qquad (8\text{-}28)$$

When σ_1 and σ_2 are known to have a common value, say σ, the 95% confidence interval becomes:

$$(\mu_1 - \mu_2) = (\overline{X}_1 - \overline{X}_2) \pm z_{.025} \, \sigma \sqrt{\frac{1}{n_1} + \frac{1}{n_2}} \qquad (8\text{-}29)$$

(b) Population Variances Unknown

The variances of the two populations, σ_1^2 and σ_2^2 in (8-28), usually are not known; the best the statistician can do is to estimate them, with the variances s_1^2 and s_2^2 that he can calculate from the two samples. Provided that the samples are large, these estimated values can be plugged into (8-28), making that equation a reasonable enough approximation to use. But if the samples are small, a new source of error is introduced; as before, we compensate for this by using the t distribution. Thus:

> 95% confidence interval, in independent samples, when the population variances are unequal and unknown,
>
> $$(\mu_1 - \mu_2) = (\overline{X}_1 - \overline{X}_2) \pm t_{.025} \sqrt{\frac{s_1^2}{n_1} + \frac{s_2^2}{n_2}}$$

(8-30)
like (8-28)

We would expect that the degrees of freedom for t would be $(n_1 - 1) + (n_2 - 1)$, the total degrees of freedom on which s_1^2 and s_2^2 are calculated; in fact, however, the d.f. is somewhat less than this, for technical reasons we need not consider here. (In practice, we always can be very safe in using d.f. $= (n_1 - 1)$ or $(n_2 - 1)$, whichever is smaller.)[10]

Now turn to (8-29). If the common σ^2 is unknown, how should this equation be modified? The answer is exactly as we might (by now) predict:

> 95% confidence interval, in independent samples, when the population variances are equal and unknown,
>
> $$(\mu_1 - \mu_2) = (\overline{X}_1 - \overline{X}_2) \pm t_{.025}\, s_p \sqrt{\frac{1}{n_1} + \frac{1}{n_2}}$$

(8-32)
like (8-29)

---

[10]If this safe estimate is unsatisfactory, the following complex formula gives a better approximation:

$$\text{d.f.} \approx \frac{(s_1^2/n_1 + s_2^2/n_2)^2}{\dfrac{(s_1^2/n_1)^2}{n_1 - 1} + \dfrac{(s_2^2/n_2)^2}{n_2 - 1}}$$

(8-31)

If n_1 and n_2 are reasonably large, say $n_1 = n_2 = 30$, then the original rule (using d.f. $= (n_1 - 1) + (n_2 - 1) = 58$) gives $t_{.025} = 2.00$; the safe rule (using d.f. $= n_1 - 1 = 29$) gives $t_{.025} = 2.045$; and the more correct rule (8-31) gives $t_{.025}$ somewhere in between. But there is so little difference that usually it is best to keep things simple and just use the safe rule.

with t having:

$$\text{d.f.} = (n_1 - 1) + (n_2 - 1) \qquad (8\text{-}33)$$

The only question remaining is how to calculate s_p^2, the "pooled variance," which is estimated from the information provided by both samples. The answer is to add up all the squared deviations from both samples, and then divide by the degrees of freedom in both samples, $(n_1 - 1) + (n_2 - 1)$. That is:

$$s_p^2 = \frac{1}{(n_1 + n_2 - 2)}\left[\sum^{n_1}(X_1 - \overline{X}_1)^2 + \sum^{n_2}(X_2 - \overline{X}_2)^2\right] \qquad (8\text{-}34)$$

where X_1 (or X_2) represents the typical observation in the first (or second) sample.[11]

Example 8-3.   From a large class, a sample of four grades were drawn: 64, 66, 89, and 77. From a second large class, a sample of three grades were drawn: 56, 71, 53. Calculate a 95% confidence interval for the difference between the two class means, $\mu_1 - \mu_2$.

Solution.   Unless there is evidence to the contrary, it is customary with small samples to assume that the underlying population variances are about equal, and thus to use (8-32).

In Table 8-2, we calculate the sample means and the squared deviations. Thus, from (8-34):

$$s_p^2 = \frac{1}{(4 + 3 - 2)}[398 + 186] = \frac{584}{5} = 117 \quad (8\text{-}35)$$

Since d.f. $= 5$, from Table V, $t_{.025} = 2.571$. Substituting into (8-32), we obtain:

$$(\mu_1 - \mu_2) = (74.0 - 60.0) \pm 2.571 \sqrt{117}\sqrt{\frac{1}{4} + \frac{1}{3}}$$

$$= 14 \pm 21 \qquad (8\text{-}36)$$

---

[11] For simplicity, we have dropped the subscript i. Strictly speaking, we should have written X_{1i} (or X_{2i}) for the typical ith observation in the first (or second) sample.

Thus, we see that the large difference in sample means is obscured by an even larger sampling allowance (due primarily to the smallness of the samples).

TABLE 8-2
Analysis of 2 Independent Samples

	Class 1			Class 2		
Observed X_1	$X_1 - \overline{X}_1$	$(X_1 - \overline{X}_1)^2$		Observed X_2	$(X_2 - \overline{X}_2)$	$(X_2 - \overline{X}_2)^2$
64	-10	100		56	-4	16
66	-8	64		71	11	121
89	15	225		53	-7	49
77	3	9				
$\overline{X}_1 = \dfrac{296}{4}$ $= 74.0$	$0\checkmark$	398		$\overline{X}_2 = \dfrac{180}{3}$ $= 60.0$	$0\checkmark$	186

The confidence interval (8-32) is appropriate only if the two samples are drawn independently[12]—in other words, only if the second class of students is sampled independently of the first. As another example, consider just one class of students, examined at two different times—say fall and spring terms. The fall and spring population of grades then could be sampled and compared, using (8-32)—provided of course, that the two samples are drawn independently (e.g., that we do not make a point of canvassing the same students twice).

(c) Paired Samples

Suppose, in a comparison of fall and spring grades, that we *do* wish to use the same students twice in both samples; then (8-32), which requires independent samples, no longer is applicable. Instead, we proceed as follows.

The paired values for a sample of four students are set out in Table 8-3. The natural first step is to see how each student changed; that is, calculate the difference $D = X_1 - X_2$ for each student.[13] Once we have calculated these differences (in color), then we can discard the original

---

[12]This is because (8-32) is derived from (8-29), (8-28), and (8-26), which in turn used (5-40)—and this crucially required independence (or zero correlation, at least).

[13]Or, leading to the same conclusion, the difference $X_2 - X_1$.

data, which has served its purpose. We now treat the differences D as a *single sample*, and we analyze them just as we would analyze any other single sample (for example, the sample in Table 8-1). First, we calculate the average sample difference $\overline{D}$. Then we use $\overline{D}$ appropriately in (8-21) to construct a confidence interval for the average population difference Δ, obtaining:

95% confidence interval, for paired samples,

$$\Delta = \overline{D} \pm t_{.025} \frac{s_D}{\sqrt{n}}$$ (8-37)

TABLE 8-3
Analysis of 2 Paired Samples

Student	Observed Grades X_1 (spring)	X_2 (fall)	Difference $D = X_1 - X_2$	$(D - \overline{D})$	$(D - \overline{D})^2$
A	64	54	10	−4	16
B	66	54	12	−2	4
C	89	70	19	5	25
D	77	62	15	1	1
			$\overline{D} = \dfrac{56}{4}$ $= 14.0$	$0 \checkmark$	$s_D^2 = \dfrac{46}{3}$ $= 15.3$

For our example, where $\overline{D} = 14$, $s_D = \sqrt{15.3}$, and $n = 4$, we find that d.f. $= n - 1 = 3$, so that $t_{.025} = 3.182$. Substituting into (8-37), we obtain:

$$\Delta = 14.0 \pm 3.182 \frac{\sqrt{15.3}}{\sqrt{4}}$$

$$= 14.0 \pm 6.2$$ (8-38)

Of course, we also recognize Δ as the difference in the two population means,[14] $\mu_1 - \mu_2$, so that our problem is solved.

(d) Why Paired Samples?

It is interesting to compare the case of paired samples (8-38) with the case of independent samples (8-36). The sampling fluctuation for the

---

[14]This can be seen in the samples as well as the populations: the average difference $\overline{D} = 14$ is just the difference in averages $\overline{X}_1 - \overline{X}_2 = 74 - 60 = 14$.

paired data is much reduced (± 6.2 vs. ± 21). Of the several possible reasons for this, the most important is intuitively clear: the differences in the two samples are not obscured by the second sample of students being entirely different (independent) from the first. In other words, it is desirable to design this pairing (matching) of observations into the sampling experiment whenever possible. It gives us more leverage on the problem at hand (difference in fall vs. spring) because we have been able to keep "other things equal" (i.e., the students the same).

Example 8-4.   To measure the effect of a fitness campaign, a university randomly sampled five employees before the campaign, and another five employees after. The weights were as follows (along with the person's initials):

Before

J. H. 168, K. L. 195, M. M. 155, T. R. 183, M. T. 169

After

L. W. 183,   V. G. 177,   E. P. 148,   J. C. 162,   M. W. 180

(a) Calculate a 95% confidence interval for the population mean weight:

   (i) Before the campaign.

   (ii) After the campaign.

   (iii) Change during the campaign.

(b) It was decided that a better sampling design would be to measure the same people after, as before. Their figures were:

After

K. L. 197, M. T. 163, T. R. 180, M. M. 150, J. H. 160

On the basis of these people, calculate a 95% confidence interval for the weight change during the campaign.

Solution. (a)

X_1	$X_1 - \overline{X}_1$	$(X_1 - \overline{X}_1)^2$	X_2	$X_2 - \overline{X}_2$	$(X_2 - \overline{X}_2)^2$
168	-6	36	183	13	169
195	21	441	177	7	49
155	-19	361	148	-22	484
183	9	81	162	-8	64
169	-5	25	180	$+10$	100
$\overline{X}_1 = \dfrac{870}{5}$ $= 174$	$0\checkmark$	944	$\overline{X}_2 = \dfrac{850}{5}$ $= 170$	$0\checkmark$	866

By (8-21):

$$\mu_1 = 174 \pm 2.776 \frac{\sqrt{944/4}}{\sqrt{5}}$$

$$= 174 \pm 19$$

$$\mu_2 = 170 \pm 2.776 \frac{\sqrt{866/4}}{\sqrt{5}}$$

$$= 170 \pm 18$$

To measure the change, it is most convenient to take the latest value (μ_2) minus the preceding reference value (μ_1). So we construct a confidence interval for $\mu_2 - \mu_1$, using (8-32):

$$\mu_2 - \mu_1 = (170 - 174) \pm 2.306 \sqrt{\frac{944 + 866}{5 + 5 - 2}} \sqrt{\frac{1}{5} + \frac{1}{5}}$$

$$= -4 \pm 22$$

The *negative* sign indicates a *decrease* (weight loss) of four pounds, ± 22 pounds.

(b) We must be sure to list the people in the same matched order so that it is meaningful to calculate the individual weight changes.

Person	X_1	X_2	$D = X_2 - X_1$	$(D - \overline{D})$	$(D - \overline{D})^2$
J. H.	168	160	-8	-4	16
K. L.	195	197	$+2$	$+6$	36
M. M.	155	150	-5	-1	1
T. R.	183	180	-3	$+1$	1
M. T.	169	163	-6	-2	4
			$D = -4$	$0\checkmark$	58

By (8-37):

$$\Delta = -4 \pm 2.776 \frac{\sqrt{58/4}}{\sqrt{5}} = -4 \pm 5$$

Remarks.   The matched samples gave a much more precise interval for the weight change (± 5 vs. ± 22). This is because the matched sample kept "other things equal": in using exactly the same five individuals, we kept constant sex, age, race, etc.[15]

---

[15] Of course, we could not keep *everything* constant. Amount of stress, the number of tempting meals, etc., could not be kept constant without a much more rigid experimental regime. These varying factors are what produce the random fluctuations that appear in the column of differences D.

PROBLEMS

8-11 Two samples of seedlings were grown with 2 different fertilizers. The first sample, with 200 seedlings, had an average height of 10.9 inches and a standard deviation of 2.0 inches. The second sample, with 100 seedlings, had an average height of 10.5 inches and a standard deviation of 5.0 inches. Construct a confidence interval for the difference between the average population heights ($\mu_1 - \mu_2$):
(a) At the 95% level of confidence.
(b) At the 90% level of confidence.

8-12 A random sample of 120 workers in one large plant took an average of 22.0 minutes to complete a task, with a variance of 4. A random sample of 120 workers in a second large plant took an average of 19.0 minutes to complete the task, with a variance of 10. Construct a 95% confidence interval for the difference between the two population mean completion times.

8-13 Independent random samples of adult Whites and Blacks were taken in 1972. They gave the following years of school completed:[16]

$$\text{Whites} \qquad 8, \quad 18, \quad 10, \quad 10, \quad 14$$

$$\text{Blacks} \qquad 9, \quad 12, \quad 5, \quad 10, \quad 14$$

Construct a 95% confidence interval for:
(a) The White population mean.
(b) The Black population mean.
(c) The mean difference between Whites and Blacks.

8-14 Five people selected at random had their breathing capacity measured before and after a certain treatment, obtaining the data below. Let μ_X (and μ_Y) be the mean capacity of the whole population before (and after) treatment. Construct a 95% confidence interval for $(\mu_Y - \mu_X)$.

	Breathing Capacity	
Person	Before (X)	After (Y)
A	2750	2850
B	2360	2380
C	2950	2930
D	2830	2860
E	2250	2320

8-15 In a large American university in 1969, the men and women professors were sampled independently, yielding the following annual salaries (in thousands of dollars, rounded):[17]

Men	Women
12	9
11	12
19	8
16	10
22	16

Calculate a 95% confidence interval for:
(a) The mean salary difference between men and women.
(b) The mean salary for men.

---

[16]From the 1974 *American Almanac,* p. 116, Table 177.

[17]Same source as Problem 2-1.

(c) The mean salary for women.

(d) Repeat parts (a), (b), and (c), using the complete data given in Problems 2-1 and 2-2, namely, independent random samples with the following characteristics:

Men	Women
$n_1 = 25$	$n_2 = 5$
$\overline{X}_1 = 16.0$	$\overline{X}_2 = 11.0$
$s_1^2 = 16$	$s_2^2 = 10$

8-16 A random sample of 180 students was taken in each of two different universities. The first sample had an average mark of 77 and a standard deviation of 6. The second sample had an average mark of 68 and a standard deviation of 10.

(a) Find a 95% confidence interval for the difference between the mean marks in the two universities.

(b) What increase in the sample size would be necessary to cut the sampling allowance by $\frac{1}{2}$?

(c) What increase in the sample size would be necessary to reduce the sampling allowance to 1.0?

8-17 To determine which of two seeds was better, a state agricultural station chose 7 two-acre plots of land randomly within the state. Each plot was split in half, and a coin was tossed to determine in an unbiased way which half would be sown with seed A, and which with seed B. The yields, in bushels, were as follows:

County	Seed A	Seed B
B	82	88
R	68	66
T	109	121
S	95	106
A	112	116
M	76	79
C	84	92

Which seed do you think is better? To back up your answer, construct a 95% confidence interval.

8-18 Given the following random samples from two populations,

$$n_1 = 25 \qquad \overline{X}_1 = 60.0 \qquad s_1 = 12$$

$$n_2 = 15 \qquad \overline{X}_2 = 68.0 \qquad s_2 = 10$$

and assuming (reasonably enough), that $\sigma_1 = \sigma_2$, find a 95% confidence interval for $(\mu_1 - \mu_2)$.

8-4 PROPORTIONS

(a) Large Samples

In Example 6-8, we saw that a sample proportion P is just a disguised sample mean $\overline{X}$ drawn from a 0–1 population. For example, if we observe 7 Democrats in a sample of 10, then:

$$P = \overline{X} = (1 + 0 + 1 + 1 + 0 + 1 + 1 + 0 + 1 + 1)/10 = 7/10$$

Similarly, the population proportion π is just the disguised mean μ in this same sort of population. Therefore, the simplest method of deriving an interval estimate for a proportion is to modify the interval estimate for a mean. We substitute P for $\overline{X}$, and $\sqrt{\pi(1 - \pi)}$ for σ, according to (6-16). Then (8-14) becomes:

$$\pi = P \pm 1.96 \sqrt{\frac{\pi(1 - \pi)}{n}} \qquad (8\text{-}39)$$

What can we do about the unknown π that appears in the right-hand side of (8-39)? Fortunately, we can substitute the sample P for π. We used this strategy before, when we substituted s for σ in the confidence interval for μ. This approximation introduces another source of error; but with a large sample size, this is no problem. Thus:

> 95% confidence interval for the proportion, for large n,
> $$\pi = P \pm 1.96 \sqrt{\frac{P(1 - P)}{n}} \qquad (8\text{-}40)$$

For example, the voter poll in Chapter 1 used this formula.

(b) Moderately Large Samples

For moderately large samples (roughly, $25 < n < 100$), there are several options. The crudest is to use (8-40) with $t_{.025}$ in place[18] of 1.96.

A more conservative alternative is to allow for the worst that could happen in (8-39). Since the maximum value[19] of $\pi(1 - \pi)$ is $1/4$, the maximum value of the sampling allowance in (8-39) is:

$$\pm 1.96 \sqrt{\frac{1/4}{n}} = \pm \frac{.98}{\sqrt{n}} \tag{8-41}$$

Thus, the conservative 95% confidence interval is:

$$\boxed{\pi = P \pm \frac{.98}{\sqrt{n}}} \tag{8-42}$$

But this is assuming the worst: if, in fact, π is not $1/2$, then $\pi(1 - \pi)$ is less than $1/4$, and (8-42) is too wide. That is, (8-42) may have more than a 95% level of confidence. But this is an error on the safe side.

The simple formula (8-42) is used, for example, in political polls, where it is known on the basis of historical experience that the proportion of Democrats is close to $1/2$. In these circumstances, it becomes a very accurate approximation.

(c) Small Samples

There is yet another way to find an interval estimate for π, which works for all sample sizes, large and small. This method, based on Figure 8-5, is

---

[18]This is like the use of t that we first met in (8-22). In the standard error, when a population parameter (such as π) is replaced with a sample estimator (such as P), we allow for this by replacing $z_{.025}$ with the vaguer $t_{.025}$.

This method is not exactly valid, because the t distribution, strictly speaking, assumes a normal population, rather than a 0–1 population. However, this is still better than using the value $z_{.025} = 1.96$.

[19]The simplest proof is with calculus, setting the derivative of $\pi(1 - \pi)$ equal to zero. To show it without calculus, we simply graph $f(\pi) = \pi(1 - \pi)$, as follows:

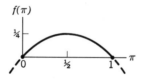

Note that for either extreme value of π (1 or 0), the value of $f(\pi)$ is zero; and if $\pi = \frac{1}{2}$, then $\pi(1 - \pi)$ reaches its maximum value because of symmetry of the parabola.

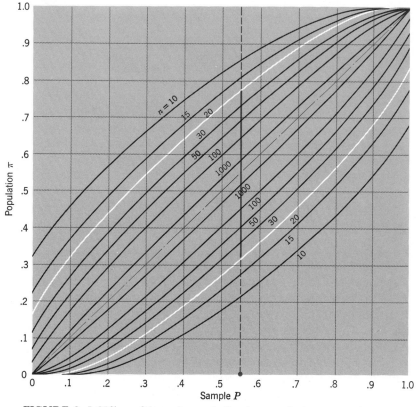

FIGURE 8–5 95% confidence intervals for the population proportion π

extremely easy to use. For example, suppose that we observe 11 Democrats in a sample of 20. We first calculate $P = 11/20 = .55$, as usual. Then, on Figure 8-5, we locate the two curves labeled $n = 20$, and see where they contain the vertical line through $P = .55$. This is our confidence interval, shown in color. It may be expressed numerically, if desired, as:

$$.31 < \pi < .77 \tag{8-43}$$

Although this method is easy to use, it is rather complicated to justify. Nevertheless, we shall work through the derivation, not only for its own sake, but also to shed light on the meaning of confidence intervals in general.

The first step is the mathematical deduction of how the variable estimator P is distributed. This is shown in Figure 8-6 for a sample size

Probability

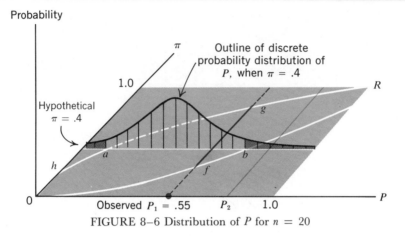

FIGURE 8–6 Distribution of P for $n = 20$

$n = 20$ and for $\pi = .4$, for example. Thus, there is a 95% probability that any P calculated from a random sample of 20 will lie in the interval ab. For each possible value of π, such a probability distribution of P defines two critical points like a and b. When all such points are joined, the result is the two white curves enclosing a 95% probability band.

This deduction of how the statistic P is related to the population π is now turned around to draw a statistical inference about π from a given sample P. For example, if we observe a sample proportion $P_1 = 11/20 = .55$, then the 95% confidence interval for π is defined by the interval fg contained within this probability band above P_1:

$$.31 < \pi < .77 \qquad\qquad \text{(8-43) repeated}$$

Whereas the (deduced) probability interval is defined in the horizontal direction of the P axis, the (induced) confidence interval is defined in the vertical direction of the π axis.

To see why this works, suppose, for example, that the true value of π is .4. Then the probability is 95% that a sample P will fall between a and b. *If and only if it does* (e.g., P_1) will the confidence interval we construct bracket the true $\pi = .4$. We therefore are using an estimation procedure that is 95% likely to bracket the true value of π, and thus yield a correct statement. But we must recognize the 5% probability that the sample P will fall beyond a or b (e.g., P_2); in this case, our interval estimate will not bracket $\pi = .4$, and our conclusion will be wrong.

This is a more general theory of confidence intervals than we have encountered before. In previous instances, such as estimating μ, we constructed a confidence interval *symmetrically* about the point estimate $\overline{X}$.

But in estimating π, no such symmetry is generally involved.[20] For example, the confidence interval for π in (8-43) was *not* quite symmetric about the point estimate .55.

(d) Difference in Two Proportions, for Large Samples

In the same way that we derived (8-40), we could derive the confidence interval to compare two population proportions:

> 95% confidence interval for the difference in proportions, for large n_1 and n_2,
>
> $$(\pi_1 - \pi_2) = (P_1 - P_2) \pm 1.96 \sqrt{\frac{P_1(1 - P_1)}{n_1} + \frac{P_2(1 - P_2)}{n_2}} \quad (8\text{-}44)$$

PROBLEMS

8-19 In a random sample of tires produced by a certain company, 20% did not meet the company's standards. Construct a 95% confidence interval for the proportion π (in the whole population of tires) that does not meet the standards:
(a) If the sample size $n = 10$.
(b) If $n = 25$.
(c) If $n = 2,500$.

8-20 In a poll of 1,063 college students in 1970, the Gallup Poll found that 49% of those interviewed believed that change in America is likely to occur in the next 25 years through relatively peaceful means (rather than through a revolution). Although the poll was not exactly a simple random sample,[21] treat it as such in order to get an approximate 95% confidence interval for the population proportion π.

8-21 In response to the question in 8-20, the same affirmative reply was given by 50% of students 18 years old and under, and 69% of those 24 years old and over.

---

[20]You may wonder why, in Figure 8-6, the 95% probability band does not converge on the two end-points O and R. It is true that one half of this band (made up of all points similar to b) does intersect the P axis at 0; this means that if π is zero (e.g., no Socialists in the U.S.), then any sample P also must be zero (no Socialists in the sample). But the other half of this band does not intersect the π axis at 0; instead it intersects at h. This means that an observed P of zero (e.g., no Socialists in a sample) does *not* necessarily imply that π is zero (no Socialists in the U.S.).

[21]The sample was a combination of stratified sampling (which is more reliable than simple random sampling), and quota sampling (which is less reliable).

Construct a 99% confidence interval for the difference in these two subpopulation proportions:

(a) Assuming that $n_1 = 300$ and $n_2 = 300$.

(b) Assuming that $n_1 = 500$ and $n_2 = 100$.

8-22 In a 1974 poll,[22] 1,650 American were asked for their opinions on the following issue:

> The U.S. Supreme Court has ruled that a woman may go to a doctor to end pregnancy at any time during the first three months of pregnancy. Do you favor or oppose this ruling?

A week later, an independent sample of 1,650 Americans were asked the same question, except that the words "to end pregnancy" were changed to "for an abortion." The responses were as follows:

Response ↓ Wording →	In Favor	Opposed	No Opinion[23]
"To end pregnancy"	46%	39%	15%
"For an abortion"	41%	49%	10%

(a) Let the proportion of voters in favor be denoted by π_1 for the first wording ("to end pregnancy") and by π_2 for the second wording ("for an abortion"). Construct a 95% confidence interval for the difference $\pi_2 - \pi_1$.

(b) Repeat (a) for the proportion of voters *who had an opinion*. (Now, for example, $P_1 = .46/(.46 + .39) = .54$.)

(c) Would you agree with the following conclusion of the Committee: "If you want to know what Americans really think about abortion, *you have to ask them about abortion.*"

8-23 There is another factor that we should have mentioned in Problem 8-22— the time factor. The second poll was taken a week later. To try to clarify the effect of time, a third poll was taken yet another week later (again, an independent sample of size 1,650). The total results were:

---

[22]Conducted by Sindlinger and Company, Swarthmore, Pa., commissioned by the Ad Hoc Committee in Defense of Life, Inc. (henceforth referred to as the "Committee"). As advertised in the *New York Times,* June 2, 1974, p. E6.

[23]"No opinion" covers those who refused to answer, as well as those who were undecided.

Week	↓Wording Response →	In Favor	Opposed	"No Opinion"
1	"To end pregnancy"	46%	39%	15%
2	"For an abortion"	41%	49%	10%
3	"For an abortion"	43%	54%	3%

(a) What would you conclude is the effect of time?

(b) Would you agree with the following conclusion of the Committee: "Clearly more and more people are making up their minds on the issue— very likely because abortion is *finally* getting major coverage from the media."

8-24 In a survey of U.S. consumer intentions, 30 families in a random sample of 150 indicated that they intended to buy a new car within a year. Construct a 95% confidence interval for the proportion of all U.S. families who intend to buy a new car within a year:

(a) Using the usual formula (8-40).

(b) Using the simplified formula (8-42).

8-25 If $\pi = .2$, what is the percentage error introduced by using (8-42)? Does this suggest that (8-42) is a reasonable approximation, provided that $.2 \leq \pi \leq .8$?

8-26 To measure the effect of an aerial-spray treatment against a certain insect, 300 trees were selected at random from a stand of timber. Each tree was measured both before and after treatment, and then classified in one of the 4 cells of the following table.

TABLE 1

Before \ After	Not Infected	Infected	Subtotals
Not infected	220	3	223
Infected	32	45	77
Subtotals	252	48	300 total

If we let π_X be the infected proportion in the population of trees *before* treatment, and let π_Y be the infected proportion in the population of trees *after* treatment, find a 95% confidence interval for $(\pi_Y - \pi_X)$, the change in infection rates. (*Hint:* For each of the 300 trees, let X be a counting variable to measure whether or not the tree is infected *before* treatment, and Y a counting variable to measure *after* treatment. Then, as an alternative to Table 1, the data could be recorded as follows:

TABLE **2**
Tree-by-Tree
Detail of Table 1

Tree #	X	Y
1	1	0
2	0	0
3	0	0
4	1	1
5	1	0
6	0	0
⋮	⋮	⋮

Note that trees #2, 3, 6, . . . , will be among the 220 trees classified in the upper left cell of Table 1; trees #1, 5, . . . , will be among the 32 trees classified in the lower left cell, etc.

Recording the data in the detail of Table 2 yields a conceptual advantage: in this form, it is obviously a matched sample, so that (8-37) may be used.

8-27 A student who wishes to enter a certain university is required to have an interview at either the West Coast Board (W) or the East Coast Board (E). In order to test whether the two boards have the same standards, the university subjected 100 candidates (chosen at random from the large population of all candidates) to an interview by both boards. The following frequency table summarizes the result.

Board E \\ Board W	Accepted	Rejected	Totals
Accepted	48	5	53
Rejected	12	35	47
Totals	60	40	100

At first glance, Board E seems to have stricter standards, because its rejection rate in the sample is 47%, while the rejection rate of Board W is only 40%.

(a) Analyze further.

(b) Can you suggest improvements or possible criticisms of the experimental design?

8-5   ONE-SIDED CONFIDENCE INTERVALS

(a) Simplest Case

There are occasions when, in order to prove a point, we wish to make a statement that a parameter is *at least as large* as a certain minimum value. The appropriate technique is then a one-sided confidence interval:[24]

$$
\boxed{
\begin{array}{l}
95\% \text{ confidence interval (one-sided),} \\[2mm]
\mu > \overline{X} - z_{.05}\dfrac{\sigma}{\sqrt{n}}
\end{array}
}
\qquad
\begin{array}{l}
(8\text{-}45) \\[2mm]
\text{like } (8\text{-}14)
\end{array}
$$

Example 8-5.   (Different analysis of the data in Example 8-1). Sixteen marks were sampled from a very large class that had a standard deviation of 12; these 16 marks had a mean of 58. The instructor wants to say that the class mean is at least a certain value, with 95% confidence. Construct the appropriate confidence interval.

Solution.   Substitute $n = 16$, $\sigma = 12$, and $\overline{X} = 58$ into (8-45), and note from Table IV that $z_{.05} = 1.64$:

$$
\mu > 58 - (1.64)\frac{12}{\sqrt{16}} = 58 - 5
$$

$$
\mu > 53 \qquad\qquad\qquad\qquad (8\text{-}46)
$$

Remarks.   The confidence interval covers all values above the sample mean $\overline{X} = 58$, and also a sufficient range of values below it to ensure 95% confidence of being correct.

---

[24]To establish (8-45), we start from the distribution of $\overline{X}$ shown in Figure 8-1. We put all the 5% error probability into *one* tail (the upper tail):

$$
\Pr\left(\overline{X} < \mu + z_{.05}\frac{\sigma}{\sqrt{n}}\right) = .95
$$

When we solve the bracketed inequality for the unknown parameter μ, we obtain (8-45).

Note that the one-sided confidence interval (8-45) does give a better lower bound than the two-sided confidence interval (8-14) (53 instead of 52). We must pay a pretty high price, however: the one-sided confidence interval has no upper bound at all.

(b) Other Cases

Any two-sided confidence interval may be similarly adjusted to give a one-sided confidence interval. As examples:

$$\mu > \overline{X} - t_{.05}\frac{s}{\sqrt{n}} \qquad (8\text{-}47)$$
$$\text{like (8-21)}$$

$$(\mu_1 - \mu_2) > (\overline{X}_1 - \overline{X}_2) - t_{.05}\, s_p \sqrt{\frac{1}{n_1} + \frac{1}{n_2}} \qquad (8\text{-}48)$$
$$\text{like (8-32)}$$

There may be occasions when we want to state that a parameter is *below* a certain value, in which case the confidence interval would take on a form such as:

$$\mu < \overline{X} + t_{.05}\frac{s}{\sqrt{n}} \qquad (8\text{-}49)$$
$$\text{like (8-47)}$$

PROBLEMS

8-28 In planning a dam, suppose that the government wishes to estimate μ, the mean annual irrigation benefit per acre. They therefore take a random sample of 25 one-acre plots and find that the benefit averages $8.10, with a standard deviation of $2.40.

　　To promote the dam, the government wishes to make a statement of the form, "μ is at least as large as. . . ." Yet to avoid political embarrassment, they want 99% confidence in this statement. What value should they put into the blank?

8-29 A manufacturing process has produced millions of TV tubes, with a mean life $\mu = 1,200$ hours, and $\sigma = 300$ hours. A new process produced a sample of 100 tubes with $\overline{X} = 1,265$.
(a) To state that the new process is at least as good as a certain value, calculate a one-sided 95% confidence interval
(b) Does the evidence indicate that the new process is better than the old?

8-30 Construct a one-sided 95% confidence interval for:
(a) Problem 8-13(c).
(b) Problem 8-14.
(c) Problem 8-15(a).

*8-6   VARIANCE OF A NORMAL POPULATION

We give one further example of a confidence interval, not so much for its practical value as for the insight that it provides.

Consider a normal population $N(\mu, \sigma^2)$ with both μ and σ^2 unknown. So far, we have estimated σ^2 with s^2 as a means of finding a confidence interval for μ. Now, suppose that our primary interest is in σ^2 rather than μ. For example, we may ask, "How much variance is there in Sweden's balance of payments?" in order to get some indication of the country's requirement of foreign exchange reserves. Or, we may ask, "What is the variance of farm income?" in order to evaluate whether a policy aimed at stabilizing farm income is necessary.[25]

We already have noted, following (7-4), that s^2 is an unbiased estimator of σ^2; to construct an interval estimate for σ^2 we must further ask: "How is the estimator s^2 distributed around σ^2?" To answer this, it is customary to define a variable, called the modified chi-square:

$$C^2 \triangleq \frac{s^2}{\sigma^2} \tag{8-50}$$

Of course, when $s^2 = \sigma^2$, this ratio is 1; thus our question can be rephrased: "how is C^2 distributed around 1?" If the parent population is normal,[26] it has been proven that the distribution of C^2 is given by Figure 8-7, with critical values displayed in Appendix Table VI(b) (tabulated according to d.f. $= n - 1$, like Student's t).[27]

Since its numerator s^2 and denominator σ^2 both are positive, the variable C^2 also is always positive, with its distribution falling to the right of zero in Figure 8-7. For small sample values, it also is skewed to the right;

---

[25] Income stabilization policies almost always are designed to stabilize income around a reasonably high level. Thus they aim both at reducing variance σ^2 *and* raising average income μ. Here we concentrate only on the variance problem.

[26] One reason that the confidence interval for σ^2 is of limited practical use is that it is not robust; i.e., if the parent population fails to satisfy the assumption of normality, the confidence interval (8-54) may have far less than 95% confidence. By constrast, confidence intervals for means are robust.

[27] C^2 is comprised of the constant parameter σ^2, and the variable s^2. Thus it has the same degrees of freedom as s^2 (explained in the footnote to equation (8-17)).

Modified chi-square (C^2, in Table VI(b)) is related to ordinary chi-square (χ^2, in Table VI(a)) according to:

$$C^2 = \frac{\chi^2}{\text{d.f.}} \tag{8-51}$$

Historically, the χ^2 distribution was used first because of its many applications to goodness-of-fit tests. This separate topic, and other χ^2 applications, are described in Chapter 17.

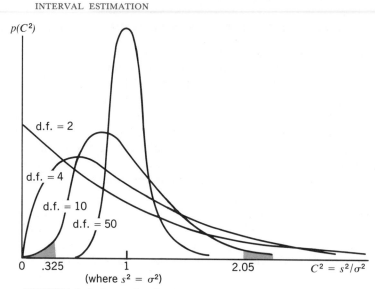

FIGURE 8–7 Some distributions of the modified chi square, C^2

but as n gets large, this skewness disappears and the C^2 distribution approaches normality. Since s^2 is an unbiased estimator of σ^2, this implies that the expected value of each of these C^2 distributions is 1. Moreover, as sample size increases, C^2 becomes more and more heavily concentrated near 1, indicating that s^2 is becoming an increasingly accurate estimator of σ^2.

With this deduction of how the estimator s^2 is distributed around its target σ^2, we now may infer a 95% confidence interval for σ^2, using our now-familiar technique. We illustrate with sample size $n = 11$ (d.f. $= 10$). From Figure 8-7 (or more precisely from Table VI(b)), we find the critical points cutting off $2\frac{1}{2}\%$ of the distribution in each tail; thus:

$$\Pr\left(.325 < \frac{s^2}{\sigma^2} < 2.05\right) = 95\%$$

Solving for σ^2, we obtain the equivalent statement:

$$\Pr\left(\frac{s^2}{2.05} < \sigma^2 < \frac{s^2}{.325}\right) = 95\% \tag{8-52}$$

If the observed value of s^2 turns out to be 3.6, for example, then the 95% confidence interval for σ^2 is:

$$1.76 < \sigma^2 < 11.1 \tag{8-53}$$

This is another example of an asymmetrical confidence interval.
The confidence interval (8-52) obviously may be generalized:

$$
\boxed{
\begin{array}{l}
\text{95\% confidence interval for the variance} \\
\text{of a normal population,} \\[2mm]
\dfrac{s^2}{C^2_{.025}} < \sigma^2 < \dfrac{s^2}{C^2_{.075}}
\end{array}
}
\qquad (8\text{-}54)
$$

where $C^2_{.025}$ and $C^2_{.975}$ are the upper and lower critical values of C^2.

PROBLEMS

Assume normal populations in the following problems.

*8-31 If a sample of 25 IQ scores from a certain population has $s^2 = 120$, construct a 95% confidence interval for the population σ^2.

*8-32 Using the data in Table 8-2, construct a 95% confidence interval for σ^2 based on:
(a) The first sample.
(b) The second sample.
(c) Both samples (use s_p^2 with d.f. $= n_1 + n_2 - 2$).

REVIEW PROBLEMS

8-33 Select the best interpretation, and criticize the others:
In a statistical report, the statement is made that the 95% confidence interval for the percentage of babies who are boys is between 51% and 55% (i.e., 53% $\pm$ 2%). This means that if, in the future, a 95% confidence interval is computed in the same way for each of a large number of random samples of the same size:
(a) 95% of such intervals will cover (contain) the midpoint 53%.
(b) 95% of such intervals will cover (contain) the population percentage of boys.
(c) 95% of such intervals will overlap (intersect) the interval 51% to 55%.
(d) 95% of such intervals will completely cover (contain) the interval 51% to 55%.

8-34 (Monte Carlo) The children in a large school system were randomly divided into two groups—a group given an enriched program (E), and a control group given the standard program (C). After two years the achievement scores of the E group and the C group were exactly the same—normally distributed with $\mu = 60$ and $\sigma = 10$.

(a) Using the random normal numbers in Table IIb, simulate a random sample of three observations from each population. From this data, and (8-32), construct a 95% confidence interval for the difference in population means.

(b) Have the instructor record and graph the confidence intervals obtained by several students in your class. If millions of such confidence intervals could be graphed:

   (i) What proportion would correctly cover the true value $\mu_E - \mu_C$?

   (ii) What proportion would be statistically discernible? Answer as well as you can, if an exact answer is not possible.

8-35 Repeat Problem 8-34, if the E population is now better—with a mean $\mu = 75$—while everything else remained the same.

8-36 Suppose that a sample of 100 men at an American university had the following distribution of IQ scores:

Midpoint	IQ Range	Frequency
100	93–107	29
115	108–122	38
130	123–137	20
145	138–152	10
160	153–167	3

(a) Graph the frequency distribution.

(b) Calculate a 95% confidence interval for the mean IQ of all men at this university.

*(c) Suppose that an independent sample of 100 women had a mean IQ of 126, and standard deviation of 12. If the student body is 70% men and 30% women, construct a 95% confidence interval for the mean IQ of this whole population.

(d) Construct a 95% confidence interval for the proportion of men who have an IQ over 137.5.

8-37 Two machines are used to produce the same good. In 400 articles produced by machine A, 16 were substandard. In the same length of time, the second machine produced 600 articles, and 60 were substandard. Construct 95% confidence intervals for:

(a) π_1, the true proportion of substandard articles from the first machine.

(b) π_2, the true proportion of substandard articles from the second machine.

(c) The difference between the two proportions $(\pi_1 - \pi_2)$.

8-38 Suppose that a psychologist runs 6 people through a certain experiment. In order to find the effect on heart rate, he collects the following data:

Heart Rate / Person	Before Experiment	After Experiment
Smith	71	84
Jones	67	72
Gunther	71	70
Wilson	78	85
Pestritto	64	71
Seaforth	70	81

Suppose that it is known that people as a whole have an average heart rate approximately normally distributed, with mean 73. Calculate a 95% confidence interval for the effect of the experiment on heart rate.

8-39 To determine whether or not it should change the design of its cereal box, a company pretested the old box versus the new: the marketing research department used a panel of 2,500 families, each receiving the two boxes with identical contents, but identified as Cereal O and Cereal N.

Of the panel, 830 preferred Cereal O, 1,220 preferred Cereal N, and the remaining 450 could see no difference between them. To what extent does this establish the superiority of the new box (Cereal N)?

8-40 Samples of eggs are taken from the nests of a species of bird in two localities, and their weights are recorded. The following are the frequency distributions.

Locality	Weight in Grams (Cell Midpoints)							Total Eggs	$\bar{X}$	s^2
	2.5	3.5	4.5	5.5	6.5	7.5	8.5			
A	3	61	84	196	80	33	0	457	5.35	1.22
B	1	72	88	174	102	46	5	488	5.45	1.47

Represent these data graphically, and judge whether the localities differ in their average egg weights.[28]

8-41 A certain scientist concluded his study in fertility control as follows:[29] "So far one result has emerged from the before-and-after survey, and it is a key measure of the outcome: at the end of 1962, 14.2% of the women in the sample were pregnant, and at the end of 1963 (after the birth-control campaign) 11.4% of the women (in a second independent sample) were pregnant, a decline of about one fifth."

If the samples (both before and after) included 2,500 women, what statistical qualification would you add to the above statement in order to make its meaning clearer?

8-42 Soon after he took office in 1963, President Johnson was approved by 160 out of a sample of 200 Americans.[30] With growing disillusionment over his Vietnam policy, by 1968 he was approved by only 70 out of a sample of 200 Americans.

(a) Construct 95% confidence intervals for the percentage of all Americans who approved of Johnson:

(i) In 1963.

(ii) In 1968.

(b) Construct a confidence interval for the change in opinion between 1963 and 1968.

8-43 In 3,576 jury trials, the presiding judge was asked how he would have decided the case without a jury, with the following results:

Judge would have / Jury	Acquitted	Convicted
Acquitted	14%	3%
Convicted	19%	64%

To what extent can a jury be said to be more lenient than a judge? Report your answer as a 95% confidence interval for the whole population (from which these 3,576 trials may be regarded as a sample).[31]

---

[28]Oxford, Trinity term, 1964.

[29]From *Scientific American,* May 1964, pp. 33–34.

[30]Based, with modification of sample size, on the Gallup Polls that recorded the high and low points of Johnson's presidency.

[31]Condensed from H. Kalven, Jr. and H. Zeisel, *The American Jury* (Boston: Little, Brown and Co., 1966). Actually, the sample was not random, and has been criticized for being unrepresentative.

8-44 Of the 453 boys released from a detention home for delinquents, there were 150 boys for whom sufficient information was available, both for the period before and the period after detention. The behavior record of these 150 boys was as follows:[32]

Before Detention \ After Detention	Good	Fair	Bad	Totals
Good	18	9	2	29
Fair	49	33	15	97
Bad	10	4	10	24
Totals	77	46	27	150

(a) We would like to know, on the average, how the stay at the detention home changes behavior. One possible way to analyze the data is, for each boy, to ascribe to bad behavior the value 0, to good behavior the value 1, and to fair behavior some intermediate value, say $\frac{1}{2}$. For each boy, therefore, a numerical improvement can be calculated. The 150 improvements may be considered a sample from a population; calculate a 95% confidence interval for the mean of this population. (Although this coding is somewhat arbitrary, it allows a simple and powerful analysis that can't be too far wrong. The analysis, then, is very similar to Problem 8-26.)
(b) In what way does your analysis in part (a) show, or fail to show, how the detention home changes behavior?

---

[32]Abridged from the Oxford University examinations, Trinity term, 1964.

CHAPTER 9

Hypothesis Testing

Say not, "I have found the truth," but rather, "I have found a truth."

Kahlil Gibran

Traditionally, hypothesis testing has been treated as a separate topic in statistics courses. It is closely related, however, to the interval estimation that we just discussed in Chapter 8. Therefore, this section starts with an overview of hypothesis testing as a rewording of confidence intervals.[1]

9-1 HYPOTHESIS TESTING USING CONFIDENCE INTERVALS

(a) A Modern Approach[2]

To illustrate the close relation between hypothesis testing and confidence intervals, we begin with an example. (Recall that in the format of this

---

[1] *Reminder to instructors:* This chapter is designed so that you may spend as much or as little time on hypothesis testing as you judge appropriate. At the end of any section you may skip ahead to the next chapters and thus have more time to cover ANOVA and regression.

[2] For this we are indebted to a number of practicing statisticians, especially John Tukey. Actually, this approach is not all that modern—a good number of statisticians have been advocating it for many years. But since it has not yet commonly been emphasized in textbooks, we shall refer to it as a "modern" approach.

book, an example is a problem that the student should work on first; the solution in the text should only be consulted later as a check.)

Example 9-1. In a large American university in 1969, the male and female professors were sampled independently, yielding the following annual salaries (in thousands of dollars, rounded):[3]

Men (X_1)		Women (X_2)
12,	20	9
11,	14	12
19,	17	8
16,	14	10
22,	15	16
$\overline{X}_1 = 16$		$\overline{X}_2 = 11$

These sample means give a rough estimate of the underlying population means μ_1 and μ_2. Perhaps they can be used to settle the following argument. A husband claims that there is no difference between μ_1 and μ_2. That is, if we denote the difference as $\Delta = \mu_1 - \mu_2$, he claims that:

$$\Delta = 0 \tag{9-1}$$

His wife, however, claims that the difference is as large as seven thousand dollars:

$$\Delta = 7 \tag{9-2}$$

Settle this argument by constructing a confidence interval.

Solution. The 95% confidence interval is, from (8-32):

$$\Delta = (\overline{X}_1 - \overline{X}_2) \pm t_{.025}\, s_p \sqrt{\frac{1}{n_1} + \frac{1}{n_2}}$$

$$= (16 - 11) \pm 2.16 \sqrt{\frac{152}{13}} \sqrt{\frac{1}{10} + \frac{1}{5}}$$

$$= 5.0 \pm 2.16(1.87) \tag{9-3}$$

$$= 5.0 \pm 4.0 \tag{9-4}$$

Thus, with 95% confidence, Δ is estimated to be between 1 and 9. Thus the claim $\Delta = 0$ seems implausible, because it falls outside this confidence interval.

[3] From D. A. Katz, "Faculty Salaries, Promotions, and Productivity at a Large University," *American Economic Review,* June 1973.

In general, any hypothesis that lies outside the confidence interval may be judged *implausible* or *rejected.* On the other hand, any hypothesis that lies within the confidence interval may be judged *plausible,* or *acceptable.* Thus:

> a confidence interval may be regarded as just the set of acceptable hypotheses. (9-5)

Since a 95% confidence interval is being used, it would be natural to speak of an hypothesis like (9-1) as being tested at a 95% confidence level. In conforming to tradition, however, we use 5%—the complement of 95%—and simply call it the level of the test. Thus, we formally conclude that the hypothesis $\Delta = 0$ is rejected at the 5% level. In other words, in (9-4) we see that there is sufficient data (a small enough sampling error) to allow us to discern a real difference between men's and women's salaries.[4] We therefore call this difference *statistically discernible* at the 5% level.

Example 9-2. Suppose that the confidence interval (9-4) had been based on smaller samples, and so had been more vague:

$$\Delta = 5 \pm 8 \tag{9-6}$$

[4] Although we have shown (at the 5% level) that men's and women's salaries are different, we have *not* necessarily shown that discrimination exists. There are many alternative explanations: men may have a longer average period of education than women. What we really should do then, is compare men and women who have the *same qualifications.* This will, in fact, be done in Problem 13-5, using multiple regression analysis.

that is:

$$-3 < \Delta < 13 \qquad (9\text{-}7)$$

Since the hypothesis $\Delta = 0$ falls within this interval, it cannot be rejected; in other words, the difference between men and women's salaries is no longer statistically discernible, so we may call it *statistically indiscernible.* Which of the following interpretations are true or false? If false, correct it.

(a) The true *population* difference may well be 0, with the difference in *sample* means $(\overline{X}_1 - \overline{X}_2 = 5)$ representing only random fluctuation.

(b) In (9-7), we see that the plausible population differences include both negative and positive values; i.e., we cannot even decide whether men's salaries on average are better or worse than women's.

(c) In (9-6), we see that the sampling allowance (± 8) dominates the estimate (5). Whenever there is this much sampling error, we call the result statistically indiscernible.

Solution. Each of these statements is a reasonable interpretation.

In summary, if a confidence interval already has been calculated, then it can be used immediately, without any further calculations, to test any hypothesis.

(b) The Traditional Approach

The hypothesis $\Delta = 0$ in (9-1) is of particular interest; since it represents no difference whatsoever, it is called the *null hypothesis* H_0. In rejecting it because it lies outside the confidence interval (9-3), we establish the important claim that there is indeed a difference between men's and women's income. Such a result traditionally has been called *statistically significant* at the 5% *significance level.*

There is a problem with this terminology. When the term "statistical significance" is used in this way, it simply means that enough data have been collected to establish that a difference does exist. It does *not* mean that the difference is necessarily important. For example, in another test based on very large samples from nearly identical populations, the 95%-confidence interval, instead of (9-3), might be:

$$\Delta = .005 \pm .004 \qquad (9\text{-}8)$$

This difference is so miniscule that we could dismiss it as being of no real interest, even though it is statistically as significant as (9-4). In other words, *statistical significance* is a technical term with a far different meaning than *ordinary* significance.[5]

Unfortunately but understandably, many people tend to confuse statistical significance with ordinary significance.[6] To reduce the confusion, we prefer the word "discernible" to the word "significant." In conclusion, therefore, the traditional phrase "statistically significant at the 5% significance level" technically means exactly the same thing as the more modern phrase, "statistically discernible at the 5% level." We prefer the modern phrase, because it is less likely to be misinterpreted.

PROBLEMS

9-1 For each of Problems 8-39 to 8-44, state and test the null hypothesis, indicating whether the results are statistically discernible at the 5% level.

9-2 A manufacturing process has produced millions of TV tubes with a mean life $\mu = 1,200$ hours, and standard deviation $\sigma = 300$ hours. A new process is tried on a sample of 100 tubes, producing a sample average $\overline{X} = 1,265$ hours. Will this new process produce a *long-run* average that is different from the null hypothesis $\mu = 1,200$? Specifically, is the sample mean $\overline{X} = 1,265$ statistically discernible (i.e., discernibly different from the H_0 value of 1,200) at a level of:

(a) 0.1%?

(b) 1%?

(c) 5%?

(d) 10%?

(e) 20%?

---

[5]There also is a problem with the term "5% significance level": it sounds like the higher this value (say 10% rather than 5%), the better the hypothesis test. However, precisely the reverse is true (our level of confidence would only be 90% rather than 95%).

[6]To make matters worse, some writers simply use the single word "significant." While they may mean "statistically significant," their readers may interpret this to mean "ordinarily significant," i.e., "important."

As an example of this, a recent journal article used the words "highly significant" to refer to relatively minor findings uncovered in a huge sample of 400,000 people (for example, one such finding was that the youngest child in a family had an IQ that was 1 or 2 points lower on average than the second youngest child).

From this example, or from (9-8), it may be concluded that a large sample is like a large magnifying glass that allows us to discern the smallest molehill (smallest difference between two populations). It is unwise to use a phrase (such as "significant" or "highly significant") that makes this molehill sound like a mountain.

9-2   PROB-VALUE

(a) What is Prob-Value?

In Section 9-1, we developed a simple way to test *any* hypothesis by examining whether or not it falls within the confidence interval. Now we shall take a new perspective by concentrating on just one hypothesis— the null hypothesis H_0. We shall calculate just how much (or how little) it is supported by the data.

Example 9-3. A traditional manufacturing process has produced millions of TV tubes, with a mean life $\mu = 1,200$ hours and a standard deviation $\sigma = 300$ hours. A new process, recommended as better by the engineering department, produces a sample of 100 tubes, producing an average $\overline{X} = 1,265$. Although this sample makes the new process look relatively good, is this just a sampling fluke? Is it possible that the new process is really no better than the old, but that we have just turned up an unlucky sample?

To formulate this problem more specifically, we state the null hypothesis: the new process would produce a population that is no different from the old; that is, $H_0 : \mu = 1,200$. This is sometimes abbreviated to:

$$\mu_0 = 1,200 \tag{9-9}$$

The claim of the engineering department may be called the *alternative* hypothesis,[7] $H_1 : \mu > 1,200$. This is sometimes abbreviated to

$$\mu_1 > 1,200 \tag{9-10}$$

Whether H_0 or H_1 is true, we still shall assume that $\sigma = 300$, which may be called a *background* hypothesis.[8]

How inconsistent is the data $\overline{X} = 1,265$ with the null hypothesis $\mu_0 = 1,200$? Specifically, if the null hypothesis were true, what is the probability that $\overline{X}$ would be as large as 1,265?

Solution. In Figure 9-1, we show the hypothetical distribution of $\overline{X}$. (Here, for the first time, we show the convention of drawing hypothetical distributions in white.) By the central limit theorem, this distribution of $\overline{X}$ is normal, with mean μ_0 = 1,200, and standard deviation $\sigma/\sqrt{n}$ = $300/\sqrt{100}$. Thus:

$$\Pr(\overline{X} \geq 1,265) = \Pr\left(\frac{\overline{X} - \mu_0}{\sigma/\sqrt{n}} \geq \frac{1,265 - 1,200}{300/\sqrt{100}}\right) \qquad (9\text{-}11)$$

$$= \Pr(z \geq 2.17) \simeq .015 \qquad (9\text{-}12)$$

Remarks. This means that if, in fact, the new process is no better, there would be only a $1\frac{1}{2}\%$ chance of observing $\overline{X}$ as large as 1,265. We call $1\frac{1}{2}\%$ the prob-value[9] for H_0. It summarizes very clearly how much agreement there is between the data and H_0. Clearly, in this example the data provides little support for H_0. If $\overline{X}$ had been observed closer to H_0 in Figure 9-1, however, the prob-value would have been larger.

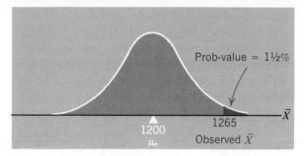

FIGURE 9–1 Prob-value $\overset{\Delta}{=}$ $\Pr(\overline{X}$ would be as large as the value actually observed, if H_0 were true)

---

[7]In practice, the null and alternative hypotheses often are determined as follows: H_1 is the claim that we want to prove (9-10). Then H_0 is really everything else, $\mu \leq 1,200$. But we need not use all of this; instead, in (9-9), we use just the boundary point, $\mu = 1,200$. Any other μ below 1,200 would be easier to distinguish from H_1 since it is "further away;" we therefore need not bother with it.

[8]The background hypotheses include other things, which seldom are stated explicitly. For example, we assume that the sample is random. These background hypotheses are not themselves subject to test; they will be maintained no matter what choice is made between μ_0 and μ_1.

[9]Although customarily it is called the *p-value*, we chose the name *prob-value* to distinguish it from the many other p's appearing in this text. Also, "prob-value" suggests that $\overline{X}$ is a probe to measure the credibility of H_0.

Unless otherwise stated, by "prob-value" we mean the one-sided prob-value, such as shown in Figure 9-1. (The two-sided prob-value is less common, and is not discussed until Figure 9-9.)

In general, for any hypothesis being tested, we define the prob-value for H_0 as:

$$\text{prob-value} \triangleq \Pr\left(\begin{array}{c}\text{The sample value would be as large}\\ \text{as the value actually observed,}[10]\\ \text{if } H_0 \text{ is true}\end{array}\right) \quad (9\text{-}13)$$

The prob-value is an excellent way to *summarize what the data says*[11] *about the credibility of H_0.*

(b) More Examples

While the statistic:

$$z = \frac{\overline{X} - \mu_0}{\sigma/\sqrt{n}}$$

in (9-11) is appropriate when σ is known, it is much more common for σ to be unknown and hence estimated by s. Then, of course, we must use the analogous t statistic:

$$t = \frac{\overline{X} - \mu_0}{s/\sqrt{n}} \qquad \text{like (8-16)}$$

or, more generally:

$$t = \frac{\text{estimate} - \text{population mean assuming } H_0}{\text{standard error}[12]} \qquad (9\text{-}14)$$

---

[10]For brevity, we use the term "as large as" to mean "*at least* as large as."

The prob-value in Figure 9-1 is calculated in the right-hand tail, because the alternative hypothesis is on the right side ($\mu > 1,200$). On the other hand, if the alternative hypothesis were on the left side ($\mu < 1,200$), then the prob-value would be calculated in the left-hand tail; i.e.:

prob-value = Pr(the sample statistic would be *as small as* the value actually observed, assuming H_0)

[11]Of course, the data is not the only thing to be considered if we want to make a final judgment on the credibility of H_0; common sense, or what sometimes is more formally called "personal prior probability," must be considered too, especially when the sample is small and hence unreliable. For example, if a penny found on the street were flipped ten times and showed eight heads, the prob-value for H_0 (fair coin) would be .05 (from the binomial Table III(c)). But obviously it would be inappropriate to conclude from this that the coin was unfair. We know that a coin picked up on the street is almost certain to be fair; thus our common sense tells us that our sample result (of eight heads) was just "the luck of the draw," and we discount it accordingly.

[12]Both $\sigma/\sqrt{n}$ and its estimate $s/\sqrt{n}$ are commonly called the standard error. The context usually will clarify which is intended—in this case, of course, $s/\sqrt{n}$ is intended. (If necessary, the terms "theoretical standard error" and "estimated standard error" could be used to distinguish between $\sigma/\sqrt{n}$ and $s/\sqrt{n}$.)

Customarily, assuming H_0, the population mean is zero, in which case (9-14) takes a very simple form:[13]

$$t = \frac{\text{estimate}}{\text{standard error}} \qquad (9\text{-}15)$$

The prob-value is obtained by calculating the t statistic in (9-14) and referring it to the Appendix Table V to find the tail probability— just as we did for the z statistic in (9-12).

Example 9-4. In Example 9-1, we calculated the following 95% confidence interval for the mean difference in men's and women's salaries (thousands of dollars):

$$(\mu_1 - \mu_2) = (\overline{X}_1 - \overline{X}_2) \pm t_{.025} (\text{standard error})$$
$$= 5.0 \pm t_{.025} (1.87) \qquad (9\text{-}16)$$
$$(9\text{-}3) \text{ repeated}$$

where there were 13 degrees of freedom for the t distribution.

Calculate the prob-value for the null hypothesis that men's and women's salaries are equal on average. Use the alternative hypothesis that men's salaries are larger.

Solution. The null hypothesis is $\mu_1 - \mu_2 = 0$, so that (9-15) yields:

$$t = \frac{\text{estimate}}{\text{standard error}} = \frac{(\overline{X}_1 - \overline{X}_2)}{\text{standard error}}$$

Of course, the values we need here are those that we already calculated for the confidence interval (9-16). Thus:

$$t = \frac{5.0}{1.87} = 2.67$$

The question is: what is the probability that t will equal or

----

[13]The form (9-15) makes good intuitive sense: we simply are calculating how large the estimate is relative to its standard error. Thus, a large standard error will make the t ratio too small for the result to be statistically discernible.

exceed 2.67? When we refer to Table V, we find that it unfortunately does not have the same detail as the z table. So we have to interpolate. We scan along the row where d.f. $= 13$, until we bracket the observed $t = 2.67$. This bracketing occurs between $t_{.010} = 2.65$ and $t_{.005} = 3.01$. Thus we conclude that:

$$.005 < \text{prob-value} < .010$$

If we are lucky, we may be able to get an even more precise estimate. For example, in this case, the calculated $t = 2.67$ is very close to the tabulated $t_{.010} = 2.65$. Thus:

$$\text{prob-value} \simeq .01$$

Accordingly, we conclude that H_0 is a very implausible hypothesis, just as we did earlier in Example 9-1.

To illustrate the diversity of distributions for which the prob-value may be calculated, we give a binomial distribution approximated by the normal as our last example.

Example 9-5.   To investigate whether children a generation ago showed racial awareness and prejudice,[14] a group of 252 black children was studied. Each child was told to choose a doll to play with from among a group of 4 dolls, 2 white and 2 nonwhite. A white doll was chosen by 169 of the 252 children.

What is the prob-value for the null hypothesis that the children ignore color? Use the alternative hypothesis that the children may be prejudiced in favor of white.

Solution.   First we formulate the problem mathematically. The 252 children may be regarded as a random sample from a large hypothetical population whose proportion π choosing white is under investigation. The null hypothesis is:

$$\pi_0 = .5$$

while the alternative hypothesis is

$$\pi_1 > .5$$

The sample proportion is $P = 169/252 = .67$. What is the probability of getting such a large proportion by chance? Since P is approximately normal, we standardize it:

$$\Pr(P \geq .67) = \Pr\left(\frac{P - \pi_0}{\sqrt{\dfrac{\pi_0(1 - \pi_0)}{n}}} \geq \frac{.67 - .50}{\sqrt{\dfrac{.5(.5)}{252}}}\right)$$

$$= \Pr(z \geq 5.40) \simeq \text{one in a million}$$

Thus the prob-value is miniscule, indicating almost no credibility for the null hypothesis. Therefore it may be concluded that the children were prejudiced in favor of white.

---

[14]From K. B. Clark and M. P. Clark, "Racial Identification and Preference in Negro Children," in *Readings in Social Psychology*, E. E. Maccoby et al., eds. (New York: Holt, Rinehart and Winston, 1958).

PROBLEMS

9-3 For Problem 8-39, state the null and alternative hypotheses. Then calculate the prob-value for H_0. Illustrate with a diagram like Figure 9-1.

For Problems 8-40 through 8-44, briefly state the null and alternative hypotheses. Then calculate the prob-value for H_0.

9-4 Of the people eligible for jury duty in Boston in 1968, about 29% were women.[15] The judge who tried Dr. Benjamin Spock for conspiracy to violate the Selective Service Act had an interesting record: of the 700 people he had selected for jury duty in his past few trials, only 15% were women. How would you judge his fairness?

9-5 A doctor tested the effectiveness of a certain drug by giving it to a group of rats, while a second group of rats was kept under identical conditions as a control. The difference (in mean weight increases) between the treated group (T) and the control group (C) yielded a t value of 2.71, and hence a prob-value of about .013, for the null hypothesis ($\mu_T = \mu_C$) against the alternative hypothesis ($\mu_T > \mu_C$). Select the best interpretation, and criticize the others:

---

[15]This is a somewhat simplified version, condensed from H. Zeisel and H. Kalven, "Parking Tickets and Missing Women," in *Statistics, A Guide to the Unknown*, J. Tanner, et al., eds. (San Francisco: Holden-Day, 1972).

(a) $\Pr(\overline{X}_T = \overline{X}_C) = .013$ whereas $\Pr(\overline{X}_T > \overline{X}_C) = .987$.

(b) The conditional probability of H_0, given the data, is .013.

(c) The probability of H_0 is .013.

(d) If we repeated the experiment, and if H_0 were true, the probability is .013 of getting a t value at least as large as the one we observed.

(e) If we repeated the experiment, and if H_1 were true, the probability is .987 of getting a t value at least as large as the one we observed.

9-3   CLASSICAL HYPOTHESIS TESTING

(a) What is a Classical Test?

Suppose that we have the same data as in Example 9-3. Recall that the traditional manufacturing process had produced millions of TV tubes, with a mean life $\mu = 1,200$ hours, and a standard deviation $\sigma = 300$ hours. Applying a classical hypothesis test to this problem involves the following three steps:

1. The null hypothesis ($H_0: \mu = 1,200$) and the alternative hypothesis ($H_1: \mu > 1,200$) are stated formally.

2. Now assume that the null hypothesis is true. What can we expect of a sample mean drawn from this sort of world? Its distribution again is shown in Figure 9-2, just as it was in Figure 9-1. But now comes the new twist. In the right-hand tail, we cut off a small probability called the *level of the test*, α, which usually is set arbitrarily[16] at 5%.

If the observed $\overline{X}$ falls in the tail in Figure 9-2, then it will be judged to be sufficiently in conflict with the null hypothesis to allow the rejection of H_0 in favor of H_1. Otherwise, H_0 will be judged acceptable. (Accordingly, the tail region is commonly called the "rejection region," while the rest is called the "acceptance region.")

The critical value $\overline{X}_c$ that marks off this tail is calculated by noting from Appendix Table IV that a z value of 1.64 cuts 5% off the tail of the normal distribution; that is:

---

[16]This arbitrary choice is equivalent to the arbitrary choice of a confidence level, which usually is set at 95%.

Note in Figure 9-2 that we are cutting off a probability only in one tail. Accordingly, this is a "one-tailed test," consistent with the one-tailed alternative hypothesis. Under certain circumstances a two-tailed test may be required, but this discussion is deferred to Section 9-6.

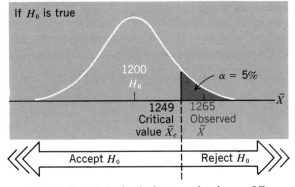

FIGURE 9–2 A classical test at level $\alpha = 5\%$

$$\text{critical } z = \frac{\overline{X}_c - \mu_0}{\sigma/\sqrt{n}} = 1.64 \tag{9-17}$$

$$\frac{\overline{X}_c - 1200}{300/\sqrt{100}} = 1.64$$

$$\overline{X}_c = 1{,}200 + 1.64\left(\frac{300}{\sqrt{100}}\right)$$

$$\overline{X}_c = 1{,}249 \tag{9-18}$$

This calculation completes our simple decision rule: we shall reject H_0 if $\overline{X}$ falls above $\overline{X}_c = 1{,}249$.

3. Finally, we take our sample. Suppose, as before, that $\overline{X}$ turns out to be 1,265. Then, since this falls above 1,249, we reject H_0.

In summary, there is another way of looking at this testing procedure. If we get an observed $\overline{X}$ exceeding 1,249, there are two explanations: (a) H_0 is true, but we have been exceedingly unlucky and got a very improbable sample $\overline{X}$. (We're born to be losers; even when we bet with odds of 19 to 1 in our favor, we still lose); or
(b) H_0 is *not* true after all; thus it is no surprise that the observed $\overline{X}$ was so high.

Since the second explanation is reasonable, we opt for that. Although the first explanation is conceivable, it is not as plausible as the second. But we are left in some doubt; it is just possible that the first explanation is the correct one. For this reason we qualify our conclusion "to be at the 5% level."[17]

(b) Classical Hypothesis Testing and Prob-Value

A comparison of the two procedures is set out in Figure 9-3. Since the prob-value ($1\frac{1}{2}\%$ from (9-12)) is less than α, the observed $\overline{X}$ is correspondingly in the rejection region; that is:

$$\boxed{\text{reject } H_0 \text{ iff prob-value} \leq \alpha} \tag{9-19}$$

To restate this, we recall that the prob-value is a measure of the credibility of H_0; if this credibility sinks below α, then H_0 is rejected.[18]

Applied statisticians increasingly prefer prob-values to classical tests, because classical tests involve setting α arbitrarily (usually at 5%). Rather than introducing such an arbitrary element, it often is preferable just to quote the prob-value, allowing the reader to pass his own judgement on H_0. (Formally, by determining whatever level of α he deems appropriate for his purpose, the reader may reach an individual decision, using (9-19).)

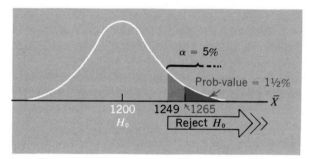

FIGURE 9–3 Classical hypothesis testing and prob-value

---

*[17] These calculations for a one-sided classical test are very similar to the calculations for a one-sided confidence interval in Problem 8-29. In both cases we reject H_0 if the difference between the observed mean $\overline{X}$ and the hypothetical mean μ_0 exceeds 1.64 standard errors. (A similar equivalence between a two-sided classical test and a two-sided confidence interval is proven in detail in Section 9-6). In view of this equivalence, we use the same terminology (such as "reject H_0" or "5% level") in this section as we did in Section 9-1.

[18] Figure 9-3 provides another useful interpretation of prob-value. Note that if we had happened to set the level of the test at the prob-value of $1\frac{1}{2}\%$ (rather than 5%), it would have been just barely possible to reject H_0. Accordingly,

$$\boxed{\text{prob-value is the lowest that we could push the level of the test and still be able (barely) to reject } H_0.} \tag{9-20}$$

(c) Type I and Type II Errors

In the decision-making process illustrated in Figure 9-2, we risk committing two distinct kinds of errors. The first is shown in Figure 9-4(a) (a reproduction of Figure 9-2), which shows what the world looks like if

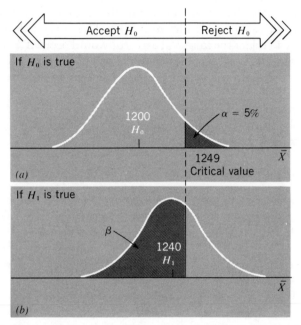

FIGURE 9–4 The two kinds of error that can occur in a classical test. (a) If H_0 is true, then α = probability of erring by rejecting the true hypothesis H_0. (b) If H_1 is true, then β = probability of erring by accepting the false hypothesis H_0.

TABLE 9-1
Possible Results of an Hypothesis Test[a]

State of the World \ Decision	Accept H_0	Reject H_0
If H_0 is true	Correct decision Probability $= 1 - \alpha$ $\quad$ = confidence level	Type I error. Probability $= \alpha$ $\quad$ = level of the test
If H_0 is false	Type II error. Probability $= \beta$	Correct decision. Probability $= 1 - \beta$ $\quad$ = power of the test

[a]Derived from Figure 9-4.

H_0 is true. In this event, there is a 5% probability that we will observe $\overline{X}$ in the shaded region, and thus erroneously reject the true H_0. Rejecting H_0 when it is true is called a type I error; its probability, of course, is α, the level of the test.

But suppose that the null hypothesis is false—i.e., that the alternative hypothesis H_1 is true—and to be specific, suppose that $\mu = 1{,}240$. Then we are living in a different sort of world; the quite different distribution of $\overline{X}$ around H_1 is shown in Figure 9-4(b). The correct decision in this case would be to reject the false H_0; an error would occur if $\overline{X}$ fell in the H_0 acceptance region. Such acceptance of H_0 when it is false is called a type II error; its probability is called β, and is shown as the shaded area in Figure 9-4(b).

The terminology of hypothesis testing is reviewed in Table 9-1. Note that the probabilities in each row must sum to 1; this must follow as long as we decide either to accept or reject H_0.[19]

PROBLEMS

9-6 (Acceptance Sampling.) The prospective purchaser of several new shipments of waterproof gloves hopes they are as good as the old shipments, which had a 10% rate of defective pairs. But he fears that they may be worse. So for each shipment, he takes a random sample of 100 pairs and counts the proportion P of defective pairs so that he can run a classical test at the level[20] $\alpha = .09$.
(a) What are the null and alternative hypotheses?
(b) How large must P be in order to reject the null hypothesis (i.e., reject the shipment)?
(c) Suppose that for 6 shipments, the values of P turned out to be 12%, 25%, 8%, 16%, 24%, 21%. Which of these shipments should be rejected?

9-7 In Problem 9-6, suppose that the purchaser tries to get away with a small sample of only 10 pairs. Suppose that instead of setting $\alpha = .09$, he arbitrarily sets the rejection region to be where $P \geq 20\%$. (That is, he will reject the shipment if there are 2 or more defective pairs among the

---

[19]Of course, other more complicated decision rules may be used. For example, the statistician may decide to suspend judgement if the observed $\overline{X}$ is in the region around 1,250 (say $1{,}240 < \overline{X} < 1{,}260$). If he observes an ambiguous $\overline{X}$ in this range, he then would undertake a second stage of sampling—which might yield a clear-cut decision, or might lead to further stages (i.e., "sequential sampling").

[20]One of the main criticisms of the classical test is that α is set arbitrarily at 5%. Instead, the level should be determined by such relevant factors as the cost of a bad shipment, the cost of an alternative supplier, the new shipper's reputation, etc. In Problem 9-6, we suppose that when all these factors have been taken into account, the appropriate α is .09.

10 pairs in the sample.) For this test, what is α? (*Hint:* Rather than using the normal approximation, the binomial distribution is easier and more accurate.)

9-8 Records show that in a random sample of 100 hours, a machine produced an hourly average of 678 articles with a standard deviation of 25. After a control device was installed, in a random sample of 500 hours the machine produced an hourly average of 674 articles with a standard deviation of 5. Management claimed that the control device was reducing production. The union countered that the drop of 4 articles in the sample mean was "merely statistical fluctuation."
(a) To objectively summarize the evidence on whether production is left unchanged, calculate the prob-value.
(b) If the arbitration board decides that a fair level is $\alpha = 1\%$, would they rule in favor of management or the union?

9-9 Fill in the blanks.
    Consider the problem facing a radar operator whose job is to detect enemy aircraft. When something irregular appears on the screen, he must decide between:
H_0: all is well; only a bit of interference on the screen.
H_1: an attack is coming.
    A "false alarm" is a type _____ error, and its probability is denoted by _____. A "missed alarm" is a type _____ error, and its probability is denoted by _____. By making the electronic equipment more sensitive and reliable, it is possible to reduce both _____ and _____.

9-10 An appliance shop sells an average of 320 appliances per day with a standard deviation of 40. After an extensive advertising campaign, management will calculate the average sales $\overline{X}$ for the next 25 days to see whether an improvement has occurred.
(a) What are the null and alternative hypotheses?
(b) For a classical test at level $\alpha = .05$, how large must $\overline{X}$ be in order to reject H_0—i.e., in order to achieve statistical discernibility?
(c) If, in fact, the average sales turn out to be $\overline{X} = 335$, do you reject H_0 at the 5% level?
(d) Calculate the prob-value for H_0, based on the data $\overline{X} = 335$.
(e) Suppose that the classical rejection region in part (b) had not been calculated. Use the prob-value in part (d) instead, to determine whether or not to reject H_0 at the 5% level. Is your answer the same as in part (c)?
(f) What assumptions have you implicitly made about the sampling? Under what conditions are they questionable?

9-11 Repeat Problem 9-10(b), (c), and (e), for a level $\alpha = .01$.

9-12 In a sample of 1,500 Americans in 1971, 42% smoked; in an independent sample of 1,500 Americans a year later, 43% smoked.[21]
(a) Construct a one-sided 95% confidence interval for the increase in the proportion of cigarette smokers in the population.
(b) Calculate the prob-value for the null hypothesis of no increase.
(c) Is the change from 42% to 43% statistically discernible at the level $\alpha = 5\%$?

9-13 In a sample of 750 men in 1974, 45% smoked; in an independent sample of 750 women, 36% smoked. For the difference between men and women, repeat Problem 9-12.

⇒9-14 Consider the classical test shown in Figure 9-2.
(a) If the observed $\overline{X}$ is 1,245, do you reject H_0?
(b) Suppose that you had designed this new process, and you were convinced that it was better than the old; specifically, on the basis of sound engineering principles, you really believed that $H_1 : \mu = 1,240$. Common sense suggests that H_0 should be rejected in favor of H_1. In this conflict of common sense with classical testing, which path would you follow? Why?
(c) Suppose that sample size is doubled to 200, while you keep $\alpha = 5\%$ and you continue to observe $\overline{X}$ to be 1,245. What would be the classical test decision now? Would it, therefore, be true to say that your problem in part (a) may have been inadequate sample size?
(d) Suppose now that your sample size n was increased to a million, and in that huge sample you observed $\overline{X} = 1,201$. Such an improvement of only 1 unit over the old process is of no economic significance (i.e., does not justify retooling, etc.); but is it statistically significant (discernible)? Is it therefore true to say that a sufficiently large sample size may provide the grounds for rejecting any specific H_0—no matter how nearly true it may be—simply because it is not *exactly* true?

9-4   CLASSICAL TESTS RECONSIDERED

(a) Reducing α and β

In Figure 9-5(a), we repeat the information of Figure 9-4—namely, the two error probabilities: α, if H_0 is true, and β, if H_1 is true. In Figure 9-5(b), we illustrate how decreasing α (by moving the critical point to the

---

[21] Problems 9-12 and 9-13 are from the *Gallup Opinion Index,* June 1974, p. 21.

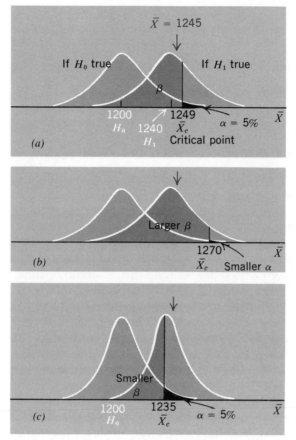

FIGURE 9–5 (a) Hypothesis test of Figure 9–4, showing α and β. (b) How a reduction in α increases β, other things being equal. (c) How an increase in sample size allows one error probability (β) to be reduced, while holding the other (α) constant.

right—say to 1,270) simultaneously will increase β. In statistics, as in economics, the problem involves trading off conflicting objectives. There is an interesting legal analogy: in a murder trial, the jury is being asked to decide between H_0, the hypothesis that the accused is innocent, and the alternative H_1, that he is guilty. A type I error is committed if an innocent man is condemned, while a type II error occurs if a guilty man is set free. The judge's admonition to the jury that "guilt must be proved beyond a reasonable doubt" means that α should be kept very small.

There have been many legal reforms (for example, limiting the power of the police to obtain a confession) that have been designed to reduce α, the probability that an innocent man will be condemned. But these same reforms have increased β, the probability that a guilty man will evade punishment. There is no way of pushing α down to 0 (insuring absolutely against convicting an innocent man) without raising β to 1 (letting every defendant go free and making the trial meaningless). The one way that α and β *both* can be reduced is by increased evidence. Figure 9-5(c) illustrates this same issue: how increased evidence, in the form of a larger sample, can reduce both α and β.

(b) Some Difficulties with Classical Tests

Figure 9-5 and Problem 9-14 illustrate some of the difficulties that we may encounter in applying a classical reject-or-accept hypothesis test at a prespecified level α. In Figure 9-5(a), an observed value of $\overline{X} = 1,245$ is not quite extreme enough to allow rejection of H_0, so it is accepted. But if we had set $\alpha = 10\%$, then H_0 would have been rejected. This illustrates once again how an arbitrary specification of α leads to an arbitrary decision. But the problem is deeper than this: it would be most unfortunate if we found H_0 acceptable, if we had had prior grounds for believing H_1, i.e., expecting that the new process would yield 1,240. In this case, our prior belief in the new process would be strongly supported by the sample observation of 1,245. Yet we have used this confirming sample result (in this classical hypothesis test) to conclude that the new process is no better![22] This serves as a warning of the serious problem that may exist in a classical test if a small sample[23] is used to accept a null hypothesis.[24]

Accordingly, it is wise, if possible, to stop short of explicitly accepting H_0; for this reason, we prefer the more reserved phrase "H_0 is acceptable," or even the phrase, "H_0 is not rejected." This means that the type II error

---

[22]This is an extreme example of how badly a classical hypothesis test may go astray. But many classical tests do not deserve such harsh criticism. In fact, in Chapter 13, we discuss cases in which accepting H_0 in a classical hypothesis test is exactly the right decision.

[23]In Figure 9-5, it is the small sample that led to the decision to accept H_0 in panel (a). By contrast, in panel (c), a larger sample (with $\overline{X}$ still 1,245) would allow us to reject H_0.

Yet at the other extreme, a huge sample size may lead us into another kind of error (the one encountered in Problem 9-14(d)). This is the error of rejecting an H_0 which, although essentially true, is not exactly true. This difficulty arises because a huge sample may reduce the standard error to the point where even a miniscule observed difference becomes statistically discernible, i.e., H_0 is rejected, although it is practically true.

[24]Although statistical theory provides a rationale for rejecting H_0, it provides no formal rationale for accepting H_0. The null hypothesis may sometimes be uninteresting, and one that we neither believe nor wish to establish; it is selected because of its simplicity. In such cases, it is the alternative H_1 that we are trying to establish, and we prove H_1 by rejecting H_0. We can see now why statistics sometimes is called "the science of disproof." H_0 cannot be proved, and H_1 is proved by disproving (rejecting) H_0. It follows that if we wish to prove some proposition, we often will call it H_1 and set up the contrary hypothesis H_0 as the "straw man" we hope to destroy. And of course if H_0 is only such a straw man, then it becomes absurd to accept it in the face of a small sample result that really supports H_1.

in its worst form may be avoided, but it also means we may be leaving the scene of the evidence with little in hand. It is for this reason that we prefer a confidence interval or prob-value, since both these provide a very informative summary of the sample.

(c) Why is Classical Testing Ever Used?

In the light of all these accumulated reservations about a classical accept-or-reject hypothesis test, why do we even bother to discuss it? There are three reasons:

1. It is helpful in guiding the student through the classical statistical literature.

2. It is helpful in clarifying certain theoretical issues, like type I and type II error.

3. A classical hypothesis test may be preferred to the calculation of prob-value if the test level α can be determined rationally, and if many samples are to be classified.

To illustrate this last issue, consider again the familiar Example 9-3; but now suppose that we are considering 5 new production processes, rather than just 1. If a sample of 100 tubes is taken in each case, suppose that the results are as follows:

TABLE **9-2**

New Process	$\overline{X}$	Is Process Really Different from Old Process (Where $\mu = 1,200$)?
1	1,265	?
2	1,240	?
3	1,280	?
4	1,150	?
5	1,210	?

We now have two options. We can calculate five prob-values for these five processes, just as we calculated it for the first process in Example 9-3. But this will involve a lot more work than a classical testing approach (which would require only that we specify α (at say 5%) and then calculate one single figure, the cut-off point $\overline{X}_c = 1,249$ derived in (9-18)). Then all five of the sample values can be evaluated immediately without any further calculation. (Note that H_0 is rejected only for processes 1 and 3; these are the only two that may be judged superior to the old method.)

But if a classical hypothesis test is to be used in this way, the level α should not be set arbitrarily. Rather, α should be determined rationally (by a technique called *Bayesian* hypothesis testing) on the basis of two considerations:

1. *Prior belief.* To again use our example, how much confidence do we have in the engineering department that assured us that these new processes are better? Is their vote divided? Have they ever been wrong before?

2. *Losses involved in making a wrong decision.* What are the costs of needlessly retooling for a new process that actually is no better (type I error)? What are the costs of failing to detect and utilize the new process if it is indeed better (type II error)?

PROBLEMS

9-15 True or false? If false, correct it.

There are two disadvantages in arbitrarily specifying α at 5%, in a classical accept-or-reject test:

(a) If the sample is very small, we may find H_0 acceptable even when H_0 is quite false.

(b) If the sample is very large, we may reject H_0, even when H_0 is approximately correct and hence a good working hypothesis.

9-16 In the legal analogy that we posed in Section 9-4(a), what would be the effect on α and β of reintroducing capital punishment on a large scale? (If the issue is not clear for murder trials, consider rape or robbery trials.)

9-17 In Problem 9-7, suppose that the alternative hypothesis is that the shipment is 30% defective. What is β?

9-18 (a) In Problem 4-34, we counted the number of persons S (out of a total of eight persons) who improved after a treatment. The null hypothesis is that the treatment is useless, i.e., the probability that an individual improves is $\pi = .50$. If the decision rule is to reject H_0 if $S \geq 6$, calculate α.

(b) If the alternate hypothesis is that $\pi = .80$, calculate β.

9-19 Consider a very simple example, in order to keep the philosophical issues clear. Suppose that you are gambling with a die, and lose whenever the die shows one (ace). After 100 throws, you notice that you have suffered a few too many losses—20 aces. This makes you suspect that your opponent is using a loaded die; specifically, you begin to wonder whether this is one of the crooked dice recently advertised as giving aces one-quarter of the time.

(a) Find the critical proportion of aces beyond which you would reject H_0 at the 5% level.

(b) Illustrate this test with a diagram similar to Figure 9-4. From this diagram, roughly estimate α and β.

(c) With your observation of 20 aces, what is your decision? Suppose that you are playing against a strange character you have just met on a Mississippi steamboat. A friend passes you a note indicating that this stranger cheated him at poker last night; and you are playing for a great deal of money. Are you happy with your decision?

(d) If you double α, use the diagram in (b) to roughly estimate what happens to β.

*9-5   THE β FUNCTION AND POWER FUNCTION

Now that we have defined and discussed the type-II error probability β at some length, it is time to actually calculate it.

Example 9-6.   Consider the test shown in Figure 9-4, where the cutoff value $\overline{X}_c$ is 1,249. The standard error of $\overline{X}$ is $\sigma/\sqrt{n}$ $= 300/\sqrt{100} = 30$. The alternate hypothesis is $\mu_1 = 1,240$.

Calculate $\beta =$ the probability that when H_1 is true, we will make the error of finding H_0 acceptable.

Solution.   According to Figure 9-4b (where H_1 is true), the error of accepting H_0 occurs when $\overline{X}$ falls below the cutoff value $\overline{X}_c = 1,249$. The probability of this is simply:

$$\Pr(\overline{X} < 1,249) = \Pr\left(\frac{\overline{X} - \mu_1}{\sigma/\sqrt{n}} < \frac{1,249 - 1,240}{300/\sqrt{100}}\right)$$

$$= \Pr(z < .30) = .62 = \beta$$

In many situations, it is unrealistic to pin down the alternative hypothesis to one specific value. For example, a more realistic version of Example 9-6 would include several possible alternatives, as follows.

Example 9-7.   Continuing Example 9-6, calculate the type II error probability β if we change the alternative hypothesis to:

(a) $\mu_1 = 1,280$.
(b) $\mu_1 = 1,320$.

Solution.   We use exactly the same method as in Example 9-6:

(a) $\beta = \Pr(\overline{X} < 1{,}249)$

$$= \Pr\left(\frac{\overline{X} - \mu_1}{\sigma/\sqrt{n}} < \frac{1{,}249 - 1{,}280}{300/\sqrt{100}}\right)$$

$$= \Pr(z < -1.03) = .15$$

(b) $\beta = \Pr(\overline{X} < 1{,}249)$

$$= \Pr\left(\frac{\overline{X} - \mu_1}{\sigma/\sqrt{n}} < \frac{1{,}249 - 1{,}320}{300/\sqrt{100}}\right)$$

$$= \Pr(z < -2.37) = .009$$

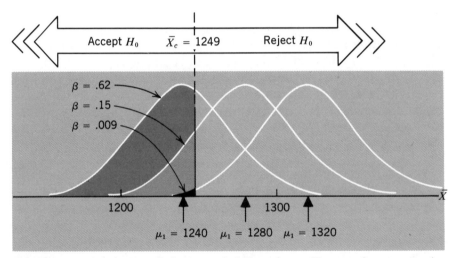

FIGURE 9–6 Calculation of β, the probability of type-II error, for a composite hypothesis

The values of β in Example 9-6 and 9-7 (corresponding to $\mu_1 = 1{,}240$, 1,280, and 1,320) are shown in Figure 9-6 as the three shaded areas (to the left of the cutoff point, where we accept H_0). Of course, an even more realistic alternative hypothesis would include *any* number larger than $\mu_0 = 1{,}200$:

$$H_1 : \mu_1 > 1{,}200 \qquad\qquad (9\text{-}10)\text{ repeated}$$

In other words:

$$H_1 : \mu_1 = 1{,}201$$

or

$$\mu_1 = 1{,}202$$

or

$$\mu_1 = 1{,}203$$

$$\vdots$$

An hypothesis such as this, which is composed of many possibilities, is called a *composite* hypothesis. For each possible μ_1, we can calculate the corresponding β—as we already have for $\mu_1 = 1{,}240, 1{,}280,$ and $1{,}320$. These results are set out in Table 9-3.

In the last column of Table 9-3, we give the power $(1 - \beta)$. This is the probability of correctly rejecting the false null hypothesis, i.e., of detecting the true alternate hypothesis. As we approach the null hypothesis of 1,200, the power approaches 5%, the value of α. On the other hand, as we proceed up the table and μ_1 becomes very large, the power to detect it approaches 100%. In other words, the further μ_1 gets from μ_0, the easier it becomes to discriminate between the two.

TABLE **9-3**

β Function and Power Function for a Test at Level $\alpha = 5\%$

(1) Possible Values of μ_1	(2) Probability of Erroneously accepting H_0 β	(3) Probability of Correctly rejecting H_0 Power $= 1 - \beta$
$\vdots$	$\vdots$	$\vdots$
1,320	0.9%	99.1%
$\vdots$	$\vdots$	$\vdots$
1,280	15%	85%
$\vdots$	$\vdots$	$\vdots$
1,240	62%	38%
$\vdots$	$\vdots$	$\vdots$
1,202	94.2%	5.8%
1,201	94.6%	5.4%
limit (1,200)[a]	(95.0%)	(5.0%)

[a]The limiting figures in the last row are bracketed as a warning. In this special case where $\mu = 1{,}200, H_0$ is true, and the 95% in column (2) represents the probability of *correctly* accepting H_0.

In Figure 9-7, we graph the information in Table 9-3—the β function and the power function. Clearly, we desire a power function that begins very close to the baseline, since its initial height is α, which we wish to keep low. At the same time, the power function should climb very steeply; the more rapidly it rises, the greater our power to distinguish between competing hypotheses.

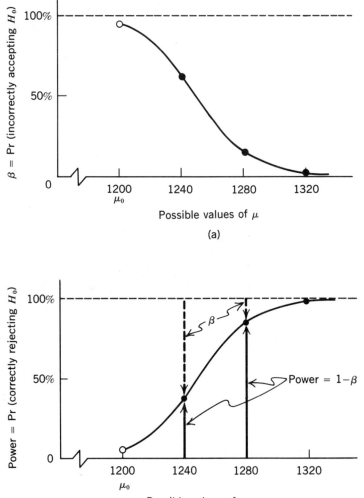

(a)

(b)

FIGURE 9–7 Graph of the functions of Table 9–3, for a test with level $\alpha = 5\%$. (a) β function (OCC.). (b) Power function.

The β function also is called the *operating characteristic curve* (OCC). Since it is the complement of the power function, our desire to have a steeply rising power function could be rephrased as a desire to have a steeply falling OCC, or β function.

Another example will further illustrate these concepts.

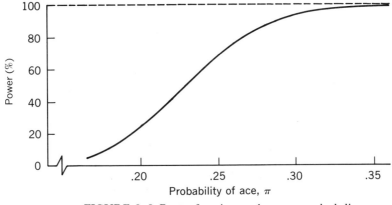

FIGURE 9–8   Power function to detect a crooked die

Example 9-8.   (Refer to Problem 9-19.) Suppose that you are gambling with a die and that you lose whenever the die shows ace. After 100 throws, you notice that you have suffered an inordinate number of losses—20 aces. Since your opponent is a stranger whom you have no particular reason to trust, you begin to wonder whether he is using a loaded die. So you set up an hypothesis test of $H_0: \pi = 1/6$ versus $H_1: \pi > 1/6$, at the 5% level. In Problem 9-19, we calculated that the region for rejecting H_0 is where the sample proportion P exceeds .228. Since the actual P was $20/100 = .20$, H_0 could not be rejected.

When the power function is calculated, it turns out to be Figure 9-8. Now answer true or false; if false, correct it.

Although you did not have enough evidence to accuse him, you can nevertheless feel satisfied with your decision procedure for three reasons:

(a) If he actually is honest, you stood only a 5% chance of falsely accusing him of cheating.

(b) If he actually is dishonest, and used a badly loaded die where $\pi = 1/4$, for example, then there is about a 95% chance that you would have uncovered him as a cheat. And if his die was loaded even more, you would have had an even better chance of uncovering him.

(c) If you continue gambling for another 100 throws, you then could run a new test on the basis of all 200 throws. This test would have an even better power function; if your opponent "passed" it, you could continue with even more confidence that he was not badly cheating you.

Solution. (a) True.

(b) Correction: . . . where $\pi = 1/4$, for example, then there is about *a 70% chance* that you would have uncovered him. . . .

(c) True.

Remarks. An unscrupulous gambler may develop a good, rough-and-ready understanding of the power function. He knows that he shouldn't use a die that *always* turns up aces. He knows that if he does this he will be found out quickly, and the game will be abandoned. The more crooked the die he uses, the greater your "power" to uncover him as a cheat. The less crooked the die, the more difficult it becomes for you to uncover him. The dishonest gambler will recognize this, and will prefer to get you to play against a slightly crooked die; then your test has little power, and it becomes almost impossible to distinguish between the two conflicting hypotheses.

PROBLEMS

*9-20 A certain type of seed has always grown to a mean height of 8.5 inches, with a standard deviation of 1 inch. A sample of 100 seeds is grown under new enriched conditions to see whether the mean height might be improved.

(a) At the 5% level, calculate the cutoff value $\overline{X}_c$ above which H_0 should be rejected.

(b) If the sample of 100 seeds actually turns out to have a mean height $\overline{X} = 8.8$ inches, do you reject H_0?

(c) Roughly graph the β curve (OCC) for this test. (*Hint:* Calculate β for a rough grid of values, then sketch the curve joining these values.)

(d) What would be the approximate chance of failing to detect a mean height improvement, if the sample of 100 seeds were to come from a population whose mean was:

  (i) 8.65 inches?
  (ii) 8.80 inches?
  (iii) 9 inches?

*9-21 By actual calculation, verify the power function in Example 9-8.

*9-22 (a) For the acceptance sampling in Problem 9-6, sketch the OCC curve and the power curve.
(b) Repeat (a), for Problem 9-7.
(c) By comparing (a) with (b), note how a large sample gives a better OCC or power curve.

9-6   TWO-SIDED TESTS

In the previous four sections, we discussed only the one-sided test, in which the alternative hypothesis and consequently the rejection region and prob-value are just on one side. The one-sided test often may be recognized by key asymmetrical phrases, such as, "more than," "less than," "better than," "worse than," "at least," etc.

However, there are occasions when it is more appropriate to use a two-sided test, which often may be recognized by key symmetrical phrases such as, "different from," "changed for better or worse," "unequal," etc. This section discusses the minor modifications required for such two-sided tests.[25]

(a) Two-Sided Prob-Value

Example 9-9. Consider again the testing of TV tubes in Example 9-3. Suppose that the null hypothesis remains as:

$$H_0: \mu = 1,200$$

But now change the alternative hypothesis by supposing that our engineers cannot advocate the new process as better but

---

[25] Since the one-sided prob-value is commonest, often it is called simply the "prob-value" (as in Section 9-2). To avoid confusion, therefore, whenever we mean a two-sided prob-value, we shall always include the phrase "two-sided."

must concede that it may be worse. Then the alternative hypothesis would be:

$$H_1 : \mu > 1{,}200 \quad or \quad \mu < 1{,}200$$

that is:

$$H_1 : \mu \neq 1{,}200 \tag{9-21}$$

In other words, we now are testing whether the new process is *different* (whereas in Example 9-3, we were testing whether it was *better*). Thus, even before we collect any data, we can agree that a value of $\overline{X}$ well below 1,200 would be just as strong evidence against H_0 as a value of $\overline{X}$ well above 1,200; that is, what counts is how far away $\overline{X}$ is, on *either side*.

If the sample mean $\overline{X} = 1{,}265$, what is the two-sided prob-value? That is, what is the probability that $\overline{X}$ will be at least as far away as 1,265 from the null hypothesis of 1,200, on either side?

Solution. To measure how far away $\overline{X}$ is from the null hypothesis, we take $|\overline{X} - \mu_0|$. The observed value of this is $|1{,}265 - 1{,}200|$. Thus we calculate:

$$Pr(|\overline{X} - \mu_0| > |1{,}265 - 1{,}200|)$$

$$= Pr\left(\left|\frac{\overline{X} - \mu_0}{\sigma/\sqrt{n}}\right| > \left|\frac{1{,}265 - 1{,}200}{300/\sqrt{100}}\right|\right) \tag{9-22}$$

$$= Pr(|z| > 2.17)$$

$$= 2\,Pr(z > 2.17) = 2(.015) = .03 \tag{9-23}$$

Remarks. This 3% is called the *two-sided prob-value* for H_0. It is just double the one-sided value of $1\frac{1}{2}\%$ given in (9-12). This is illustrated in Figure 9-9.

In general, whenever the alternative hypothesis is two-sided, it is appropriate to calculate the two-sided prob-value for H_0:

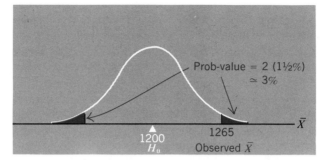

FIGURE 9–9 Prob-value $\overset{\Delta}{=} \Pr(\overline{X}$ would be as extreme as the value actually observed, if H_0 were true). Compare with Figure 9–1.

$$\text{Two-sided prob-value} \overset{\Delta}{=} \Pr\begin{pmatrix} \text{the sample value would be as} \\ \text{extreme as the value actually} \\ \text{observed, assuming } H_0 \end{pmatrix} \quad (9\text{-}24)$$

Whenever the distribution is symmetric, the two-sided prob-value is just double the one-sided prob-value.[26]

(b) Classical Tests and Confidence Intervals

It is easy to modify a one-sided classical test to a two-sided classical test; we merely reject H_0 if the two-sided prob-value falls below the specified level α. For instance, in Example 9-9 where the prob-value was only 3%, H_0 could be rejected at the 5% level.

Of course, if the two-sided prob-value has not been calculated and we wish to run a two-sided classical test over and over, we would calculate a two-sided rejection region that cuts off a total of $\alpha = 5\%$ (i.e., $2\frac{1}{2}\%$ in each tail around μ_0). Figure 9-10(a) illustrates this for the TV-tube data of Example 9-9. We note that an observed $\overline{X} = 1,265$ would mean H_0 is rejected (which agrees with what we concluded earlier when we used prob-value).

Whenever a two-sided test is appropriate, an ordinary two-sided confidence interval also is appropriate, as shown in Figure 9-10(b). Since the null hypothesis $\mu_0 = 1,200$ lies outside this confidence interval, it can be rejected. It is clear why the confidence interval and the test are exactly

---

[26]Since the two-sided prob-value is very difficult to define for asymmetric distributions, we do not deal with it here.

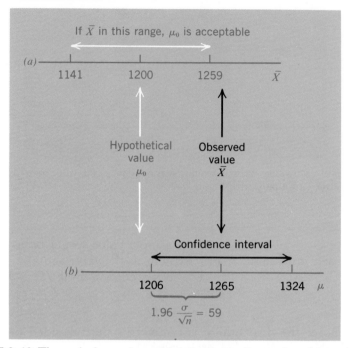

FIGURE 9–10 The equivalence of two-sided classical testing and confidence intervals. (a) The classical test at level $\alpha = 5\%$. Since $\overline{X}$ falls beyond the cutoff points, we can reject H_0. (b) The 95%-confidence interval. Since μ_0 falls outside the confidence limits, we can reject H_0.

equivalent: in both cases, we simply check whether the magnitude of the difference $|\overline{X} - \mu|$ exceeds 1.96 standard errors. The only difference is that the classical test uses the null hypothesis μ_0 as its reference point, whereas the confidence interval uses the observed $\overline{X}$ as its reference point.[27] Thus, (9-5) finally is confirmed.[28]

---

*[27] Here is the algebraic detail: to construct both a classical test and a confidence interval, we begin with standardizing $\overline{X}$:

$$z = \frac{\overline{X} - \mu_0}{\sigma/\sqrt{n}}$$

95% of the time, z will be between ± 1.96, and in these cases H_0 is judged acceptable (thus keeping our error rate to 5%). That is, H_0 is acceptable iff:

$$\left| \frac{\overline{X} - \mu_0}{\sigma/\sqrt{n}} \right| < 1.96 \qquad (9\text{-}25)$$

(cont'd)

PROBLEMS

9-23 Three different sources claimed that the mean G.P. physician's salary μ in 1970 was \$30,000, \$35,000, and \$37,500, respectively. A random sample of 25 G.P.s gave a mean[29] of \$38,000 and a standard deviation of \$8,000.

(a) At the 5% level, run a two-sided classical test on each of the three hypotheses. (That is, for each hypothesis, calculate the region for $\overline{X}$ that calls for the rejection of H_0, and then make the decision on the basis of $\overline{X} = \$38,000$. Or, alternatively, calculate the two-sided prob-value, and then decide whether or not to reject H_0.)

(b) Construct a 95% two-sided confidence interval for μ. Then test each of the hypotheses simply by noting whether it is included in the confidence interval. Is this easier than (a)?

(c) Repeat (a) for one-sided tests against the alternative hypothesis that $\mu = \$40,000$.

(d) Repeat (b) for a one-sided confidence interval of the form "μ is at least such and such."

9-24 True or false? If false, correct it.

(a) A 95% confidence interval may be regarded as the set of hypotheses acceptable at level $\alpha = 5\%$. (If the alternative hypothesis is two-sided, then the classical test and confidence interval should be two-sided.)

(b) To test the hypothesis:

$$H_0 : \pi = \tfrac{1}{2} \quad \text{against} \quad H_1 : \pi = \tfrac{1}{4}$$

we should use a two-sided test, rejecting H_0 when π turns out to be large.

(c) Comparing hypothesis testing to a jury trial, we may say that the type I error is like the error of condemning a guilty man.

(d) Suppose, in a certain test, that the prob-value turns out to be .013. Then H_0 would be acceptable at the 5% level and also at the 1% level.

For a classical test, we continue with (9-25) by solving for $\overline{X}$:

$$\mu_0 - 1.96\sigma/\sqrt{n} < \overline{X} < \mu_0 + 1.96\sigma/\sqrt{n}$$

which is the interval in Figure 9-10(a).

For a confidence interval, we continue with (9-25) by solving for μ_0:

$$\overline{X} - 1.96\sigma/\sqrt{n} < \mu_0 < \overline{X} + 1.96\sigma/\sqrt{n}$$

which is the interval in Figure 9-10(b). The point is, in both the classical test and the confidence interval, H_0 is acceptable iff (9-25) is true. Thus their perfect equivalence is established.

[28] Of course, we could confirm the one-sided analogy to (9-5) similarly: the one-sided confidence interval is the set of hypotheses that are acceptable when subjected to a one-sided test.

[29] From the *1974 American Almanac,* Table 103, with some modifications.

(e) To make a confidence interval 25 times as narrow (precise), it is necessary to make the sample 25 times as large.

9-25 For the data in Problem 8-15(a) comparing men's and women's salaries:
(a) Calculate the one-sided prob-value. Is the difference statistically discernible at the 5% level?
(b) Calculate the two-sided prob-value. Is the difference statistically discernible at the 5% level?
(c) Which of (a) or (b) is consistent with the confidence interval in Problem 8-15(a)?
(d) Which do you think is more appropriate, (a) or (b)?

9-26 Suppose that a scientist concludes that a difference in sample means is "statistically significant (discernible) at the 1% level." Answer true or false; if false, correct it.
(a) There is at least a 99% chance that there is a real difference in the population means.
(b) The prob-value for H_0 (population means are exactly equal) is 1% or less.
(c) If there were no difference in the population means, the chance of getting such a difference (or more) in the sample means is 1% or less.
(d) The scientist's conclusion is sound evidence that a difference in population means exists. Yet in itself it gives no evidence whatever that this difference is large enough to be of practical importance. This illustrates that "statistical significance" and "practical significance" are two different concepts.
(e) In the 99% confidence interval for $\mu_1 - \mu_2$, the estimate $(\overline{X}_1 - \overline{X}_2)$ would be overwhelmed by the sampling allowance.

REVIEW PROBLEMS

9-27 Let us rework one of the elementary problems from Chapter 4.
(a) Suppose that the birth of boys and girls is equally likely. In a family of five children, what is the chance that there will be a 3–2 split (three boys and two girls, or vice-versa)?
(b) Recent statistics (1969) in the U.S. showed 1,829,000 male births, and 1,742,000 female births. How well does this evidence support the "equally likely births" hypothesis of part (a)?
(c) In view of (b), do you want to change your answer to (a)?
(d) What other assumptions besides the "equally likely births" did you make in the model of part (a)?

9-28 Four different claims were made about the mean annual salaries of pediatricians and of general practitioners (G.P.s) in 1970:

Claim 1: G.P.s and pediatricians earn the same, on average.

Claim 2: G.P.s earn $500 more than pediatricians, on average.

Claim 3: G.P.s earn $1,000 more than pediatricians, on average.

Claim 4: G.P.s earn $2,000 more than pediatricians, on average.

To settle the issue, both populations were sampled randomly, yielding the following data:[30]

	G.P.s	Pediatricians
n	250	200
$\overline{X}$	$38,000	$36,000
s	$ 8,000	$10,000

(a) Without any heavy calculations, can you immediately verify any of these claims?

(b) Calculate the 95% confidence interval (two-sided) for $\mu_G - \mu_P$. Is the difference discernible at the 5% level?

(c) If a classical test (two-sided) were made on each of the four hypotheses, which could be rejected at the 5% level?

(d) Several hypotheses in (c) were found acceptable (could not be rejected). If you had to actually accept just one hypothesis to publish as a conclusion to this study, which would it be?

9-29 Continuing Problem 9-28:

(a) What would be the advantages and disadvantages of a one-sided confidence interval?

(b) Calculate the one-sided 95% confidence interval.

(c) If a classical one-sided test were made, which hypotheses could be rejected at the 5% level?

*(d) State formally the null and alternative hypotheses in (c), for each of the four claims being tested.

9-30 In a Gallup poll of 1,500 Americans in 1975,[31] 45% answered "yes" to the question, "Is there any area right around here—that is, within a mile—

---

[30] From the *1974 American Almanac*, Table 103, with some modifications.

[31] From the *Gallup Opinion Index*, October 1975, p. 16.

where you would be afraid to walk alone at night?" In an earlier poll of 1,500 Americans in 1972, only 42% had answered "yes."

(a) State a reasonable null hypothesis and alternative hypothesis, in words and symbols.

(b) Calculate the prob-value for H_0.

(c) Calculate a 95% confidence interval for the population change. (Make it one-sided or two-sided, to be consistent with parts (a) and (b).)

(d) If a classical test were made, could the null hypothesis be rejected at the 5% level? Make sure that your answer is consistent with both (b) and (c).

9-31 Continuing Problem 9-30, answer true or false; if false correct it.

(a) The change in opinion is statistically indiscernible.

(b) If we imagine that many other polls were taken of the same population, in about 5% of the cases we would find that opinion changed by three percentage points or more.

(c) The probability that the population opinion is unchanged is about 5%.

CHAPTER 10

Analysis of Variance

Nothing is good or bad but by comparison.

Thomas Fuller

10-1 ONE-FACTOR ANALYSIS OF VARIANCE

In Chapter 8, we made inferences about one population mean, and then compared two means. Now we shall compare several means, using techniques called analysis of variance.

(a) Testing for Differences

As an illustration, suppose that three machines are to be compared. Because these machines are operated by men, and because of other inexplicable reasons, output per hour is subject to chance fluctuation. In the hope of "averaging out" and thus reducing the effect of chance fluctuation, a random sample of five different hours is obtained from each machine and set out in Table 10-1, where each sample mean is calculated.

The first question is, "Are the machines really different?" That is, are the sample means $\overline{X}_i$ in Table 10-1 different because of differences in the underlying population means μ_i (where μ_i represents the lifetime performance of machine i)? Or may these differences in $\overline{X}_i$ be reasonably attributed to chance fluctuations alone? To illustrate, suppose that we collect three samples from just *one* machine, as shown in Table 10-2. As

TABLE **10-1**

Sample Outputs of Three Machines

Machine, or Sample Number	Sample from Machine i					$\overline{X}_i$
$i = 1$	47	53	49	50	46	49
$= 2$	55	54	58	61	52	56
$= 3$	54	50	51	51	49	51

Average $\overline{X} = \overline{\overline{X}} = 52$

expected, sampling fluctuations cause small differences in $\overline{X}_i$ even though the μ_i in this case are identical. So the question may be rephrased, "Are the differences in $\overline{X}_i$ of Table 10-1 of the same order as those of Table 10-2 (and thus attributable to chance fluctuation), or are they large enough to indicate a difference in the underlying μ_i?" The latter explanation seems more plausible; but how do we develop a formal test?

TABLE **10-2**

Three Samples of the Output of the
Same Machine

Sample Number	Sample Values					$\overline{X}_i$
$i = 1$	49	55	51	52	48	51
$= 2$	52	51	55	58	49	53
$= 3$	55	51	52	52	50	52

$\overline{\overline{X}} = 52$

As usual, the hypothesis of "no difference" in the population means is called the null hypothesis:

$$H_0 : \mu_1 = \mu_2 = \mu_3 \tag{10-1}$$

A test of this hypothesis first requires a numerical measure of the degree to which the sample means differ. We therefore take the three sample means in the last column of Table 10-1 and calculate their variance. Using (2-6(a))—and carefully noting that we are calculating the variance of the sample means and *not* the variance of all values in the table—we have:

$$s_{\overline{X}}^2 = \frac{1}{(r-1)} \sum_{i=1}^{r} (\overline{X}_i - \overline{\overline{X}})^2 \tag{10-2}$$

$$= \frac{1}{2}[(49 - 52)^2 + (56 - 52)^2 + (51 - 52)^2]$$

$$= 13.0 \tag{10-3}$$

where r = number of rows (the number of sample means), and:

$$\overline{\overline{X}} = \text{average of } \overline{X}_i = \frac{1}{r} \sum_{i=1}^{r} \overline{X}_i = 52 \tag{10-4}$$

Yet $s_{\overline{X}}^2$ does not tell the whole story. For example, consider the data of Table 10-3, which has the same $s_{\overline{X}}^2$ as Table 10-1, yet more erratic machines that produce large chance fluctuations within each row. The implications of this are shown in Figure 10-1. In panel (b), the machines are so erratic that all samples could be drawn from the same population—i.e., the differences in sample means may be explained by chance. On the other hand, the (same) differences in sample means hardly can be explained by chance in panel (a), because the machines in this case are *not* so erratic.

TABLE **10-3**
Sample Outputs of Three Machines

Machine	Sample Output from Machine i					$\overline{X}_i$
$i = 1$	57	42	53	38	55	49
$= 2$	46	59	64	61	50	56
$= 3$	57	59	48	46	45	51

$$\overline{\overline{X}} = 52$$

We now have our standard of comparison. In panel (a) we conclude that the μ_i are different—reject H_0—because the variance in sample means $(s_{\overline{X}}^2)$ is large *relative* to the chance fluctuation.

How can we measure this chance fluctuation? Intuitively, it seems to be the spread (or variance) of observed values *within* each sample. Thus we compute the variance within the first sample in Table 10-1:

$$s_1^2 = \frac{1}{(n-1)} \sum_{j=1}^{n} (X_{1j} - \overline{X}_1)^2$$

$$= \frac{1}{4}[(47 - 49)^2 + (53 - 49)^2 + \cdots] = 7.5 \tag{10-5}$$

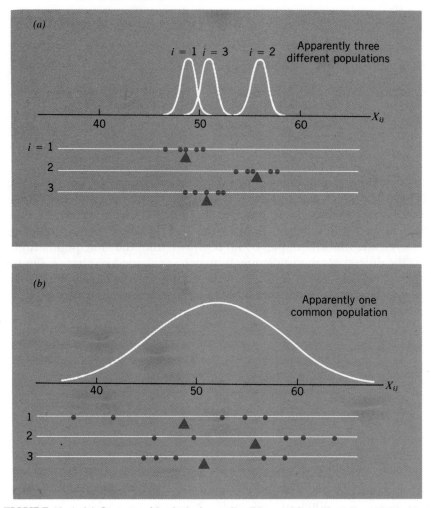

FIGURE 10–1 (a) Outputs of 3 relatively predictable machines (data from Table 10–1 —modified slightly to clarify the message). (b) Outputs of 3 erratic machines (data from Table 10–3). They have the same sample means $\overline{X}_i$, hence the same $s_{\overline{X}}^2$, as the machines in (a).

where X_{1j} is the jth observed value in the first sample.

Similarly, we compute the variance or chance fluctuation within the second (s_2^2) and third samples (s_3^2). The average of these is a measure of the total chance fluctuation, and is called the *pooled variance:*[1]

---

[1]This pooled variance is just an extension of the pooled variance in the two-sample case in (8-34).

$$s_p^2 = \frac{1}{r} \sum_{i=1}^{r} s_i^2 = \frac{1}{3}(7.5 + 12.5 + 3.5) = 7.83 \qquad (10\text{-}6)$$

From each of the r samples, we have a sample variance with $(n - 1)$ degrees of freedom, so that the pooled variance s_p^2 has $r(n - 1)$ degrees of freedom.

The key question now can be stated. Is $s_{\overline{X}}^2$ large relative to s_p^2? That is, what is the ratio $s_{\overline{X}}^2/s_p^2$? It is customary to examine a slightly modified ratio:[2]

$$F = \frac{ns_{\overline{X}}^2}{s_p^2} \qquad (10\text{-}7)$$

where n has been introduced into the numerator simply to ensure that, whenever H_0 is true, this F ratio will have a value near 1. However, because of statistical fluctuation, F sometimes will be above 1, and sometimes below.

If H_0 is not true (and the μ's are not the same), then $ns_{\overline{X}}^2$ will be relatively large compared with s_p^2, and the F ratio in (10-7) will tend to be much greater than 1. Thus, we will be able to judge whether H_0 is true or not, depending on whether F is near 1 or much greater than 1.[3]

---

[2]The F ratio is named in honor of Sir Ronald A. Fisher (1890–1962), the most influential statistician of his time.

[3]To interpret the F ratio in (10-7) further, suppose that our samples are drawn from three normal populations with the same variance. If, in addition, H_0 is true, and the three population means are the same, then the division of our data into three samples is rather artificial—all observations could be viewed as one large sample drawn from a single population. Now consider two alternative ways of estimating σ^2, the variance of that population:

1. Average the variance within each of the three samples. This is the s_p^2 in the denominator of (10-7).

2. Or, infer σ^2 from $s_{\overline{X}}^2$, the observed variance of sample means. Recall how the variance of sample means is related to the variance of the population:

$$\sigma_{\overline{X}}^2 = \frac{\sigma^2}{n} \qquad (6\text{-}7) \text{ repeated}$$

Thus:

$$\sigma^2 = n\sigma_{\overline{X}}^2$$

This suggests estimating σ^2 with $ns_{\overline{X}}^2$, which is recognized as the numerator of (10-7).

If H_0 is true, we could estimate σ^2 by either of these methods; since the two will be about equal, their ratio F will fluctuate around 1.

But if H_0 is not true, then the numerator of (10-7) will blow up because the difference in population means will result in a large spread in the sample means (large $s_{\overline{X}}^2$). At the same time, the denominator will still reflect only chance fluctuation. Consequently, the F ratio will be large.

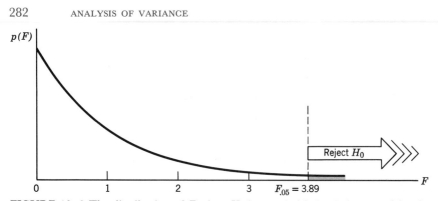

FIGURE 10–2 The distribution of F when H_0 is true (with 2, 12 degrees of freedom)

More formally, to test H_0 we must know the distribution of the test statistic F. When H_0 is true, the exact distribution is shown in Figure 10-2, along with the critical value of $F_{.05}$ that cuts off 5% of the upper tail of the distribution. To test at the 5% level, we reject H_0 if F exceeds this critical value 3.89. Thus, if H_0 is true, there is only a 5% probability that we would observe an F value exceeding 3.89 and consequently would reject H_0 erroneously.

To illustrate this procedure, let us reconsider Tables 10-1, 10-2, and 10-3. In each table, let us ask whether the machines exhibit differences that are statistically discernible (significant). In other words, in each table, we test $H_0 : \mu_1 = \mu_2 = \mu_3$ against the alternative that they are not equal. For the data in Table 10-1, an evaluation of (10-7) yields:

$$F = \frac{ns_{\bar{X}}^2}{s_p^2} = \frac{5(13.0)}{7.83} = 8.3$$

Since this exceeds $F_{.05}$, H_0 is rejected. In this case, the difference in sample means is very large relative to the chance fluctuation.

However, for the data in Table 10-2:

$$F = \frac{5(1.0)}{7.83} = .64$$

Since this is below the critical $F_{.05}$ value of 3.89, H_0 cannot be rejected. In this case, the differences in sample means can be explained reasonably by chance fluctuations. This is no surprise; we generated these three samples in Table 10-2 from the same machine. Similarly, for the data in Table 10-3:

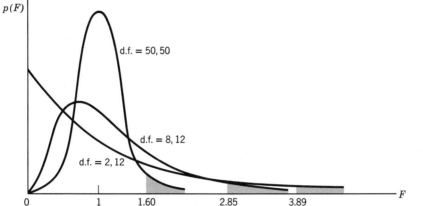

FIGURE 10–3 The F distribution, with various degrees of freedom in numerator and denominator. Note how the 5% critical point (beyond which we reject H_0) moves toward 1 as degrees of freedom increase.

$$F = \frac{5(13.0)}{57.5} = 1.13$$

Since this is also below $F_{.05}$, H_0 again cannot be rejected. In this case, the difference in sample means is swamped by the chance fluctuation in the denominator.

These three formal tests confirm our earlier intuitive conclusions. Table 10-1 provides the only case in which we conclude (at the 5% level) that the underlying populations have different means.

(b) The F Distribution

The F distribution shown in Figure 10-2 is only one of many; there is a different distribution depending on degrees of freedom $(r - 1)$ in the numerator, and degrees of freedom $r(n - 1)$ in the denominator. Critical points for the F distributions are tabulated in Table VII of the Appendix. From this table, we can confirm the critical points in Figure 10-3 as well as in Figure 10-2.

(c) The ANOVA Table

This section merely presents a convenient and customary way to lay out the calculations already described. But first, we summarize the model in Table 10-4. We confirm in the second column that all samples are assumed to be drawn from normal[4] populations with the same variance

---

[4] Just as for the t distribution in Section 8-2, our results usually remain approximately true even if the populations are nonnormal.

σ^2—but, of course, with means that may differ. (Indeed, it is the possible difference in means that is being tested).

The ensuing calculations are laid out conveniently in Table 10-5, called the ANOVA table—an obvious shorthand for ANalysis Of VAriance. This is mostly a bookkeeping arrangement, with the first row showing calculations of the numerator of the F ratio, and the second row the denominator. In part (b) of this table, we evaluate the specific example of the three machines in Table 10-1.

TABLE **10-4**
Summary of Assumptions

Population	Assumed Distribution	Observed Sample Values
1	$N(\mu_1, \sigma^2)$	X_{1j} $(j = 1 \ldots n)$
2	$N(\mu_2, \sigma^2)$	X_{2j} $(j = 1 \ldots n)$
$\vdots$	$\vdots$	$\vdots$
i	$N(\mu_i, \sigma^2)$	X_{ij} $(j = 1 \ldots n)$
$\vdots$	$\vdots$	$\vdots$
r	$N(\mu_r, \sigma^2)$	X_{rj} $(j = 1 \ldots n)$

null hypothesis $H_0 : \mu_1 = \mu_2 = \cdots = \mu_i = \cdots \mu_r$
alternate hypothesis $H_1 : \mu_i \neq \mu_I$ for some pair, $i \neq I$

In addition, this table provides two handy checks on our calculations. One check is on degrees of freedom in the third column. The other check is on sums of squares in the second column; the sum of squares *between* rows plus the sum of squares *within* rows should add up to the total sum of squares.[5] Each sum of squares is also called *variation;* when divided by the appropriate degrees of freedom, it yields *variance* in the fourth column.

---

[5]**Proof:** The deviation of any observed value X_{ij} from the mean of all observed values $\overline{\overline{X}}$ can be broken down into two parts:

$$(X_{ij} - \overline{\overline{X}}) = (\overline{X}_i - \overline{\overline{X}}) + (X_{ij} - \overline{X}_i) \tag{10-8}$$

$$\underset{\text{deviation}}{\text{total}} = \underset{\text{deviation}}{\text{explained}} + \underset{\text{deviation}}{\text{unexplained}}$$

Thus, using Table 10-1 as an example, the third observation in the second sample (58) is greater than $\overline{\overline{X}} = 52$. This total deviation can be broken down:

$$(58 - 52) = (56 - 52) + (58 - 56)$$
$$6 = 4 + 2 \tag{cont'd}$$

The variance *between* rows is "explained" by the fact that the rows may come from different parent populations (e.g., machines that perform differently). The variance *within* rows is "unexplained" because it is the random or chance variation that cannot be systematically explained (by differences in machines). Thus F sometimes is referred to as the variance ratio:

$$F = \frac{\text{explained variance}}{\text{unexplained variance}} \tag{10-11}$$

This suggests a possible means of strengthening the F test. Suppose, for example, that these three machines are sensitive to differences in the men operating them. Why not introduce the operator explicitly into the analysis? If some of the previously unexplained variation could thus be explained by differences in operator, the denominator of (10-7) would be reduced. With the larger F value that results, we would have a more

---

Thus, most (4) of this total deviation is explained by the machine, while less (2) is unexplained, due to random fluctuations. Clearly, (10-8) must always be true, since the two occurrences of $\overline{X}_i$ cancel.

Square both sides of (10-8) and sum over all i and j:

$$\sum_i \sum_j (X_{ij} - \overline{\overline{X}})^2 = \sum_i \sum_j (\overline{X}_i - \overline{\overline{X}})^2 + 2 \sum_i \sum_j (\overline{X}_i - \overline{\overline{X}})(X_{ij} - \overline{X}_i) + \sum_i \sum_j (X_{ij} - \overline{X}_i)^2 \tag{10-9}$$

On the right side, the middle (cross product) term may be written as:

$$2 \sum_{i=1}^{r} \left[(\overline{X}_i - \overline{\overline{X}}) \underbrace{\sum_{j=1}^{n} (X_{ij} - \overline{X}_i)}_{} \right] = 0$$

the algebraic sum of deviations about the mean is always zero.

Furthermore, the first term on the right side of (10-9) is:

$$\sum_{j=1}^{n} \underbrace{\left[\sum_{i=1}^{r} (\overline{X}_i - \overline{\overline{X}})^2 \right]}_{\text{independent of } j} = n \sum_{i=1}^{r} (\overline{X}_i - \overline{\overline{X}})^2$$

Substituting these two conclusions back into (10-9), we have, finally:

$$\sum_i \sum_j (X_{ij} - \overline{\overline{X}})^2 = n \sum_i (\overline{X}_i - \overline{\overline{X}})^2 + \sum_i \sum_j (X_{ij} - \overline{X}_i)^2 \tag{10-10}$$

$$\begin{array}{ccccc} \text{total} & = & \text{explained} & + & \text{unexplained} \\ \text{variation} & & \text{variation} & & \text{variation} \end{array}$$

TABLE 10-5

(a) ANOVA Table, General

Source of Variation	Variation; Sum of Squares (SS)	d.f.	Variance; Mean Sum of Squares (MSS)	F ratio
Between rows; "EXPLAINED" by differences in $\bar{X}_i$	$n\displaystyle\sum_{i=1}^{r}(\bar{X}_i - \bar{\bar{X}})^2 = SS_r$	$(r-1)$	$MSS_r = SS_r/(r-1)$ $= ns_{\bar{X}}^2$	$\dfrac{\text{explained variance}}{\text{unexplained variance}}$
Within rows; residual variation, resulting from chance fluctuation, "UNEXPLAINED"	$\displaystyle\sum_{i=1}^{r}\sum_{j=1}^{n}(X_{ij} - \bar{X}_i)^2 = SS_u$	$r(n-1)$	$MSS_u = SS_u/r(n-1)$ $= s_p^2$	
Total	$\displaystyle\sum_i\sum_j(X_{ij} - \bar{\bar{X}})^2$	$(rn-1)$		

(b) ANOVA Table, for Observations Given in Table 10-1

Source of Variation	Variation	d.f.	Variance	F ratio	prob-value
Between machines; "EXPLAINED"	130	2	65	$\dfrac{65}{7.83} = 8.3$	$.001 < p < .01$
Within machines; "UNEXPLAINED"	94	12	7.83		
Total	224✓	14✓			

TABLE 10-6
Modification of ANOVA Table 10-5 for Unequal Sample Sizes, $n_1, n_2, \ldots n_i, \ldots$

Source of Variation	Variation; Sum of Squares (SS)	d.f.	Variance; Mean Sum of Squares (MSS)	F ratio
Between rows; "EXPLAINED" by differences in $\bar{X}_i$	$\displaystyle\sum_{i=1}^{r} n_i(\bar{X}_i - \bar{\bar{X}})^2 = SS_r$	$(r-1)$	$MSS_r = SS_r/(r-1)$	$\dfrac{\text{explained variance}}{\text{unexplained variance}} = F$
Within rows; residual variation, resulting from chance fluctuation, "UNEXPLAINED"	$\displaystyle\sum_{i=1}^{r}\sum_{j=1}^{n_i}(X_{ij} - \bar{X}_i)^2 = SS_u$	$\displaystyle\sum_{i=1}^{r}(n_i - 1)$	$MSS_u = SS_u\big/\displaystyle\sum(n_i - 1) \\ = s_p^2$	
Total	$\displaystyle\sum^{r}\sum^{n_i}(X_{ij} - \bar{\bar{X}})^2$	$\displaystyle\sum_{i=1}^{r} n_i - 1$		

where $\bar{\bar{X}}$ = the grand average of all the X_{ij}:

$$\bar{\bar{X}} = \frac{\sum\sum X_{ij}}{\sum n_i} = \frac{\sum n_i \bar{X}_i}{\sum n_i}$$

287

powerful test of the machines (i.e., we would be in a stronger position to reject H_0). Thus, our ability to detect whether one factor (machine) is important would be strengthened by introducing another factor (operator) to help explain variance. In fact, we shall carry out precisely this calculation in the next section, two-factor ANOVA.

Although H_0 is sometimes tested at the arbitrary level of 5%, we already pointed out in Chapter 9 that it may be more meaningful to state the prob-value. This is the tail probability, which can be approximated from the F table. For example, consider $F = 8.3$ in Table 10-5(b), having 2 and 12 d.f. We find that the closest critical values of F in Table VII are $F_{.01} = 6.93$ and $F_{.001} = 13.0$. We therefore conclude that the prob-value for H_0 is:

$$.001 < \text{prob-value} < .010 \qquad (10\text{-}12)$$

Or, by crude interpolation, we would guess that:

$$\text{prob-value} \simeq .006$$

In other words, the null hypothesis (that the machines of Table 10-1 are all the same) has very little credibility.

*(d) Unequal Sample Sizes

The most efficient way to collect observations is to make all samples the same size n. However, when this is not feasible, it still is possible to modify the ANOVA calculations. Table 10-6 provides the necessary modifications of Table 10-5 to take into account different sample sizes n_1, n_2, etc. Note especially that the definition of $\overline{X}$ is no longer the simple average of $\overline{X}_i$, but rather a *weighted* average with weights n_i.

PROBLEMS

10-1 (a) (Monte Carlo) Simulate a random sample of 5 observations from a normal population with $\mu = 50, \sigma = 10$. Call it sample A. Then simulate a second and third sample (B and C) from the same population. From the array of 15 observations, evaluate the ANOVA table to test whether these three samples are from the same population.

(b) Have the instructor record from every student the F statistic found in part (a). Graph the resulting distribution of F. Does the graph resemble Figure 10-2? What proportion of the F values exceed $F_{.05} = 3.89$? What proportion would exceed this value if the class were very large?

(c) Repeat (a) and (b), using three different populations: $\mu_1 = 40$, $\mu_2 = 50$, and $\mu_3 = 60$ (while σ remains 10).

For Problems 10-2 to 10-6, calculate the ANOVA table, including the approximate prob-value for the null hypothesis.

10-2 Twelve plots of land are divided randomly into three groups. The first is held as a control group, while fertilizers A and B are applied to the other two groups. Yield is observed to be:

Control, C	60	64	65	55
A	75	70	66	69
B	74	78	72	68

10-3 In a large American university in 1969, the male and female professors were sampled independently, yielding the following annual salaries (in thousands of dollars, rounded).[6]

Men	Women
12	9
11	12
19	8
16	10
22	16

10-4 A sample of American homeowners in 1970 reported the following home values (in thousands of dollars) by region:[7]

Region	Population Size	Sample of Home Values					$\overline{X}$	s^2
Northeast	16,000,000	23,	18,	12,	15,	27	19	36.5
Northcentral	19,000,000	17,	32,	12,	13,	11	17	75.5
South	21,000,000	13,	10,	16,	12,	19	14	12.5
West	12,000,000	13,	39,	17,	25,	17	22	99.0

---

[6]From D. A. Katz, "Faculty Salaries, Promotions, and Productivity at a Large University," *American Economic Review*, June 1973.

[7]From the 1974 *American Almanac*, Table 1168.

*10-5 A sample of American families in 1971 reported the following annual incomes (in thousands of dollars) by region.[8]

Northeast	8, 14
Northcentral	13, 9
South	7, 14, 8, 7
West	7, 7, 16

10-6 From each of 4 very large classes, 50 students were sampled, with the following results:

Class	Average Grade $\overline{X}$	Standard Deviation, s
A	68	11
B	74	12
C	70	8
D	68	10

*10-2   CONFIDENCE INTERVALS

(a) Simple Confidence Intervals

The limitations of hypothesis tests that we cited in Chapter 9 also hold in ANOVA. It may not be too enlightening to ask *whether* population means differ; by increasing sample size enough, statistical discernibility (significance) can nearly always be established—even though the population difference may be too small to be of any practical or economic importance. It is therefore more important to find out, "By *how much* do population means differ?"

It is easy to compare only two machines in Table 10-1 by constructing a confidence interval for $(\mu_1 - \mu_2)$ using $(\overline{X}_1 - \overline{X}_2)$:

$$(\mu_1 - \mu_2) = (\overline{X}_1 - \overline{X}_2) \pm t_{.025}\, s_p \sqrt{\frac{1}{n_1} + \frac{1}{n_2}} \qquad \begin{array}{c} (10\text{-}13) \\ (8\text{-}32) \text{ repeated} \end{array}$$

In (8-32), s_p^2 was the variance pooled from the two samples. However, it is more reasonable to use all the information available and pool the

---

[8] From the 1974 *American Almanac*, Table 534.

variance from all three samples, as in (10-6), obtaining $s_p^2 = 7.83$ with $4 + 4 + 4 = 12$ degrees of freedom. Thus, the 95% confidence interval is:

$$(\mu_1 - \mu_2) = (49 - 56) \pm 2.179\sqrt{7.83}\,\sqrt{\tfrac{1}{5} + \tfrac{1}{5}} = -7.0 \pm 3.9$$

Similar confidence intervals for $(\mu_1 - \mu_3)$ and for $(\mu_2 - \mu_3)$ may be constructed, for a total of three intervals; in our example, these intervals all are the same width, since the n_i all are the same:

$$
\left.
\begin{aligned}
(\mu_1 - \mu_2) &= -7.0 \pm 3.9 \quad \text{(a)}\\
(\mu_1 - \mu_3) &= -2.0 \pm 3.9 \quad \text{(b)}\\
(\mu_2 - \mu_3) &= +5.0 \pm 3.9 \quad \text{(c)}
\end{aligned}
\right\}
\qquad (10\text{-}14)
$$

(b) Simultaneous Confidence Intervals

There is just one difficulty with the above approach. Although we can be 95% confident of each individual statement such as (10-14(a)), we can be far less confident that the whole *system* of statements (10-14) is true. There are three ways (three statements) where this could go wrong.[9]

The level of confidence in the system (10-14) would be reduced to $(.95)^3 = .857$, if the three individual statements were independent.[10] But in fact they are not; for example, they all use the same s_p. Thus, if the observed s_p is high, all three interval estimates in (10-14) will be broad as a consequence. The problem is how to allow for this dependence in order to obtain the correct *simultaneous* confidence level for the whole system. In fact, this problem usually is stated the other way around: how much wider must the *individual* intervals in (10-14) be in order to yield a 95% level of confidence that all are simultaneously true?

One very simple solution is to cut the error rate of each interval from 5% to $5\%/3 = 1.67\%$.[11] This can be achieved by using (10-13), with a

---

[9]To emphasize this point, suppose that there were 100 individual confidence intervals comprising (10-14), instead of merely 3. Then we would expect, with average luck, for 95 of them to be right, and 5 of them to be wrong; hence the system of statements as a whole would be wrong.

[10]According to (3-30), for independent events we simply multiply the individual probabilities.

[11]Then the overall error rate will be at most $3(1.67\%) = 5\%$, as required. To prove this formally, let E_i represent the probability that an error occurs on the ith confidence interval. Then we begin with a simple probability formula:

$$\Pr(E_1 \text{ or } E_2) = \Pr(E_1) + \Pr(E_2) - \Pr(E_1 \text{ and } E_2) \le \Pr(E_1) + \Pr(E_2) \qquad \begin{matrix}(3\text{-}17)\\ \text{repeated}\end{matrix}$$

Similarly:

$$\Pr(E_1 \text{ or } E_2 \text{ or } E_3) \le \Pr(E_1) + \Pr(E_2) + \Pr(E_3) \le 1.67\% + 1.67\% + 1.67\% = 5\%$$

i.e.:

$$\Pr(\text{any error at all}) \le 5\%$$

more stringent t value. Since the two-tailed probability we require is .0167, we use $t_{.0083} = 2.8$ (found by interpolating Table V). And so we obtain, for the data of Table 10-1:

$$\left.\begin{array}{l} \mu_1 - \mu_2 = -7.0 \pm 5.0 \\ \mu_1 - \mu_3 = -2.0 \pm 5.0 \\ \mu_2 - \mu_3 = +5.0 \pm 5.0 \end{array}\right\} \qquad (10\text{-}15)$$

In other words, the interval of ± 3.9 in (10-14) must be widened to ± 5.0.

Although (10-15) is the simplest solution conceptually, there is a more efficient method available, called *Scheffé's multiple comparisons:*[12]

> With 95% confidence, *all* the following statements are true:
>
> $$(\mu_1 - \mu_2) = (\overline{X}_1 - \overline{X}_2) \pm \sqrt{(r-1)F_{.05}}\, s_p \sqrt{\frac{1}{n_1} + \frac{1}{n_2}} \quad \text{(a)}$$
>
> $$(\mu_1 - \mu_3) = (\overline{X}_1 - \overline{X}_3) \pm \sqrt{(r-1)F_{.05}}\, s_p \sqrt{\frac{1}{n_1} + \frac{1}{n_3}} \quad \text{(b)} \qquad (10\text{-}16)$$
>
> $$(\mu_2 - \mu_3) = (\overline{X}_2 - \overline{X}_3) \pm \sqrt{(r-1)F_{.05}}\, s_p \sqrt{\frac{1}{n_2} + \frac{1}{n_3}} \quad \text{(c)}$$

where $F_{.05}$ = the critical value of F (with $r - 1$ and $r(n - 1)$ d.f.) leaving 5% in the upper tail.

s_p^2 = the pooled sample variance, as calculated in Table 10-6 or equation (10-6).

r = number of rows (means) to be compared.

n_i = sample sizes.

For the machines in Table 10-1, (10-16(a)) yields:

$$\mu_1 - \mu_2 = (49 - 56) \pm \sqrt{(2)3.89}\, \sqrt{7.83}\, \sqrt{\frac{1}{5} + \frac{1}{5}} = -7.0 \pm 4.9$$

---

[12]A proof may be found in H. Scheffe, *The Analysis of Variance* (New York: John Wiley & Sons, 1959).

Note that each formula in (10-16) is just like (10-13), except that $t_{.025}$ has been replaced by $\sqrt{(r-1)F_{.05}}$. In the special case $r = 2$, this latter term reduces to $\sqrt{F_{.05}}$, which is just $t_{.025}$ (as will be shown in Problem 10-13).

Similar intervals for $(\mu_1 - \mu_3)$ and $(\mu_2 - \mu_3)$ may be calculated; in this example, the intervals all are the same width, since the n_i are the same. Thus:

$$
\left.
\begin{aligned}
\mu_1 - \mu_2 &= -7.0 \pm 4.9 \quad \text{(a)} \\
\mu_1 - \mu_3 &= -2.0 \pm 4.9 \quad \text{(b)} \\
\mu_2 - \mu_3 &= +5.0 \pm 4.9 \quad \text{(c)}
\end{aligned}
\right\} \qquad (10\text{-}17)
$$

As expected, (10-17) is similar to (10-14), only slightly wider (compare 4.9 versus 3.9). Indeed, it is this increased width (vagueness) that makes us 95% confident that *all* statements are true.

The results in (10-17) are summarized in Table 10-7. We star the confidence intervals where the null hypothesis lies outside and so can be rejected. Note how we can quickly evaluate a machine by examining its row: thus the positive differences in row 2 indicate how machine 2 outperforms the other two machines.[13] The same information can be represented graphically, as in Figure 10-4.

TABLE **10-7**

Differences in means for the data in Table 10-1[a]

i / I	1	2	3
1		−7.0*	−2.0
2	7.0*		5.0*
3	2.0	−5.0*	

[a]To get 95% simultaneous confidence intervals for $(\mu_i - \mu_I)$, take the listed value of $(\overline{X}_i - \overline{X}_I)$ and add ± 4.9. We star (*) the statistically discernible (significant) differences that exceed this ± 4.9 allowance.

---

[13]Any column shows these comparisons in reverse: thus, column 1 indicates how machine 1 is *outperformed by* the other two machines.

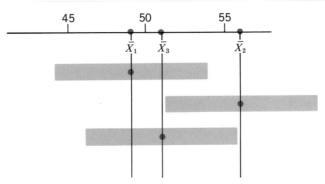

FIGURE 10–4 Differences in means for the data in Table 10–1. The colored bars show the simultaneous 95% confidence allowance of ± 4.9. Thus, for example, the mean $\overline{X}_2$ that lies outside the top bar is statistically discernible from $\overline{X}_1$; the means $\overline{X}_1$ and $\overline{X}_3$ that lie outside the second bar are discernible from $\overline{X}_2$; and so on.

(c) Single Contrasts

This idea can best be introduced with a concrete illustration.

Example 10-1. Suppose that the company that collected the data in Table 10-1 presently uses 8 machines of type 1 and 2 machines of type 2. There is a proposal to replace them with 10 machines of type 3. Will production go up or down? Find a 95% confidence interval for the increase in production, per machine-hour.

Solution. The new mean production will be μ_3. The old mean production was $.8\mu_1 + .2\mu_2$. What we want, therefore, is a confidence interval for the difference:

$$\mu_3 - (.8\mu_1 + .2\mu_2)$$

that is:

$$\mu_3 - .8\mu_1 - .2\mu_2 \tag{10-18}$$

Proceeding as in Section 8-3, we estimate (10-18), using the corresponding sample means:

$$\overline{X}_3 - .8\overline{X}_1 - .2\overline{X}_2 \tag{10-19}$$

Since this is a linear combination of independent random variables, its variance can be easily found by extending (5-40):

$$\text{var} = 1^2 \text{ var } \overline{X}_3 + (-.8)^2 \text{ var } \overline{X}_1 + (-.2)^2 \text{ var } \overline{X}_2$$

$$= \frac{\sigma^2}{n_3} + (-.8)^2 \frac{\sigma^2}{n_1} + (-.2)^2 \frac{\sigma^2}{n_2}$$

$$= \sigma^2 \left(\frac{1}{n_3} + \frac{(-.8)^2}{n_1} + \frac{(-.2)^2}{n_2} \right)$$

Thus the 95% confidence interval is:

$$(\mu_3 - .8\mu_1 - .2\mu_2) = (\overline{X}_3 - .8\overline{X}_1 - .2\overline{X}_2)$$

$$\pm t_{.025}\, s_p \sqrt{\frac{1^2}{n_3} + \frac{(-.8)^2}{n_1} + \frac{(-.2)^2}{n_2}}$$

$$\text{(10-20)}$$

$$= 51 - .8(49) - .2(56)$$

$$\pm 2.179 \sqrt{7.83} \sqrt{\frac{1}{5} + \frac{.64}{5} + \frac{.04}{5}}$$

$$(\mu_2 - .8\mu_1 - .2\mu_2) = .60 \pm 3.5 \qquad \text{(10-21)}$$

Note that the coefficients in (10-18) sum to zero $(1 - .8 - .2 = 0)$. Any linear combination of means that satisfy this condition is called a *contrast of means*, which may be written formally as:

$$\sum C_i \mu_i, \quad \text{where} \quad \sum C_i = 0 \qquad \text{(10-22)}$$

Then the 95% confidence interval will be a straightforward generalization of (10-20):

$$\boxed{\sum C_i \mu_i = \sum C_i \overline{X}_i \pm t_{.025}\, s_p \sqrt{\sum \left(\frac{C_i^2}{n_i} \right)}} \qquad \text{(10-23)}$$

Our earlier comparison of means in (10-13) represented the special case where $C_1 = 1$, $C_2 = -1$, and all the other $C_i = 0$.

(d) Multiple Contrasts

The preceding example gave a confidence interval for one particular contrast of interest. Suppose now that many possible contrasts are being contemplated in preliminary planning of factory renovations. Or, alternatively, suppose that our interest is scientific rather than technological, and that we are interested in every contrast that the data indicate may be interesting. Is it possible to get a formula that would allow us to make confidence intervals for all possible contrasts, which are *all* simultaneously true with 95% confidence? This sounds like a tall order, but in fact such a formula does exist:

---

with 95% simultaneous confidence, *all possible* contrasts are bracketed by the bounds:

$$\sum C_i \mu_i = \sum C_i \overline{X}_i \pm \sqrt{(r-1)F_{.05}} \, s_p \sqrt{\sum \left(\frac{C_i^2}{n_i}\right)} \qquad (10\text{-}24)$$

This is a remarkable formula. It includes not only all the statements[14] in (10-16), but also an infinite number of other contrasts that could be constructed. It is natural, of course, to wonder how we can be 95% confident of an infinite number of statements. It is because these statements are dependent. Thus, for example, once we have made the first two statements in (10-16), our intuition tells us that the third is likely to follow. Moreover, once we have made these three basic confidence intervals, other contrasts tend to follow, which can be added with little damage to our level of confidence. As the number of statements or contrasts grows and grows, each new statement tends to become simply a restatement of contrasts that already have been made, and essentially does no damage to our level of confidence. Thus, it can be mathematically confirmed that the entire (infinite) set of contrasts in (10-24) all are estimated simultaneously at a 95% level of confidence.

Example 10-2. For the data in Table 10-1, find confidence intervals that contrast each mean with the average of the other two means. Make the intervals broad enough so that there is 95% confidence that they all are true (and indeed that all possible contrasts like them also would be true).

Solution. We use the formula (10-24) for multiple contrasts. To contrast μ_1 with $(\mu_2 + \mu_3)/2$, we write it as:

---

[14]Each statement in (10-16) involves setting one coefficient in (10-24) equal to 1, another equal to -1, and the rest equal to 0.

$$(\mu_1 - .5\mu_2 - .5\mu_3) = (\overline{X}_1 - .5\overline{X}_2 - .5\overline{X}_3)$$

$$\pm \sqrt{(r-1)F_{.05}}\, s_p \sqrt{\frac{1^2}{5} + \frac{(-.5)^2}{5} + \frac{(-.5)^2}{5}}$$

$$= 49 - .5(56) - .5(51)$$

$$\pm \sqrt{(2)3.89}\, \sqrt{7.83}\, \sqrt{.30}$$

$$\text{Similarly:} \quad \left. \begin{array}{l} (\mu_1 - .5\mu_2 - .5\mu_3) = -4.5 \pm 4.3 \\[4pt] (\mu_2 - .5\mu_1 - .5\mu_3) = 6.0 \pm 4.3 \\[4pt] (\mu_3 - .5\mu_1 - .5\mu_2) = -1.5 \pm 4.3 \end{array} \right\} \quad (10\text{-}25)$$

In summary, (10-23) is appropriate for considering *one specific* linear contrast of means. On the other hand, (10-24) is appropriate for considering *all* possible contrasts simultaneously. Since (10-24) includes contrasts that may be formulated *after* the data has been studied, it sometimes is called the set of *posterior contrasts*. On the other hand, a specific contrast of the form (10-23) that is prespecified *before* any data is collected sometimes is called a *prior contrast*.

We might say that (10-24) is a "hunting license" to go after any contrast that looks interesting when we are in the "data forest." The "cost" of this license is that (10-24) necessarily must be somewhat vaguer than (10-23) ($\sqrt{(r-1)F_{.05}}$ is larger than $t_{.025}$, except when $r = 2$ and they are equal.)

PROBLEMS

*10-7 (a) For the simulated data in Problem 10-1(c), construct 95% confidence intervals for the three differences in means, using (10-13).

(b) Have the instructor record from every student whether or not he succeeded in having all three confidence intervals correct. What proportion of the students succeeded? What proportion would succeed if the class were very large?

(c) Repeat (a) and (b), using the 95% simultaneous confidence intervals (10-16).

*10-8 (a) For each of Problems 10-2 to 10-4, construct a set of 95% simultaneous intervals like Table 10-7. Also, graph them like Figure 10-4.

(b) For Problem 10-5, construct a set of 95% simultaneous confidence intervals like Table 10-7.

*10-9 Refer to the machine example of Table 10-1 and its ANOVA Table 10-5(b).

Suppose that one factory is to be outfitted entirely with machines of the first type. Suppose that a second factory is to be outfitted with machines of the second and third types, in the proportions 70% and 30%. Find a confidence interval for the difference in mean production for the two factories:

(a) That has 95% confidence, by itself.

(b) That can be included with all other contrasts (including all the pairwise comparisons of means in Table 10-7) at the 95% simultaneous-confidence level.

*10-10 Repeat Problem 10-9, if the first factory is to be outfitted instead with machines of types 1 and 2, in the proportions 50% and 50% (while the second factory remains outfitted with machines of types 2 and 3, in the proportions 70% and 30%).

*10-11 In Problem 10-4, suppose that we wanted to find out the average home value in the South, compared with the rest of the country. Construct the appropriate confidence interval:

(a) That has 95% confidence, by itself.

(b) That can be included with all other contrasts, at the 95% simultaneous confidence level.

*10-12 (a) Is the prob-value in Problem 10-3 the same as when you used the 2-sample t-test in Problem 9-25(b)?

(b) Is the confidence interval in Problem 10-8 (relating to Problem 10-3) the same as when you used the two-sample t in Problem 8-15(a)?

*10-13 (a) Generalize Problem 10-12: whenever there are just two rows to be compared, must the confidence intervals based on the ANOVA F test and the t test always coincide?

(b) Answer true or false; if false, correct it.

The one-factor ANOVA that we have studied so far may be regarded as an extension of the two-sample t test (unmatched, with σ_1^2 and σ_2^2 assumed equal). That is, ANOVA gives the same answer as t for 2 samples, and goes on to give answers for any number of samples (where t cannot be applied).

10-3   TWO-FACTOR ANALYSIS OF VARIANCE

(a) The ANOVA Table

We have suggested that the F test on the differences in machines given in (10-11) would be strengthened if the unexplained variance could be reduced by taking other factors into account. Suppose, for example, that the sample outputs given in Table 10-1 were produced by five different operators—with each operator producing one of the sample observations on each machine. This data, reorganized according to a two-way classification (by machine *and* operator), is shown in Table 10-8. It is necessary to complicate our notation somewhat. We now are interested in the average of each operator (each column average $\overline{X}_{.j}$) as well as the average of each machine (each row average $\overline{X}_{i.}$).[15]

Now the picture is clarified; some operators are efficient (the first and fourth), and some are not. The machines are not that erratic, after all; there is just a wide difference in the efficiency of the operators. If we can adjust for this explicitly, it will reduce the unexplained (or chance) variance in the denominator of (10-11); since the numerator will remain unchanged, the F ratio will be larger as a consequence, perhaps allowing us to reject H_0. To sum up, it appears that another factor (difference in operators) was responsible for a lot of extraneous noise in our simple one-way analysis in the previous section. By removing this noise, we hope to get a much more powerful test of the machines.

The analysis is an extension of the one-factor ANOVA, and is sum-

TABLE **10-8**

Samples of Production (X_{ij}) of 3 Different Machines
(As given in Table 10-1, but now arranged according to operator)

Machine \ Operator	$j = 1$	2	3	4	5	Machine Means $\overline{X}_{i.}$
$i = 1$	53	47	46	50	49	49
2	61	55	52	58	54	56
3	51	51	49	54	50	51
						$\overline{\overline{X}} = 52$
Operator Means $\overline{X}_{.j}$	55	51	49	54	51	$\overline{\overline{X}} = 52$

---

[15]The dot indicates the subscript over which summation occurs. For example, the dot suppresses the subscript j in $\overline{X}_{i.} = (1/n)\sum_{j} X_{ij}$.

TABLE 10-9(a)
Two-Way ANOVA, General

Source	Variation; Sum of Squares (SS)	d.f.	Variance; Mean Sum of Squares (MSS)	F
Between rows; EXPLAINED by differences in machines, i.e., differences in $\bar{X}_{i\cdot}$.	$SS_r = c\sum_{i=1}^{r}(\bar{X}_{i\cdot} - \bar{\bar{X}})^2$	$r-1$	$MSS_r = \dfrac{SS_r}{r-1} = cs^2_{\bar{X}_{i\cdot}}$	$\dfrac{MSS_r}{MSS_u}$
Between columns; EXPLAINED by differences in operators, i.e., differences in $\bar{X}_{\cdot j}$	$SS_c = r\sum_{j=1}^{c}(\bar{X}_{\cdot j} - \bar{\bar{X}})^2$	$c-1$	$MSS_c = \dfrac{SS_c}{c-1} = rs^2_{\bar{X}_{\cdot j}}$	$\dfrac{MSS_c}{MSS_u}$
UNEXPLAINED, i.e., residual variation, resulting from chance fluctuation.	$SS_u = \sum_{i=1}^{r}\sum_{j=1}^{c}(X_{ij} - \bar{X}_{i\cdot} - \bar{X}_{\cdot j} + \bar{\bar{X}})^2$	$(r-1)(c-1)$	$MSS_u = \dfrac{SS_u}{(r-1)(c-1)} = s^2$	
Total	$SS = \sum_{i=1}^{r}\sum_{j=1}^{c}(X_{ij} - \bar{\bar{X}})^2$	$rc-1$		

marized in Table 10-9; panel (a) is general, and panel (b) is specific to the data of Table 10-8. We note several features: the number of columns (operators) is denoted by c, and replaces n in Table 10-5. The column (operator) variation is defined like the row (machine) variation. The residual (unexplained) variation is a little more complicated, and will be explained in the next section. But for now, let us note that the unexplained variation (94) in Table 10-5(b) has been broken down in Table 10-9(b) into a major component that is explained by operators (72) and a minor residual component (22). So finally, in Table 10-9(b), all the component sources of variation still must sum to the total variation in the last line,[16] just as they did in Table 10-5(b).

(b) Testing Hypotheses

Now that we have broken the total variation down into its components, we can test whether there is a discernible difference in machines. We also can test whether there is a discernible difference in operators. In either test, the extraneous influence of the other factor will be taken into account.

On the one hand, we test for differences in machines by constructing the F ratio:

$$F = \frac{\text{variance explained by machines}}{\text{unexplained variance}}$$

(10-28)
like (10-11)

TABLE **10-9(b)**
Two-Way ANOVA, for Observations Given in Table 10-8

Source	Variation	d.f.	Variance	F Ratio	Prob-value for H_0
Between machines	130	2	65	23.6	$p < .001$
Between operators	72	4	18	6.5	$.01 < p < .05$
Residual	22	8	2.75		
Total	224 ✓	14 ✓			

---

[16]This may be proved just as in one-factor ANOVA. We begin with the identity:

$$(X_{ij} - \overline{\overline{X}}) = (\overline{X}_{i.} - \overline{\overline{X}}) + (\overline{X}_{.j} - \overline{\overline{X}}) + (X_{ij} - \overline{X}_{i.} - \overline{X}_{.j} + \overline{\overline{X}})$$

(10-26)
like (10-8)

By squaring and summing and noting how all the cross-product terms drop out, we finally obtain:

$$\sum_{i=1}^{r} \sum_{j=1}^{c} (X_{ij} - \overline{\overline{X}})^2 = c \sum_{i=1}^{r} (\overline{X}_{i.} - \overline{\overline{X}})^2 + r \sum_{j=1}^{c} (\overline{X}_{.j} - \overline{\overline{X}})^2 + \sum_{i=1}^{r} \sum_{j=1}^{c} (X_{ij} - \overline{X}_{i.} - \overline{X}_{.j} + \overline{\overline{X}})^2$$

(10-27)
like (10-10)

| Total variation | = | machine (row) variation | + | operator (column) variation | + | unexplained variation |

If H_0 is true, this F ratio has an F distribution. Specifically, from Table 10-9(b) we have:

$$F = \frac{\text{variance explained by machines}}{\text{unexplained variance}} = \frac{65}{2.75} = 23.6 \quad (10\text{-}29)$$

Referring to Table VII with 2 and 8 d.f., the closest critical value is $F_{.001} = 18.5$. Thus, the prob-value for H_0 (equal machines) is less than .001, and H_0 may be rejected.

This is much sharper evidence against H_0 than we got from the one-factor ANOVA in (10-12). The numerator has remained unchanged, but the chance variation in the denominator is much smaller, since the effect of the operators' differences has been netted out. Thus, our statistical leverage on H_0 has been increased.[17]

Similarly, we can get a powerful test of the null hypothesis that the operators perform equally well. Once again, F is the ratio of explained to unexplained variance; but this time, of course, the numerator is the variance between operators. Thus:

$$F = \frac{\text{variance explained by operators}}{\text{unexplained variance}} = \frac{18}{2.75} = 6.5 \quad (10\text{-}30)$$

Referring to Table VII with 4 and 8 d.f., the closest critical values are $F_{.05} = 3.84$ and $F_{.01} = 7.01$. Thus the prob-value for H_0 (equal operators) is:

$$.01 < \text{prob-value} < .05$$

Accordingly, we conclude that at the 5% level, there is a discernible (significant) difference in operators.

There is one issue that we passed over quickly that still requires clarification. In the one-factor test, we calculated unexplained variation by looking at the spread of n observed values within a category—i.e., within a whole row in Table 10-1. But in the two-factor Table 10-8, we split our observations column-wise, as well as row-wise; this leaves us with only one observation within each category. Thus, for example, there is only a single observation (61) of how much output is produced by operator 1 on machine 2. Variation no longer can be computed within that cell. What should we do?

---

[17] Strictly speaking, we have a stronger test because we have gained more by reducing unexplained variance than we have lost because our degrees of freedom in the denominator have been reduced by 4. (If we already are short of degrees of freedom—i.e., if we are near the top of Table VII—loss of degrees of freedom may be serious.)

Well, if there were no random residual, how would we predict the output of operator 1 on machine 2? This is a relatively good machine ($\overline{X}_{2.} = 56$) and a relatively good operator ($\overline{X}_{.1} = 55$). On both counts we would predict output to be above average. This strategy easily can be formalized to predict $\hat{X}_{21}$. We can do this for each cell, estimating the random residual as the difference between the observed value X_{ij} and the corresponding predicted value $\hat{X}_{ij}$. This yields a whole set of residuals, whose sum of squares is precisely the unexplained variation[18] SS_u (the last term in (10-27), also appearing in the second column of Table 10-9(a)). Divided by d.f., this becomes the unexplained variance used in the denominator of both F tests.

One final warning: in computing predicted output $\hat{X}_{ij}$, we assume that there is no interaction between the two factors—as would occur, for example, if certain operators liked some machines, but other operators disliked them. Such interaction would require a more complex model and several observations per cell. The two-way analysis of variance developed in this section is based on the assumption that interaction does not exist—the so-called *simple additive model*.

---

[18]Predicted value $\hat{X}_{ij}$ is defined as:

$$\hat{X}_{ij} = \overline{\overline{X}} + \text{adjustment reflecting machine performance} + \text{adjustment reflecting operator}$$
$$= \overline{\overline{X}} + (\overline{X}_{i.} - \overline{\overline{X}}) + (\overline{X}_{.j} - \overline{\overline{X}}) \tag{10-31}$$

Specifically, in our example:

$$\hat{X}_{21} = 52 + (56 - 52) + (55 - 52) = 52 + 4 + 3 = 59$$

Thus, our prediction of the performance of operator 1 on machine 2 is calculated by adjusting average performance (52) by the degree to which this machine is above average (4) and the degree to which this operator is above average (3).

Cancelling $\overline{\overline{X}}$ values in (10-31), we get:

$$\hat{X}_{ij} = \overline{X}_{i.} + \overline{X}_{.j} - \overline{\overline{X}}$$

and the random element, being the difference between the observed and expected, becomes:

$$X_{ij} - \hat{X}_{ij} = X_{ij} - \overline{X}_{i.} - \overline{X}_{.j} + \overline{\overline{X}} \tag{10-32}$$

We emphasize that this random element is output left unexplained after adjustment for both machine i and operator j.

In our example,

$$X_{21} - \hat{X}_{21} = 61 - 59 = 2$$

Thus, this observed output is 2 units above what we expected, and must be left unexplained—the result of random influences.

Unexplained variation (SS_u) is recognized to be the sum of squares of all random elements as defined in (10-32).

*(c) Multiple Comparisons

Turning from hypothesis tests to confidence intervals, we may write a statement for two-factor ANOVA similar to one-factor ANOVA:

> With 95% simultaneous confidence, all possible contrasts in row means are bracketed by the bounds:
>
> $$\sum C_i \mu_i = \sum C_i \overline{X}_{i.} \pm \sqrt{(r-1)F_{.05}} \; s \sqrt{\frac{\Sigma(C_i^2)}{c}} \qquad (10\text{-}33)$$

where

$F_{.05}$ = the critical value of F, with $(r-1)$ and $(r-1)(c-1)$ d.f.

$s = \sqrt{\mathrm{MSS}_u}$, as calculated in Table 10-9(a).

r = number of rows

c = number of columns

Formula (10-33) differs from (10-24) primarily because its unexplained variance s^2 now is smaller, making the confidence intervals more precise. To illustrate, consider the machines of Table 10-8, analyzed in ANOVA Table 10-9(b). With 95% confidence, all the following statements are true:

$$\mu_1 - \mu_2 = (49 - 56) \pm \sqrt{(2)4.46} \; \sqrt{2.75} \sqrt{\frac{1}{5} + \frac{1}{5}}$$

that is:

Similarly:
$$\left.\begin{array}{l} \mu_1 - \mu_2 = -7.0 \pm 3.1 \\ \mu_1 - \mu_3 = -2.0 \pm 3.1 \\ \mu_2 - \mu_3 = +5.0 \pm 3.1 \end{array}\right\} \qquad (10\text{-}34)$$

and all other possible contrasts. Note how the confidence intervals indeed are reduced relative to one-factor ANOVA (compare ± 3.1 in (10-34) with ± 4.9 in (10-17)).

Of course, we could contrast the column means just as well simply by interchanging r and c in (10-33). For example, how do the operators compare? With 95% confidence, all the following statements are true:

$$\mu_1 - \mu_2 = (\overline{X}_{.1} - \overline{X}_{.2}) \pm \sqrt{(c-1)F_{.05}}\, s\sqrt{\frac{1}{r} + \frac{1}{r}}$$

$$= (55 - 51) \pm \sqrt{(4)3.84}\,\sqrt{2.75}\,\sqrt{\frac{1}{3} + \frac{1}{3}}$$

that is:

$$\left.\begin{aligned}
\mu_1 - \mu_2 &= 4.0 \pm 5.3 \\
\mu_1 - \mu_3 &= 6.0 \pm 5.3 \\
\mu_1 - \mu_4 &= 1.0 \pm 5.3
\end{aligned}\right\} \qquad (10\text{-}35)$$

Similarly:

and all other possible contrasts.

PROBLEMS

10-14 To refine the experimental design of Problem 10-2, suppose that the 12 plots of land are on 4 farms (3 plots on each). You suspect that there may be a difference in fertility between farms. Retabulate the data in Problem 10-2, according to fertilizer *and* farm, as follows.

Fertilizer \ Farm	1	2	3	4
Control C	60	64	65	55
A	69	75	70	66
B	72	74	78	68

(a) Calculate the ANOVA table to determine whether fertilizers differ, and whether farms differ.

*(b) Construct a table of differences in fertilizers similar to Table 10-7. Do the same for differences between farms.

10-15 Three men work on an identical task of packing boxes. The number of boxes packed by each in three selected hours is shown in the table below.

Hour \ Man	A	B	C
11–12 A.M.	24	19	20
1–2 P.M.	23	17	14
4–5 P.M.	25	21	17

(a) Calculate the ANOVA table.

*(b) For the factors that are statistically discernible (significant) at the 5% level, construct a table of 95% simultaneous confidence intervals, as in Table 10-7.

10-16 A sample of eight American families in 1971 reported the following annual incomes (in thousands of dollars) by region and race.[19]

Race \ Region	Northeast	Northcentral	South	West
White	11.3	11.0	9.7	10.8
Black	7.7	7.6	5.5	7.6

(a) Calculate the ANOVA table.

*(b) For the factors that are statistically discernible (significant) at the 5% level, construct a table of simultaneous confidence intervals, as in Table 10-7.

(c) Construct a confidence interval for the mean difference between races, using the t-test (8-37). Also calculate the 2-sided prob-value for H_0.

(d) Do your answers in (c) agree with (a) and (b)?

(e) In view of your experience in (d), guess whether the following is true or false; if false, correct it:

Two-factor ANOVA may be regarded as an extension of the two-sample matched t. That is, ANOVA gives the same confidence intervals as t for 2 samples, and goes on to give simultaneous confidence intervals for any number of samples (where t cannot be applied).

REVIEW PROBLEMS (CHAPTERS 6-10)

10-17 Suppose that a 95% confidence interval for a population mean was calculated to be $\mu = 170 \pm 20$. From the following, select the best interpretation and criticize the others.

(a) Any hypothesis in the interval $150 < \mu < 190$ is called the null hypothesis, while any hypothesis outside this interval is called the alternate hypothesis.

(b) The population mean is a random variable with expectation 170 and standard deviation 20.

(c) If this sampling experiment were repeated many times, and if each

---

[19] From the 1974 *American Almanac*, Table 534. In this problem, as in all others, we are assuming no interaction. That is, we assume that the population differential between whites and blacks is the same for all 4 regions. To the extent that this assumption is true, ANOVA is valid.

time a confidence interval were similarly constructed, 95% of these confidence intervals would cover $\mu = 170$.

(d) The sample mean is a random variable with expectation 170 and standard deviation 10.2

(e) Any hypothesis in the interval $150 < \mu < 190$ may be called acceptable at the 5% level.

10-18 An anthropologist collected a random sample of 100 men from among the large population of men on a certain island. Their heights in inches were as follows:

Height x (Cell Midpoints)	Frequency
75	1
72	6
69	20
66	39
63	24
60	8
57	0
54	2

Does this throw any light on the theory of some anthropologists that the island men have the same heights as North Americans (as given in Table 2-3)?

10-19 A random sample of 100 people was asked, both before and after a major policy speech, whether or not they approved of the president's policies. The number of responses in each category were:

After ＼ Before	Not Approve	Approve
Not Approve	57	13
Approve	2	28

To what extent has approval in the population changed? Answer with a 95% confidence interval.

10-20 An instructor wished to compare grades in his large statistics class of 810 students. From the 220 students who attended class less than half the time (the "absentees," A), he took a random sample of 10 students. From the remaining 590 students who attended at least half the time (the "participants," P), he also took an independent random sample of 10 students. The following data were obtained:

	Population Size	Sample Size	Sample Mean	Sample Variance
Absentees A	220	10	53.2	280
Participants B	590	10	71.2	170

Construct a 95% confidence interval for the difference Δ in the mean grade between the two groups of students.

10-21 In a certain freshman statistics course, there are four sections. From each section, a sample of six grades was drawn, with the following results:

Section	Random Sample of 6 Grades from Each Section	Sample Mean $\overline{X}$
A	63, 87, 64, 69, 78, 65	71
B	77 etc.	70
C		50
D		81

ANOVA Table (partial)

Source	Variation (Sum of Squares)
Between sections	3,036
Within sections	2,080
Total	5,116

(a) Construct 95% simultaneous confidence intervals for all differences in section means.

(b) Sections A and C were taught by Professor Jones, while sections B and D were taught by Professor Koestler. Therefore, the contrast:

$$\left[\frac{\mu_B + \mu_D}{2}\right] - \left[\frac{\mu_A + \mu_C}{2}\right]$$

is of interest. Construct a 95% confidence interval for it.

(c) Does answer (b) prove (statistical proof at a 95% level of confidence, not logical proof) that Professor Koestler is better than Professor Jones, in the restricted sense that Koestler can get students to perform better on exams?

10-22 (Simpson's Paradox.) Hundreds of students were sampled randomly at Hypothetical U. over a ten year period. The failure rates, broken down by sex and faculty, were as follows:

	Arts		Science		University as a Whole (Totals)	
	Men	Women	Men	Women	Men	Women
Failed	150	100	50	200	200	300
Passed	450	300	50	200	500	500
Totals	600	400	100	400	700	800

(a) Construct a 95% confidence interval for the difference in failure rates between men and women for each of the following populations:

(i) All students.

(ii) Arts students.

(iii) Science students.

(b) How do you answer the following statement of a puzzled friend: "When I look at the university as a whole, women seem dumber. But when I look at which women are dumber, the women in Arts and the women in Science are all as good as the men! Did I miscalculate? Is it sampling fluctuations?"

10-23 Admissions to Graduate Division, Berkeley, Fall 1973, were as follows:[20]

	Men	Women
Admitted	3,700	1,500
Rejected	4,600	2,800
Totals	8,300	4,300

[20] From P. J. Bickel, et al., "Sex Bias in Graduate Admissions: Data from Berkeley," *Science* 187, February 7, 1975.

(a) Suppose that the above data can be regarded as a random sample of 8,300 men (or 4,300 women) from an infinite hypothetical population seeking admission to Berkeley. Let π_M (or π_W) be the proportion of men (or women) in this hypothetical population who are admitted.

Construct a 95% confidence interval for $\pi_M - \pi_W$. Is there evidence of bias? If so, is it directed against men or women?

(b) Let us disaggregate the data of (a) by faculty, in order to find the source of the "bias."[21]

	Humanities		Science	
	Men	Women	Men	Women
Admitted	700	900	3,000	600
Rejected	1,600	2,300	3,000	500
Totals	2,300	3,200	6,000	1,100

Repeat (a) for the subpopulation applying to Humanities, and to Science.

(c) Answer true or false; if false, correct it.

(i) If faculty is kept constant, men and women are admitted about equally. However, there is a tendency for women to apply to the tougher faculty, which explains why their overall admission rate is considerably lower.

(ii) Problems 10-22 and 10-23 illustrated how easy it is to leap to false conclusions with a naive analysis of aggregated data.

10-24 We introduced the notion of a biased sample in Chapter 1, and the notion of a biased estimator in Chapter 7. In order to see how these two ideas are related, consider the following vastly simplified example:

A certain population consists of two classes of people:

70% are poor people earning $4,000 each;

30% are rich people earning $20,000 each.

A politician records the incomes of 100 visitors to his office during a week when each rich person is twice as likely as each poor person to visit the office, thus making the sample mean a biased estimator of the population mean.

---

[21] Although this disaggregation is hypothetical, it preserves the spirit of the problem while at the same time simplifying the computations.

(a) How much is this bias, in dollars?

(b) What would the bias in part (a) be if:

   (i) The sample size were 25 (instead of 100)?

   (ii) The split between poor and rich were 60%–40% (instead of 70%–30%)?

10-25 In a certain county, suppose that the electorate is 80% urban and 20% rural; 70% of the urban voters, and only 25% of the rural voters, vote for D in preference to R. In a certain straw vote conducted by a small-town newspaper editor, a rural voter has 6 times the chance of being selected as an urban voter. This bias in the sampling will cause the sample proportion to be a biased estimator of the population proportion in favor of R.

(a) How much is this bias?

(b) Is the bias large enough to cause the average sample to be wrong (in the sense that the average sample "elects" a different candidate from the one the population elects)?

10-26 Turning now from the omniscient viewpoint of 10-25 to the more limited viewpoint of the newspaper editor, realistically we cannot suppose that the population proportion π favoring D is known. However, we can suppose that the 80%–20% urban–rural split in the population is known— through census figures, for example. Suppose that the editor then obtains the following data from a biased sample of 700 voters:

Location \ Vote	For D	For R	Totals
Urban	210	92	302
Rural	80	318	398
Totals	290	410	700

The simple-minded and biased estimate of π is the simple proportion $290/700 = 41\%$. Calculate an *unbiased* estimate of π. Incidentally, a technique like this, which is based on several population strata (urban, rural) whose proportions are known and allowed for, is an example of *stratified sampling*.

PART III

Regression: Relating Two or More Variables

CHAPTER 11

Fitting a Line

The cause is hidden, but the result is known.

Ovid

11-1   INTRODUCTION

In our first example of statistical inference, we estimated the mean of a single population. Then we compared two population means. Finally, we compared r population means, using analysis of variance. Now we ask whether we could improve the analysis if we are able to rank the r populations numerically rather than in unordered categories.

For example, we can use the analysis of variance (as in Problem 10-2) to show how wheat yield depends on several different *kinds* of fertilizer. If we wish to consider how yield depends on several different *amounts* of fertilizer, we define fertilizer application on a numerical scale. If we plot the yield Y that follows from various fertilizer applications X, a scatter similar to Figure 11-1 might be observed. From this scatter, it seems clear that fertilizer does affect yield. Moreover, it should be possible to describe *how*, by an equation relating Y to X. Estimating an equation is, of course, equivalent geometrically to fitting a curve through this scatter. This is called the "regression" of Y on X; as a simple mathematical model, it will be useful as a brief and precise description, or as a means of predict-

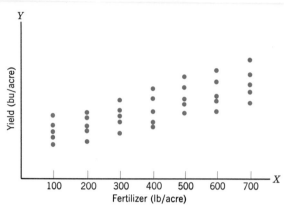

FIGURE 11-1 Observed relation of wheat yield to fertilizer application

ing the yield Y for a given amount of fertilizer X. This chapter is devoted exclusively to how a straight line may best be fitted.

Since yield depends on fertilizer, yield is called the "dependent variable" or "response" Y. Since fertilizer application is not dependent on yield, but instead is determined independently by the experimenter, we refer to it as an "independent variable" or "factor," or "regressor" X.

Example 11-1. Suppose, in a study of how wheat yield depends on fertilizer, that funds are available for only seven experimental observations. So the experimenter sets X at seven different values, taking only one observation Y in each case, as shown in Table 11-1. Graph these points, and roughly fit a line by eye.

TABLE **11-1**
7 Observations of Fertilizer
and Yield

X Fertilizer (Pound/Acre)	Y Yield (Bushel/Acre)
100	40
200	50
300	50
400	70
500	65
600	65
700	80

Solution.

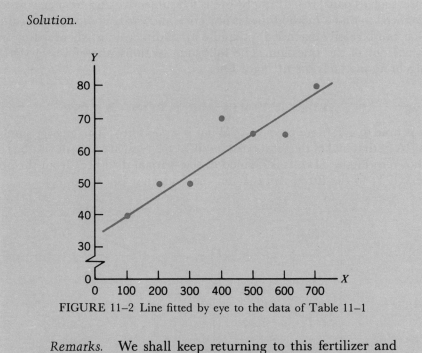

FIGURE 11–2 Line fitted by eye to the data of Table 11–1

Remarks. We shall keep returning to this fertilizer and yield example throughout the next few chapters, to illustrate what regression means.

How good is a rough fit by eye, such as we used in Example 11-1? In Figure 11-3, we note in panel (a) that if all the points were exactly in a line, then the fitted line could be drawn in with a ruler, by eye, perfectly accurately. Even if the points were *nearly* in a line, such as in panel (b), fitting by eye would be reasonably satisfactory. But in the highly scattered

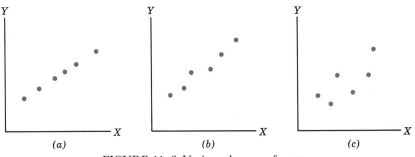

FIGURE 11–3 Various degrees of scatter

case, such as panel (c), fitting by eye is too subjective and too inaccurate. We need to find a method that is objective and easily computerized, and that can be easily extended to handle more dimensions where fitting by eye is out of the question. The following sections, therefore, set forth algebraic methods for fitting a line.

11-2   POSSIBLE CRITERIA FOR FITTING A LINE

It is time to ask, more precisely, "What is a good fit?" The answer surely is, "A fit that makes the total error small." One typical error (deviation) is shown in Figure 11-4. It is defined as the vertical distance[1] from the observed Y_i to the fitted value $\hat{Y}_i$ on the line, that is, $(Y_i - \hat{Y}_i)$.

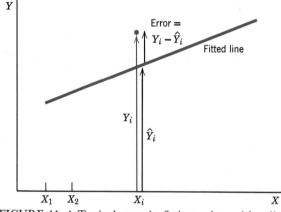

FIGURE 11–4 Typical error in fitting points with a line

We note that this error is positive when the observed Y_i is above the line, and negative when the observed Y_i is below the line.

1. As our first tentative criterion, consider a fitted line that minimizes the sum of all these errors:

$$\sum_{i=1}^{n} (Y_i - \hat{Y}_i) \tag{11-1}$$

---

[1]Vertical distance is used as the measure of error because our objective is to minimize the error in explaining Y, and Y is measured vertically.

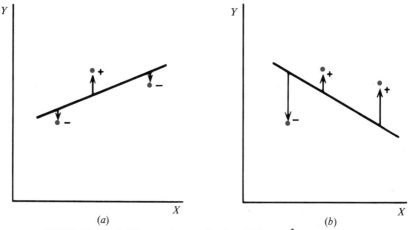

(a)   (b)

FIGURE 11–5 The weakness of using $\Sigma(Y_i - \hat{Y}_i)$ to fit a line

Unfortunately, this works badly. Using this criterion, the two lines shown in Figure 11-5 fit the observations equally well, even though the fit in panel (a) is intuitively a good one, and the fit in panel (b) is a very bad one. The problem is one of sign; in both cases, positive errors just offset negative errors, leaving their sum equal to zero. This criterion must be rejected, since it provides no distinction between bad fits and good ones.

2. There are two ways of overcoming the sign problem.[2] The first is to minimize the sum of the *absolute* values of the errors:

$$\sum |Y_i - \hat{Y}_i| \tag{11-2}$$

Since large positive errors are not allowed to offset large negative ones, this criterion would rule out bad fits like Figure 11-5(b). However, it still has a drawback. In Figure 11-6, the fit in panel (b) satisfies this criterion better than the fit in panel (a), since $\Sigma |Y_i - \hat{Y}_i|$ is 3, rather than 4. In fact, you can satisfy yourself that the line in panel (b) joining the two end-points satisfies this criterion better than *any* other line. But perhaps it is not the best solution to the problem, because it pays no attention whatever to the middle point. The fit in panel (a) may be preferable because it takes account of all points.

---

[2]You may recognize that (11-2) and (11-3) are close analogies with (2-4) and (2-5).

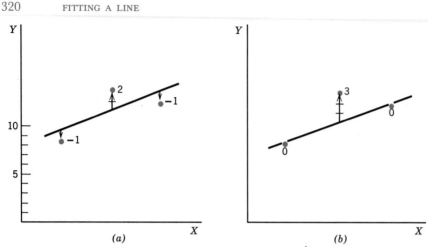

FIGURE 11–6 The weakness of using $\Sigma |Y_i - \hat{Y}_i|$ to fit a line

3. As a second way to overcome the sign problem, we finally propose to minimize the sum of the squares of the errors:

$$\sum (Y_i - \hat{Y}_i)^2 \qquad (11\text{-}3)$$

This is the famous "least squares" criterion; its justifications include the following:

(a) Squaring overcomes the sign problem by making all errors positive.

(b) The algebra of least squares is very manageable

(c) There are two important theoretical justifications for least squares: the Gauss–Markov theorem (discussed in Chapter 12), and the maximum likelihood criterion (discussed in Chapter 18).

11-3   THE LEAST SQUARES SOLUTION

The scatter of observed X and Y values from Table 11-1 is graphed again in Figure 11-7(a). Our objective is to fit a line:

$$\hat{Y} = a_0 + bX \qquad (11\text{-}4)$$

(The geometry of lines and planes, including the concepts of intercept, slope, etc., is reviewed in Appendix 13-A.) The fitting of the line (11-4) involves three steps:

STEP 1.   Translate X into deviations from its mean; that is, define a new variable x:

$$x \overset{\Delta}{=} X - \overline{X} \qquad (11\text{-}5)$$

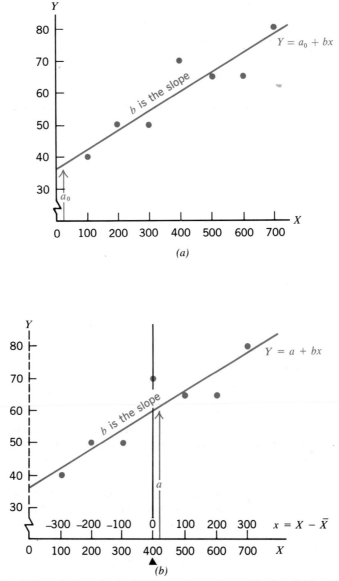

FIGURE 11–7 Translation of axis. (a) Regression using original variables. (b) Regression after translating X.

This is equivalent to a geometric translation of axis, as shown in Figure 11-7(b); the Y axis has been shifted right from 0 to $\bar{X}$. The new x value becomes positive or negative, depending on whether X was above or below $\overline{X}$. There is no change in the Y values. The intercept a differs from the original a_0, but the slope b remains the same.

Measuring X as a deviation from $\overline{X}$ will simplify the mathematics because the sum of the new x values equals zero,[3] that is:

$$\sum x_i = 0 \tag{11-6}$$

STEP 2.   Fit the line in Figure 11-7(b), that is, the line:

$$\hat{Y} = a + bx \tag{11-7}$$

Fit it to the scatter by selecting the values for a and b that satisfy the least squares criterion; select those values of a and b that:

$$\text{minimize } \sum (Y_i - \hat{Y}_i)^2 \tag{11-8}$$
$$\text{(11-3) repeated}$$

Since each fitted value $\hat{Y}_i$ is on the estimated line (11-7):

$$\hat{Y}_i = a + bx_i \tag{11-9}$$

When this is substituted into (11-8), the problem becomes one of selecting a and b to minimize the sum of squares, that is:

$$\text{minimize }\ S(a, b) = \sum (Y_i - a - bx_i)^2 \tag{11-10}$$

The notation $S(a, b)$ is used to emphasize that this expression depends on a and b. As a and b vary (as various lines are tried), $S(a, b)$ will vary too, and we ask at what value of a and b it will be a minimum. This will give us our optimum (least squares) line.

The simplest minimization technique is calculus, and we will use it in the next paragraph. (Readers without calculus can minimize (11-10) with the more cumbersome algebra of Appendix 11-A, and rejoin us

---

[3]Although this was proved in (2-3) essentially, we shall repeat the proof because of its simplicity and importance:

$$\sum x_i = \sum (X_i - \overline{X})$$
$$= \sum X_i - n\overline{X}$$

Noting that $\overline{X}$ is defined as $\sum X_i/n$, it follows that $\sum X_i = n\overline{X}$ and hence:

$$\sum x_i = n\overline{X} - n\overline{X} = 0 \tag{11-6 proved}$$

where the resulting theorems (11-13) and (11-16) are set out in the box below.)

Minimizing $S(a, b)$ requires setting its partial derivatives equal to zero. So we first set the partial derivative with respect to a equal to zero:

$$\frac{\partial}{\partial a} \sum (Y_i - a - bx_i)^2 = \sum 2(-1)(Y_i - a - bx_i)^1 = 0 \qquad (11\text{-}11)$$

Dividing through by -2 and rearranging:

$$\sum Y_i - na - b \sum x_i = 0 \qquad (11\text{-}12)$$

Noting that $\Sigma x_i = 0$ by (11-6), we can solve for a:

$$a = \frac{\sum Y_i}{n} = \overline{Y} \qquad (11\text{-}13)$$

Thus, the least squares estimate of a is simply the average value of Y. Referring back to Figure 11-7, we see that this ensures that the fitted regression line must pass through the point $(\overline{X}, \overline{Y})$, which may be interpreted as the center of gravity of the sample of n points.

It also is necessary in (11-10) to set the partial derivative with respect to b equal to zero:

$$\frac{\partial}{\partial b} \sum (Y_i - a - bx_i)^2 = \sum 2(-x_i)(Y_i - a - bx_i)^1 = 0 \qquad (11\text{-}14)$$

Cancelling -2:

$$\sum x_i(Y_i - a - bx_i) = 0 \qquad (11\text{-}15)$$

Rearranging:

$$\sum x_i Y_i - a \sum x_i - b \sum x_i^2 = 0$$

Noting that $\Sigma x_i = 0$, we can solve for b:

$$b = \frac{\sum x_i Y_i}{\sum x_i^2} \qquad (11\text{-}16)$$

TABLE 11-2

Fitting a Least Squares Line to the Data of Example 11-1

(a) Calculations for the Slope b and Intercept a (b) Calculations for the Residual Variance s^2

X	Y	$x = X - \bar{X}$ $= X - 400$	xY	x^2	$\hat{Y} = a + bx$ $= 60$ $+ .059x$	$(Y - \hat{Y})$	$(Y - \hat{Y})^2$
100	40	−300	−12,000	90,000	42.3	−2.3	5.29
200	50	−200	−10,000	40,000	48.2	1.8	3.24
300	50	−100	−5,000	10,000	54.1	−4.1	16.81
400	70	0	0	0	60.0	10.0	100.00
500	65	100	6,500	10,000	65.9	−0.9	.81
600	65	200	13,000	40,000	71.8	−6.8	46.24
700	80	300	24,000	90,000	77.7	2.3	5.29

$\sum X = 2,800 \quad \sum Y = 420 \quad \sum x = 0 \quad \sum xY = 16,500 \quad \sum x^2 = 280,000 \quad \sum(Y - \hat{Y})^2 = 177.68$

$\bar{X} = \dfrac{1}{n}\sum X \qquad \bar{Y} = \dfrac{1}{n}\sum Y$

$= \dfrac{2,800}{7} \qquad\quad = \dfrac{420}{7} = 60$

$= 400 \qquad\qquad \boxed{a = 60}$

$b = \dfrac{\sum xY}{\sum x^2} = \dfrac{16,500}{280,000}$

$\boxed{b = .059}$

$s^2 = \dfrac{1}{n-2}\sum(Y - \hat{Y})^2$

$= \dfrac{177.68}{5} = 35.5$

and $s = 5.96$

Our results[4] in (11-13) and (11-16) are important enough to restate:

> With x values measured as deviations from their mean, the least-squares values of a and b are:
>
> $$a = \overline{Y}$$
>
> $$b = \frac{\sum x_i Y_i}{\sum x_i^2}$$

(11-13) repeated

(11-16) repeated

For the data in Table 11-1, a and b are calculated in Table 11-2(a). It follows that the least squares equation is:

$$\hat{Y} = 60 + .059x \tag{11-17}$$

This fitted line is graphed in Figure 11-7(b).

STEP 3. If desired, the regression can now be translated back into our original frame of reference in Figure 11-7(a). Express (11-17) in terms of the original X values:

$$\hat{Y} = 60 + .059(X - \overline{X})$$
$$= 60 + .059(X - 400)$$
$$= 60 + .059X - 23.6$$
$$\hat{Y} = 36.4 + .059X \tag{11-18}$$

This fitted line is graphed in Figure 11-7(a).

A comparison of (11-17) and (11-18) confirms that the slope of our fitted regression ($b = .059$) remains the same; the only difference is in the intercept. Moreover, we note how easily the original intercept ($a_0 = 36.4$) was recovered.

An estimate of yield for any given fertilizer application is now easily derived from the least squares equation (11-18). For example, if 350 lbs. of fertilizer is to be applied:

$$\hat{Y} = 36.4 + .059(350) = 57$$

---

[4]To be rigorous, we could examine second derivatives to prove that we actually do have a minimum sum of squares—rather than a maximum or saddle point. (See, for example, chapter 5 of T. H. Wonnacott, *Calculus, an applied approach*, Wiley, New York, 1977.)

The alternative least squares equation (11-17) yields exactly the same result. When $X = 350$, then $x = -50$, and:

$$\hat{Y} = 60 + .059(-50) = 57$$

PROBLEMS

11-1 Suppose that four randomly chosen plots were treated with various levels of fertilizer, resulting in the following yields of corn:

Fertilizer X (Pounds/Acre)	Yield Y (Bushels/Acre)
100	70
200	70
400	80
500	100

(a) Calculate the regression line of yield on fertilizer.
(b) Plot the four points and the regression line. Check that the line fits the data reasonably well.
(c) Explain what the intercepts a and a_0 represent.
(d) Estimate how much the yield is increased for every pound of fertilizer applied—that is, what is the marginal physical product of fertilizer?
(e) If the value of the crop is $2 per bushel, estimate how much revenue is increased for every pound of fertilizer applied—that is, what is the marginal *revenue* product of fertilizer? If fertilizer costs $.10 per pound, would it be economical to apply?

11-2 Suppose that a random sample of five families had the following annual income and saving (in thousands of dollars):

Family	Income X	Saving S
A	8	.6
B	11	1.2
C	9	1.0
D	6	.7
E	6	.3

(a) Plot S on X for each of the families.
(b) Calculate and graph the regression line of saving S on income.
(c) Interpret the intercepts a and a_0.

11-3 Use the data of Problem 11-2 to regress consumption C on income X, where $C = X - S$. Then compare this slope with the slope of S on X in Problem 11-2.

11-4 Suppose that four firms had the following profits and research expenditures:

Profit, P (Thousands of Dollars)	Research Expenditure, R (Thousands of Dollars)
50	40
60	40
40	30
50	50

(a) Fit a regression line of P on R.
(b) Graph the data and the fitted line.
(c) Does this regression line show how research generates profits?

⇒11-5 Suppose that we translate Y as well as X into deviation form (so that $y = Y - \overline{Y}$ and $x = X - \overline{X}$).
(a) Prove that $\Sigma x_i y_i = \Sigma x_i Y_i$. Hence, instead of (11-16), we may alternatively calculate b as:

$$b = \frac{\sum x_i y_i}{\sum x_i^2} \tag{11-19}$$

(b) Use (11-19) to calculate b in Problem 11-1. Check that it agrees with the value calculated previously using (11-16).
(c) Draw a figure similar to Figure 11-7 to show the translation of the y-axis as well as the x-axis. What is the new y-intercept? the new slope? Does this mean that the equation of the fitted line is simply:

$$y = bx \tag{11-20}$$

*11-6 (Requires calculus.) Suppose that both X and Y are left in their original form, rather than being translated into deviation form.
(a) Write out the sum of squared deviations as in (11-10), in terms of a_0 and b.
(b) Set equal to zero the partial derivatives with respect to a_0 and b, thus obtaining two so-called "normal" equations.

(c) Write out these two normal equations using the data in Problem 11-1, and solve for a_0 and b.

(d) Does the regression line calculated in (c) correspond to the regression previously obtained in Problem 11-1? Which method do you think is easier?

APPENDIX 11-A

Least Squares Without Calculus

First, we must solve the general problem of minimizing an ordinary quadratic function of one variable b, of the form:

$$f(b) = k_2 b^2 + k_1 b + k_0 \qquad (11\text{-}21)$$

where k_2, k_1, k_0 are constants, with $k_2 > 0$. With a little algebraic manipulation, (11-21) may be written as:

$$f(b) = k_2 \left(b + \frac{k_1}{2k_2} \right)^2 + \left(k_0 - \frac{k_1^2}{4k_2} \right)$$

Note that b appears in the first term, but not in the second. Therefore, our hope of minimizing the expression lies in selecting a value of b to minimize the first term. Being a square and hence never negative, the first term will be minimized when it is zero, that is, when:

$$b + \frac{k_1}{2k_2} = 0$$

that is:

$$b = \frac{-k_1}{2k_2} \qquad (11\text{-}22)$$

This result is shown graphically in Figure 11-8. To restate: *a quadratic function of the form (11-21) is minimized by setting:*

$$b = - \frac{(\text{coefficient of first power})}{2(\text{coefficient of second power})} \qquad (11\text{-}23)$$

With this theorem in hand, let us return to the problem of selecting values for a and b to minimize:

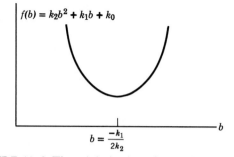

FIGURE 11–8 The minimization of a quadratic function

$$S(a, b) = \sum [(Y_i - a) - bx_i]^2 \qquad (11\text{-}24)$$
$$(11\text{-}10) \text{ repeated}$$

It will be useful to multiply out the square:

$$S(a, b) = \sum [(Y_i - a)^2 - 2bx_i(Y_i - a) + b^2 x_i^2] \qquad (11\text{-}25)$$
$$= \sum (Y_i - a)^2 - 2b \sum x_i(Y_i - a) + b^2 \sum x_i^2 \qquad (11\text{-}26)$$

In the middle term, consider:

$$\sum x_i(Y_i - a) = \sum x_i Y_i - a \sum x_i$$
$$= \sum x_i Y_i + 0$$

Using this to rewrite the middle term of (11-26), we have:

$$S(a, b) = \sum (Y_i - a)^2 - 2b \sum x_i Y_i + b^2 \sum x_i^2 \qquad (11\text{-}27)$$

This is a useful recasting of (11-24), because the first term contains a alone, while the last two terms contain b alone. To find the value of b that minimizes (11-27), therefore, we need consider only the last two terms. According to (11-23), this is minimized when:

$$b = \frac{-\left(-2 \sum x_i Y_i\right)}{2 \sum x_i^2} = \frac{\sum x_i Y_i}{\sum x_i^2} \qquad (11\text{-}16) \text{ proved}$$

To find the value of a that minimizes (11-27), we need consider only the first term. This may be written

$$\sum (Y_i - a)^2 = \sum Y_i^2 - 2a \sum Y_i + na^2$$

According to (11-23) (with a in the role of b), this is minimized when:

$$a = \frac{-\left(-2 \sum Y_i\right)}{2n} = \frac{\sum Y_i}{n} = \overline{Y} \qquad \text{(11-13) proved}$$

CHAPTER 12

Regression Theory

Models are to be used, but not to be believed.

Henri Theil

12-1   THE MATHEMATICAL MODEL

So far, our treatment of a sample of points has only involved mechanically fitting a line. Now we wish to make inferences about the parent population from which this sample was drawn. Specifically, we must consider the mathematical model that allows us to construct confidence intervals and test hypotheses.

(a) Simplifying Assumptions

Consider again the fertilizer-yield example in Chapter 11. Suppose that the experiment could be repeated many times at a fixed level of fertilizer x. Even though fertilizer application is fixed from experiment to experiment, we would not observe exactly the same yield each time. Instead, there would be statistical fluctuation of the Y values, clustered about a central value. We can think of the many possible values of Y forming a population; the probability distribution of Y for a given x we shall call $p(Y/x)$. Moreover, there will be a similar probability distribution for Y at any other experimental level of x. One possible sequence of Y populations is shown in Figure 12-1(a). There obviously would be great problems

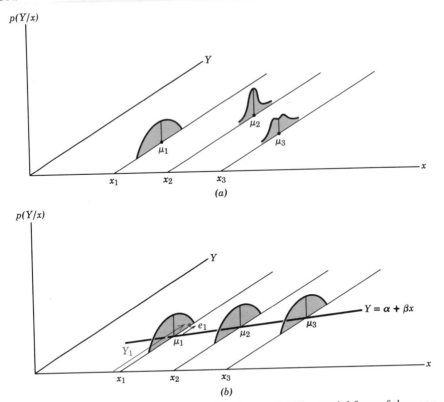

FIGURE 12–1 (a) General populations of Y, given x. (b) The special form of the populations of Y assumed in simple linear regression.

in analyzing populations as peculiar as these. To keep the problem manageable, therefore, we make several assumptions about the regularity of the populations; as shown in Figure 12-1(b), we assume that:

1. The probability distributions $p(Y_i/x_i)$ have the same variance σ^2 for all x_i.

2. The means $E(Y_i)$ lie on a straight line, known as the true (population) regression line:

$$E(Y_i) = \mu_i = \alpha + \beta x_i \qquad (12\text{-}1)$$

The population parameters α and β specify the line; they are to be estimated from sample information.

3. The random variables Y_i are statistically independent. For example, a large value of Y_1 does not tend to make Y_2 large; that is, Y_2 is "unaffected" by Y_1.

These assumptions may be written more concisely as:

> The random variables Y_i are statistically independent, with:
>
> $$\text{mean} = \alpha + \beta x_i$$
>
> and
>
> $$\text{variance} = \sigma^2$$

(12-2)

On occasion, it is useful to describe the deviation of Y_i from its expected value as the error or disturbance term e_i, so that the model alternatively may be written as:

> $$Y_i = \alpha + \beta x_i + e_i$$
>
> where the e_i are independent random variables, with:
>
> $$\text{mean} = 0$$
>
> and
>
> $$\text{variance} = \sigma^2$$

(12-3)

We note that the distributions of Y and e are identical, except that their means differ. In fact, the distribution of e is just the distribution of Y translated onto a zero mean. To emphasize this, in Figure 12-1(b) the observed value of Y_1 is shown in color (our usual convention for observations), along with the corresponding value for the error term, e_1.

No assumption is made about the *shape* of the distribution of e (normal or otherwise), provided that it has a finite variance. We therefore refer to assumptions (12-3) as the "weak set." We shall derive as many results as possible from these before we add a more restrictive normality assumption later.

(b) The Nature of the Error Term

Now let us consider in more detail the "purely random" part of Y_i, the error or disturbance term e_i. Where does it come from? Why doesn't a precise and exact value of Y_i follow, once the value of x_i is given? The error may be regarded as the sum of two components:

1. *Measurement error.* There are various reasons why Y may be measured incorrectly. In measuring crop yield, an error may result from sloppy harvesting or inaccurate weighing. If the example is a study of the consumption of families at various income levels, the measurement error in consumption might consist of budget and reporting inaccuracies.

2. *Stochastic error* occurs because of the inherent irreproducibility of biological and social phenomena. Even if there were no measurement error, continuous repetition of an experiment using exactly the same amount of fertilizer would result in different yields; these differences are unpredictable, and are called "stochastic" or "random." They may be reduced by tighter experimental control—for example, by holding constant soil conditions, amount of water, etc. But *complete* control is impossible—for example, seeds cannot be duplicated. Stochastic error may be regarded as the influence on Y of many omitted variables, each with an individually small effect.

In the social sciences, controlled experiments usually are not possible. For example, an economist cannot hold U.S. national income constant for several years while he examines the effect of interest rate on investment. Since he cannot neutralize extraneous influences by holding them constant, his best alternative is to take them into account explicitly, by regressing Y on x *and* the extraneous factors. This is a useful technique for reducing stochastic error; it is called "multiple regression" and is discussed fully in the next chapter.

(c) Estimating α and β

Suppose that the true regression $E(Y) = \alpha + \beta x$ is the black line shown in Figure 12-2. This is unknown to the statistician, who must estimate it as best he can by observing x and Y. At the first level x_1, if the random error e_1 takes on a negative value, as shown in the diagram, he will observe Y_1 on the low side. Similarly, suppose that he has two more randomly disturbed observations, Y_2 and Y_3. The statistician would estimate the true line by applying the least-squares method of Chapter 11 to the only information he has—the sample values Y_1, Y_2 and Y_3. He would come up with the colored estimating line in this figure, $\hat{Y} = a + bx$. To emphasize that this line provides estimates, we rename it:

$$\hat{Y} = \hat{\alpha} + \hat{\beta}x \tag{12-4}$$

where $\hat{\beta}$ is read "β hat" or "β estimate" (and $\hat{\alpha}$ and $\hat{Y}$ similarly are read "α hat" and "Y hat").[1]

---

[1] This notation is just like the notation $\hat{\theta}$ in (7-1) and $\hat{Y}$ in (11-9).

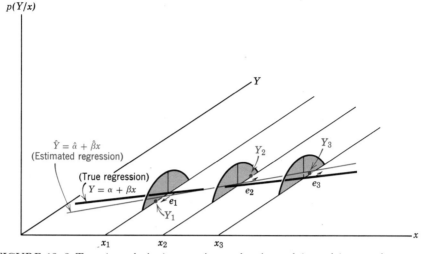

FIGURE 12–2 True (population) regression and estimated (sample) regression

Figure 12-2 is a critical diagram. Before proceeding, you should be sure that you can distinguish clearly between:

1. The true regression and its surrounding e distribution. Since these are population values and cannot be observed, they are shown in black.

2. The Y observations and the resulting fitted regression line. Since these are sample values, they are known to the statistician, and are shown in color.

Unless the statistician is very lucky indeed, it is obvious that his estimated line will not coincide exactly with the true population line. The best that he can hope for is that the least squares method of estimation will be close to the target.

PROBLEMS

12-1 (Monte Carlo.) Suppose that the true (long-run average) relation of corn yield Y to fertilizer X is given by:

$$E(Y) = 60 + .10X$$

(a) Graph this line for $0 \le X \le 400$.
(b) Even when fertilizer is kept constant, suppose that the yield randomly varies from one experimental plot of land to another, with a standard

deviation $\sigma = 10$ and a distribution that is normal. Simulate a sample of five yields, one each from $X = 0, 100, 200, 300, 400$.
(c) Calculate the line that best fits the sample.
(d) Graph the sample and the fitted line.
(e) Have the instructor graph several of the lines such as found in (d).
(f) Have the instructor record the slope estimator $\hat{\beta}$ from every student, and graph the resulting frequency distribution of $\hat{\beta}$. What are the mean and standard error of $\hat{\beta}$, approximately?

12-2   THE MEAN AND VARIANCE OF â AND $\hat{\beta}$

How close does the estimated line come to the true population line? Specifically, how is the estimator $\hat{\alpha}$ distributed around its target α, and $\hat{\beta}$ around its target β?

We shall show that $\hat{\alpha}$ and $\hat{\beta}$ have the following moments:

$$E(\hat{\alpha}) = \alpha \tag{12-5}$$

$$\text{var}\,(\hat{\alpha}) = \frac{\sigma^2}{n} \tag{12-6}$$

$$E(\hat{\beta}) = \beta \tag{12-7}$$

$$\text{var}\,(\hat{\beta}) = \frac{\sigma^2}{\sum x_i^2} \tag{12-8}$$

where σ^2 is the variance of the error e (the variance of Y around the regression line).

Because of its greater importance, we shall concentrate on the slope estimator $\hat{\beta}$, rather than $\hat{\alpha}$, for the rest of the chapter.

proof of (12-7) and (12-8)   The formula for $\hat{\beta}$ in (11-16) may be rewritten as:

$$\hat{\beta} = \sum \left(\frac{x_i}{k}\right) Y_i \tag{12-9}$$

where:

$$k = \sum x_i^2 \tag{12-10}$$

Thus:

$$\hat{\beta} = \sum w_i Y_i = w_1 Y_1 + w_2 Y_2 + \cdots + w_n Y_n \qquad (12\text{-}11)$$

where:

$$w_i = \frac{x_i}{k} \qquad (12\text{-}12)$$

Since each x_i is a fixed constant, so is each w_i. Thus, from (12-11), we establish the important conclusion:

> $\hat{\beta}$ is a weighted sum (i.e., a linear combination) of the random variables Y_i

$(12\text{-}13)$

Hence, by (5-32) we may write:

$$E(\hat{\beta}) = w_1 E(Y_1) + w_2 E(Y_2) + \cdots + w_n E(Y_n) = \sum w_i E(Y_i) \quad (12\text{-}14)$$

Moreover, noting that the varibles Y_i were assumed to be independent, it follows from (5-40) that:

$$\text{var } (\hat{\beta}) = w_1^2 \text{ var } Y_1 + \cdots + w_n^2 \text{ var } Y_n = \sum w_i^2 \text{ var } Y_i \qquad (12\text{-}15)$$

For the mean, substitute (12-1) into (12-14):

$$E(\hat{\beta}) = \sum w_i (\alpha + \beta x_i) \qquad (12\text{-}16)$$
$$= \alpha \sum w_i + \beta \sum w_i x_i \qquad (12\text{-}17)$$

and, noting (12-12):

$$= \frac{\alpha}{k} \sum x_i + \frac{\beta}{k} \sum (x_i) x_i \qquad (12\text{-}18)$$

But $\sum x_i$ is zero, according to (11-6). Thus:

$$E(\hat{\beta}) = 0 + \frac{\beta}{k} \sum x_i^2$$

Finally, from (12-10):

$$E(\hat{\beta}) = \beta \qquad\qquad \text{(12-7) proved}$$

Thus $\hat{\beta}$ is an unbiased estimator of β.

For the variance, substitute (12-2) into (12-15):

$$\text{var}\,(\hat{\beta}) = \sum w_i^2 \sigma^2 \qquad\qquad \text{(12-19)}$$

and, noting (12-12):

$$= \sum \frac{x_i^2}{k^2}\,\sigma^2 \qquad\qquad \text{(12-20)}$$

$$= \frac{\sigma^2}{k^2} \sum x_i^2$$

From (12-10) again, we finally obtain:

$$\text{var}\,(\hat{\beta}) = \frac{\sigma^2}{\sum x_i^2} \qquad\qquad \text{(12-8) proved}$$

A similar derivation of the mean and variance of $\hat{\alpha}$ is left as an exercise.

We observe from (12-12) that in calculating $\hat{\beta}$, the weight w_i attached to the Y_i observation is proportional to the deviation x_i. Hence, outlying observations exert a relatively heavy influence in the calculation of $\hat{\beta}$.

12-3 THE GAUSS-MARKOV THEOREM

The major justification for using the least squares method to estimate a linear regression is the following:

> **Gauss-Markov Theorem**
> Within the class of linear unbiased estimators of β, the least squares estimator $\hat{\beta}$ has minimum variance (is most efficient). $\qquad$ (12-21)
> Similarly, $\hat{\alpha}$ is the minimum variance estimator of α.

This theorem is important because it follows even from the weak set of assumptions (12-3), and hence requires no assumption about the shape of the distribution of the error term. Instead of a proof, we shall give an interpretation of the Gauss–Markov theorem.

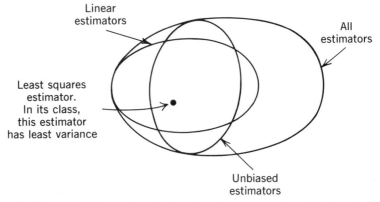

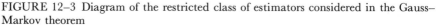

FIGURE 12–3 Diagram of the restricted class of estimators considered in the Gauss–Markov theorem

We already have seen in (12-13) that $\hat{\beta}$ is a linear estimator, and we restrict ourselves to linear estimators because they are easy to compute and analyze. We restrict ourselves even further, as shown in Figure 12-3; within this set of linear estimators, we consider only the limited class that is unbiased. Not only is the least squares estimator in this class, according to (12-7), but, of all the estimators in this class, it has the minimum variance. Therefore, it is often referred to as BLUE, the "best linear unbiased estimator."

The Gauss–Markov theorem has an interesting corollary. As a special case of regression, we might ask what happens if we are explaining Y, but $\beta = 0$ in (12-2), so that no independent variable x comes into play. From (12-2), α is the mean μ of the Y population. Moreover, from (11-13), its least squares estimator is $\overline{Y}$. Thus:

Gauss–Markov Corollary
Within the class of linear unbiased estimators of a population mean μ, the sample mean $\overline{Y}$ has minimum variance.   (12-22)

It must be emphasized that the Gauss–Markov theorem is restricted; it only applies to estimators that are both linear and unbiased. It follows that there may be a nonlinear estimator that has smaller variance than the least squares estimator. For example, to estimate a population mean, the sample median is a *nonlinear* estimator that has smaller variance than the sample mean for certain kinds of nonnormal populations, as we mentioned in Section 7-4.

PROBLEMS

12-2 Suppose that the experimenter of Example 11-1 was in a hurry to analyze his data, and so drew a line joining the first and last points. We shall denote the slope of this line by $\tilde{\beta}$.

(a) Calculate $\tilde{\beta}$ for the data in Table 11-1.

(b) Write out a formula for $\tilde{\beta}$ in terms of X_1, Y_1, X_7, Y_7.

(c) Is $\tilde{\beta}$ a linear estimator?

(d) Is $\tilde{\beta}$ an unbiased estimator of β?

(e) Without doing any calculations, can you say how the variance of $\tilde{\beta}$ compares to the variance of the least squares estimator $\hat{\beta}$?

(f) Verify your answer in (e) by actually calculating the variance of $\tilde{\beta}$ and $\hat{\beta}$ for the data of Example 11-1. (Express your answer in terms of the unknown error variance σ^2.)

12-3 In Problem 12-2, we considered the two extreme values, X_1 and X_7. Now let us consider an alternative pair of less extreme values—say X_2 and X_6. We shall denote the slope of this line by $\bar{\beta}$. Like $\tilde{\beta}$, it can easily be shown to be linear and unbiased.

(a) Calculate the variance of $\bar{\beta}$. How does it compare with the variance of $\tilde{\beta}$?

(b) Answer true or false; if false, correct it. $\tilde{\beta}$ has less variance than $\bar{\beta}$, which illustrates a general principle: The more a pair of observations are spread out, the more "statistical leverage" they exert, hence the more efficient they are.

12-4 How does Problem 7-4 illustrate the Gauss–Markov theorem?

12-4   THE DISTRIBUTION OF $\hat{\beta}$

Now that we have established the mean and variance of $\hat{\beta}$ in (12-7) and (12-8), we ask about the shape of the distribution of $\hat{\beta}$. Let us add (for the first time) the strong assumption that the Y_i are normal. Since $\hat{\beta}$ is a linear combination of the Y_i, it follows from (6-9) that $\hat{\beta}$ also will be normal. But even without assuming that the Y_i are normal, we know that, as sample size increases, the distribution of $\hat{\beta}$ usually will approach normality. This can be justified by a generalized form of the central limit theorem (6-10).[2]

Our objective is to develop a clear intuitive picture of how this estimator varies from sample to sample. First, of course, we note that (12-7)

---

[2]The central limit theorem (6-10) proved the large sample normality of the sample mean $\overline{X}$. It applies also to a *weighted sum* of random variables such as $\hat{\beta}$ in (12-11), under most conditions. Similarly, the normality of $\hat{\alpha}$ is justified.

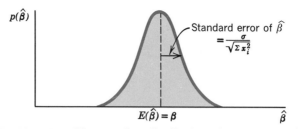

FIGURE 12-4 The sampling distribution of the estimator $\hat{\beta}$

established that $\hat{\beta}$ has its distribution centered on its target β, so that $\hat{\beta}$ is unbiased.

The interpretation of its variance (12-8) is more subtle, with some interesting implications for experimental design. Suppose that the experiment has been designed badly, with the X_i close together. This makes the deviations x_i small; hence $\Sigma\, x_i^2$ is small. Therefore, the variance of $\hat{\beta}$ in (12-8) is large, and $\hat{\beta}$ is a comparatively unreliable estimator. To check the intuitive validity of this, consider the scatter diagram in Figure 12-5(a). The bunching of the X_i means that the small part of the line being investigated is obscured by the error e_i, making the slope estimate $\hat{\beta}$ very unreliable. In this specific instance, our estimate has been pulled badly out of line by the errors—in particular by the one indicated by the arrow.

By contrast, in Figure 12-5(b), we show the case where the X_i are reasonably spread out. Even though the errors e_i remain the same, the estimate $\hat{\beta}$ is much more reliable, because errors no longer exert the same leverage.

As a concrete example, suppose that we wish to examine how sensitive Canadian imports (Y) are to the international value of the Canadian dollar (x). A much more reliable estimate should be possible using the periods when the Canadian dollar was floating (and took on a range of values) than during the periods when this dollar was fixed (and only allowed to fluctuate within very narrow limits).

12-5   CONFIDENCE INTERVALS AND HYPOTHESIS TESTS FOR β

(a) Standard Error of $\hat{\beta}$

Now that we have established the mean, variance, and normality of $\hat{\beta}$, statistical inferences about β are in order. But first we have one remaining problem: σ^2, the variance about the population line, is generally unknown, and so it must be estimated. A natural estimator is to use the

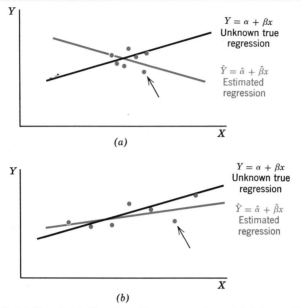

FIGURE 12–5 (a) Unreliable fit when X_i are very close. (b) More reliable fit when X_i are spread out.

deviations about the *fitted* line:

$$s^2 = \frac{1}{n-2} \sum (Y_i - \hat{Y}_i)^2 \qquad (12\text{-}23)$$

like (11-3)

where $\hat{Y}_i$ is the fitted value on the estimated regression line; that is:

$$\hat{Y}_i = \hat{\alpha} + \hat{\beta} x_i \qquad (12\text{-}24)$$

like (12-4)

s^2 often is referred to as "residual variance," a term that is used similarly in analysis of variance. In (12-23), the divisor $(n - 2)$ is used rather than n, in order to make s^2 unbiased.[3]

_____

[3] Unbiasedness may be proved with an argument similar to the one that we used in the footnote to (8-17). But in the present calculation of s^2, *two* estimators, $\hat{\alpha}$ and $\hat{\beta}$, are required; thus two degrees of freedom are lost for s^2. Hence $(n - 2)$ is the divisor in s^2, and also the degrees of freedom for the subsequent t distribution.

An alternative viewpoint may be helpful. If there were only $n = 2$ points in the data, the least-squares line could be fitted. It would turn out to have a perfect fit (through any two given points, a line can always be drawn that goes through them exactly). Thus, although $\hat{\alpha}$ and $\hat{\beta}$ would be determined easily enough, there would be no information at all about the variance of the observations about the line, σ^2. Only to the extent that n exceeds 2 can we get information about

(cont'd)

When s^2 is substituted for σ^2 in (12-8), and when we take the square root, we obtain the estimated standard error:

$$s_{\hat{\beta}} = \frac{s}{\sqrt{\sum x_i^2}}$$ (12-25)

Now we can make statistical inferences.

(b) Confidence Intervals

We can derive the 95% confidence interval for β easily, arriving at a result[4] analogous to (8-22):

$$\beta = \hat{\beta} \pm t_{.025}\, s_{\hat{\beta}}$$ (12-26)

Substituting $s_{\hat{\beta}}$ from (12-25) yields:

95% confidence interval for the slope,

$$\beta = \hat{\beta} \pm t_{.025}\, \frac{s}{\sqrt{\sum x_i^2}}$$ (12-27)

where the degrees of freedom for t are, as always, the same as the divisor in s^2:

$$\text{d.f.} = n - 2$$ (12-28)

Using a similar argument, and noting (12-6), we could easily derive:

95% confidence interval for the intercept,[5]

$$\alpha = \hat{\alpha} \pm t_{.025}\, \frac{s}{\sqrt{n}}$$ (12-29)

σ^2. Thus it is customary to say, "Two degrees of freedom are used up in calculating the estimators for the line, $\hat{\alpha}$ and $\hat{\beta}$, leaving the remaining $n - 2$ degrees of freedom for the estimate of the variance about the line, s^2."

[4]For strict validity, this use of the t distribution requires the strong assumption that the distribution of Y_i is normal; but even when it is not, t often remains a good approximation.

[5]This intercept α of course is for the y-axis placed over the center of the data (as in Figure

(cont'd)

Example 12-1.   Find the 95% confidence interval for the slope in (11-17) that relates wheat yield to fertilizer.

Solution.   To evaluate (12-27), first s^2 is calculated according to (12-23), in the last three columns of Table 11-2 on p. 324. The critical t value then has $n - 2 = 7 - 2 = 5$ degrees of freedom (the same as the divisor in s^2); from Table V, $t_{.025}$ is found to be 2.571. Finally, note that $\hat{\beta}$ and $\Sigma\, x_i^2$ already were calculated in Table 11-2. When these values all are substituted into (12-27):

$$\beta = .059 \pm 2.571\, \frac{5.96}{\sqrt{280,000}}$$

$$= .059 \pm 2.571(.0113) \tag{12-31}$$

$$= .059 \pm .029$$

$$.030 < \beta < .088 \tag{12-32}$$

(c) Testing Hypotheses

The hypothesis typically tested is the null hypothesis,

$$H_0 : \beta = 0$$

Should this be a one-tailed or a two-tailed test? This question must be answered on the basis of prior theoretical reasoning. To illustrate, suppose that we wish to investigate the effect of wages W on the national price level P, using the simple relationship,

$$P = \alpha + \beta\, W$$

On theoretical grounds, it may be concluded *a priori* that if wages affect prices at all, this relation will be a positive one. In this case, H_0 is tested against the one-sided alternative:

11-7(b)). When the y-axis is left in its original position (as in Figure 11-7(a)), the intercept is denoted α_0; in Problem 12-16 we show that its 95% confidence interval is:

$$\alpha_0 = (\overline{Y} - \hat{\beta}\overline{X}) \pm t_{.025}\, s\, \sqrt{\frac{1}{n} + \frac{\overline{X}^2}{\sum x_i^2}} \tag{12-30}$$

$$H_1 : \beta > 0 \qquad\qquad (12\text{-}33)$$

On the other hand, as an example of a case in which such clear prior guidelines do not exist, consider an equation explaining saving. How does it depend on the interest rate? It is not clear, on theoretical grounds, whether this effect is a positive or a negative one. Since interest is the reward for saving, a high interest rate should provide an incentive, leading us to expect interest to affect saving positively. But if individuals save in order to accumulate some target sum (perhaps for their retirement, or to buy a house), then the higher the interest rate, the more rapidly any saving will accumulate, hence the less they need to save to reach this target. In this case, interest affects saving negatively, and so it is appropriate to test H_0 against the two-sided alternative:

$$H_1 : \beta \neq 0 \qquad\qquad (12\text{-}34)$$

In such a two-tailed test,[6] H_0 can be rejected if it is excluded from the corresponding confidence interval for β.

> *Example 12-2.* In Example 12-1, test at the 5% level the null hypothesis that yield is unrelated to fertilizer; i.e., test $H_0 : \beta = 0$ against the two-sided alternative $H_1 : \beta \neq 0$.
>
> *Solution.* Since $\beta = 0$ is excluded from the confidence interval (12-32), the null hypothesis $\beta = 0$ is rejected. Thus, yield really does have a relation to fertilizer.

*(d) Prob-Value

A more appropriate form for a test is the calculation of the prob-value. We first calculate the t statistic:

$$t = \frac{\hat{\beta}}{s_{\hat{\beta}}} \qquad\qquad \begin{array}{l}(12\text{-}35)\\ \text{like } (9\text{-}15)\end{array}$$

Then we calculate the probability in the tail beyond this realized value of t.

---

[6]This may be viewed as testing H_0 (that the two effects cancel out) against H_1 (that one or the other dominates).

Example 12-3. Suppose, in Example 12-1, that chemists know that the fertilizer will not burn or otherwise harm the crop. Then we should test against the one-sided alternative hypothesis $\beta > 0$. What is the prob-value for H_0?

Solution. In the confidence interval (12-31), we already have calculated $\hat{\beta}$ and its standard error $s_{\hat{\beta}}$, which we now can substitute into (12-35):

$$t = \frac{.059}{.0113} = 5.2$$

In Table V, we find that $t = 5.2$ lies between $t_{.005} = 4.032$ and $t_{.001} = 5.893$, so that:

$$.001 < \text{prob-value} < .005 \qquad (12\text{-}36)$$

Remarks. Since the prob-value for H_0 is so very low, we would reject H_0 at the 5% level, or at any reasonable level (according to (9-19)). But the calculation of the prob-value is much more informative than a mere statement of rejection of H_0.

PROBLEMS

12-5 (a) For each of Problems 11-1 to 11-4, construct a 95% confidence interval for the true slope β.

(b) For Problem 11-2, construct a 95% confidence interval for α.

12-6 Suppose that a random sample of 4 families had the following annual income and saving:

Family	Income X (Thousands of \$)	Saving S (Thousands of \$)
A	12	1.0
B	8	1.0
C	7	.6
D	17	2.2

(a) Estimate the regression line $S = \alpha + \beta x$.

(b) Construct a 95% confidence interval for the slope β.

(c) Graph the 4 points and the fitted line, and then indicate as well as you can the acceptable slopes given by the confidence interval in (b).

(d) Construct a 95% confidence interval for α.

12-7 Which of the following hypotheses do the data of Problem 12-6 reject at the 5% level? Use a two-sided test. (*Hint:* (9-5).)

(i) $\beta = 0$
(ii) $\beta = .05$
(iii) $\beta = .10$
(iv) $\beta = .50$

*12-8 Using the data in Problem 12-6,

(a) Suppose that we are interested in making a statement of the form, "The marginal propensity to save (β) is at least as large as such and such." Construct a one-sided 95% confidence interval of this form.

(b) Calculate the prob-value for the null hypothesis $\beta = 0$ (against a one-sided alternative hypothesis: $\beta > 0$).

(c) At the 5% level, can we reject the null hypothesis $\beta = 0$? Use the one-sided alternative hypothesis $\beta > 0$, and test in two ways:

(i) Is $\beta = 0$ excluded from the confidence interval?
(ii) Is the prob-value less than 5%?

*12-9 Repeat Problem 12-8, using instead the data of Problem 11-2.

12-6   INTERPOLATION (INTERVAL ESTIMATES)

In the previous section, we considered the broad aspects of the model, namely, the position of the whole line (determined by α and β). In this section, we shall consider two narrower problems:

(a) For a given value x_0, what is the interval that will predict the corresponding *mean* value of Y_0 (i.e., the confidence interval for $E(Y_0)$ or μ_0)? For example, in our fertilizer problem, we may want an interval estimate of the mean yield resulting from the application of 550 lbs. of fertilizer. (Note that we are not deriving an interval estimate of mean yield by observing repeated applications of 550 lbs. of fertilizer; in that case, we could apply the simpler technique of estimating a population mean with the sample mean. Instead we are observing only seven *different* applications of fertilizer. This is clearly a more difficult problem.)

(b) What is the interval that will predict a *single observed* value of Y_0 (referred to as the prediction interval for an individual Y_0)? Again using our example, what would we predict a single yield to be from an application of 550 lbs. of fertilizer? This individual value clearly is less predictable than the mean value in (a). We now consider both in detail.

(a) The Confidence Interval for the Mean μ_0

First we shall find the point estimator $\hat{\mu}_0$, and then construct an interval estimate around it. The appropriate estimator $\hat{\mu}_0$ is just the point on the estimated regression line above x_0:

$$\hat{\mu}_0 = \hat{\alpha} + \hat{\beta}x_0 \tag{12-37}$$

But as a point estimate, this almost certainly will involve some error, because of errors made in the estimates $\hat{\alpha}$ and $\hat{\beta}$. Figure 12-6 illustrates the effect of these errors. In panel (a), the true regression is shown, along with an estimated regression. Note how $\hat{\mu}_0$ underestimates in this case. In panel (b), the true regression again is shown, but now with several estimated regressions fitted from several possible sets of sample data. The fitted colored dot is sometimes too low, sometimes too high; but on average, it seems just right. To verify this, we note that $\hat{\mu}_0$ in (12-37) is a linear combination of the random variables $\hat{\alpha}$ and $\hat{\beta}$. Thus, from (5-32):

$$E(\hat{\mu}_0) = E(\hat{\alpha}) + x_0 E(\hat{\beta})$$

From (12-5) and (12-7):

$$= \alpha + x_0\beta = \mu_0$$

$$\boxed{E(\hat{\mu}_0) = \mu_0} \tag{12-38}$$

Thus $\hat{\mu}_0$ is indeed an unbiased estimator of μ_0.

Consider next its variance. Because $\hat{\alpha}$ and $\hat{\beta}$ are uncorrelated,[7] from (5-40):

$$\text{var}(\hat{\mu}_0) = \text{var}\,\hat{\alpha} + x_0^2\,\text{var}\,\hat{\beta}$$

From (12-6) and (12-8):

$$= \frac{\sigma^2}{n} + x_0^2 \frac{\sigma^2}{\sum x_i^2}$$

---

[7] One reason for redefining the x variable as a deviation from the mean was to make the covariance of $\hat{\alpha}$ and $\hat{\beta}$ zero. The proof, straightforward but tedious, is omitted.

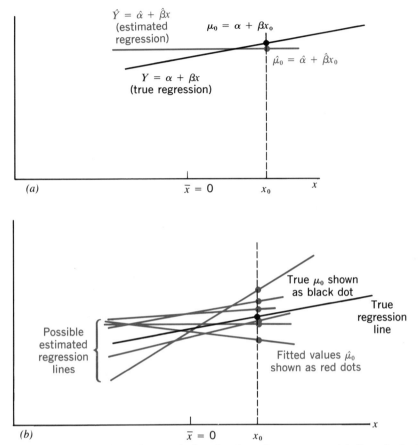

FIGURE 12–6 How the estimator $\hat{\mu}_0$ is related to the target μ_0. (a) One single $\hat{\mu}_0$. (b) A whole series of possible $\hat{\mu}_0$.

$$\text{var}(\hat{\mu}_0) = \sigma^2 \left(\frac{1}{n} + \frac{x_0^2}{\sum x_i^2} \right) \qquad (12\text{-}39)$$

where, of course, x_0 is the new specified x value, while the x_i are the originally observed x values. To interpret (12-39), we note that this variance (uncertainty) of $\hat{\mu}_0$ has two components, due to the variance (uncertainty) of $\hat{\alpha}$ and of $\hat{\beta}$, respectively. The uncertainty term resulting from $\hat{\beta}$ increases as x_0^2 increases, that is, the further x_0 is from the central value 0. This also can be seen from Figure 12-6(b): if x_0 were further to the right, then the estimates $\hat{\mu}_0$ would be spread out over an even wider range.

Finally, we note that $\hat{\alpha}$ and $\hat{\beta}$ are normal (as we established at the beginning of Section 12-4). It follows from (6-9) that $\hat{\mu}_0$ also is normal. Thus, the 95% confidence interval for μ_0 is:

$$\boxed{\begin{array}{l} \text{95\% confidence interval,} \\[2mm] \mu_0 = \hat{\mu}_0 \pm t_{.025}\, s\, \sqrt{\dfrac{1}{n} + \dfrac{x_0^2}{\sum x_i^2}} \end{array}}$$

$$\begin{array}{l} \text{(12-40)} \\ \text{like (8-22)} \end{array}$$

where, of course, s has been substituted for the unknown σ, and the t distribution (with d.f. $= n - 2$) has been used correspondingly. When x_0 is set at its central value of 0, note how this confidence interval reduces simply to the confidence interval for α in (12-29). In this case, there is no uncertainty introduced by estimating β.

(b) The Prediction Interval for an Individual Y_0

In predicting a single observed Y_0, once again the best estimate is the point on the estimated regression line above x_0. In other words, the best *point* prediction for Y_0 is:

$$\hat{Y}_0 = \hat{\alpha} + \hat{\beta} x_0 = \hat{\mu}_0 \tag{12-41}$$

When we try to find the *interval* estimate for Y_0, we will face all the problems involved in the interval for the mean, μ_0. And we have an additional problem because we are trying to estimate only one observed Y, rather than the more stable average of all the possible Ys. Hence, to the previous variance (12-39), we now must add the inherent variance σ^2 of an individual Y observation, obtaining:

$$\sigma^2\left(\frac{1}{n} + \frac{x_0^2}{\sum x_i^2}\right) + \sigma^2 = \sigma^2\left(\frac{1}{n} + \frac{x_0^2}{\sum x_i^2} + 1\right) \tag{12-42}$$

Except for this larger variance, the prediction interval for Y_0 is the same as the confidence interval for μ_0:

$$\boxed{\begin{array}{l} \text{95\% prediction interval for an individual } Y \text{ observation,} \\[2mm] Y_0 = \hat{\mu}_0 \pm t_{.025}\, s\, \sqrt{\dfrac{1}{n} + \dfrac{x_0^2}{\sum x_i^2} + 1} \end{array}} \tag{12-43}$$

with the t distribution again having $(n - 2)$ degrees of freedom.

Example 12-4. If 550 lbs. of fertilizer are applied in the wheat example (11-17), find a 95% interval estimate for:
(a) The mean wheat yield that we would obtain if we planted many many plots (μ_0).
(b) The wheat yield on just one plot (Y_0).

Solution. The point estimate is the same in both cases. We simply calculate:

$$x_0 = X_0 - \overline{X}$$
$$= 550 - 400 = 150 \tag{12-44}$$

and substitute it into the equation of the estimated line (11-17):

$$\hat{\mu}_0 = \hat{Y}_0 = 60 + .059(150)$$
$$= 68.8 \tag{12-45}$$

(a) For an interval estimate for μ_0, substitute (12-44) and (12-45) into (12-40) along with s^2 and $\Sigma\, x_i^2$ from Table 11-2:

$$\mu_0 = 68.8 \pm 2.571(5.96)\sqrt{\frac{1}{7} + \frac{150^2}{280,000}}$$
$$= 68.8 \pm 7.2 \tag{12-46}$$

(b) For an interval estimate for Y_0, (12-43) yields the same calculation except for an extra 1 under the square-root sign:

$$Y_0 = 68.8 \pm 2.571(5.96)\sqrt{\frac{1}{7} + \frac{150^2}{280,000} + 1}$$
$$= 68.8 \pm 16.9 \tag{12-47}$$

This interval is more than twice as wide as (12-46), which shows how much more difficult it is to predict an *individual* observation than a *mean*.

The relationship of prediction and confidence intervals is shown in Figure 12-7. The two potential sources of error in a confidence interval for the mean are shown in panels (a) and (b); these are combined to form the dark band in panel (c). The wider, lighter band in (c) gives the predic-

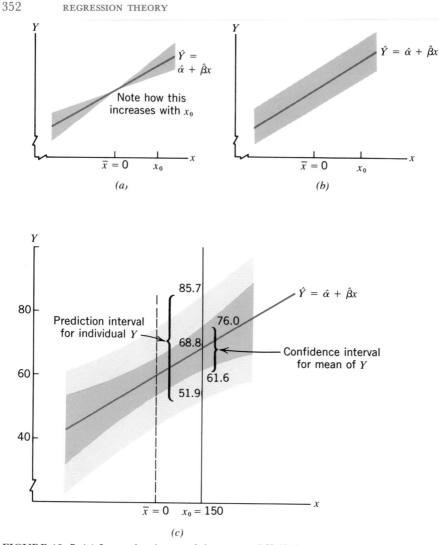

FIGURE 12–7 (a) Interval estimate of the mean of Y, if there were error only in estimating β. (b) Interval estimate of the mean of Y, if there were error only in estimating α. (c) Interval estimate for mean of Y, and prediction interval for individual Y, when there is error in estimating both α and β.

tion intervals for individual Y observations. Note how both bands expand as x_0 moves farther away from its central value of zero; this reflects the fact that x_0^2 appears in both variances.

PROBLEMS

12-10 Using the data of Problem 11-2, what is the 95% prediction interval for the saving of a family with the following income (in thousands of dollars):

(a) 6?

(b) 8?

(c) 10?

(d) 12?

(e) Which of these four intervals is least precise? Most precise? Why?

(f) How is the answer in (b) related to the confidence interval for α found in Problem 12-5(b)?

12-11 Repeat Problem 12-10, calculating instead a 95% confidence interval for the *average* saving of all families at each given income level.

12-12 (a) Suppose that you are trying to explain how the interest rate affects investment. Would you prefer to take observations over a period in which the Federal Reserve is trying to stabilize the rate, or a period in which it is allowed to vary widely?

(b) Suppose that you have estimated the regression of saving on income using data from families in the $8,000 to $16,000 income range. Would you feel more confident predicting the future saving of a family with a $16,000 income, or a family with a $10,000 income? Why?

12-7   DANGERS OF EXTRAPOLATION

We emphasize that, in Formulas (12-40) and (12-43), x_0 may be *any* value of x. If x_0 lies *among* the observed values $x_1 \ldots x_n$, the process is called interpolation. (If x_0 *is* one of the observed values $x_1 \ldots x_n$, the process might be called, "using also the other values of x to sharpen our knowledge of this one population at x_0.") If x_0 is out beyond the observed values $x_1 \ldots x_n$, then the process is called extrapolation. The techniques developed in Section 12-6 may be used for extrapolation, but only with great caution, as we shall see.

There is no sharp division between safe interpolation and dangerous extrapolation. Rather, there is *continually* increasing danger of misinterpretation as x_0 gets further and further from its central value.

(a) Statistical Risk

We emphasized in the previous section that prediction intervals get larger as x_0 moves away from the center. This is true, even if all the assumptions underlying our mathematical model hold exactly.

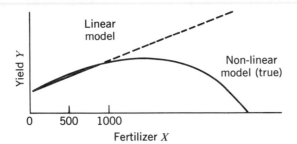

FIGURE 12–8 Comparison of linear and nonlinear models

(b) Risk of Invalid Model

In practice, we must recognize that a mathematical model is never absolutely correct. Rather, it is a useful approximation. In particular, we cannot take seriously the assumption that the population means are strung out in an *exactly* straight line. If we consider the fertilizer example, it is likely that the true relation increases initially, but then bends down eventually as a "burning point" is approached, and the crop is over-dosed. This is illustrated in Figure 12-8, which is an extension of Figure 11-2 with the scale appropriately changed. In the region of interest, from 0 to 700 lbs., the relation is *practically* a straight line, and no great harm is done in assuming the linear model. However, if the linear model is extrapolated far beyond this region of experimentation, the result becomes meaningless. In such cases, a nonlinear model should be considered (as in Chapter 15).

12-8   CONCLUDING OBSERVATIONS

Two points warrant emphasis. First, most of the theory of this chapter—and, in particular, the Gauss–Markov justification of least squares—requires no assumption of normality of the error term (i.e., normality of Y). The one exception occurred in Sections 12-4 to 12-6, where the normality assumption was required only for small sample estimation—and this because of a quite general principle that small sample estimation requires a normally distributed parent population to strictly validate the t distribution. But even here, t is often a reasonably good approximation in nonnormal populations.

Second, we have assumed that the independent variable x has taken on a given set of fixed values (for example, fertilizer application was set at certain specified levels). But in many cases, x cannot be controlled in this way. For example, if we are examining the effect of rainfall on yield, we must recognize that x (rainfall) is a random variable that is completely outside our control. The surprising thing is that most of this chapter remains valid whether x is fixed *or* a random variable, provided that we assume, as well as (12-3), that:

$$\boxed{\begin{array}{l} \sigma^2 \text{ (and } \alpha \text{ and } \beta) \text{ are independent of } x, \text{ and} \\ \text{the error } e \text{ is statistically independent of } x \end{array}} \qquad \begin{array}{l} (12\text{-}48) \\ (12\text{-}49) \end{array}$$

This greatly generalizes the application of the regression model.

REVIEW PROBLEMS

12-13 (a) Let us define the variance of the X values as:

$$\sigma_X^2 \triangleq \frac{1}{n} \sum x^2 \qquad (12\text{-}50)$$

Then show that the standard error of $\hat{\beta}$ can be written as:

$$\frac{\sigma/\sigma_X}{\sqrt{n}}. \qquad (12\text{-}51)$$

This formula has the advantage of showing explicitly how the reliability of $\hat{\beta}$ depends upon the following three components:
   (i) σ^2, the residual variance about the line.
   (ii) σ_X^2, the variance of the X values.
   (iii) n, the sample size.
(b) Suppose that an agricultural economist cannot change the inherent inaccuracy of the individual observation (σ^2), but she does have a chance to change her experimental design. State how much the standard error of $\hat{\beta}$ will change if she takes:
   (i) Four times as many observations, spread over the same X range.
   (ii) The same number of observations, spread over 4 times the former X range.
   (iii) Half as many observations, spread over twice the former X range.

12-14 A class of 150 registered students wrote two tests, for which the grades were denoted X_1 and X_2. The instructor calculated the following summary statistics:

$$\overline{X}_1 = 60 \qquad \overline{X}_2 = 70$$

$$\sum (X_1 - \overline{X}_1)^2 = 36{,}000 \qquad \sum (X_2 - \overline{X}_2)^2 = 24{,}000$$

$$\sum (X_1 - \overline{X}_1)(X_2 - \overline{X}_2) = 15{,}000$$

residual variances about the fitted lines (X_1 on X_2, and X_2 on X_1):

$$s^2_{X_1/X_2} = 180, \qquad s^2_{X_2/X_1} = 120$$

The instructor then discovered that there was one more student, who was unregistered; worse yet, one of this student's grades (X_1) was lost, although the other grade was discovered ($X_2 = 55$). The dean told the instructor to estimate the missing grade X_1 as closely as possible.

(a) Calculate the best estimate you can. (*Hint:* use (11-19).)

(b) Calculate an interval about your estimate in (a) that you have 95% confidence will contain the true grade.

(c) What assumptions did you have to make implicitly in (a) and (b)?

12-15 In order to estimate this year's inventory, a tire company sampled 6 dealers, in each case getting inventory figures for both this year and last:

X = Inventory Last Year	Y = Inventory This Year
70	60
260	320
150	230
100	120
20	50
60	60

Summary statistics are $\overline{X} = 110$, $\overline{Y} = 140$.

$$\sum x^2 = 36{,}400 \qquad \sum y^2 = 61{,}800 \qquad \sum xy = 46{,}100$$

residual variances about the fitted lines (X on Y, and Y on X):

$$s^2_{X/Y} = 500 \qquad s^2_{Y/X} = 850$$

(a) Calculate the least squares line showing how this year's inventory Y is related to last year's X. (*Hint:* use (11-19).)

(b) Suppose that a complete inventory of all dealers is available for last year (but not for this year). Suppose also that the mean inventory for last year was found to be $\mu_X = 180$ tires per dealer. On the graph below, we show this population mean μ_X and sketch the population scatter (although this scatter remains unknown to the company, because Y values are not available yet). On this graph, plot the six observed points, along with $\overline{X}$, $\overline{Y}$, and the estimated regression line.

(c) Indicate on the graph how μ_Y should be estimated. Construct a 95% confidence interval for μ_Y.

(d) Construct a 95% confidence interval for μ_Y, if last year's data X had been unavailable or ignored (i.e., using only Y values).

(e) Comparing (c) to (d), state in words the value of exploiting prior knowledge about last year's inventory.

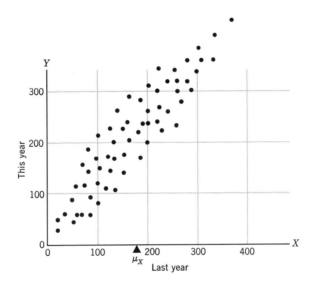

12-16 (a) Prove (12-30). (*Hint:* (12-40))

(b) For the fertilizer data in Table 11-2 calculate a 95% confidence interval for α_0.

(c) For the same data, calculate a 95% confidence interval for α (the y-intercept, using the deviation form x.). Is this confidence interval wider or narrower than in (b)? Why?

12-17 (a) In Problem 5-15 (e), we calculated $E(Y/X)$, for $X = 10, 15$, and 20. Graph these 3 points. Is the linearity assumption (12-1) satisfied?

(b) Continue Problem 5-15: calculate var (Y/X) for $X = 10, 15$, and 20. Is the assumption of constant variance preceding equation (12-1) satisfied?

CHAPTER 13

Multiple Regression

"The cause of lightning," Alice said very decidedly, for she felt quite sure about this, "is the thunder—no, no!" she hastily corrected herself, "I meant it the other way."

"It's too late to correct it," said the Red Queen, "When you've once said a thing, that fixes it, and you must take the consequences."

Lewis Carroll

13-1   INTRODUCTION

Multiple regression is the extension of simple regression, to take account of more than one independent variable X. It is obviously the appropriate technique when we want to investigate the effects on Y of several variables simultaneously. Yet, even if we are interested in the effect of only one variable, it usually is wise to include the other variables influencing Y in a multiple regression analysis, for two reasons:

1. To reduce stochastic error (as we discussed in Section 12-1). The objective here, as in introducing a second factor in ANOVA, is to reduce the residual variance s^2, and hence increase the strength of our statistical tests.

2. Even more important, to eliminate bias that might result if we just ignored a variable that substantially affects Y.

Example 13-1. Suppose that the fertilizer and yield observations in Example 11-1 were taken at seven different agricultural experiment stations across the country. If soil conditions and temperature were essentially the same in all these areas, we still might ask whether part of the fluctuation in Y (i.e., the disturbance term e) can be explained by varying levels of rainfall in different areas. A better prediction of yield may be possible if *both* fertilizer and rainfall are examined. The observed levels of rainfall are therefore given in Table 13-1, along with the original observations of yield and fertilizer from Table 11-1.

TABLE **13-1**
Observed Yield, Fertilizer Application,
and Rainfall

Y *Wheat Yield* *(Bushels/Acre)*	X *Fertilizer* *(Pounds/Acre)*	Z *Rainfall* *(Inches)*
40	100	10
50	200	20
50	300	10
70	400	30
65	500	20
65	600	20
80	700	30

(a) On Figure 11-2, tag each point with its value of rainfall Z. Then, considering just those points with low rainfall ($Z = 10$), roughly fit a line by eye. Next, repeat for the points with moderate rainfall ($Z = 20$), and then for the points with high rainfall ($Z = 30$).

(b) Now, if rainfall were kept constant, roughly estimate what the slope of yield on fertilizer would be. That is, what would be the increase in yield per pound of fertilizer?

(c) If fertilizer were kept constant, roughly estimate what would be the increase in yield per inch of rainfall.

(d) What would you estimate would be the yield if fertilizer were 400 pounds, and rainfall were 10 inches?

Solution.

(a)

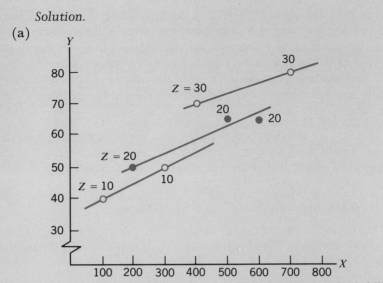

FIGURE 13–1 How yield depends on 2 variables (fertilizer and rainfall)

(b) The largest slope in Figure 13-1 is $10/200 = .05$ for the line $Z = 10$, while the smallest slope is $10/300 = .033$ for the line $Z = 30$; on average, these slopes are about .04 bushels per pound of fertilizer.

(c) The vertical distance between the line where $Z = 10$ and the line where $Z = 30$ is about 15 bushels. Since this increase of 15 bushels comes from an increase of 20 inches of rain, this means that rain increases yield by about $15/20 = .75$ bushels per inch of rainfall.

(d) On Figure 13-1, we use the line where $Z = 10$, at the point where $X = 400$, obtaining a yield of 55 bushels.

Example 13-1 illustrates the two reasons why the additional variable Z improves our analysis:

1. We have a better fit of the data. Note how the fluctuations about the fitted lines in Figure 13-1 are less than in Figure 11-2. This should allow us to make more precise statistical conclusions about how X affects Y.

2. Figure 13-1 shows the relationship of yield to fertilizer while rainfall is held constant. If rainfall is not held constant and is ignored, we obtain the slope in Figure 11-2; this slope is larger because high rainfall tends to accompany high fertilizer. Thus our estimate in Figure 11-2 was biased because we were erroneously attributing to fertilizer the effects of both fertilizer *and* rainfall. (Any estimate that tends to undershoot or overshoot like this is by definition called "biased.")

Example 13-1 was vastly oversimplified. We need to develop a more objective, easily computerized method that will handle the more complicated cases where fitting by eye is out of the question.

13-2   THE MATHEMATICAL MODEL

Yield Y now is to be regressed on the two independent variables, or "regressors": fertilizer X and rainfall Z. Let us suppose that the relationship is of the form:

$$E(Y_i) = \alpha + \beta x_i + \gamma z_i \qquad (13\text{-}1)$$
$$\text{like (12-1)}$$

where both regressors x and z are measured as deviations from their means. Geometrically, this equation is a plane in the three-dimensional space shown in Figure 13-2. (The geometry of planes and lines is reviewed

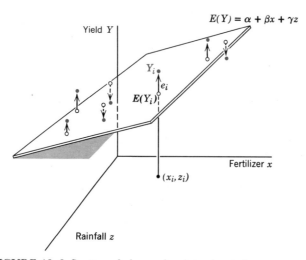

FIGURE 13–2 Scatter of observed points about the true regression plane

in Appendix 13-A.) For any given combination of rainfall and fertilizer (x_i, z_i), the expected yield $E(Y_i)$ is the point on this plane directly above, shown as a hollow dot. Of course, the observed value of Y_i is very unlikely to fall precisely on this plane. For example, the observed Y_i at this particular fertilizer-rainfall combination is somewhat greater than its expected value, and is shown as the colored dot lying directly above this plane.

The difference between the observed and expected value of Y_i is the stochastic or error term e_i. Thus, any observed value Y_i may be expressed as its expected value plus this disturbance term:

$$Y_i = \alpha + \beta x_i + \gamma z_i + e_i \qquad (13\text{-}2)$$

with the assumptions about e_i the same as in Chapter 12.

like (12-3)

β is interpreted geometrically as the slope of the plane as we move in the x-direction, keeping z constant; thus β is the marginal effect of fertilizer x on yield Y. Similarly, γ is the slope of the plane as we move in the z-direction, keeping x constant; thus γ is the marginal effect of z on Y.

13-3 LEAST SQUARES ESTIMATION

Least squares estimates again are derived by selecting the estimates $\hat{\alpha}$, $\hat{\beta}$, and $\hat{\gamma}$ that minimize the sum of the squared deviations between the observed Y_i and the fitted $\hat{Y}_i$; that is:

$$\text{minimize} \sum (Y_i - \hat{\alpha} - \hat{\beta} x_i - \hat{\gamma} z_i)^2 \qquad (13\text{-}3)$$

like (11-10)

This is done with calculus by setting the partial derivatives with respect to $\hat{\alpha}$, $\hat{\beta}$, and $\hat{\gamma}$ equal to zero (or algebraically, by a technique similar to that used in Appendix 11-A). The result is the following three estimating equations (sometimes called normal equations):

$$\hat{\alpha} = \overline{Y}$$
$$\sum x_i Y_i = \hat{\beta} \sum x_i^2 + \hat{\gamma} \sum x_i z_i \qquad (13\text{-}4)$$
$$\sum z_i Y_i = \hat{\beta} \sum x_i z_i + \hat{\gamma} \sum z_i^2$$

Again, note that the intercept estimate $\hat{\alpha}$ is the average $\overline{Y}$. The second and third equations may be solved for $\hat{\beta}$ and $\hat{\gamma}$. These calculations are shown in Table 13-2, and yield the fitted multiple regression equation:

TABLE **13-2**
Least Squares Multiple Regression of Y on X and Z

Y	X	Z	$x = X - \bar{X}$	$z = Z - \bar{Z}$	xY	zY	x^2	z^2	xz
40	100	10	−300	−10	−12,000	−400	90,000	100	3,000
50	200	20	−200	0	−10,000	0	40,000	0	0
50	300	10	−100	−10	−5,000	−500	10,000	100	1,000
70	400	30	0	10	0	700	0	100	0
65	500	20	100	0	6,500	0	10,000	0	0
65	600	20	200	0	13,000	0	40,000	0	0
80	700	30	300	10	24,000	800	90,000	100	3,000
$\sum Y$ $= 420$ $\bar{Y} = 60$	$\sum X$ $= 2800$ $\bar{X} = 400$	$\sum Z$ $= 140$ $\bar{Z} = 20$	$0 \checkmark$	$0 \checkmark$	$\sum xY$ $= 16,500$	$\sum zY$ $= 600$	$\sum x^2$ $= 280,000$	$\sum z^2$ $= 400$	$\sum xz$ $= 7,000$

Estimating Equations (13-4) $\begin{cases} 16,500 = 280,000\hat{\beta} + 7,000\hat{\gamma} \\ 600 = 7,000\hat{\beta} + 400\hat{\gamma} \end{cases}$

Solution $\begin{cases} \hat{\beta} = .0381 \\ \hat{\gamma} = .833 \end{cases}$

$$\hat{Y} = 60 + .0381x + .833z \qquad (13\text{-}5)$$
$$= 60 + .0381(X - 400) + .833(Z - 20)$$
$$= 28.1 + .0381X + .833Z$$

Note that the slope of Y on x is .0381, which closely agrees with the eyeball estimate of .04 in Example 13-1(b). Also note that the slope of Y on z is .833, which closely agrees with the eyeball estimate of .75 in Example 13-1(c).[1]

PROBLEMS

13-1 Suppose that a random sample of five families yielded the following data (an extension of Problem 11-2, where everything is measured in thousands of dollars):

Family	Saving S	Income X	Assets W
A	.6	8	12
B	1.2	11	6
C	1.0	9	6
D	.7	6	3
E	.3	6	18

(a) Estimate the multiple regression equation of S on X and W.
(b) Does the coefficient of Y differ from the answer to Problem 11-2?
(c) For a family with assets of 5 thousand and income of 8 thousand dollars, what would you predict saving to be?
(d) If a family had a 2 thousand dollar increase in income, while assets remained constant, estimate by how much their saving would increase.
(e) If a family had a 1 thousand dollar increase in income, and a 3 thousand dollar increase in assets, estimate by how much their saving would increase.

(13-2) Suppose that a random sample of five families yielded the following data (another extension of Problem 11-2):

---

[1] Fitting a plane to a scatter can be interpreted as fitting three evenly spread and *parallel* lines (with common slope $\hat{\beta}$) to the scatter in Figure 13-1. This gives a result slightly different from our original, less precise method of fitting the three lines freely, and then taking the average of the three slopes.

Family	Saving S	Income X	Number of Children N
A	.6	8	5
B	1.2	11	2
C	1.0	9	1
D	.7	6	3
E	.3	6	4

(a) Estimate the multiple regression of S on X and N.

*b) For a family with 5 children and income of 6 thousand dollars, what would you predict saving to be?

13-3 Suppose that the data in Problems 13-1 and 13-2 apply to exactly the same families. Then this can be combined to obtain the following table:

Family	Saving S	Income X	Assets W	Number of Children N
A	.6	8	12	5
B	1.2	11	6	2
C	1.0	9	6	1
D	.7	6	3	3
E	.3	6	18	4

Measuring the independent variables as deviations from the mean, we want the estimated equation:

$$\hat{S} = \hat{\alpha} + \hat{\beta}x + \hat{\gamma}w + \hat{\psi}n$$

(a) Generalizing (13-4), use the least squares criterion to derive the system of 4 equations in the 4 unknown estimates, $\hat{\alpha}$, $\hat{\beta}$, $\hat{\gamma}$, and $\hat{\psi}$.

*(b) Using a table such as Table 13-2, calculate the 4 estimates.

13-4   MULTICOLLINEARITY

(a) In Simple Regression

In Figure 12-5(a), we showed how the estimate $\hat{\beta}$ became unreliable if the X_i were closely bunched, that is, if the regressor X had little variation. It will be instructive to consider the limiting case, where the X_i are concentrated on one single value $\overline{X}$, as in Figure 13-3. Then $\hat{\beta}$ is not determined at all. There are any number of differently sloped lines passing

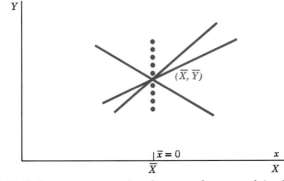

FIGURE 13–3 Degenerate regression, because of no spread (variation) in X

through $(\overline{X}, \overline{Y})$ that fit equally well: for each line in Figure 13-3, the sum of squared deviations is the same, since the deviations are measured vertically from $(\overline{X}, \overline{Y})$. This geometric fact has an algebraic counterpart. If all $X_i = \overline{X}$, then all $x_i = 0$, and the term involving $\hat{\beta}$ in (11-10) is zero; hence, the sum of squares does not depend on $\hat{\beta}$ at all. It follows that any $\hat{\beta}$ will do equally well in minimizing the sum of squares. Another way of looking at the same problem is that, since all x_i are zero, Σx_i^2 in the denominator of (11-16) is zero, and $\hat{\beta}$ is not defined.

In conclusion, when the values of X show little or no variation, then the effect of X on Y can no longer be sensibly investigated. But if the problem is *predicting* Y—rather than investigating Y's dependence on X—this bunching of the X values does not matter *provided that* we limit our prediction to this same value of X. All the lines in Figure 13-3 predict Y equally well. The best prediction is $\overline{Y}$, and all these lines give us that result.

(b) In Multiple Regression

Again consider the limiting case where the values of the independent variables X and Z are completely bunched up on a line L, as in Figure 13-4. This means that all the observed points in our scatter lie in the vertical plane running up through L. You can think of the three-dimensional space as a room in a house; our observations are not scattered throughout this room, but instead lie embedded in an extremely thin pane of glass standing vertically on the floor.

In explaining Y, multicollinearity makes us lose one dimension. In the earlier case of simple regression, our best fit for Y was not a line, but rather a point $(\overline{X}, \overline{Y})$; in this multiple regression case, our best fit for Y is not a plane, but rather the line F. To get F, we just fit the least squares

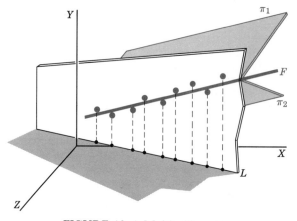

FIGURE 13–4 Multicollinearity

line through the points on the vertical pane of glass. The problem is identical to the one shown in Figure 11-2; in one case, a line is fitted on a flat pane of glass; in the other case, on a flat piece of paper. This regression line F is therefore our best fit for Y. As long as we stick to the same *combination* of X and Z—that is, as long as we confine ourselves to predicting Y values on that pane of glass—no special problems will arise. We can use the regression F on the glass to predict Y in exactly the same way as we did in the simple regression analysis of Chapter 11.

But there is no way to examine how X affects Y. Any attempt to define β, the marginal effect of X on Y (holding Z constant), involves moving off that pane of glass; and we have no sample information whatsoever on what the world out there looks like. Or, to put it differently, if we try to explain Y with a plane—rather than a line F—we find that there are any number of planes running through F (e.g., π_1 and π_2) that do an equally good job. Since each passes through F, each yields an identical sum of squared deviations; thus each provides an equally good fit. This is confirmed in the algebra in the estimating equations (13-4). When X is a linear function of Z (i.e., when x is a linear function of z), it may be shown that the last two equations are dependent, and cannot be solved uniquely[2] for $\hat{\beta}$ and $\hat{\gamma}$.

---

[2]Two linear equations usually can be solved for two unknowns, but not always. For example, suppose that John's age (X) is twice Harry's (Y). Then we can write:

$$X = 2Y$$

$$5X = 10Y$$

These two equations tell us the same thing. We have two equations in two unknowns, but they don't generate a unique solution, because they don't provide independent information.

Now let us be less extreme in our assumptions and consider the near-limiting case where Z and X are almost on a line (i.e., where all our observations in the room lie very close to a vertical pane of glass). In this case, a plane may be fitted to our observations, but the estimating procedure is very unstable; it becomes very sensitive to random errors, reflected in large variances of the estimators $\hat{\beta}$ and $\hat{\gamma}$. Thus, even though the relation of Y to X may be real, it may be indiscernible because the standard error of $\hat{\beta}$ is so large. This is analogous to the argument in the simple regression case in Section 12-4.

When the independent variables X and Z are collinear, or nearly so (i.e., highly correlated), it is called the *problem of multicollinearity*.[3] As we indicated earlier, this does not raise problems in predicting Y, provided there is no attempt to predict for values of X and Z removed from their line of collinearity. But structural questions cannot be answered—the influence of X alone (or Z alone) on Y cannot be investigated sensibly.

Example 13-2.   (Continuation of Example 11-1.) Suppose that X is still the amount of fertilizer, measured in pounds per acre. But now suppose that the statistician makes the incredibly foolish error of also measuring fertilizer in ounces per acre, and using it as another regressor, Z. Since any weight measured in ounces must be sixteen times its measurement in pounds:

$$Z = 16X \qquad (13\text{-}6)$$

Thus, all combinations of X and Z must fall on this straight line, and we have an example of perfect collinearity. Now if we try to fit[4] a regression plane to the observations of yield and fertilizer given in Table 11-1, one possible answer would be the original regression given in (11-18):

$$\hat{Y} = 36.4 + .059X + 0Z \qquad (13\text{-}7)$$

What other answers are possible?

Solution.   An equally satisfactory solution would follow from substituting $X = \frac{1}{16}Z$ into (13-7):

$$\hat{Y} = 36.4 + 0X + .0037Z$$

---

[3]Actually, it would be more accurate to call it the problem of *collinearity*, and reserve the term *multicollinearity* if there are many regressors $X, Z, \ldots$ that are highly correlated.

Any number of solutions could be obtained by making a partial substitution for X into (13-7) with arbitrary weight λ:

$$\hat{Y} = 36.4 + .059[\lambda X + (1 - \lambda)X]$$

$$= 36.4 + .059[\lambda X + (1 - \lambda)\tfrac{1}{16}Z]$$

$$\hat{Y} = 36.4 + .059\lambda X + .0037(1 - \lambda)Z \qquad (13\text{-}8)$$

By assigning successive values to λ, we could generate successive planes (such as π_1 and π_2 in Figure 13-3). But all these three-dimensional planes are just equivalent expressions for the simple two-dimensional relationship between fertilizer and yield. While all give the same correct prediction of Y, no meaning can be attached to whatever coefficients of X and Z we may come up with.

---

[4] Since a computer program would probably break down, suppose that the calculations are done by hand.

While the previous extreme example may have clarified some of the theoretical issues, no statistician would make such a naive error in practice. Instead, more subtle difficulties arise. For example, suppose that demand for a group of goods is being related to prices and income, with the overall price index being the first regressor. Suppose that aggregate income measured in money terms is the second regressor. If this is real income multiplied by the same price index, the problem of multicollinearity may become a serious one. The solution is to use real income, rather than money income, as the second regressor. This is a special case of a more general warning: in any multiple regression where price is one regressor, beware of other regressors that are measured in prices.

This completes our discussion of the method and problems involved in simply fitting a plane to a three dimensional scatter of observations. Now we can turn to questions of statistical inference.

13-5   CONFIDENCE INTERVALS AND STATISTICAL TESTS

(a) Standard Error

As in simple regression, the true relation of Y to X is measured by the unknown population parameter β; we estimate it with the sample estimator $\hat{\beta}$. Although the unknown β is fixed, our estimator $\hat{\beta}$ is a random variable, differing from sample to sample. The properties of $\hat{\beta}$ may be established, just as in the previous chapter. Thus $\hat{\beta}$ may be shown to be normal—again provided that the sample size is large, or that the error term is normal. $\hat{\beta}$ can also be shown to be unbiased, with its mean equal to β. The fluctuation of $\hat{\beta}$ is measured by its standard error; this is very complicated to calculate, even in the relatively simple case of only two regressors.[5] Customarily, it is calculated by an electronic computer, using a library program; then the user does not need to know the formula. Instead, he needs to understand the *meaning* of the standard error, which is quite analogous to the simple regression case.

For example, in Table 13-3 we produce the computer output for the wheat-yield data of Table 13-2. In successive columns, the computer prints, for each regressor:

1. The name.
2. The estimated coefficient.
3. The standard error.

TABLE **13-3**
Computer Output for Multiple
Regression of Wheat Yield[a]

| Multiple Regression Equation | | |
Variable	Coefficient	Std. Error
Const	28.09524	2.49148
Fert	.03810	.00583
Rain	.83333	.15430
Residual Variance = 5.35714		

[a]Data in Table 13-1.

From these basic quantities, we can easily go on to calculate confidence intervals and tests.

---

[5]With only two regressors X and Z, the standard error of $\hat{\beta}$ is:

$$s_{\hat{\beta}} = \frac{s}{\sqrt{\sum x_i^2 - \left(\sum x_i z_i\right)^2 / \sum z_i^2}}$$ (13-9)

where s^2, as usual, is the residual variance, the analogue of (12-23) with a divisor of $n - 3$ (since 3 estimates are now required, $\hat{\alpha}$, $\hat{\beta}$, and $\hat{\gamma}$). With more regressors, the formula for $s_{\hat{\beta}}$ would be longer. The notation (but not the calculations) could then be greatly simplified by using matrices. (See, for example, chapter 13 of R. J. Wonnacott and T. H. Wonnacott, *Econometrics,* 2nd edition, Wiley, New York, 1978.) Matrices are also used in computer programs.

(b) Confidence Intervals

The formula for the 95% confidence interval is of the standard form:

$$\boxed{\beta = \hat{\beta} \pm t_{.025} \, s_{\hat{\beta}}}$$

(13-10)

like (12-26)

With k regressors, the degrees of freedom[6] for t are:

$$\boxed{\text{d.f.} = n - k - 1}$$

(13-11)

like (12-28)

Example 13-3. From the computer output in Table 13-3, calculate 95% confidence intervals for the two regression coefficients.

Solution. From (13-11), d.f. $= 7 - 2 - 1 = 4$, so that Table V gives $t_{.025} = 2.776$. Also substitute $\hat{\beta}$ and $s_{\hat{\beta}}$ from Table 13-3, and then (13-10) yields:

for the fertilizer coefficient:

$$\beta = .03810 \pm 2.776(.00583)$$
$$\simeq .038 \pm .016$$

(13-12)

for the rainfall coefficient:

$$\gamma = .83333 \pm 2.776(.15430)$$
$$\simeq .83 \pm .43$$

(13-13)

(c) Prob-Value

The t ratio to test $\beta = 0$ is, as usual:

$$\boxed{t = \frac{\hat{\beta}}{s_{\hat{\beta}}}}$$

(13-14)

like (12-35)

---

[6] Just as in (12-28), the degrees of freedom for t are always the same as the divisor in s^2. This is just the number of observations n reduced by the number of estimates $(\hat{\alpha}, \hat{\beta}_1, \hat{\beta}_2, \ldots, \hat{\beta}_k)$. That is, d.f. $= n - (k + 1)$.

Example 13-4. (a) From the computer output in Table 13-3, calculate the prob-value for the null hypothesis $\beta = 0$ (fertilizer does not improve yield).
(b) Repeat for the coefficient of rainfall, γ.

Solution. (a) From (13-14):

$$t = \frac{.03810}{.00583} = 6.53$$

From Table V:

$$.001 < \text{prob-value}[7] < .005 \qquad (13\text{-}15)$$

Since the prob-value, which measures the credibility of H_0, falls below .05, according to (9-19) we could reject H_0 at the 5% level.
(b) For the null hypothesis $\gamma = 0$, (13-14) gives:

$$t = \frac{.83333}{.15430} = 5.40$$

From Table V:

$$.001 < \text{prob-value} < .005 \qquad (13\text{-}16)$$

In other words, the null hypothesis has such a small prob-value in both cases that it may be concluded that yield *is* related to both fertilizer and rainfall.

----

[7] The null hypothesis is that "fertilizer does not improve yield" ($\beta = 0$, or strictly speaking, $\beta \leq 0$). The alternative hypothesis is that fertilizer does improve yield ($\beta > 0$). The prob-value is therefore one-sided too.

For a two-sided prob-value, we would double the numbers in (13-15). Then for a two-sided test, we would reject H_0 at the 5% level if the two-sided prob-value fell below 5%—or equivalently, if $\beta = 0$ fell outside the 95% confidence interval.

It is customary to summarize the calculations of Example 13-4, along with the original information of Table 13-3, by arranging them in equation form, as follows:

$$\boxed{\text{YIELD} = 28.1 + .038 \text{ FERTILIZER} + .833 \text{ RAINFALL}} \quad (13\text{-}17)$$

standard error	.0058	.154
t ratio	6.5	5.4
95% CI	± .016	± .43

13-6   HOW MANY REGRESSORS SHOULD BE RETAINED?

The t ratio for fertilizer or rainfall in (13-17) would lead us to reject H_0 at the 5% level (since the prob-value is less than 5%). We therefore should retain fertilizer and rainfall as statistically discernible variables (or, to use the traditional phrase, "statistically significant variables"); in this case, there are no problems.

But now suppose that we had weaker data (perhaps because of a smaller sample); accordingly, suppose that the standard error for rainfall Z was .55 (instead of .15). Then the t ratio would be $t = .833/.55 = 1.51$, which does *not* let us reject H_0 at the 5% level. If we use this evidence to actually accept H_0 (no effect of rainfall), and thus drop rainfall as a regressor, we may encounter the same difficulty that we discussed in Section 9-4. Since this is so important in regression analysis, let us review the argument briefly.

Although it is true that a t ratio of 1.51 for rainfall Z is statistically indiscernible, this *does not* prove that there is no relationship between Z and Y. It is easy to see why. We have strong biological grounds for believing that yield Y is positively related to rainfall Z. In (13-17), this belief is confirmed by the positive coefficient $\hat{\gamma} = .833$. Thus our statistical evidence is consistent with our prior belief, even though it is a weaker confirmation than we would like.[8] To actually accept the null hypothesis $\gamma = 0$, and to conclude that Z does not affect Y, would be to contradict directly both the (strong) prior belief and the (weak) statistical evidence. We would be reversing a prior belief, even though the statistical evidence weakly confirmed it. And this would remain true for any positive t ratio— although, as t became smaller, our statistical confirmation would become weaker. Only if $\hat{\gamma}$ is zero or negative do the statistical results contradict our prior belief.

It follows from this that, if we had strong prior grounds for believing that Z is related positively to Y, Z should not be dropped from the regres-

---

[8] Perhaps the confirmation is weak because of too small a sample. Thus, .833 possibly may be an accurate description of how Y is related to Z; but our t value may be statistically indiscernible because the sample is small, causing the standard error to be relatively large.

sion equation; instead, it should be retained, with all the pertinent information on its confidence interval, t ratio, etc.

On the other hand, what if our prior belief is that H_0 is approximately true? Then the decision to drop or retain a variable would be different. For example, a weak observed relationship (such as $t = 1.51$) would be in some conflict with our prior expectation of no relationship. But it is so minor a conflict that it is easily explained by chance (prob-value $\simeq .10$). Hence, resolving it in favor of our prior expectation and continuing to use H_0 as a working hypothesis might be a reasonable judgment. Under such circumstances, this regressor would be dropped from the equation.

In the case of regression, there is another argument that may lead a statistician with very weak prior belief (in either H_0 or H_1) to accept H_0 when the test yields a statistically indiscernible result: it keeps the model simple, and conserves degrees of freedom to strengthen tests on other regressors.[9] (When a regressor is dropped, the equation involving the fewer remaining regressors must be recalculated; then, since the number of regressors k is reduced, d.f. $= n - k - 1$ is increased.)

We conclude once again that classical statistical theory alone does not provide absolutely firm guidelines for accepting H_0; acceptance must be based also on extrastatistical judgment. Thus, prior belief plays a key role, not only in the initial specification of which regressors should be in the equation, but also in the decision about which ones should be dropped in the light of the statistical evidence, as well as in the decision on how the model eventually will be used.

Prior belief plays a less critical role in the rejection of an hypothesis, but it is by no means irrelevant. Suppose, for example, that although you believed Y to be related to three variables, you didn't really expect it to be related to a fourth; someone had just suggested that you "try on" the fourth at a 5% level. This means that if H_0 (no relation) is true, there is a 5% chance of ringing a false alarm (and erroneously concluding that a relation does exist). If this is the *only* variable that is "tried on," then this is a risk that you can live with. However, if many similar variables are included in a multiple regression by someone who is "bag-shaking"

---

[9] A more sophisticated argument for accepting H_0 in these circumstances goes like this. Even though we know that $\beta = 0$ (i.e., accepting H_0 and dropping this regressor) must be somewhat wrong, we know that $\beta = \hat{\beta}$ (rejecting H_0, and retaining the regressor) is somewhat wrong too, but in a different way. Whereas the estimate 0 has some bias but no variance (it was arbitrarily prespecified), the estimate $\hat{\beta}$ has no bias but some variance. Which is more wrong? The best criterion, which combines bias and variance, is the mean squared error (7-8). This leads to accepting H_0:

(a) If we think that the bias is small (if H_0 was considered near the truth, a priori); and/or

(b) If $\hat{\beta}$ has a lot of variance, either because of a small sample, or multicollinearity problems.

These are recognized as the two conditions for accepting H_0 that we already encountered above.

(i.e., trying on everything in sight),[10] then the chance of a false alarm increases dramatically.[11] Of course, this risk can be kept small by reducing the level for each t test from 5% to 1% or less. This has led some statisticians to suggest a 1% level with the variables just being "tried on," and a 5% level with the other variables that are expected to affect Y.

To sum up, hypothesis testing should not be done mechanically. It requires:

1. Good judgment and good prior understanding of the model being tested.

2. An understanding of the assumptions and limitations of the statistical techniques.

PROBLEMS

13-4 Suppose that a multiple regression of Y on three independent variables yields the following estimates, based on a sample of $n = 30$:

$$\hat{Y} = 25.1 + 1.2X_1 + 1.0X_2 - .50X_3$$

standard error	(2.1)	(1.5)	(1.3)	(.06)
t ratio	(11.9)	( )	( )	( )
95% CI	(±4.3)	( )	( )	( )

(a) Fill in the brackets.
(b) Answer true or false; if false, correct it.
    (i) If there were strong prior reasons for believing that X_1 is unrelated to Y, it is reasonable to reject the null hypothesis $\beta_1 = 0$ at the 5% level.
    (ii) If there were strong prior reasons for believing that X_2 is positively related to Y, it is reasonable to use the estimated coefficient 1.0 rather than accept the null hypothesis $\beta_2 = 0$.

13-5 A recent study of several hundred professors' salaries in a large American university in 1969[12] yielded the following multiple regression (in the

---

[10]This procedure of trying on irrelevant variables is also referred to as the "kitchen sink" approach (i.e., tossing everything into the equation including the kitchen sink), or simply, "garbage in, garbage out" (the fact that we can expect to screen out 95% of the garbage that goes in is little consolation, since the 5% that comes out may still be garbage).

[11]Suppose, for simplicity, that the t tests for the several variables (say k of them) were independent. Then the probability of no error at all is $(.95)^k$. For $k = 10$, for example, this is .60, which makes the probability of some error (some false alarm) as high as .40.

[12]From D. A. Katz, "Faculty Salaries, Promotions, and Productivity at a Large University," *American Economic Review*, June 1973. We referred to this same article in Problems 2-1, 8-15, and 10-3, but now we are using more information than just sex.

interests of brevity, we omit many terms, some of which we will discuss later):

$$\hat{S} = 230B + 18A + 100E + 490D + 190Y + 50T + \cdots$$

standard error	(86)	(8)	(28)	(60)	(17)	(370)
t ratio	( )	( )	( )	( )	( )	( )
95% CI	( )	( )	( )	( )	( )	( )

where S = the professor's annual salary (dollars)
B = number of books he has written
A = number of ordinary articles
E = number of excellent articles
D = number of Ph.D.s supervised
Y = number of years' experience
T = teaching score as measured by student evaluations, severely rounded: the best half of the teachers were rounded up to 100% (i.e., 1), the worst half were rounded down to 0.

(a) Fill in the brackets below the equation.

(b) Answer true or false; where false, correct it. (Actually, since this is real data, you may find that certain points are controversial, rather than simply true or false. In such cases, clarify and support your own point of view as well as you can.)

(i) The coefficient of B is estimated to be 230. Other social scientists might collect other samples from the same population and calculate other estimates. The distribution of these estimates would be centered around the true population value of 230. Therefore, the estimator is called unbiased.

(ii) Since there is strong prior reason to believe that T does not affect S, it is reasonable to accept the null hypothesis that its coefficient is zero, and so drop it from the equation. This will, fortunately, make the equation briefer.

(iii) Repeat (ii), substituting Y for T.

(c) For someone who knows no statistics, briefly summarize the influences on professors' incomes, by indicating where strong evidence exists and where it does not.

13-6 Give an example of a multiple regression of Y on X_1 and X_2, in which you would retain X_1 in the equation but drop X_2, even though its coefficient had a higher t ratio.

13-7 Suggest possible additional regressors that might be used to improve the multiple regression analysis of wheat yield.

13-8 True or false? If false, correct it.

(a) Multicollinearity occurs when the regressors are linearly related, or nearly so.

(b) This means that some regression coefficients will have large standard errors.

(c) Some regressors therefore may be statistically indiscernible; if these regressors also are regarded *a priori* as unimportant, they may be dropped from the model.

(d) Then, when the regression equation is recalculated, the multicollinearity problem will be reduced.

13-9 Suppose that your roommate is a bright student, but that he has studied no economics, and little statistics. (Specifically, he understands only simple—but not multiple—regression.) In trying to explain what influences the U.S. price level, he has regressed U.S. prices on 100 different economic variables one at a time (i.e., in 100 simple regressions). Moreover, he apparently selected these variables in a completely haphazard way without any idea of potential cause-and-effect relations. He discovered 5 variables that were statistically discernible at the level $\alpha = 5\%$, and concluded that each of these has an influence on U.S. prices.

(a) Explain to him in simple terms what reservations, if any, you have about his conclusion.

(b) If he had uncovered 20 statistically discernible variables, would your criticism remain the same? How would you suggest that he improve his analysis?

13-7   INTERPRETATION OF REGRESSION: "OTHER THINGS BEING EQUAL"

The coefficients in a linear regression model have a very simple but important interpretation, which we shall now consider. (The same interpretation is reviewed geometrically in Appendix 13-A.)

(a) Simple Regression Reviewed

Recall the simple regression model of Chapter 12:

$$Y = \alpha + \beta x \qquad (13\text{-}18)$$

(In this section we will ignore the error term e, since we are interested in interpreting β for models with or without a stochastic error term.) It often is very useful to interpret β as:

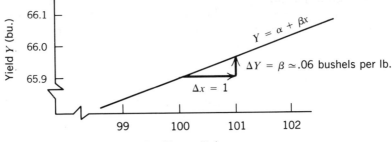

FIGURE 13–5 Regression coefficient β = slope = increase in Y that accompanies a unit increase in x.

$$\boxed{\beta = \text{increase in } Y \text{ if } x \text{ is increased by one unit}} \qquad (13\text{-}19)$$

For example, in the relation of wheat yield Y to fertilizer x, β is the increase in yield when fertilizer is increased one pound (called "marginal physical product" of fertilizer). This is illustrated in Figure 13-5.

To prove (13-19) algebraically, let us consider what happens when x is increased by 1 unit, say from the initial x_0 to $(x_0 + 1)$. Then the increase in yield Y can be found by solving for Y in (13-18), both before and after the increase in x:

$$\begin{aligned}
\text{initial } Y &= \alpha + \beta x_0 \\
\text{new } Y &= \alpha + \beta(x_0 + 1) \\
\hline
\text{difference} = \text{increase in } Y &= \beta
\end{aligned} \qquad (13\text{-}19) \text{ proved}$$

(b) Nonlinear Regression

To appreciate the linear model, it is useful to contrast it with a more complicated model, for example, the quadratic model:

$$Y = \alpha + \beta x + \gamma x^2$$

When the marginal product of x is calculated as before, by increasing x one unit, then:

$$\begin{aligned}
\text{initial } Y &= \alpha + \beta x_0 + \gamma x_0^2 \\
\text{new } Y &= \alpha + \beta(x_0 + 1) + \gamma(x_0 + 1)^2 \\
\hline
\text{difference} = \text{increase in } Y &= \beta + 2\gamma x_0 + \gamma
\end{aligned} \qquad (13\text{-}20)$$

In this case, the marginal productivity of x is no longer simply the coefficient β. It also involves the coefficient γ, and the level x_0. Thus, a major advantage of the linear model is that β has such a clear and direct interpretation.

(c) Multiple Regression

Consider again the multiple regression model:

$$Y = \alpha + \beta x + \gamma z \tag{13-21}$$

For example, wheat yield Y may depend on both fertilizer x and rainfall z. The interpretation of β now is:

$$\beta = \text{the increase in } Y \text{ if } x \text{ is increased}$$
$$\text{one unit, } \textit{while } z \textit{ is held constant} \tag{13-22}$$

To prove this, we keep z constant (say at z_0), while we increase x from x_0 to $(x_0 + 1)$. Then from (13-21):

$$\text{initial } Y = \alpha + \beta x_0 + \gamma z_0$$
$$\underline{\text{new } Y = \alpha + \beta(x_0 + 1) + \gamma z_0}$$
$$\text{increase in } Y = \beta \qquad \text{(13-22) proved}$$

For the general linear model:

$$Y = \alpha + \beta_1 x_1 + \beta_2 x_2 + \cdots + \beta_k x_k \tag{13-23}$$

it may be confirmed that the interpretation of each coefficient is similar:

$$\boxed{\begin{array}{l} \beta_i = \text{the increase in } Y \text{ if } x_i \text{ is increased one unit,} \\ \textit{while all other } x \textit{ variables are held constant} \end{array}} \tag{13-24}$$

13-8   SIMPLE AND MULTIPLE REGRESSION COMPARED

In order to evaluate the benefits of a proposed irrigation scheme in a certain region, suppose that the relation of yield Y to rainfall R is investigated over several years. From the data set out in the first 3 columns of Table 13-4, we could calculate the simple regression equation:

TABLE 13-4

Yield, Rainfall and Temperature Over
Several Years

Year	Yield, Y (Bu./Acre)	Total Spring Rainfall R (Inches)	Average Spring Temperature T (° Fahr.)
1963	60	8	56
1964	50	10	47
1965	70	11	53
1966	70	10	53
1967	80	9	56
1968	50	9	47
1969	60	12	44
1970	40	11	44

$$\hat{Y} = 60 - 1.67\, r \qquad (13\text{-}25)$$

standard error 4.0

But the negative coefficient (indicating that rainfall *reduces* yield!) strongly suggests that something in this analysis has gone very wrong. Actually, even before we calculated the regression (13-25), we should have known that it might be wrong, because it only measures the simple relation of Y to R. What we really need to know is how yield is related to rainfall, *while all other important variables are held constant.* According to (13-24), therefore, we should carry out a multiple regression of yield on rainfall *and* the other important variables such as temperature, obtaining:

$$\hat{Y} = 60 + 5.71\, r + 2.95\, t \qquad (13\text{-}26)$$

standard error 2.68   .69

This regression yields a much more reasonable conclusion. Rainfall R does have the expected effect of increasing yield, other things being equal (i.e., when T is constant).[13] This is the positive direct relation, and we show it in Figure 13-6.

---

[13] The magnitude of the rainfall coefficient (5.71) also is important: it estimates the productivity of water, which then can be compared with the cost of water in order to determine whether irrigation is worthwhile.

Although temperature was introduced into the regression equation primarily to clarify the relation of yield to rainfall, the temperature coefficient is of some interest in its own right. We note that it is positive, and even more discernible than rainfall.

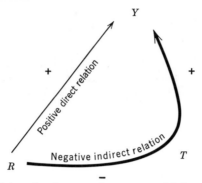

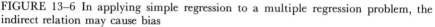

FIGURE 13–6 In applying simple regression to a multiple regression problem, the indirect relation may cause bias

To see why (13-25) yielded the wrong sign, we must realize that in the data from which it was calculated, T is *not* held constant. In fact, T and R are inversely related; for example, in 1970, high rainfall occurs with low temperature. Low temperature in turn results in low yields (because of the positive coefficient of t in (13-26)). Thus, high rainfall, because it lowers temperature, *indirectly* lowers yield. This is the negative indirect relation, shown in Figure 13-6. Whereas multiple regression sorts out and isolates the direct relation, simple regression does not; instead, the simple regression coefficient reflects both direct and indirect effects (in our example, positive direct effect of rainfall on yield, and its negative indirect effect). If the indirect relation predominates (as in our example), then it will determine the sign of the simple regression coefficient. But even in situations less spectacular than this, the simple regression coefficient still will be biased to some extent by the indirect effect of an omitted regressor.[14] This confirms our conclusion that we stated at the beginning of this chapter: the great advantage of multiple regression is to eliminate bias.[15]

---

[14]Unless, of course, the omitted regressor is totally unrelated to the other regressors.

It may seem that we are giving contradictory advice about related regressors. Didn't we warn earlier that highly related regressors might cause a problem of multicollinearity? Here is how we resolve the paradox:

To keep multicollinearity to a minimum, we should design an experiment or collect data that has as little relation as possible among the regressors (for example, as we suggested at the end of Section 13-4, if one regressor is price, the second regressor should be real income, rather than money income). In other words, we try to get our regressors as unrelated as possible; but having done this, we then must live with them. Multicollinearity should not make us simply omit a regressor that we believe to be important, since this omission would introduce bias.

[15]To illustrate this issue geometrically, it would be interesting to graph the data of Table 13-4. We would plot Y against R, and label each point according to its value of T—just as in Example 13-1.

(cont'd on page 383)

Another advantage of multiple regression is to reduce the residual variance;[16] this reduces the standard error of the coefficient of r, from 4.0 in (13-25) to 2.68 in (13-26). Therefore, statistical tests and confidence intervals are strengthened. Of course, the addition of other regressors (as well as R and T) might further reduce bias and residual variance.

PROBLEMS

13-10 Suppose that a psychologist computed the following multiple regression on the basis of a random sample of 60 people from a large population:

$$\hat{Y} = 64 + 16X_1 - 1.2X_2$$

standard error	1.06	3.0	1.7
95% CI	±2.1	±6.0	±3.4

Select the statement that is most appropriate for the coefficient $\hat{\beta}_1 = 16$ (and criticize the other statements):

(a) $\hat{\beta}_1$ estimates the total increase in Y that would accompany a unit increase in X_1 and the associated increase in X_2.

(b) $\hat{\beta}_1$ estimates the total increase in Y that would be caused by a unit increase in X_1 (while X_1 simultaneously caused an estimated decrease of 1.2 units in X_2).

(c) $\hat{\beta}_1$ estimates the increase in Y that would accompany a unit increase in X_1, if X_2 were held constant.

(d) $\hat{\beta}_1$ is a fixed parameter that estimates the sample coefficient β_1, with a mean of 16 and variance of 3.

(e) The null hypothesis ($\hat{\beta}_1 = 0$) should be accepted at the 5% level.

13-11 In Problem 13-5, recall that the salary regression was

If you carry this out, you will see why the simple regression coefficient (ignoring T) was negative in (13-25), although the multiple regression coefficients were positive in (13-26).

Incidentally, we can express precisely how an (inappropriate) simple regression coefficient may be biased by the indirect effect, by the following theorem:

simple regression coefficient = direct effect + indirect effect (bias)

$$\hat{\beta}_{Yr} = \hat{\beta}_{Yr/t} + \hat{\beta}_{Yt/r}\hat{\beta}_{tr} \tag{13-27}$$

where

$\hat{\beta}_{Yr}$ = simple regression coefficient of Y on r
$\hat{\beta}_{tr}$ = simple regression coefficient of t on r
$\hat{\beta}_{Yr/t}$ = multiple regression coefficient of Y on r (holding t fixed)
$\hat{\beta}_{Yt/r}$ = multiple regression coefficient of Y on t (holding r fixed).

[16]The computer showed that the residual variance was reduced from $s^2 = 194$ in (13-25) to $s^2 = 50$ in (13-26).

$$\hat{S} = 230B + 18A + 100E + 490D + 190Y + 50T + \cdots$$

(a) Answer true or false; if false, correct it.

(i) Other things being equal, we estimate that a professor who has written one or more books earns $230 more annually.

(ii) Or, to draw an analogy with the fertilizer-yield relation of Problem 11-1(d), we might say that $230 estimates the value (in terms of a professor's salary) of writing one or more books.

(iii) Other things being equal, we estimate that a professor who is one year older earns $190 more annually. In other words, the annual salary increase averages $190.

(b) Similarly, interpret all the other coefficients for someone who knows no statistics.

13-12 (a) Should your previous answer to Problem 13-8(c) and (d) be changed, if the regressors are regarded *a priori* as important? If so, how?

(b) What further criticism do you now have of your roommate's analysis in Problem 13-9(a)?

13-13 Cigarette smokers have a life expectancy of about 5 years less than non-smokers.[17] Is all this necessarily caused by their smoking? Which way do you think the bias is (if any)?

13-14 Answer true or false; if false, correct it.

(a) The simple regression equation (13-25) occasionally can be useful. For example, in the absence of any information on temperature, it would correctly lead us to hope for a year with low rainfall rather than high.

(b) In view of the positive multiple regression coefficients in (13-26), however, it would be even better to hope for a year with low rainfall, and then irrigate.

13-15 Referring to Table 13-4, suppose that we wished to know the relation of Y to R (other things being equal). Would there be a bias in the simple regression of Y on R:

(a) If R and T were positively related? If so, in which direction?

(b) If R and T had no relation? If so, in which direction?

(c) Now answer true or false; if false, correct it.

Applying simple regression to a multiple regression problem will introduce bias if the independent variables are unrelated.

---

[17] A simplified version from "11–12–13, World Conference on Smoking and Health," New York, 1967, p. 23. Actually, life expectancy depended on the amount smoked. For 25-year-old American men, the reduction in life expectancy ranged from 4.6 years for light smokers (<10 cigarettes daily) to 8.3 years for heavy smokers (≥ 40 cigarettes daily).

*13-16 (a) From Table 13-4, calculate the regression coefficient of T on R.

(b) Using also the regression coefficients calculated in (13-25) and (13-26), verify (13-27).

13-9   DUMMY (0–1) VARIABLES

(a) Including a Dummy Variable

Suppose that we wish to investigate how the public purchase of government bonds (B) is related to national income (Y). A hypothetical scatter of annual observations of these two variables is shown for Canada in Figure 13-7 and in Table 13-5. It immediately is evident that the relationship of bonds to income follows two distinct patterns—one for wartime (1940–1945), the other for peacetime.

The normal relation of B to Y (shown as the line L_0) is subject to an upward shift (to L_1) during wartime; heavy bond purchases in those years is explained not by Y alone, but also by selling campaigns whose appeal was based on patriotism. B therefore should be related to Y *and* another variable—war W. W does not have a whole range of values, but only two: we set its value at 1 for all wartime years and at 0 for all peacetime years. (W is a 0–1 or dummy variable of the kind that we encountered in Section 6-5.)[18] Therefore:[19]

$$E(B) = \alpha_0 + \beta Y + \gamma W \tag{13-29}$$

where

$$W = 0 \text{ for peacetime years}$$
$$= 1 \text{ for wartime years.} \tag{13-30}$$

By substituting (13-30) into (13-29), we obtain the following equivalent pair of equations:

---

[18]Once again, note that this 0–1 coding is not entirely arbitrary; it allows an easy verbal interpretation of W:

$$W = \text{the number of wars Canada was fighting in the given year.} \tag{13-28}$$

This 0–1 definition also is motivated by the simplicity that it brings to the multiple regression analysis. In particular, when the coefficient of W is given the customary interpretation (13-24), it gives the increase in response (bond sales) if W is increased one unit (as we go from peace 0 to war 1), if the other variable (income) is held constant. This is just the distance between lines L_0 and L_1 in Figure 13-7, as we soon will confirm by comparing (13-32) with (13-31).

[19]In equation (13-29), since Y and W are measured as original values (rather than deviations), we call the constant α_0 instead of α. This issue, which we first raised in Figure 11-7, is relatively trivial and will not occur again.

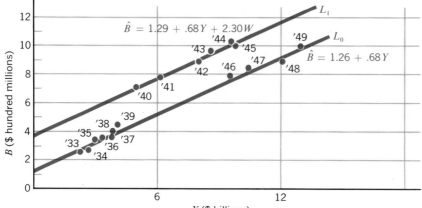

FIGURE 13–7 Hypothetical scatter of public purchases of bonds (B) and national income (Y)

$$E(B) = \alpha_0 + \beta Y \qquad \text{for peacetime} \qquad (13\text{-}31)$$

$$E(B) = \alpha_0 + \beta Y + \gamma \qquad \text{for wartime} \qquad (13\text{-}32)$$

We note that γ represents the effect of wartime on bond sales, and that β represents the effect of income changes. (The latter is assumed to remain the same in war or peace.) The important point is that one multiple regression of B on Y and W, as in (13-29), will yield the *two* estimated lines shown in Figure 13-7; L_0 is the estimate of the peacetime function (13-31), and L_1 is the estimate of the wartime function (13-32).

Complete calculations for this example are set out in Table 13-5, and the procedure is interpreted in Figure 13-8. Since all observations are at $W = 0$ or $W = 1$, the scatter is confined to the two vertical planes π_0 and π_1. The estimated regression plane,

$$\hat{B} = \hat{\alpha}_0 + \hat{\beta} Y + \hat{\gamma} W \qquad (13\text{-}33)$$

may be viewed as a plane resting on two supporting buttresses L_0 and L_1; some of the observed dots, of course, lie above this fitted plane, and others below it. The slopes of L_0 and L_1 are (by assumption) equal to the common[20] value $\hat{\beta}$, and the estimated wartime shift is $\hat{\gamma}$.

---

[20]This means that L_0 and L_1 are not fitted independently. In other words, the least squares plane (13-33) has a slope $\hat{\beta}$ that tries to fit *all* the data as well as possible. Thus, $\hat{\beta}$ cannot be the best

(cont'd on page 388)

TABLE 13-5

Calculations for Regression of B on Y and W, where W is a Dummy Variable

Year	B	Y	W	$y = Y - \bar{Y}$	$w = W - \bar{W}$	yw	By	Bw	y^2	w^2
1933	2.6	2.4	0	−4.44	−.35	1.55	−11.54	−.91	19.71	.12
1934	3.0	2.8	0	−4.04	−.35	1.41	−12.12	−1.05	16.32	.12
1935	3.6	3.1	0	−3.74	−.35	1.31	−13.46	−1.26	13.99	.12
1936	3.7	3.4	0	−3.44	−.35	1.20	−12.73	−1.29	11.83	.12
1937	3.8	3.9	0	−2.94	−.35	1.03	−11.17	−1.33	8.64	.12
1938	4.1	4.0	0	−2.84	−.35	0.99	−11.64	−1.43	8.07	.12
1939	4.4	4.2	0	−2.64	−.35	0.92	−11.62	−1.54	6.97	.12
1940	7.1	5.1	1	−1.74	.65	−1.13	−12.35	4.62	3.03	.42
1941	8.0	6.3	1	−.54	.65	−.35	−4.32	5.20	.29	.42
1942	8.9	8.1	1	1.26	.65	.82	11.21	5.78	1.59	.42
1943	9.7	8.8	1	1.96	.65	1.27	19.01	6.30	3.84	.42
1944	10.2	9.6	1	2.76	.65	1.79	28.15	6.63	7.62	.42
1945	10.1	9.7	1	2.86	.65	1.86	28.89	6.56	8.18	.42
1946	7.9	9.6	0	2.76	−.35	−.97	21.80	−2.77	7.62	.12
1947	8.7	10.4	0	3.56	−.35	−1.25	30.97	−3.05	12.67	.12
1948	9.1	12.0	0	5.16	−.35	−1.81	46.96	−3.19	26.63	.12
1949	10.1	12.9	0	6.06	−.35	−2.12	61.21	−3.53	36.72	.12

(War years bracket 1940–1945)

$$\sum B = 115 \qquad \sum Y = 116.3 \qquad \sum W = 6$$

$$\bar{B} = \frac{115}{17} \qquad \bar{Y} = \frac{116.3}{17} \qquad \bar{W} = \frac{6}{17}$$

$$= 6.76 \qquad\qquad = 6.84 \qquad\qquad = .35$$

$$\sum yw \qquad \sum By \qquad \sum Bw \qquad \sum y^2 \qquad \sum w^2$$

$$= 6.55 \qquad = 147.2 \qquad = 13.74 \qquad = 193.7 \qquad = 3.88$$

Estimating equations (13-4)
$$\begin{cases} \sum By = \hat{\beta} \sum y^2 + \hat{\gamma} \sum yw \\ \sum Bw = \hat{\beta} \sum yw + \hat{\gamma} \sum w^2 \end{cases}$$

or
$$\begin{cases} 147.2 = 193.7\hat{\beta} + 6.55\hat{\gamma} \\ 13.74 = 6.55\hat{\beta} + 3.88\hat{\gamma} \end{cases}$$

Solution:
$$\begin{cases} \hat{\beta} = .681 \approx .68 \\ \hat{\gamma} = 2.30 \end{cases}$$

Thus our estimated regression is: $B = 6.76 + .68y + 2.30w$

Or, in terms of the original variables: $B = 6.76 + .68(Y - \bar{Y}) + 2.30(W - \bar{W})$

$$= 6.76 + .68(Y - 6.84) + 2.30(W - .35)$$

$$B = 1.29 + .68Y + 2.30W$$

387

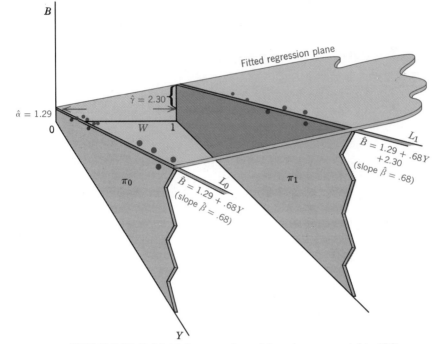

FIGURE 13–8 Multiple regression with a dummy variable (W)

(b) Bias Caused by Excluding the Dummy Variable

In this dummy variable model—as in any regression model—we can see how ignoring one variable would invite bias, as well as increase residual variance. For example, consider what happens if W is ignored, so that the scatter involves only the two dimensions B and Y. Geometrically, this involves projecting the three-dimensional scatter in Figure 13-8 onto the two-dimensional B–Y plane, as in Figure 13-9(a). This immediately is

---

possible least squares fit to the peacetime data alone, nor to the wartime data alone; instead, $\hat{\beta}$ is a compromise.

By contrast, the uncompromising model that fits L_0 and L_1 independently, is:

$$E(B) = \alpha_1 + \beta_1 Y \quad \text{for wartime}$$
$$E(B) = \alpha_2 + \beta_2 Y \quad \text{for peacetime}$$

where the slopes β_1 and β_2 are not constrained to be equal. So four parameters are required for this model, rather than the three parameters in the dummy variable model (13-29).

In this model, to independently estimate the wartime slope with only 5 observations may yield a very unreliable estimator $\hat{\beta}_2$. This was a good reason for our pooling the peacetime and wartime observations to obtain just one slope estimator $\hat{\beta}$ in (13-33).

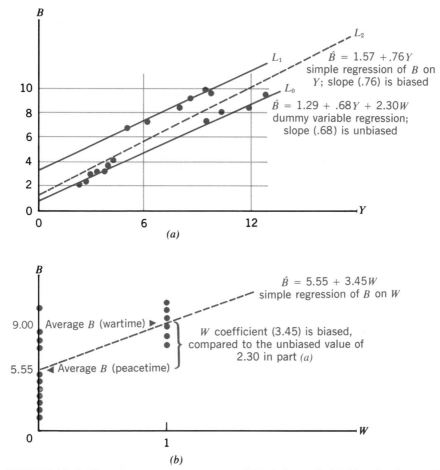

FIGURE 13–9 Bias when one explanatory variable is ignored. (a) Biased estimate of slope (the effect of Y) because the dummy variable W is ignored. (b) Biased estimate of the effect of W because the numerical variable Y is ignored.

recognized as the same scatter that we plotted in Figure 13-7. We also reproduce from that diagram L_0 and L_1, the estimated multiple regression, using W as a dummy variable. Now if we erroneously calculate the simple regression of B on Y (L_2 say), it clearly has too great a slope. This upward bias results from the fact that the war years had a slight tendency to be high-income years: thus, on the right-middle side of this scatter, higher bond sales that should be attributed in part to wartime would be attributed erroneously to income alone.

A similar bias occurs if Y is ignored in a regression of B on W. With no Y dimension, the scatter in Figure 13-8 would be projected into the

B–W plane, as in Figure 13-9(b). In this diagram, the slope of the best fit also happens to be the difference in means.[21] It is $9.00 - 5.55 = 3.45$, which is too large, compared with the unbiased estimate of wartime (2.30) obtained from the multiple regression. This upward bias comes from the same cause: higher bond sales that should be attributed in part to higher income would be attributed erroneously to wartime alone.

PROBLEMS

13-17 Referring to the bond sales in Figure 13-9(a):

(a) Estimate what the bond sales would have been in 1946 if the war had lasted until then and if national income had still been only \$9.6 billion.

(b) Suppose that the data in Table 13-5 and Figure 13-9(a) had not included the last four years. In a simple regression of *B* on *Y*, roughly estimate the slope by eye. Then what would be the bias caused by using simple regression instead of multiple regression?

(c) Repeat (b), assuming instead that the *first* seven years were missing.

13-18 The following is the result of a test of gas consumption on a sample of 6 cars:

	M Miles Per Gallon	*H* Engine Horsepower
Make *A*	21	210
	18	240
	15	310
Make *B*	20	220
	18	260
	15	320

(a) Code *X* as 0 or 1, depending on whether the car is make *A* or *B*. Then estimate the multiple regression equation of *M* as a function of *H* and *X*.

(b) Graph the data and the fitted pair of lines, as in Figure 13-7.

⇒13-19 A random sample of men and women in a large American university in 1969 gave the following annual salaries (in thousands of dollars, rounded):[22]

---

[21] This is because the range of *W* is exactly 1, and because the mean of a set of points is the position of best fit (as shown in Section 12-3). This same point will be illustrated again in Problem 13-19.

[22] Same source as Problems 8-15 and 13-5.

Men	12,	11,	19,	16,	22
Women	9,	12,	8,	10,	16

Denote income by Y and denote sex by a dummy variable X, having $X = 0$ for men and $X = 1$ for women. Then

(a) Graph Y against X.

(b) Estimate by eye the regression line of Y on X. (*Hint:* Where will the line pass through the men's salaries? through the women's salaries?)

(c) Estimate by least squares the regression line of Y on X. How well does your eyeball estimate in (b) compare?

(d) Construct a 95% confidence interval for the coefficient of X. Explain what it means in simple language.

(e) Compare (d) with the solution in Problem 8-15.

(f) Do you think that the answer to (d) is a measure of how much the university's sex discrimination affects women's salaries?

13-20 Now we can consider more precisely the regression equation for professors' salaries in Problem 13-5, by including some additional regressors:

$$\hat{S} = 230B + \cdots + 50T - 2400X + 1900P + \cdots$$

standard error	(370)	(530)	(610)
t ratio	( )	( )	( )
95% CI	( )	( )	( )

where

S = the professor's annual salary (dollars)

$$T = \begin{cases} 0 \text{ if the professor received a student evaluation score below the median} \\ 1 \text{ if above the median} \end{cases}$$

$$X = \begin{cases} 0 \text{ if the professor is male} \\ 1 \text{ if the professor is female} \end{cases}$$

$$P = \begin{cases} 0 \text{ if the professor has no Ph.D.} \\ 1 \text{ if the professor has a Ph.D.} \end{cases}$$

(a) Fill in the brackets in the equation.

(b) Answer true or false; if false, correct it.

(i) A professor with a Ph.D. earns annually $1900 more than one without a Ph.D.

(ii) Or, to draw an analogy with Problem 13-11(a), we might say that $1900 estimates the value (in terms of a professor's salary) of one more unit (in this case, a Ph.D.).

(c) Give an interpretation of the coefficient of X, and the coefficient of T.

13-21 For the raw data of Problem 13-20, the mean salaries for male and female professors were $16,100 and $11,200 respectively. By referring to the coefficient of X in Problem 13-20, answer true or false; if false, correct it.

After holding constant all other variables, women made $2,400 less than men. Therefore, $2,400 is a measure of the extent of sex discrimination and $2,500 (16,100 − 11,200 − 2,400) is a measure of the salary differential due to other factors, for example, productivity and experience.[23]

13-22 Comparing Problems 13-5 and 13-20, note that the same variable T appears in two different forms. In the first form, it is apparent that it involves a severe degree of rounding. What are the advantages and disadvantages of such rounding?

13-10   REGRESSION AND ANALYSIS OF VARIANCE

(a) Regression with Dummies is Equivalent to ANOVA

Example 13-5. A random sample of 5 home values was taken from the U.S. South, and also from the Northeast, in 1970.[24] In thousands of dollars, they were as follows:

S	NE
13	23
10	18
16	12
12	15
19	27

(a) Let us represent region by a dummy variable X, as follows:

$$X = 0 \quad \text{for } S$$
$$= 1 \quad \text{for } NE$$

---

[23]Quoted, with rounding, from the same source as Problems 13-5 and 13-20.

Also let Y denote home value. Then calculate and graph the regression of Y on X. Also calculate the 95% confidence interval for the slope β.

(b) Calculate a 95% confidence interval for the mean difference between NE and S, using (8-32).

Solution.   (a)

X	Y	$x = X - \overline{X}$	xY	x^2	$\hat{Y} = \hat{\alpha} + \hat{\beta}x$	$Y - \hat{Y}$	$(Y - \hat{Y})^2$
0	13	$-.5$	-6.5	.25	14	-1	1
0	10	$-.5$	-5.0	.25	14	-4	16
0	16	$-.5$	-8.0	.25	14	2	4
0	12	$-.5$	-6.0	.25	14	-2	4
0	19	$-.5$	-9.5	.25	14	5	25
1	23	$+.5$	11.5	.25	19	4	16
1	18	$+.5$	9.0	.25	19	-1	1
1	12	$+.5$	6.0	.25	19	-7	49
1	15	$+.5$	7.5	.25	19	-4	16
1	27	$+.5$	13.5	.25	19	8	64

$$\overline{X} = \frac{5}{10} = .5 \qquad \overline{Y} = \frac{165}{10} = 16.5 \qquad \sum xY = 12.5 \qquad \sum x^2 = 2.50 \qquad 196$$

$$\hat{\alpha} = \overline{Y} = 16.5 \qquad\qquad \text{(11-13) repeated}$$

$$\hat{\beta} = \frac{\sum xY}{\sum x^2} = \frac{12.5}{2.50} = 5.0 \qquad\qquad \text{(11-16) repeated}$$

Thus:

$$Y = 16.5 + 5.0x$$

or:
$$= 16.5 + 5.0(X - .5)$$

$$= 14 + 5.0X$$

The 95% confidence interval is:

$$\beta = \hat{\beta} \pm t_{.025} \frac{\sqrt{\sum (Y - \hat{Y})^2/(n - 2)}}{\sqrt{\sum x^2}} \qquad \text{like (12-27)}$$

$$= 5.0 \pm 2.306 \frac{\sqrt{196/8}}{\sqrt{2.50}} \qquad (13\text{-}34)$$

$$= 5.0 \pm 7.22$$

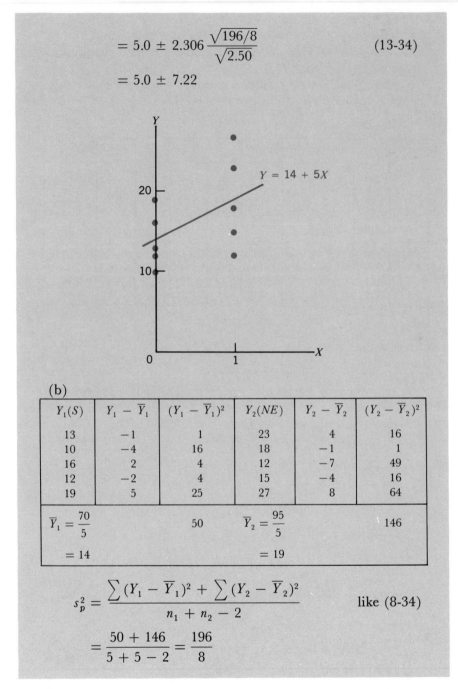

(b)

$Y_1(S)$	$Y_1 - \overline{Y}_1$	$(Y_1 - \overline{Y}_1)^2$	$Y_2(NE)$	$Y_2 - \overline{Y}_2$	$(Y_2 - \overline{Y}_2)^2$
13	-1	1	23	4	16
10	-4	16	18	-1	1
16	2	4	12	-7	49
12	-2	4	15	-4	16
19	5	25	27	8	64
$\overline{Y}_1 = \dfrac{70}{5}$		50	$\overline{Y}_2 = \dfrac{95}{5}$		146
$= 14$			$= 19$		

$$s_p^2 = \frac{\sum (Y_1 - \overline{Y}_1)^2 + \sum (Y_2 - \overline{Y}_2)^2}{n_1 + n_2 - 2} \qquad \text{like (8-34)}$$

$$= \frac{50 + 146}{5 + 5 - 2} = \frac{196}{8}$$

Thus:

$$\mu_2 - \mu_1 = (\overline{Y}_2 - \overline{Y}_1) \pm t_{.025}\, s_p \sqrt{\frac{1}{n_2} + \frac{1}{n_1}} \qquad \text{like (8-32)}$$

$$= (19 - 14) \pm 2.306\, \sqrt{196/8}\, \sqrt{\frac{1}{5} + \frac{1}{5}} \qquad (13\text{-}35)$$

$$= 5.0 \pm 7.22$$

Remarks. The confidence intervals in (a) and (b) were both the same. This is no coincidence: the calculations in (13-34) and (13-35) were entirely similar, term by term.

It is equally apparent why they must be the same, if we consider the general interpretation of the regression coefficient (13-19): $\hat{\beta} = 5$ estimates the average increase in Y if X is increased one unit (from 0 to 1; i.e., from S to NE). But this is precisely what is being estimated in (13-35) by $\overline{Y}_2 - \overline{Y}_1 = 5$.

---

[24] From the 1974 *American Almanac,* Table 1168. These data also appear in Problem 10-4.

From this example, we conclude that regression with one dummy variable is equivalent to the 2-sample t test, which, in turn, is equivalent to ANOVA with 2 categories (as shown in Problem 10-13). More generally, it similarly may be shown that regression with $(r - 1)$ appropriately chosen dummy variables is equivalent to ANOVA with r categories.[25]

(b) Analysis of Covariance (ANOCOVA)

The example of bond sales (13-29) was a regression on a numerical variable (income) and a dummy variable (wartime). This alternatively could be described as a combination of standard regression analysis and analysis of variance. Technically, this combination is referred to as analysis of covariance (ANOCOVA).

Another example of ANOCOVA might be a study of the effects of racial discrimination on income; here the major concern would be the

---

[25] Why do we need only $(r - 1)$ dummy variables to handle r categories? Because one category is used as a reference, and does not need its own dummy variable. For example, in Problem 13-24 the control treatment serves as a natural reference category and needs no dummy variable.

effect on income of the dummy variable (Black versus White), with other numerical variables (years of experience, education, etc.) being included in the regression mainly to keep these other influences from biasing the result.

(c) Review, with Examples

Multiple regression is an extremely useful tool, which has many broad applications. We summarize its three special cases, which are distinguished by the nature of the independent variables:

1. *"Standard" regression* is regression on numerical variables only.

2. *Analysis of Variance* (ANOVA) is equivalent to regression on dummy variables only.

3. *Analysis of Covariance* (ANOCOVA) is regression on both dummy and numerical variables.

These three techniques are compared using the hypothetical data of Figures 13-10 to 13-13, which show the possible ways that mortality may be analyzed.

Figure 13-10 shows mortality for several age groups of American men. Applying standard regression, we would reject the hypothesis that the slope $\beta = 0$; thus, we conclude that age *does* affect the mortality rate. In the process, we derive a useful estimate $\hat{\beta}$, of *how* age affects mortality.

If the data were grouped crudely into 3 categories, the result would be the scatter shown in Figure 13-11. Note that this is exactly the same set of mortality observations Y as in Figure 13-10. The only difference is that we are no longer as detailed about the age variable X. ANOVA

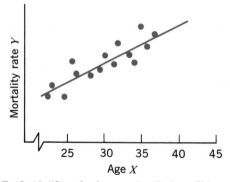

FIGURE 13–10 "Standard regression," since X is numerical

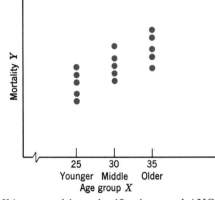

FIGURE 13–11 X is grouped into classifications, and ANOVA may be used

now can be applied[26] easily to test whether the means of these three categories differ discernably. Once again, the conclusion is that age affects mortality. However, ANOVA would not tell us how age affects mortality (unless we used multiple comparisons, which would still be less meaningful than a regression slope). As long as X is numerical, as in Figures 13-10 and 13-11, we conclude that standard regression can be applied and is usually the preferred technique. But when X is categorical, standard regression cannot be applied. For example, in Figure 13-12 we graph some

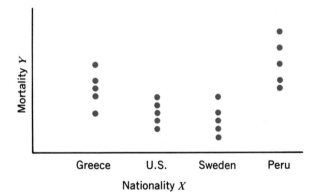

FIGURE 13–12 X is categorical, and ANOVA must be used (or, equivalently, regression on dummies)

---

[26]Standard regression also could be applied, with a line fitted to the scatter in Figure 13-11. However, if standard regression is to be applied, it is more efficient to use the ungrouped data of Figure 13-10.

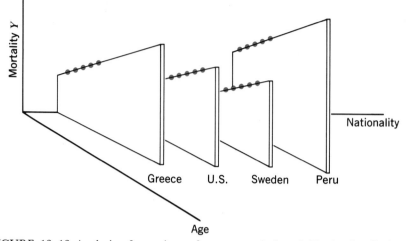

FIGURE 13–13 Analysis of covariance for a categorical variable (nationality) and numerical variable (age)

hypothetical data on how mortality depends on nationality.[27] X ranges over various categories (Greece, U.S., etc.), and there is no natural way of placing these on a numerical scale—or even of ordering them. Hence, standard regression is out of the question,[28] and ANOVA must be used (or, equivalently, regression on dummies).

If mortality is dependent on both age and nationality, the analysis of covariance shown in Figure 13-13 is appropriate. This uses nationality dummies, and introduces the numerical variable age explicitly, to eliminate the bias that it would cause if it were omitted. For example, if age were ignored in Figure 13-13, the mortality rate for Sweden would exceed that of Greece!

In summary, standard regression is the more powerful tool whenever the independent variable X is numerical. Analysis of variance is appropriate if the independent variable is a set of unordered categories.

---

[27] In Figure 13-12 we assume that all samples have been drawn from a single age group; if the ages are different, we must make the allowance shown in Figure 13-13.

[28] That is, we cannot fit a line sensibly. To confirm this, note that a standard regression line fitted to the scatter in Figure 13-12 would yield $\beta \simeq 0$ (i.e., no evidence that nationality matters). Yet if Peru were graphed first rather than last, β would have a large negative value, and we would have to conclude that nationality does matter. Thus, the conclusions would depend on the arbitrary ordering of the countries.

PROBLEMS

13-23 (a) Based on the following sample information, use analysis of covariance (i.e., multiple regression) to describe how education is related
to father's income and place of residence.
(b) Graph your results.

	Years of Formal Education (E)	Father's Income (F)
Urban sample	15	$ 8,000
	18	11,000
	12	9,000
	16	12,000
Rural sample	13	$ 5,000
	10	3,000
	11	6,000
	14	10,000

13-24 (a) Using the data in Problem 10-2, estimate the regression of yield on
fertilizer type, using the following two dummy variables:

$$D_1 = 1 \text{ if fertilizer } A \text{ is used}$$

$$= 0 \text{ otherwise}$$

$$D_2 = 1 \text{ if fertilizer } B \text{ is used}$$

$$= 0 \text{ otherwise}$$

(b) Do you get exactly the same answer as you did in Problem 10-2,
for the effect of fertilizer A? for the effect of fertilizer B?

REVIEW PROBLEMS

13-25 Consider a multiple regression model of personal income Y as a function
of age A, number of years of education E, and sex S (where sex is coded
0 for male, 1 for female):

$$\hat{Y} = \hat{\alpha} + \hat{\beta}a + \hat{\gamma}e + \hat{\delta}s$$

Answer true or false; if false, correct it.
(a) The coefficient $\hat{\delta}$ may be interpreted as estimating the amount of
income that the average man earns more than the average woman.

(b) Other things being equal, a person who is 1 year older earns $\hat{\beta}$ more income.

13-26 A sociologist collected a random sample of 1000 men to see how divorce rates are related to religion, income, region, and degree of urbanization. Outline how you would analyze the data.

13-27 A 1974 study[29] of 1072 subjects showed how lung function was related to several variables, including three hazardous occupations. The following abbreviations were used:

AIRCAP = air capacity (cubic centimeters) that the subject can expire in one second

BRONC = dummy variable for bronchitis (1 if subject has it, 0 if not)

AGE = age (years)

HEIT = height (inches)

PRSMOK = present smoking (cigarettes per day)

PASMOK = past smoking as measured by:
(number of years smoked) × (cigarettes per day)

CPSMOK = present cigar and pipe smoking (cigars + pipes, per week)

CHEMW = dummy variable to measure whether subject is a chemical worker (1 if he is, 0 if not)

FIREW = dummy variable to measure whether subject is a fireman (1 if he is, 0 if not)

FARMW = dummy variable to measure whether subject is a farm worker (1 if he is, 0 if not)

Since the fourth occupation, physician, was the reference group, it did not need its own dummy. The following 2 regressions were computed, with standard errors in brackets. (All regressors are in deviation form.)

$$AIRCAP = 3605 - 39\,AGE + 98\,HEIT - 9.0\,PRSMOK$$
$$(1.8) \qquad (7.5) \qquad (2.2)$$

---

[29]N. M. Lefcoe, and T. H. Wonnacott, "The Prevalence of Chronic Respiratory Disease in Four Occupational Groups," *Archives of Environmental Health* 29 (September 1974): 143–146.

$$- .0039 \text{ PASMOK} - 2.6 \text{ CPSMOK} - 350 \text{ CHEMW}$$
$$\quad (.070) \qquad\qquad (1.1) \qquad\qquad (46)$$

$$- 180 \text{ FIREW} - 380 \text{ FARMW}$$
$$\quad (54) \qquad\qquad (53)$$

$$\text{BRONC} = .107 + .0021 \text{ AGE} + .00037 \text{ HEIT} + .0047 \text{ PRSMOK}$$
$$\qquad\qquad (.0009) \qquad\quad (.0038) \qquad\quad (.0011)$$

$$+ .000098 \text{ PASMOK} + .00063 \text{ CPSMOK}$$
$$\quad (.000036) \qquad\qquad (.00054)$$

$$+ .065 \text{ CHEMW} - .032 \text{ FIREW} + .002 \text{ FARMW}$$
$$\quad (.024) \qquad\qquad (.027) \qquad\qquad (.027)$$

(a) Star the coefficients that are statistically discernible (significant) at the 5% level (one-sided).
Fill in the blanks:
(b) The average value of AIRCAP is _____cc., while the average incidence of bronchitis is _____%.
(c) Other things (such as _____, _____, _____) being equal, chemical workers on average have AIRCAP values that are _____cc. lower than the physicians and bronchitis rates that are _____ percentage points higher.
(d) Repeat (c), substituting "firemen" for "physicians."
(e) Other things being equal, on average a man who is 1 year older has an AIRCAP value that is _____cc. lower, and a bronchitis rate that is _____ percentage points higher.
(f) Repeat (e), substituting "presently smokes one more pack (20 cigarettes) per day" instead of "is 1 year older."
(g) As far as AIRCAP is concerned, we estimate that smoking one pack a day is roughly equivalent to aging _____ years. But this estimate may be biased because of _____.

*13-28 A sociologist computed a regression of mobility M as a function of family income X_1:

$$M = b_0 + b_1 x_1$$

Then she realized that family size X_2 also was relevant, and so she calculated the multiple regression:

$$M = c_0 + c_1 x_1 + c_2 x_2$$

Under what conditions will the coefficients of x_1 in the two regressions be equal ($b_1 = c_1$)?

APPENDIX 13-A

Lines and Planes; Elementary Geometry

(a) Lines

The definitive characteristic of a straight line is that it continues forever in the *same constant direction.* In Figure 13-14, we make this idea precise. In moving from one point P_1 to another point P_2, we denote the horizontal distance by ΔX (where Δ means change, or difference), and the vertical distance by ΔY. Then the slope[30] is defined as:

$$\text{slope} = \frac{\Delta Y}{\Delta X}$$

The characteristic of a straight line is that this slope remains the same everywhere:

$$\boxed{\frac{\Delta Y}{\Delta X} = b \text{ (a constant)}} \tag{13-36}$$

For example, the slope between P_3 and P_4 is the same as between P_1 and P_2, as calculation will verify:

$$P_1 \text{ to } P_2: \quad \frac{\Delta Y}{\Delta X} = \frac{3}{6} = .50$$

$$P_3 \text{ to } P_4: \quad \frac{\Delta Y}{\Delta X} = \frac{2}{4} = .50$$

A very instructive case occurs when X increases just one unit; then (13-36) yields:

$$\boxed{\text{when } \Delta X = 1, \quad \Delta Y = b} \tag{13-37}$$

In words, "b is the increase in Y that accompanies a unit increase in X," which agrees with the regression interpretation (13-19).

It is now very easy to derive the equation of a line, if we know its

---

[30] Slope is a concept that is useful in engineering as well as mathematics. For example, if a highway rises 12 feet over a distance of 200 feet, its slope is $12/200 = 6\%$.

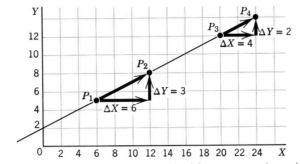

FIGURE 13–14 A straight line is characterized by constant slope, $\Delta Y/\Delta X = b$

slope b and any one point on the line. Suppose that the one point we know is P_0, the Y-intercept; since its coordinates, as shown in Figure 13-15, are 0 and a_0, it is denoted $P_0(0, a_0)$. In moving to any other point $P(X, Y)$ on the line, we may write:

$$\text{slope,}\ \frac{\Delta Y}{\Delta X} = \frac{Y - a_0}{X - 0} \tag{13-38}$$

In (13-36), we insisted that, if the line is to be straight, this slope must equal the constant b:

$$\frac{Y - a_0}{X - 0} = b$$

$$Y - a_0 = bX$$

$$\boxed{Y = a_0 + bX} \tag{13-39}$$
$$\text{(11-4) proved}$$

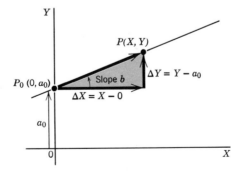

FIGURE 13–15 Derivation of the equation of a straight line

Hence the value of Y is seen to depend on the Y-intercept a_0, the slope b, and the value of X (i.e., how far along this line you are).

(b) Planes

In Figure 13-16, we show a plane in the 3-dimensional (X, Y, Z) space. Let L_{XY} denote the line where this plane cuts the XY plane; then we may think of the plane as a grid of lines parallel to L_{XY} with slope b, or, equally well, as a grid of lines parallel to L_{ZY} with slope c.

Now suppose that we start at point P_1. If we hold Z constant and move to P_2, then we are moving along one of the grid lines parallel to L_{XY} (with slope $= b$). Thus, by analogy with (13-37):

$$\boxed{\begin{aligned} &\text{If } Z \text{ is held constant,} \\ &\text{when} \quad \Delta X = 1, \quad \Delta Y = b \end{aligned}} \qquad (13\text{-}40)$$

In words, "b is the increase in Y that accompanies a unit increase in X, while Z is held constant," which agrees with the regression interpretation (13-22).

Similarly, we can interpret a move from P_1 to P_3:

$$\boxed{\begin{aligned} &\text{If } X \text{ is held constant,} \\ &\text{when} \quad \Delta Z = 1, \quad \Delta Y = c \end{aligned}} \qquad (13\text{-}41)$$

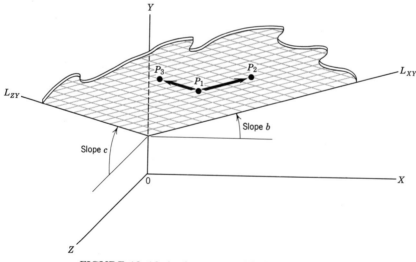

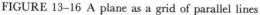

FIGURE 13–16 A plane as a grid of parallel lines

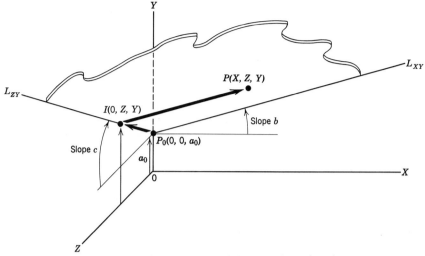

FIGURE 13–17 Derivation of the equation of a plane

In words, "c is the increase in Y that accompanies a unit increase in Z, while X is held constant."

Now it is very easy to derive the equation of a plane. Referring to Figure 13-17, let us start at the Y-intercept P_0 $(0, 0, a_0)$. Let us move to the typical point $P(X, Z, Y)$ in two steps: (1) along the line L_{ZY} to an intermediate point I; and then (2) parallel to L_{XY} to finally reach P. Then we find Y (the height of P) as follows.

1. Just as (13-39) gives the height as a function of X, so we analogously find that the height of I as a function of Z is:

$$a_0 + cZ \qquad (13\text{-}42)$$

2. To move from I to P, note that our intercept at I is now $(a_0 + cZ)$, as given in (13-42); note also that we will be moving along a grid line parallel to L_{XY} with slope b. Applying (13-39):

$$Y = (a_0 + cZ) + bX$$

$$\boxed{Y = a_0 + bX + cZ} \qquad (13\text{-}43)$$

This is the equation of the plane, which agrees with (13-1) (except for notational changes, of course).

CHAPTER 14

Correlation

14-1   SIMPLE CORRELATION

Simple regression analysis showed us *how* variables are linearly related; correlation analysis will only show us the *degree* to which variables are linearly related. In regression analysis, a whole function is estimated (the regression equation); but correlation analysis yields only one number—an index designed to give an immediate picture of how closely two variables move together. Although correlation is a less powerful technique than regression, the two are so closely related that correlation often becomes a useful aid in interpreting regression. In fact, this is the major reason for studying it.

(a) The Population Correlation ρ

We already have defined the correlation ρ, sometimes called *Pearson's correlation coefficient*, in (5-25):

$$\rho \triangleq \frac{\sigma_{XY}}{\sigma_X \sigma_Y} \tag{14-1}$$

or, noting (5-23):

$$\rho = \frac{E(X - \mu_X)(Y - \mu_Y)}{\sigma_X \sigma_Y} \tag{14-2}$$

It is useful to reexpress this in terms of the standardized X and Y:

$$\boxed{\rho = E\left(\frac{X - \mu_X}{\sigma_X}\right)\left(\frac{Y - \mu_Y}{\sigma_Y}\right)} \tag{14-3}$$

(b) The Sample Correlation r

By analogy with (14-3), the sample correlation is:

$$\boxed{r \triangleq \frac{1}{n - 1} \sum_{i=1}^{n} \left(\frac{X_i - \overline{X}}{s_X}\right)\left(\frac{Y_i - \overline{Y}}{s_Y}\right)} \tag{14-4}$$

An intuitive development of this formula for r is shown in Figure 14-1, closely parallelling the development of ρ in Section 5-3. Panel (a) shows the scatter of marks on a math (X) and verbal (Y) test that were scored by a sample of eight college students; these data are also set out in the first two columns in Table 14-1. To ensure that our resulting index will be independent of the choice of origin, we shift both axes in panel (b), now defining x and y as deviations from the mean. Values of these translated variables are given in the next two columns of Table 14-1.

Suppose that we multiply the x and y coordinate values for each student, and sum them to get Σxy. This gives us a good measure of how math and verbal results tend to move together. We can see this by referring to Figure 14-1(b): for any observation such as P_1 in the first or third quadrant, x and y agree in sign, so their product xy is positive. Conversely, for any observation such as P_2 in the second or fourth quadrant, x and y disagree in sign, so their product xy is negative. If X and Y move together, most observations will fall in the first and third quadrants; consequently, most products xy will be positive, as will their sum—a reflection of the positive relationship between X and Y. But if X and Y are related negatively (i.e., if one rises when the other falls), most observations will fall in the second and fourth quadrants, yielding a negative value for our Σxy index. We conclude that as an index of correlation, Σxy at least carries the right sign. Moreover, when there is no relation-

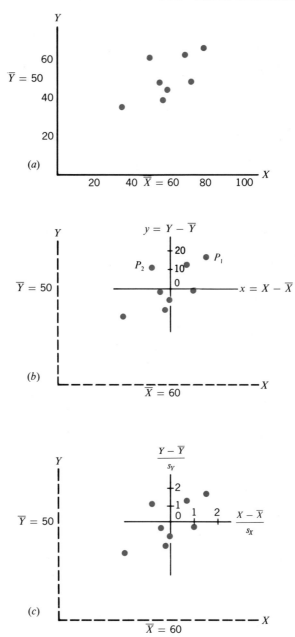

FIGURE 14–1 Scatter of math and verval scores. (a) Original observations. (b) Axes shifted. (c) Axes rescaled to standard units.

TABLE 14-1

Math Score (X) and Corresponding Verbal Score (Y) of a Sample of 8 Students Entering College.

X	Y	$x =$ $X - \bar{X}$	$y =$ $Y - \bar{Y}$	xy	x^2	y^2	Regression of Y on X $\hat{Y} =$ $\bar{Y} + \hat{\beta}x$	$Y - \hat{Y}$	$(Y - \hat{Y})^2$	Regression X on Y $\hat{X} =$ $\bar{X} + \hat{\beta}_{*}y$	$X - \hat{X}$	$(X - \hat{X})^2$
36	35	-24	-15	360	576	225	38	-3	9	48	-12	144
80	65	20	15	300	400	225	60	5	25	72	8	64
50	60	-10	10	-100	100	100	45	15	225	68	-18	324
58	39	-2	-11	22	4	121	49	-10	100	51	7	49
72	48	12	-2	-24	144	4	56	-8	64	59	13	169
60	44	0	-6	0	0	36	50	-6	36	55	5	25
56	48	-4	-2	8	16	4	48	0	0	58	-2	4
68	61	8	11	88	64	121	54	7	49	69	-1	1

$$\sum X = 480 \qquad \sum Y = 400 \qquad \sum x = 0 \qquad \sum y = 0 \qquad \sum xy = 654 \qquad \sum x^2 = 1304 \qquad \sum y^2 = 836$$

$$\bar{X} = 60 \qquad \bar{Y} = 50$$

$$\sum (Y - \hat{Y})^2 = 508 \qquad \sum (X - \hat{X})^2 = 780$$

Regression of Y on X:

$$\hat{\beta} = \frac{\sum xy}{\sum x^2} = .50$$

$$s^2 = \frac{\sum (Y - \hat{Y})^2}{n - 2} = \frac{508}{6} = 84.7 \qquad s^2_{*} = \frac{\sum (X - \hat{X})^2}{n - 2} = \frac{780}{6} = 130.0$$

$$s = 9.20$$

Regression of X on Y:

$$\hat{\beta}_{*} = \frac{\sum xy}{\sum y^2} = .78$$

$$s_{*} = 11.4$$

ship between X and Y, with the observations distributed evenly over the four quadrants, positive and negative terms will cancel, and $\Sigma\, xy$ will be zero.

There are just two ways that $\Sigma\, xy$ can be improved. First, it depends on sample size. (Suppose that we observed exactly the same sort of scatter from a sample of double the size; then $\Sigma\, xy$ also would double, even though the picture of how these variables moved together would remain the same.) To avoid this problem, we divide by the sample size—actually by $(n - 1)$—to yield the index:

$$\text{sample covariance, } s_{XY} \triangleq \frac{\sum xy}{n - 1} \tag{14-5}$$

$$s_{XY} = \frac{1}{n - 1} \sum (X_i - \overline{X})(Y_i - \overline{Y}) \tag{14-6}$$
$$\text{like (5-23)}$$

This is a highly useful concept in statistics, but it does have one remaining weakness: s_{XY} depends on the units in which x and y are measured. (Suppose that the math test had been marked out of 50 instead of 100; x values, and hence s_{XY}, would be only half as large— even though the degree to which verbal and mathematical performance are related would not have changed.) This difficulty is avoided by measuring both variables in terms of standard units; both x and y are divided by their standard deviations. This step is shown in Figure 14-1(c), with the resulting index being the sample correlation (14-4).

Finally, to simplify calculations, (14-4) may be reexpressed[1] as:

$$r = \frac{\sum (X_i - \overline{X})(Y_i - \overline{Y})}{\sqrt{\sum (X_i - \overline{X})^2 \sum (Y_i - \overline{Y})^2}}$$
$$= \frac{\sum xy}{\sqrt{\sum x^2 \sum y^2}} \tag{14-7}$$

For example, to calculate the correlation coefficient between the math and verbal scores of the sample of eight students, we substitute the appropriate sums from Table 14-1 into (14-7):

---

[1]We simply substitute into (14-4) the formula (2-6a) for s_X and s_Y. Then we cancel $(n - 1)$. Incidentally, this explains why the divisor $(n - 1)$ is used instead of n in (14-5).

$$r = \frac{654}{\sqrt{(1304)(836)}} = .63 \tag{14-8}$$

Some idea of how r behaves is given in Figure 14-2. Especially note the line in panel (b): when there is a perfect positive relation, the product

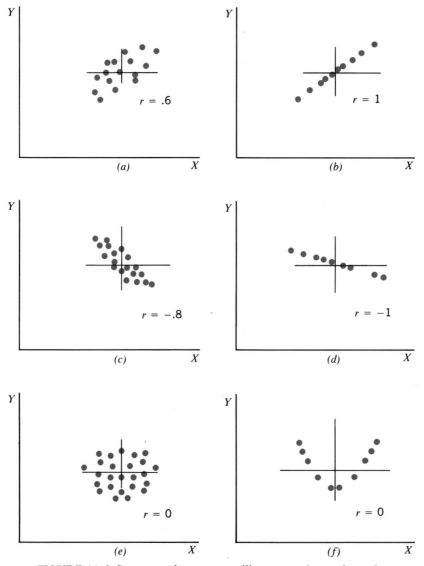

FIGURE 14–2 Some sample scatters to illustrate various values of r

xy is always positive. Thus $\Sigma\ xy$, and hence r, are as large as possible. A similar argument holds for the perfect negative relation shown in panel (d). This suggests that r has an upper limit of $+1$ and a lower limit of -1. This will be proved in Section 14-2(c), below.

Finally, consider the symmetric scatters shown in panels (e) and (f). The calculation of r in either case yields zero, because each positive product xy is offset by a corresponding negative xy in the opposite quadrant. Yet these two scatters show quite different patterns: in (e), there is no relation between X and Y; in (f), however, there is a strong relation (knowledge of X will tell us a great deal about Y).

A zero value for r therefore does not necessarily imply "no relation;" rather, it means "no *linear* relation." Thus, simple correlation is a measure of linear relation only; it is of no use in describing nonlinear relations.

(c) Inference from r to ρ

In calculating r, what can we infer about the underlying population ρ? First, we must clarify our assumptions about the underlying population itself. In our example, this would be the math and verbal marks scored by *all* college entrants. This population might appear as in Figure 14-3, except that there would be, of course, many more dots in this scatter, each representing another student. If we subdivide both X and Y axes into intervals, the area in our diagram will be divided up in a checkerboard pattern. From the relative frequency in each of the squares, the histogram in Figure 14-4 is constructed. If this histogram is rescaled to relative frequency *density*, and then approximated by a smooth surface, the result is the continuous function shown in Figure 14-5, representing the probability density of any X and Y combination.

A special kind of joint distribution of X and Y is assumed in making statistical inferences in simple correlation analysis—a bivariate normal distribution. Bivariate, because both X and Y are random variables; one

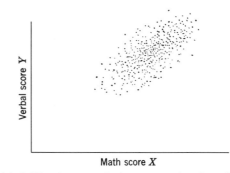

FIGURE 14–3 Bivariate population scatter (math and verbal scores)

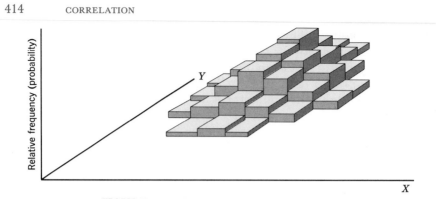

FIGURE 14–4 Bivariate population histogram

is not fixed, as fertilizer was in Chapter 11. Normal, because the conditional distribution of X or of Y is always normal. Specifically, if we slice the surface at any value of Y (say Y_0), the shape of the resulting cross-section is normal. Similarly, if we select any X value (say X_0) and slice the surface in this other direction, the resulting cross-section also is normal.

Let us pause briefly to consider the alternative way in which the bivariate normal population, shown in three dimensions in Figure 14-5, can be graphed in two dimensions. Slice the surface horizontally, as in Figure 14-6 (or, equivalently, you can think of flooding Figure 14-6 with water up to a certain level). The resulting cross-section (or shoreline) is an ellipse, representing all X, Y combinations with the same probability density. It is called an *isoprobability curve* (or *level curve*). This level curve marked c is reproduced in the two-dimensional X, Y space in Figure 14-7. Level curves at higher and lower levels also are shown. It is useful, in Figure 14-7, to mark the major axis d that is common to all these level

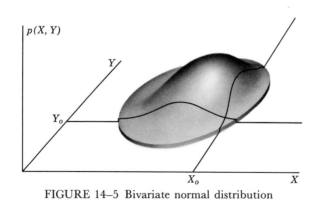

FIGURE 14–5 Bivariate normal distribution

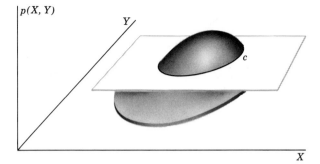

FIGURE 14–6 An isoprobability ellipse from a bivariate normal surface

curves. If the bivariate normal distribution is concentrated about this major axis, ρ has a high numerical value. Several examples of populations and their associated correlations ρ are shown in Figure 14-8.

Provided that the parent population is bivariate normal, inferences about the population ρ can easily be made from a sample correlation r. Recall the inferences about π from P in Chapter 8. Using the same reasoning that established Figure 8-5, we construct Figure 14-9. Thus, from any sample r, a 95% confidence interval for the population ρ can be found. For example, if a sample of 10 students has $r = .6$, we show in color the 95% confidence interval for ρ, read vertically as:

$$-.05 < \rho < .87 \qquad (14\text{-}9)$$

Because of space limitations, the balance of this chapter will concentrate on sample correlations and ignore the corresponding population

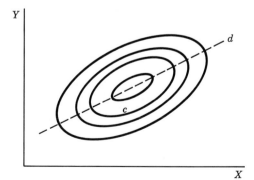

FIGURE 14–7 The bivariate normal distribution shown as a set of isoprobability or level curves

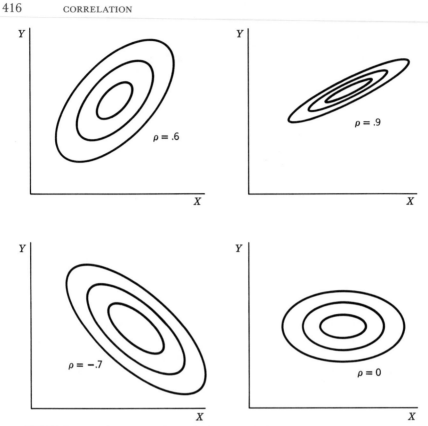

FIGURE 14–8 Examples of population correlations

correlations. But each time a sample correlation is introduced, we should recognize that an equivalent population correlation is defined similarly, and inferences may be made about it from the sample correlation.

PROBLEMS

14-1 From the following random sample of 5 son-and-father pairs:

Son's Height (Inches)	Father's Height (Inches)
68	64
66	66
72	71
73	70
66	69

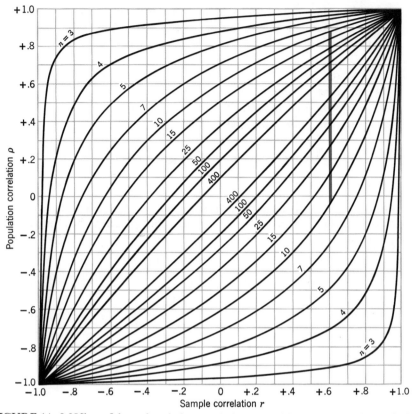

FIGURE 14–9 95% confidence bands for correlation ρ in a bivariate normal population, for various sample sizes n

(a) Calculate the sample correlation r.
(b) Find the 95% confidence interval for the population correlation ρ.
(c) At the 5% level, can you reject the hypothesis that $\rho = 0$? (Test against the two-sided alternative hypothesis $\rho \neq 0$.)

⇒14-2 From the following random sample of student grades:

Student	First Test X	Second Test Y
A	80	90
B	60	70
C	40	40
D	30	40
E	40	60

(a) Calculate r, and find a 95% confidence interval for ρ.

(b) Calculate the regression of Y on X, and find a 95% confidence interval for β.

(c) Graph the 5 students and the estimated regression line.

(d) At the 5% level, against the two-sided alternative hypothesis that X and Y have some linear relation, can you reject:

    (i) The null hypothesis $\beta = 0$?

    (ii) The null hypothesis $\rho = 0$?

14-2   CORRELATION AND REGRESSION

(a) Relation of $\hat{\beta}$ to r

If regression and correlation analysis both were applied to the same scatter of math (X) and verbal (Y) scores, how would they be related? From (11-19):

$$\hat{\beta} = b = \frac{\sum xy}{\sum x^2} \tag{14-10}$$

and repeating (14-7):

$$r = \frac{\sum xy}{\sqrt{\sum x^2}\sqrt{\sum y^2}} \tag{14-11}$$

When (14-10) is divided by (14-11):

$$\frac{\hat{\beta}}{r} = \frac{\sqrt{\sum x^2}\sqrt{\sum y^2}}{\sum x^2} = \sqrt{\frac{\sum y^2}{\sum x^2}} \tag{14-12}$$

If we divide both the numerator and denominator inside the square root sign by $(n - 1)$, we get:

$$\frac{\hat{\beta}}{r} = \frac{\sqrt{\sum y^2/(n-1)}}{\sqrt{\sum x^2/(n-1)}} = \frac{s_Y}{s_X} \tag{14-13}$$

$$\boxed{\hat{\beta} = r\frac{s_Y}{s_X}} \tag{14-14}$$

Thus $\hat{\beta}$ and r are closely related. For example, if either is zero, the other will also be zero. Similarly, if either of the population parameters β or ρ is zero, the other also will be zero. Thus it is no surprise that in Problem 14-2(d), the tests for $\beta = 0$ and $\rho = 0$ were equivalent ways of examining "no linear relation between X and Y."

(b) Explained and Unexplained Variation

In Figure 14-10, we reproduce the sample of math (X) and verbal (Y) scores, along with the fitted regression of Y on X, calculated in Table 14-1. If we wished to predict a student's verbal score Y without knowing X, then the best prediction would be the average observed value $\overline{Y}$. At x_i, it is clear from this diagram that we would make a very large error—namely $(Y_i - \overline{Y})$, the deviation of Y_i from its mean. However, if X is known and the regression of Y on X has been calculated, we predict Y to be $\hat{Y}_i$. Note how this reduces our error, since $(\hat{Y}_i - \overline{Y})$—a large part of our deviation— is now "explained." This leaves only a relatively small "unexplained" deviation $(Y_i - \hat{Y}_i)$. The total deviation of Y_i is the sum:

$$(Y_i - \overline{Y}) = (\hat{Y}_i - \overline{Y}) + (Y_i - \hat{Y}_i), \quad \text{for any } i \qquad (14\text{-}15)$$

$$\underset{\substack{\text{total} \\ \text{deviation}}}{} = \underset{\substack{\text{explained} \\ \text{deviation}}}{} + \underset{\substack{\text{unexplained} \\ \text{deviation}}}{}$$

It follows that:

$$\sum (Y_i - \overline{Y}) = \sum (\hat{Y}_i - \overline{Y}) + \sum (Y_i - \hat{Y}_i) \qquad (14\text{-}16)$$

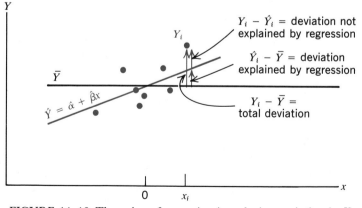

FIGURE 14–10 The value of regression in reducing variation in Y

What is surprising is that this same equality holds true when these deviations are squared, and we obtain a result[2] very similar to ANOVA in Chapter 10:

$$\sum (Y_i - \overline{Y})^2 = \sum (\hat{Y}_i - \overline{Y})^2 + \sum (Y_i - \hat{Y}_i)^2 \qquad (14\text{-}17)$$

$$\underset{\substack{\text{total} \\ \text{variation}}}{} = \underset{\substack{\text{explained} \\ \text{variation}}}{} + \underset{\substack{\text{unexplained} \\ \text{variation}}}{}$$

where variation is defined as the sum of squared deviations. We may rewrite (14-17) by substituting into its middle term:

$$(\hat{Y}_i - \overline{Y}) = \hat{y}_i = \hat{\beta} x_i \qquad \begin{array}{c} (14\text{-}19) \\ \text{like (11-20)} \end{array}$$

We then obtain:

$$\sum (Y_i - \overline{Y})^2 = \hat{\beta}^2 \sum x_i^2 + \sum (Y_i - \hat{Y}_i)^2 \qquad \begin{array}{c} (14\text{-}20) \\ \text{like (10-10)} \end{array}$$

$$\underset{\substack{\text{total} \\ \text{variation}}}{} = \underset{\substack{\text{variation} \\ \text{explained} + \\ \text{by } X}}{} \underset{\substack{\text{unexplained} \\ \text{variation}}}{}$$

This equation makes it clear that explained variation is the variation accounted for by the estimated regression coefficient $\hat{\beta}$. This procedure of analyzing or decomposing total variation into its components is called "analysis of variance for regression." The components of variance may be displayed in an ANOVA table such as Table 14-2(a). (Recall that variance is variation divided by degrees of freedom.) From this, we may formulate a test of the null hypothesis $\beta = 0$. Just as in the standard ANOVA test in Chapter 10, the question is whether the ratio of the explained variance to unexplained variance is sufficiently large to reject H_0. Specifically, we form the ratio:

---

[2]The proof of (14-17) is very similar to the proof of (10-10): Square both sides of (14-15), and sum over all values of i:

$$\sum (Y_i - \overline{Y})^2 = \sum [(\hat{Y}_i - \overline{Y}) + (Y_i - \hat{Y}_i)]^2$$
$$= \sum (\hat{Y}_i - \overline{Y})^2 + 2 \sum (\hat{Y}_i - \overline{Y})(Y_i - \hat{Y}_i) + \sum (Y_i - \hat{Y}_i)^2 \qquad (14\text{-}18)$$

The middle term can be rewritten using (14-19):

$$2\hat{\beta} \sum x_i (Y_i - \hat{Y}_i)$$

But this sum vanishes; in fact, it was set equal to zero in equation (11-15), which was used to estimate the regression line. Thus, the middle term in (14-18) disappears, and (14-17) is proved. This same theorem similarly can be proved in the general case of multiple regression.

A further justification of the least squares technique (not mentioned in Chapter 11) is that it results in this useful relation between explained, unexplained, and total variation.

$$F = \frac{\text{variance explained by regression}}{\text{unexplained variance}} \qquad (14\text{-}21)$$

$$= \frac{\hat{\beta}^2 \sum x_i^2}{s^2}$$

TABLE 14-2
Analysis of Variance Table for Linear Regression
(a) General

Source of Variation	Variation	d.f.	Variance	F ratio
Explained (by regression)	$\sum (\hat{Y}_i - \overline{Y})^2$ or $\hat{\beta}^2 \sum x_i^2$	1	$\dfrac{\hat{\beta}^2 \sum x_i^2}{1}$	$\dfrac{\hat{\beta}^2 \sum x_i^2}{s^2}$
Unexplained (residual)	$\sum (Y_i - \hat{Y}_i)^2$	$n - 2$	$s^2 = \dfrac{\sum (Y_i - \hat{Y}_i)^2}{n - 2}$	
Total	$\sum (Y_i - \overline{Y})^2$	$n - 1$		

(b) For Sample of Math and Verbal Scores (in Table 14-1)

Source of Variation	Variation[a]	d.f.	Variance	F ratio
Explained (by regression)	328	1	328	3.87
Unexplained (residual)	508	6	84.7	
Total	836 ✓	7 ✓		

[a]The explained variation is calculated most easily as $\hat{\beta}^2 \sum x_i^2 = (.5015)^2(1304) = 328$.

This is shown as the last entry in Table 14-2(a). By referring to the critical F values in Table VII, we can calculate the prob-value for H_0, and if this is sufficiently low, we can reject H_0.

The F test is just an alternative way of testing the null hypothesis that $\beta = 0$. The first method—using the t ratio as in (12-35)—is preferable if a confidence interval also is desired. The two tests are equivalent because the t statistic is related to F (with one degree of freedom in the numerator) by:

$$\boxed{t^2 = F} \qquad (14\text{-}22)$$

To sum up: there are three equivalent ways of testing the null hypothesis that the regressor has no effect on Y: the F test and the t test of $\beta = 0$, and the test of $\rho = 0$ in Figure 14-9. All three now are illustrated in an example.

> *Example 14-1.*   (a) Analyze the math and verbal scores of Table 14-1 in an ANOVA table, including the prob-value and a test of the null hypothesis $\beta = 0$ at the 5% level.
> (b) Test the same null hypothesis by alternatively using the t confidence interval.
> (c) Test the equivalent null hypothesis $\rho = 0$ using the confidence interval based on $r = .63$, as given in (14-8).
>
> *Solution.*   (a) The ANOVA table is set out in Table 14-2(b), yielding an F ratio of 3.87. Comparing this to the nearby critical values (in Table VII), we find that $F_{.10} = 3.78$ and $F_{.05} = 5.99$. Thus:
>
> $$.05 < \text{prob-value} < .10 \qquad (14\text{-}23)$$
>
> Since the prob-value is more than 5%, we cannot reject H_0.
> (b) The appropriate estimates derived in Table 14-1 are substituted into the formula for the 95% confidence interval:
>
> $$\beta = \hat{\beta} \pm t_{.025} \frac{s}{\sqrt{\sum x^2}} \qquad (12\text{-}27) \text{ repeated}$$
>
> Thus:
>
> $$\beta = .50 \pm 2.45 \frac{9.20}{\sqrt{1304}}$$
>
> $$= .50 \pm 2.45(.254) \qquad (14\text{-}24)$$
>
> $$= .50 \pm .62 \qquad (14\text{-}25)$$
>
> Since $\beta = 0$ is included in the confidence interval, we cannot reject the null hypothesis at the 5% level.
> (c) In Figure 14-9, we must interpolate to find $n = 8$ and $r = .63$. This yields the approximate 95% confidence interval:
>
> $$-.15 < \rho < +.90$$

Since $\rho = 0$ is included in the confidence interval, we cannot reject the null hypothesis at the 5% level.

(c) Coefficient of Determination, r^2

We now will relate the variations in Y to r. It follows from (14-12) that:

$$\hat{\beta} = r \sqrt{\frac{\sum y_i^2}{\sum x_i^2}} \tag{14-26}$$

Substituting this value for $\hat{\beta}$ in (14-20):

$$\sum (Y_i - \overline{Y})^2 = r^2 \sum y_i^2 + \sum (Y_i - \hat{Y}_i)^2 \tag{14-27}$$

Noting that Σy_i^2 is, by definition, $\Sigma (Y_i - \overline{Y})^2$, the solution for r^2 is:

$$\frac{\sum (Y_i - \overline{Y})^2 - \sum (Y_i - \hat{Y}_i)^2}{\sum (Y_i - \overline{Y})^2} = r^2 \tag{14-28}$$

Finally, we can reexpress the numerator using (14-17). Thus:

$$r^2 = \frac{\sum (\hat{Y}_i - \overline{Y})^2}{\sum (Y_i - \overline{Y})^2} = \frac{\text{explained variation of } Y}{\text{total variation of } Y} \tag{14-29}$$

This equation provides a clear intuitive interpretation of r^2. Note that this is the *square* of the correlation coefficient r, and often is called the *coefficient of determination. It is the proportion of the total variation in Y explained by fitting the regression.* Since the numerator cannot exceed the denominator, the maximum value of the right-hand side of (14-29) is 1; hence the limits on r are ± 1. These two limits were illustrated in Figure 14-2: in panel (b), $r = +1$ and all observations lie on a positively sloped straight line; in panel (d), $r = -1$ and all observations lie on a negatively sloped straight line. In either case, a regression fit will explain 100% of the variation in Y.

At the other extreme, when $r = 0$, then the proportion of the variation of Y that is explained is $r^2 = 0$, and a regression line explains nothing. That is, when $r = 0$, then $\hat{\beta} = 0$. These are just two equivalent ways of formally stating that there is "no observed linear relation between X and Y."

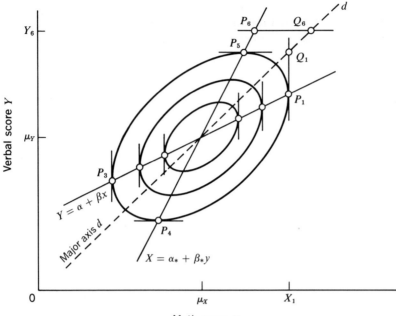

Math score X

FIGURE 14–11 The two regression lines in a bivariate normal population

(d) Regression Applied to a Bivariate Normal Population

In Table 14-1, we calculated a regression of Y on X for sample values that we assumed had been taken from a bivariate normal population. Is the $\hat{\beta}$ that we calculated an estimator of a population β, or does β even exist? That is, for a bivariate normal population, does a true regression line of Y on X exist? We now will show that the answer is yes.

As before, our assumed bivariate normal population is shown as a probability "hill," represented in Figure 14-11 by a set of level curves, with major axis d. Now consider the straight line $Y = \alpha + \beta x$, defined by joining points of vertical tangency such as P_1 and P_3. Each of these vertical tangents defines a cross-section slice of Y that is normal. Concentrating on the slice through $P_1 Q_1$, for example, we see that the mean of these Y values occurs at the point of tangency P_1; at this point, our vertical line touches its highest level curve, and the highest point on any normal distribution is at the mean. Thus, we see that the means of the Y populations lie on the straight line $Y = \alpha + \beta x$. Next, the variance of the Y populations can be shown to be constant.[3] Thus, the assumptions of the *regression* model (12-2) are satisfied by a bivariate normal (*correlation*) population. Therefore, the line $Y = \alpha + \beta x$ may be regarded as a true linear regression of Y on X.

If we knew a student's math score and we wished to predict his verbal score, this regression line would be appropriate, (e.g., if his math score were X_1, we would predict his verbal score to be at P_1). We should understand fully why we would *not* predict Q_1; that is, we do *not* use the major axis of the ellipse (line d) for prediction, even though this represents "equivalent" performance on the two tests. Since this student is far above average in mathematics, an equivalent verbal score seems too optimistic a prediction. Recall the large random element that is involved in performance. There are a lot of students who will do well in one exam, but not so well in the other; in other words, ρ is less than 1 for this population. Therefore, instead of predicting at Q_1, we will be more moderate and predict at P_1—a compromise[4] between "equivalent" performance Q_1 and "average" performance μ_Y.

This is the origin of the word "regression." Whatever a student's score in math, there will be a tendency for his verbal score to "regress" towards the population average.[5] It is evident from Figure 14-11 that this is equally true for a student with a very low math score; in this case, the predicted verbal score regresses upward toward the average.

Another interesting fact is that the correlation coefficient between X and Y is unique (i.e., ρ_{XY} is identically ρ_{YX}); but there are two regres-

---

[3]This may seem like a curious conclusion since, in Figure 14-5, the size of each cross-section slice differs, depending on the value of X_0. However, each slice $p(X_0, Y)$ must be adjusted by division by $p(X_0)$ in order to define the conditional distribution of Y, as shown in (5-10). Thus the conditional distribution is:

$$p(Y/X_0) = \frac{p(X_0, Y)}{p(X_0)}$$

This adjustment makes all the conditional distributions of Y "look alike," and thus have the same variance.

[4]P_1 is, in fact, a weighted average of Q_1 and μ_Y, with weights depending on ρ and $(1 - \rho)$. Thus, in the limiting case in which $\rho = 1$, X and Y are perfectly correlated, and we would predict Y at Q_1. At the other limit, in which $\rho = 0$, we can learn nothing about likely performance on one test from the result of the other, and we would predict Y at μ_Y. For all cases between these two limits, we predict between Q_1 and μ_Y.

[5]The classical case, encountered by Karl Pearson, et al., *Biometrika* 1903, involved trying to predict a son's height from his father's height. If the father is a giant, the son is likely to be tall; but there are good reasons for expecting him to be shorter than his father. So the prediction for the son was derived by "regressing" his father's height toward the population average.

sions, the regression of Y on X and the regression of X on Y. This immediately is evident if we ask how we would predict a student's math score (X) if we knew his verbal score (e.g., Y_6). Then equivalent performance (point Q_6 on line d) is a bad predictor; since he has done very well in the verbal test, we would expect him to do less well in math, although still better than average. Thus, the best prediction is P_6 on the regression line of X (math) on Y (verbal), $X = \alpha_* + \beta_* y$. In this case our regression is defined by joining points $(P_5, P_4$, etc.) of *horizontal*, rather than vertical, tangency. Each of these horizontal lines defines a normal conditional distribution of X, given Y; each of these distributions has the same variance, and has its mean lying on this regression line, thus satisfying our conditions for a true regression of X on Y.

Example 14-2.   From the math and verbal scores in Table 14-1:

(a) Calculate the two estimated regression lines $(Y$ on X, and X on $Y)$.

(b) For a student with a math score of 90, what is the best estimate of his verbal score?

(c) For a student with a verbal score of 10, what is the best estimate of his math score?

Solution.   (a) The appropriate estimates from Table 14-1 are substituted into the regression line:

$$Y = \hat{\alpha} + \hat{\beta}x$$

$$= \overline{Y} + \left(\frac{\sum xy}{\sum x^2}\right)x$$

$$Y = 50 + .50x \tag{14-30}$$

Similarly:

$$X = \hat{\alpha}_* + \hat{\beta}_* y$$

$$= \overline{X} + \left(\frac{\sum xy}{\sum y^2}\right)y$$

$$X = 60 + .78y \tag{14-31}$$

(b) Substitute $x = X - \overline{X} = 90 - 60 = 30$ into (14-30):

$$Y = 50 + .50(30) = 65$$

(c) Substitute $y = Y - \overline{Y} = (10 - 50) = -40$ into (14-31):

$$X = 60 + .78(-40) = 29$$

These answers are graphed in Figure 14-12.

(e) Correlation or Regression?

Both the standard regression and correlation models require that Y be a random variable. But the two models differ in the assumptions made about X. The regression model makes few assumptions about X, but the more restrictive correlation model of this chapter requires that X be a random variable, having with Y a bivariate normal distribution. We therefore conclude that the standard regression model has wider application. It may be used, for example, to describe the fertilizer-yield problem in Chapter 11 where X was fixed at prespecified levels, or the bivariate normal population of X and Y in this chapter. However, the standard correlation model describes only the latter. (It is true that r^2 can be *calculated* even when X is prespecified, as an indication of how effectively regression explains Y in (14-29); but r cannot be used for inferences about ρ in Figure 14-9.)

In addition, regression answers more interesting questions. Like correlation, it indicates whether two variables move together; but it also estimates how. Moreover, the key question in correlation analysis—whether

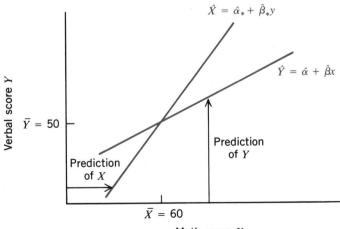

FIGURE 14–12 The two regression lines for the verbal and math scores in Table 14–1

or not any relationship exists between the two variables, i.e., whether $\rho = 0$—can be answered directly from regression analysis by testing the equivalent null hypothesis $\beta = 0$. If this is the only question, then there is no need to introduce correlation analysis at all.

In conclusion, since regression answers a broader and more interesting set of questions (including some correlation questions as well), it usually is the preferred technique. Correlation is used primarily as an aid to understanding regression.

(f) "Nonsense" Correlations

Even though correlation or regression may have established that two variables move together, no claim can be made that this necessarily indicates cause and effect. For example, the correlation of teachers' salaries and the consumption of liquor over a period of years turned out to be .9. This does not prove that teachers drink; nor does it prove that liquor sales increase teachers' salaries. Instead, both variables moved together, because both are influenced by a third variable—long-run growth in national income and population. If only third factors of this kind could be kept constant—or their effects fully discounted—then correlation would become more meaningful. This is the objective of multiple regression—or, equivalently, *partial correlation* in the next section.

Correlations such as the above are often called "nonsense" correlations. It would be more accurate to say that the correlation is real enough, but any naive inference of cause and effect is nonsense. As we already suggested in Section 13-8, the same issue occurs in simple regression analysis. For example, a regression applied to teachers' salaries and liquor sales also would yield a statistically discernable $\hat{\beta}$ coefficient; but any inference of cause and effect from this still would be nonsense.

PROBLEMS

14-3 Suppose that a shoe manufacturer takes a random sample of his production in order to examine the relationship between wearing performance and cost. (Assume that the population is approximately bivariate normal.)

$X = $ Cost of Production	$Y = $ Months of Wear
10	8
15	10
10	6
20	12
20	9

(a) Calculate and graph the regression line of Y on X.

(b) Find the prob-value for H_0. Assume that the only difference in cost is in materials—with a very costly and durable compound being used in the more expensive shoes.

(c) Find the prob-value for H_0 under a different assumption: The more expensive shoes are made out of exactly the same materials as the cheaper shoes; the only difference is that the expensive shoes have been designed at very high cost by an internationally famous Italian designer. (Continue to use this alternative hypothesis for the remaining parts (d) through (f).)

(d) Write out the ANOVA table for the regression of Y on X. What proportion of the variation in Y is explained by this regression? What proportion is left unexplained? What is the prob-value for the null hypothesis $\beta = 0$?

(e) Calculate r. Test the null hypothesis $\rho = 0$ at the 5% level.

(f) Do you get consistent answers in (c) through (e) for the question "Are X and Y linearly related?"

(g) What is the least squares estimate of:

(i) Y, if $X = 12$?

(ii) Y, if $X = 20$?

(iii) X, if $Y = 10$?

14-4 Suppose that a bivariate normal distribution of scores is perfectly symmetric in X and Y, with $\rho = .50$ and with level curves (isoprobability ellipses) as follows:

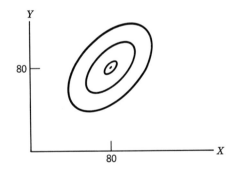

Answer true or false; if false, correct it.

(a) The regression line of Y on X is:

$$Y = 80 + .5(X - 80)$$

(b) The regression line of Y on X has the graph shown on page 430.

(c) The variance of Y is $1/4$ the variance of X.

(d) The proportion of the Y variation explained by X is only $1/4$.

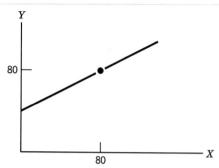

(e) Thus, the residual Y values (after fitting X) would have 3/4 the variation of the original Y values.

(f) For a student with a Y score of 70, the predicted X score is 60.

14-5 Let $\hat{\beta}$ and $\hat{\beta}_*$ be the sample regression slopes of Y on X, and X on Y for any given scatter of points. Answer true or false; if false, correct it.

(a) $\hat{\beta} = r\dfrac{s_Y}{s_X}$

(b) $\hat{\beta}_* = r\dfrac{s_X}{s_Y}$

(c) $\hat{\beta}\hat{\beta}_* = r^2$

(d) $\hat{\beta}_* = \dfrac{1}{\hat{\beta}}$

14-6.

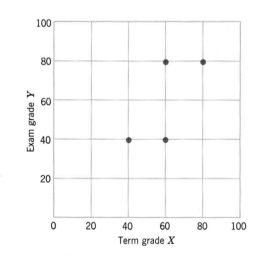

In the above graph of four students' marks, find geometrically (without doing any algebraic calculations):

(a) The regression line of Y on X.
(b) The regression line of X on Y.
(c) The correlation r (*Hint:* Problem 14-5(c)).
(d) The predicted Y for a student with $X = 70$.
(e) The predicted X for a student with $Y = 70$.

14-3   PARTIAL AND MULTIPLE CORRELATION

(a) Partial Correlation

In a multiple regression equation, the statistical discernibility (significance) of each regressor commonly is found from the t ratio (13-14). The same information can be expressed equivalently by the *partial correlation,* which is defined as the correlation of the regressor and the response, if all other regressors could be held constant.[6] From the partial correlation, the prob-value can be computed. Of course, this prob-value coincides exactly with the prob-value computed from the t ratio,[7] so that the partial correlation coefficient adds nothing here; it is quoted for the benefit of those who are familiar with correlation coefficients and their intuitive meaning.

(b) Multiple Correlation, R

Whereas the partial correlations measure how Y is related to each of the regressors one by one, the multiple correlation R measures how Y is related to all the regressors at once. R is derived by first calculating the fitted values $\hat{Y}$ using all the regressors; for example, when there are two regressors:

---

[6]If desired, the partial correlation can be expressed in terms of simple correlations. Although the general case is so complicated that it must be expressed in matrix form, the case of two regressors has a manageable formula for the partial correlation of Y and X, if Z could be held constant:

$$r_{YX/Z} = \frac{r_{YX} - r_{YZ}r_{XZ}}{\sqrt{1 - r_{XZ}^2}\,\sqrt{1 - r_{YZ}^2}} \tag{14-32}$$

where r_{YZ} represents the simple correlation of Y and Z, etc. This formula shows explicitly that there need be no close correspondence between the partial and simple correlation coefficient; however, in the special case that both X and Y are completely uncorrelated with Z (i.e., $r_{XZ} = r_{YZ} = 0$), then (14-32) reduces to:

$$r_{YX/Z} = r_{YX}$$

that is, the partial and simple correlation coefficients are the same. It also is instructive to note what happens at the other extreme when X becomes perfectly correlated with Z. In this case, $r_{YX/Z}$ cannot be calculated, since $r_{XZ} = 1$ and the denominator of (14-32) becomes zero as a consequence. This is recognized as the problem of perfect multicollinearity in Chapter 13, where the corresponding multiple regression estimate $\hat{\beta}$ could not be defined.

[7]In fact, the prob-value for the partial correlation usually is calculated by first calculating the t ratio, and then using this to read off the prob-value in the t table (Table V).

$$\hat{Y} = \hat{\alpha} + \hat{\beta}X + \hat{\gamma}Z$$

Then the multiple correlation R is defined as the ordinary (simple) correlation between the fitted $\hat{Y}$ and the observed Y:

$$R \stackrel{\Delta}{=} r_{\hat{Y}Y} \tag{14-33}$$

This has all the nice algebraic properties of any simple correlation. In particular, we note (14-29), which takes the form:

$$R^2 = \frac{\sum (\hat{Y}_i - \overline{Y})^2}{\sum (Y_i - \overline{Y})^2} = \frac{\text{variation of } Y \text{ explained by all regressors}}{\text{total variation of } Y} \tag{14-34}$$

Thus R^2 is seen to provide an overall index of how well Y can be explained by all the regressors, i.e., how well a multiple regression fits the data. Moreover, as we add additional regressors to our model, we can see how helpful they are in explaining the variation in Y, by noting how much they increase R^2.

*(c) Stepwise Regression

When there are a large number of regressors, a computer sometimes is programmed to introduce them one at a time, in a so-called "stepwise regression." The order in which the regressors are introduced may be determined in several ways. Two of the commonest are:

1. The statistician may specify a priori the order in which he wants the regressors to be introduced. For example, if there are, among many other regressors, 3 dummy variables to take care of 4 different occupations, then it would make sense to introduce these 3 dummy variables together in succession, without interruption from other regressors.

2. If there are no such clear guidelines, the statistician may want to let the data determine the order, introducing the most important regressors first. This customarily is achieved by having the computer choose the sequence that will make R^2 climb as quickly as possible. Suppose, for example, that 2 regressors already have been introduced. In deciding which of the remaining regressors should be added in the next step, the computer tries each of them in turn, and selects the one that increases R^2 the most.

Whichever method is used, the computer customarily prints the regression equation after each step (after each new regressor is introduced). Then at the end, the computer prints a summary, consisting of the list of regressors in the order they were introduced and the corresponding value of R^2 at each step.

Example 14-3.   A recent study related lung capacity Y to several variables, including three hazardous occupations.[8] Table 14-3 was computed by stepwise regression using method 1, where the statistician specified the order *a priori*.

TABLE **14-3**

Summary of stepwise regression, for lung capacity Y related to 8 regressors[a]

New Regressor Introduced	R^2 for All the Regressors Included So Far	Increase in R^2 Due to the New Regressor
Physical Variables		
X_1 = age	.363	.363
X_2 = height	.467	.104
Smoking Variables		
X_3 = present smoking	.483	.016
X_4 = past smoking	.484	.001
X_5 = smoking pipes, cigars	.485	.001
Dummy Occupation Variables[b]		
X_6 = chemical work	.491	.006
X_7 = fire fighting	.496	.005
X_8 = farming	.519	.023

[a]Based on a sample of n = 1,072 workers.
[b]These dummy variables compare chemical workers, firemen, and farmers with a reference group of physicians. For example X_6 = 1 for chemical workers, 0 otherwise.

Which steps increase R^2 the most? the least?

Solution.   From the last column of Table 14-3, we see that R^2 is increased the most by the physical variables X_1 (age) and X_2 (height).[9] R^2 is increased the least by the two minor smoking variables X_4 (past smoking) and X_5 (smoking pipes, cigars). The remaining variables seem to be of moderate importance.

---

[8]This is a continuation of Problem 13-27.

[9]Of course it doesn't require a statistical study to show that lung capacity depends upon age and height. These two regressors were included in the study primarily to avoid the bias that would have occurred if they had been omitted. That is, the age and height regressors were introduced to allow for the fact that farmers, for example, tended to be older and shorter.

Although stepwise regression is a useful device to help sort out the relative important of the regressors, there are several potential abuses to beware of:

1. Suppose that a regressor is dropped simply because it increases R^2 too little (or because it is statistically indiscernible in a formal test, as described below). This omission may bias the remaining regression coefficients, as described in Section 13-8.

2. In Table 14-3, consider the tiny increase in R^2 when X_4 (past smoking) is introduced. This does not necessarily mean that past smoking is unimportant. Rather, perhaps it could be explained by the fact that past smoking is highly correlated with present smoking (few adults change their smoking habits, either to quit or begin). When two regressors such as this are very highly correlated, they cannot be "sorted out" clearly. Thus we can expect that the first one that is introduced (present smoking) will tend to capture the effect of both, thus increasing R^2 on both accounts. Since the effect of the second variable already has been captured in this way, its formal introduction may have little additional effect[10] on R^2. This illustrates an important point: the judgment of which variables are most important in explaining Y may depend heavily on the order in which they are introduced.

*(d) Hypothesis Testing

Consider the problem of testing the null hypothesis that a *whole group* of g regressors has no relation to the response Y (i.e., that a set of g regression coefficients are all zero).[11] First, we decompose the total variation into its relevant components:

$$\text{total variation} = \text{variation explained by first regressors}$$

$$+ \text{ additional variation explained by last } g \text{ regressors}$$

$$+ \text{ unexplained variation} \qquad (14\text{-}35)$$
$$\text{like } (14\text{-}20)$$

Then we form the variance ratio:

---

[10] This is a good example of the multicollinearity problem illustrated in Figure 13-4: the effect of a regressor is less discernible when the model contains another regressor that is highly correlated with it.

[11] For a more precise statement of what actually is being tested, see a more advanced text, such as R. J. Wonnacott and T. H. Wonnacott, *Econometrics* (New York: John Wiley & Sons, 1970), Chapter 13. This text also gives confidence intervals (ellipses) as an alternative to hypothesis testing.

$$F = \frac{\text{additional variance explained by last } g \text{ regressors}}{\text{unexplained variance}} \quad (14\text{-}36)$$

Specifically, these are the six steps that are involved in (14-36):

1. Calculate the increase in R^2 due to the g regressors; call it ΔR^2.

2. Calculate the average increase in R^2 per regressor,

$$\Delta R^2 / g \quad (14\text{-}37)$$

3. Calculate the proportion of variation left unexplained after the addition of the g regressors, $(1 - R^2)$.

4. Divide by $n - k - 1$, the degrees of freedom in the unexplained variation; i.e., calculate

$$(1 - R^2)/(n - k - 1) \quad (14\text{-}38)$$

where n = number of observations, and k = total number of regressors considered—both the group of g regressors that are being tested and the earlier regressors that already were in the model.[12]

5. See how large (14-37) is relative to (14-38), by forming their ratio,

$$F = \frac{\Delta R^2 / g}{(1 - R^2)/(n - k - 1)} \quad \begin{array}{l} (14\text{-}39) \\ \text{like } (14\text{-}36) \end{array}$$

6. Assuming that the null hypothesis is true, this F ratio can be proved to have the F distribution that we already tabulated in Appendix Table VII. So we can calculate the prob-value in the right-hand tail roughly, and use this for testing the null hypothesis, just as we did in Chapter 10.

Example 14-4.   Using Table 14-3, calculate the prob-value for the null hypothesis that lung function has no relation to occupation.

---

[12] Since there is also a constant term, $k + 1$ is the number of parameters (regression coefficients) in the model; then $n - k - 1$ is the excess of observations over parameters, i.e., degrees of freedom (d.f.). Equation (14-38) represents, therefore, a kind of "average unexplained error per excess observation."

Solution. For the group of the last three (occupation) regressors in Table 14-3, the above 6 steps yield:

(1) $$\Delta R^2 = .519 - .485 = .034$$

(2) $$\Delta R^2/g = .034/3 = .0113$$

(3) $$1 - R^2 = 1 - .519 = .481$$

(4) $$(1 - R^2)/(n - k - 1) = .481/(1072 - 8 - 1)$$
$$= .481/1063 = .00045$$

(5) $$F = \frac{.0113}{.00045} = 25.0$$

(6) Referring to Table VII, the d.f. are $g = 3$ and $n - k - 1 = 1063 \simeq \infty$; we therefore consult the last line of Table VII, and find that $F_{.001} = 5.42$. Since the observed F of 25.0 far exceeds this, then:

$$\text{prob-value} \ll .001$$

and we conclude that occupation is highly discernible (statistically significant).

PROBLEMS

14-7 (a) Referring to Table 14-1, using Y in column 2 and $\hat{Y}$ in column 8, calculate the multiple correlation coefficient R according to (14-33).
(b) In this example, does R agree with the simple correlation $r = .63$ (as given in (14-8))?
*(c) Prove that R is *always* the same as r, when there is just one regressor. Thus we may think of r as a special case of R.

14-8 For the data of Problem 13-1 relating saving S to income X and assets W, find:
(a) The multiple correlation of S on X and W.
(b) The simple correlation of S on X.
(c) What is the proportion of variation in S that is:
   (i) Explained by X alone.
   (ii) Explained by X and W.
   (iii) Explained by the addition of W (after X).
   (iv) Left unexplained (after the addition of W and X).

*14-9 Use the results in Problem 14-8 to calculate the prob-value for the null hypothesis that S has no relation to W, after X has been included in the model.

*14-10 Referring to Table 14-3, calculate the prob-value for the following null hypotheses about the population:[13]
(a) The smoking variables, as a group, have no relation to lung function.
(b) The 6 smoking and occupational variables as a group have no relation to lung function.
(c) Present smoking (X_3) has no relation to lung function (here $g = 1$, a special but interesting case).
(d) Farming (X_8) has no relation to lung function (again, $g = 1$).

REVIEW PROBLEMS

14-11 In a random sample, students' aptitude score X and achievement score Y had means $\overline{X} = 60$, $\overline{Y} = 68$, variances $s_X^2 = 100$, $s_Y^2 = 150$, and a correlation $r = .40$. Calculate each of the following, if possible; if not possible, state what further information would be necessary.
(a) The estimated slope in the regression of Y on X.
(b) The prob-value for H_0 (no linear relation between X and Y).
(c) The predicted achievement score Y of a student with an aptitude score $X = 80$.
(d) The predicted aptitude score X of a student with an achievement score $Y = 90$.

14-12 Repeat Problem 14-11(a) and (b) for the following data (annual income, in thousands of dollars):

$X = $ Husband's Income	$Y = $ Wife's Income
25	12
20	12
30	18
25	14

14-13 (a) Referring to the math and verbal scores of Table 14-1, suppose that only the students with math score exceeding 65 were admitted to college. For this subsample of three students, calculate the correlation of X and Y.
(b) For the other subsample of the five remaining students, calculate the correlation of X and Y.

---

[13] In each case, the model under consideration includes, as usual, just those regressors that we have listed in the table so far.

(c) Are these two correlations in the subsamples greater or less than the correlation in the whole sample? Do you think that this will be generally true?

14-14 Suppose that all the firms in a certain industry recorded their profits P (after tax) in 1975 and again in 1976, as follows:

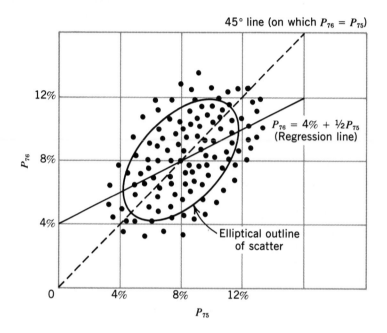

The least squares regression line also is shown in the graph. From it we would predict, for example, that a firm making a profit of 12% in 1975 would make a profit of 10% in 1976; and that a firm making a profit of 4% in 1975 would make a profit of 6% in 1976. That is, the outstandingly prosperous firms in 1975, as well as the outstandingly poor firms in 1975, tended to become less outstanding in 1976. Select the most appropriate statement (and criticize the others).

(a) This shows that the firms tended to become more homogeneous (more stable) in their profits in 1976—fewer risks were run.

(b) This indicates, but does not prove, that some factor (perhaps a too progressive taxation policy, or a more conservative outlook by business-people, etc.) caused profits to be much less extreme in 1976 than in 1975.

(c) This shows, among other things, how difficult it is to stay near the top, from year to year (from 1975 to 1976, specifically).

(d) To predict the 1976 profit for a firm making a 1975 profit of 10%, we ought to use the 45° line and hence predict $P_{76} = P_{75} = 10\%$.

14-15 Suppose that a sociologist analyzed two demographic variables X and Y, and found the correlation r and the coefficient $\hat{\beta}$ for the regression of Y on X. A second sociologist used the same data, but with rescaled variables. Instead of using X and Y, he used:

$$X' = 1,000X$$

$$Y' = 100,000Y$$

Suppose that all his subsequent calculations also are denoted by primes (r', $\hat{\beta}'$, etc.). Are the following quantities independent of scale? (Answer "yes," "no," or "can't say"):
(a) Correlation (i.e., is $r' = r$?).
(b) Regression coefficient (i.e., is $\hat{\beta}' = \hat{\beta}$?).
(c) t-ratio (i.e., is $t' = t$ in (12-35)?).
(d) Prob-value for H_0.

*14-16 Suppose that we wish to regress Y on X and Z. Considering only the regressors X and Z for the moment, note that their coefficient of determination r^2 is given by:

$$r^2 = \frac{\left(\sum x_i z_i\right)^2}{\left(\sum x_i^2\right)\left(\sum z_i^2\right)}$$

(a) Now, considering the regression of Y on X and Z, use this value of r^2 to show that (13-9) may be rewritten as:

$$s_{\hat{\beta}} = \frac{s}{\sqrt{\sum x_i^2}\,\sqrt{1 - r^2}} \qquad (14\text{-}40)$$

(b) Formula (14-40) is a very useful form, from which we may draw several conclusions. Answer true or false; if false, correct it.
  (i) If X and Z are uncorrelated, the formula (14-40) for $s_{\hat{\beta}}$ is remarkably similar to the formula (12-25) in simple regression. The only difference is that in (14-40), s^2 is the residual variance left unexplained after *two* regressors have been fitted, instead of just one.

(ii) The more X and Z are positively correlated, the larger s_β will be. For example, if $r = .99$, then s_β will be about 100 times as large as if $r = 0$.

(iii) This illustrates the problem of multicollinearity: if X and Z are highly correlated, the coefficient of X will have a much larger standard error.

CHAPTER 15

Nonlinear Regression

All Mathematics would suggest
A steady straight line as the best,
But left and right alternately
Is consonant with History.

W. H. Auden

So far, we have considered only *linear* regression. But there are many other ways in which Y may depend on X, such as the example that we gave in Figure 12-8. In this chapter, we analyze some of the basic forms that nonlinearity may take. It will become evident, even from these few examples, that our standard regression procedure can often still be applied—simply by redefining the variables or transforming the equation.

To be specific, consider two examples; the first is:

$$Y = \alpha + \beta X^2 \tag{15-1}$$

Here, the only nonlinearity is in the variable X; α and β appear in this equation in the same linear way as they did in Chapter 12. The material in Section 15-1 will confirm that standard regression analysis can be applied directly.

The second example is:

$$Y = X^\beta \tag{15-2}$$

Here, the problem is more difficult, since the nonlinearity directly involves β, the parameter to be estimated. Consequently, this equation first requires a transformation that will make it linear in β, as we will show in Section 15-2. Then we can apply standard regression analysis.

15-1   NONLINEARITY IN THE VARIABLES, BUT NOT THE PARAMETERS

(a) The Solution—Remarkably Easy

The regressions that we studied in earlier chapters have been of the linear form:

$$Y = \alpha + \beta X + \gamma Z + e \qquad (15\text{-}3)$$
$$\text{like (13-2)}$$

Equation (15-3) is linear in the variables (X, Y, Z) as well as in the parameters (α, β, γ). Now, if we closely examine the least-squares method in Chapter 13, we note that as long as α, β, γ appear in a linear way, the estimating equations (13-4) will be linear in the estimates $\hat{\alpha}$, $\hat{\beta}$, $\hat{\gamma}$. Hence, no problems are involved in their solution.

(b) The U-Shaped Cost Curve

Suppose that a firm's marginal cost Y is a function of the total quantity of its output Q. If this function is U-shaped (initially falling, then rising), then an appropriate mathematical model may be a second-degree polynomial (parabola):

$$Y = \alpha + \beta Q + \gamma Q^2 + e \qquad (15\text{-}4)$$

For example, we give some hypothetical data in Figure 15-1 and Table 15-1, columns (1) and (2).

To find the least-squares estimators $\hat{\alpha}$, $\hat{\beta}$, and $\hat{\gamma}$, we simply define:

$$X \overset{\Delta}{=} Q \qquad (15\text{-}5)$$
$$Z \overset{\Delta}{=} Q^2 \qquad (15\text{-}6)$$

When these are substituted into (15-4), we obtain the standard multiple regression equation (15-3):

$$Y = \alpha + \beta X + \gamma Z + e$$

TABLE 15-1
Multiple Regression Fit of a Parabolic Marginal Cost Curve

			Translation of Q and Q² into X and Z							Retranslation of X and Z back into Q and Q²
(1) Y_i	(2) $X_i = Q_i$	(3) $Z_i = Q_i^2$	(4) $x_i = X_i - \overline{X}$	(5) $z_i = Z_i - \overline{Z}$	(6) $Y_i x_i$	(7) $Y_i z_i$	(8) x_i^2	(9) z_i^2	(10) $x_i z_i$	(11) $\hat{Y}_i = 40.2 - 1.44x_i + .0366z_i$
37	10	100	−20	−967	−740	−35,767	400	934,451	19,333	33.6
27	15	225	−15	−842	−405	−22,725	225	708,408	12,625	31.0
31	20	400	−10	−667	−310	−20,667	100	444,449	6,667	30.2
27	25	625	−5	−442	−135	−11,925	25	195,072	2,208	31.2
36	30	900	0	−167	0	−6,000	0	27,779	0	34.1
42	35	1,225	5	158	210	6,650	25	25,068	792	38.8
45	40	1,600	10	533	450	24,000	100	284,441	5,333	45.3
55	45	2,025	15	958	825	52,708	225	918,396	14,375	53.7
62	50	2,500	20	1,433	1,240	88,866	400	2,054,435	28,667	63.8
$\sum Y_i = 362$	$\sum X_i = 270$	$\sum Z_i = 9,600$	$\sum x_i = 0$	$\sum z_i = 0$	$\sum Y_i x_i = 1,135$	$\sum Y_i z_i = 75,140$	$\sum x_i^2 = 1,500$	$\sum z_i^2 = 5,592,500$	$\sum x_i z_i = 90,000$	
$\overline{Y} = 40.22$	$\overline{X} = 30$	$\overline{Z} = 1,067$								

$$\hat{Y} = 40.2 - 1.44(X - \overline{X}) + .0366(Z - \overline{Z})$$
$$\text{or } \hat{Y} = 44.4 - 1.44X + .0366Z$$
$$\text{i.e., } \hat{Y} = 44.4 - 1.44Q + .0366Q^2$$

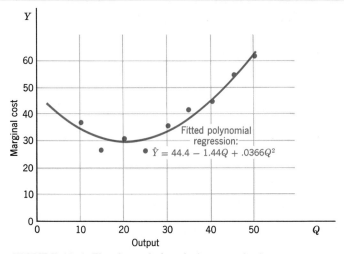

FIGURE 15–1 Fitted parabola relating marginal cost to output

We note in passing that although Y is related to only one *independent variable Q*, our fit involves regressing Y on two *regressors, Q* and Q^2. This is the first time that we have had to distinguish between independent variables and regressors, but the distinction is important now. When one variable is used to obtain several regressors, as in this instance, we may wonder if multicollinearity becomes a problem.

Although Z_i and X_i are *functionally* dependent (i.e., one is the square of the other), they are not *linearly* dependent (i.e., one is not, say, three times the other). Geometrically, the points (X_i, Z_i) do lie on a curve, as shown in Figure 15-2; however, the important point is that they do not lie on a line. Thus, we avoid the problem of complete multicollinearity. From a mathematical point of view, the physical or economic source of the X_i and Z_i values is irrelevant; just as long as X and Z are *linearly* independent, we can use the mathematical model of Chapter 13.

Therefore, we may compute the estimated coefficients $\hat{\alpha}$, $\hat{\beta}$, and $\hat{\gamma}$ using a standard computer program for multiple regression.[1] The results are:

---

[1] It is completely unnecessary, in this age of computers, to carry out the computations by hand; most canned programs not only will run the regression, but also will transform variables beforehand and retransform them afterwards. Nevertheless, in order to explicitly show how this procedure works, we lay out the hand computations in Table 15-1.

Columns (2) and (3) set out the first step: the transformation of Q into X and Z. Columns (4) to (10) set out the second step: the standard regression of Y on X and Z. Our calculations are substituted into the last two estimating equations in (13-4), obtaining:

(cont'd)

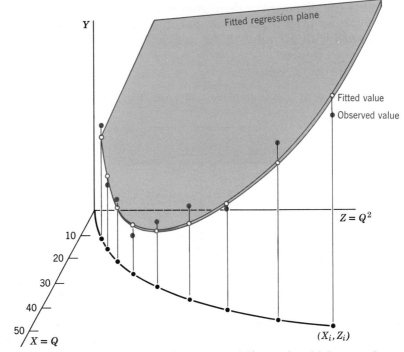

FIGURE 15–2 Polynomial regression as a special case of multiple regression

$$\hat{Y} = 44.4 - 1.44Q + .0366Q^2 \tag{15-8}$$

$$(t = -3.1) \quad (t = 4.9)$$

When this regression is plotted in Figure 15-1, it raises an interesting question: Is the parabolic model really necessary, or would a straight line suffice? The answer lies in prior knowledge and the statistical dis-

---

$$1,135 = \hat{\beta}(1,500) + \hat{\gamma}(90,000)$$
$$75,140 = \hat{\beta}(90,000) + \hat{\gamma}(5,592,500)$$

These two may be solved:

$$\hat{\beta} = -1.44$$
$$\hat{\gamma} = .0366 \tag{15-7}$$

In the last rows of Table 15-1, we then set out the third step: substituting (15-7), and retranslating X and Z back into Q. This yields (15-8).

In the last column of Table 15-1, we calculate the fitted points, which form the curve in Figures 15-1 and 15-2.

cernibility (significance) of the regressor Q^2, as we explained in Section 13-6. Since our prior expectation that this was a parabolic relationship is confirmed with a t value of 4.9 in Equation (15-8), we retain Q^2.

PROBLEMS

15-1 (a) Fit Y with a second-degree polynomial in Q for the following data:

Output Q (Thousands)	Marginal Cost Y
1	32
2	20
3	20
4	28
5	50

   (b) Graph the five points, and the fitted curve.
   (c) What is the estimated Y when $Q = 2.5$?

15-2 (a) If a parabolic curve were fitted to the whole population in Problem 11-1, would you expect it to be concave up or down? What, then, would be the sign of the coefficient of X^2?
   (b) Now fit a parabola to the data of Problem 11-1.
   (c) Does the coefficient of X^2 have the sign that you expected a priori?
   (d) Do you think that the straight line or the parabola is a better model to fit the data? Why?

15-2   NONLINEARITY IN THE PARAMETERS, REQUIRING A TRANSFORMATION

(a) A Simple Growth Model

Figure 15-3(a) shows U.S. population P during a period of sustained growth, 1860–1900.[2] It curves up in a way that suggests a constant percentage growth (like compound interest, or unrestrained biological growth). The appropriate model is therefore the exponential,

$$P = Ae^{\beta T} \tag{15-9}$$

where $e \simeq 2.718$ is the base for natural logs. The advantage of using this base is that:

---

[2] From 1974 The American Almanac, Table 1.

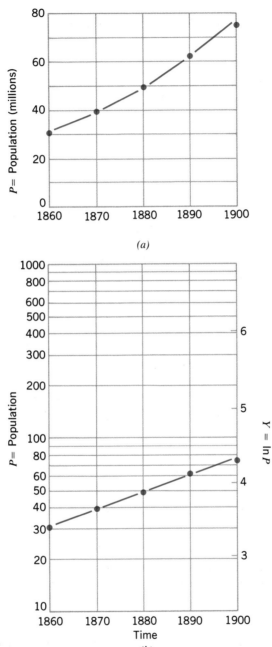

FIGURE 15–3 (a) U.S. population growth, 1860–1900, and its exponential fit. (b) Same curve plotted on semilog paper.

$$\boxed{\beta \text{ represents the rate of growth}^3} \qquad (15\text{-}10)$$

as proved in Section 15-3.

If it is reasonable to assume that large errors are associated with large values of the dependent variable P, then it is appropriate to treat the error term as multiplicative, rather than additive; then the statistical model is:

$$\boxed{P = Ae^{\beta T} \cdot u} \qquad (15\text{-}11)$$

where the errors u have a distribution that fluctuates around 1. Although (15-11) is nonlinear in the parameters A and β, it easily can be linearized by taking natural logarithms, yielding:

$$ln\ P = ln\ A + \beta T + ln\ u \qquad (15\text{-}12)$$

Let us define:

$$Y \overset{\Delta}{=} ln\ P \qquad (15\text{-}13)$$

$$\alpha \overset{\Delta}{=} ln\ A$$

$$e \overset{\Delta}{=} ln\ u \qquad (15\text{-}14)$$

This transformed error term e has a distribution that fluctuates around zero.[4] Thus (15-12) takes the standard linear form:

$$Y = \alpha + \beta T + e$$

This can be estimated by using a standard regression computer program, which yields:

$$\hat{Y} = 3.90 + .022t \qquad (15\text{-}15)$$

$$(s_{\hat{\beta}} = .0006)$$

For reference, we also show the hand calculations in detail[5] in Table 15-2. The transformed values $Y = ln\ P$ in column (3) are graphed in

---

[3] If time T is measured in years (or months, etc.), then β is the rate of growth per year (or month, etc.).

[4] See Figure 15-4 below. If $e = ln(u)$, and u has a median of 1, then e will have a median of $ln(1) = 0$.

[5] The natural logs in Table 15-2, as well as the antilog in (15-18), are obtained by multiplying 2.30 times the common logs in Appendix Table I.

Figure 15-3(b) (note the scale on the right of this diagram). This is equivalent to graphing the original values P on semilog graph paper. Note how close these observed values are to the fitted regression line (15-15), which also is graphed in this figure; this provides further confirmation of our selection of the exponential model.

To retransform (15-15) back into the exponential model, we substitute (15-13), obtaining:

$$ln\ \hat{P} = 3.90 + .022t \tag{15-16}$$

TABLE **15-2**

Calculation of Exponential Growth Curve, U.S. Population
1860–1900

						Transformation of Equation	Regression of Y on T
						Retransformation	
(1)	(2)	(3)	(4)	(5)	(6)		
Year T	Population P (Millions)	$Y = ln\ P$	$t = T - \overline{T}$	Yt	t^2		
1860	31.4	3.45	−20	−69.0	400		
1870	39.8	3.68	−10	−36.8	100		
1880	50.2	3.91	0	0	0		
1890	63.0	4.14	10	41.4	100		
1900	76.0	4.33	20	86.6	400		
$\overline{T} = 1880$		$\overline{Y} = 3.90$		$\sum Yt = 22.2$	$\sum t^2 = 1,000$		

$$\hat{\alpha} = \overline{Y} = 3.90$$

$$\hat{\beta} = \frac{\sum Yt}{\sum t^2} = \frac{22.2}{1,000} \simeq .022$$

$$\hat{Y} = 3.90 + .022t$$
$$ln\ \hat{P} = 3.90 + .022t$$
$$\hat{P} = e^{3.90}\ e^{.022t}$$
$$\hat{P} = 49.5e^{.022t}$$

Taking antilogs (exponentials),

$$\hat{P} = e^{3.90}e^{.022t} \tag{15-17}$$

i.e.:

$$\hat{P} = 49.5e^{.022t} \tag{15-18}$$

For convenience, we have left time t in deviation form. Thus, the coefficient $\hat{A} = 49.5$ (million) may be interpreted as the estimate of the population in 1880 (when $t = 0$). The coefficient $\hat{\beta} = .022 = 2.2\%$ is the approximate annual growth rate.

Although a fitted equation like (15-18) may shed a good deal of light on past growth, there is an important warning against using it for any short-term prediction of a time series, such as population. In this sort of very simple growth model, the error u is likely to be serially correlated (when population is unduly high in one period, it may consequently be high in the next period). Then (as we shall see in Chapter 21), the prediction should take account of this correlation.

The problems are even greater in a long-term projection. So many changes occur in the underlying process determining population growth (wars, recessions, the development and acceptance of birth control methods, etc.), that there is no single unchanging relationship that can be trusted for a long-term prediction. Hence, this sort of projection may be a futile exercise. It is far more fruitful to try to discover what changes may be expected in the underlying growth process as the population adjusts to the pressure of increasing numbers.

(b) Potential Advantages of a Transformation

Recall that the logarithmic transformation can only be applied provided that the error term u in (15-11) can reasonably be assumed to be multiplicative. But suppose that this assumption cannot be made; for example, suppose that we know that the error term is additive, so that our model is:

$$P = Ae^{\beta T} + u \qquad (15\text{-}19)$$

Since the log transformation cannot be applied validly, much more complicated and expensive computer-fitting techniques would be the only option.

But if the model is in a form like (15-11) that does yield to the log transformation, then this will provide a number of advantages simultaneously. Not only will it linearize the model and make the error term additive, but it also may make the error term more normally distributed. This is shown in Figure 15-4: if the distribution of u is skewed to the right, the distribution of $ln\ u$ usually is more symmetric, hence closer to normal. This provides certain advantages, such as more valid t tests.

These log transformation benefits are not restricted to growth models, of course. The log transformation is a useful tool for any multiplicative and exponential model. In fact, in many areas of the social sciences, it is more common to use the log transformation than to leave the data untransformed.

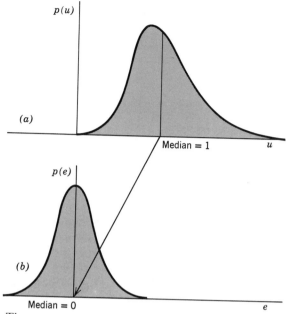

FIGURE 15–4 The error term in the multiplicative model (15-11). (a) Original skewed error. (b) The more symmetric error after the log transformation.

PROBLEMS

15-3 The population of the world since 1650 has been estimated as follows:[6]

Year	Population (Millions)
1650	470
1750	690
1850	1090
1950	2510

(a) Graph. Join the dots by a rough curve, and extrapolate to the year 2000.
(b) Fit an exponential growth by least squares. Graph it. Extrapolate it to 2000. Is this a very meaningful exercise?
(c) What is the estimated growth rate?

---

[6]From 1964 *Information Please Almanac*, p. 275.

15-4 Suppose that in 1950, a bank balance was left to accumulate at a constant interest rate i.

(a) Assuming no deposits or withdrawals, estimate the original balance B_0 in 1950, and the annual interest (growth) rate from the following observations:

Time	Balance
1952	$106.09
1953	$109.27
1954	$112.55
1955	$115.92

(b) If bank calculations are error-free (i.e., there is no random disturbance e), would you need all four observations to calculate i and B_0? How many observations are necessary?

15-5 For each of the following models, outline the method you would use to estimate the parameters (Greek letters).

(a) $Y = \alpha + \beta X + \gamma X^2 + \delta X^3$
(b) $Y = \alpha + \beta t + \gamma \sin(2\pi t/12)$ (Linear growth with cycle.)
(c) $Y = \alpha(1 + \beta)^t$
(d) $Y = \alpha \beta^t \gamma^X$
(e) $Y = \alpha + \beta X + \gamma Z + \delta X^2 + \epsilon XZ + \zeta Z^2$

15-6 In each model in Problem 15-5, what form should the error term take in order for your transformation to be most effective?

*15-3   ELASTICITY

One of the most important applications of logarithmic regression is to provide a direct estimate of elasticity, the standard measure of the responsiveness of one variable to another. For example, consider the economist's familiar concept of demand, which describes how the quantity demanded Q is related to price P. The price elasticity of demand (sometimes abbreviated to "elasticity" of demand) is defined[7] as:

---

[7]In terms of calculus, (15-20) becomes:

$$\text{elasticity} = \left| \frac{dQ}{dP} \cdot \frac{P}{Q} \right| \qquad (15\text{-}21)$$

$$
\begin{array}{rl}
\text{price elasticity} = & \dfrac{\left|\text{relative change in quantity demanded } Q\right|}{\text{relative change in price } P} \\[2mm]
= & \left|\dfrac{\Delta Q/Q}{\Delta P/P}\right|
\end{array}
\tag{15-20}
$$

with ΔQ representing the change in Q, and the absolute value sign added in elementary treatments so that elasticity will be positive. For example, an elasticity of 5 means that demand is very elastic, with the relative change in quantity demanded being 5 times as large as the relative change in price.

Now suppose that we wish to estimate the price elasticity of demand. If we can cast the relationship of Q and P into logarithmic form,[8] i.e.:

$$
\begin{array}{l}
\text{if } ln\ Q = \alpha + \beta\ ln\ P \\[2mm]
\text{then } |\beta| = \text{price elasticity}
\end{array}
\tag{15-22}
$$

Thus, a simple regression of $ln\ Q$ on $ln\ P$ will yield $|\hat{\beta}|$, the desired estimate of elasticity of demand.

Moreover, (15-22) can be extended further, as the following illustration suggests. The farm problem often is viewed as the result of low price elasticity of demand for agricultural products, and low income elasticity of demand as well (with income elasticity of demand, of course, just being (15-20), with income replacing price in the denominator). Suppose that we wish to evaluate this proposition. If demand for farm products Q may be specified as depending on price P and income Y in the logarithmic form:[9]

$$
ln\ Q = \alpha + \beta\ ln\ P + \gamma\ ln\ Y
\tag{15-23}
$$

then, by extending (15-22), a multiple regression of Q on $ln\ P$ and $ln\ Y$ will

---

[8]Although we have used natural logs to the base $e = 2.718$ in (15-22), the theorem is equally true for any other base.

[9]Note that this specification (15-23) requires the demand function to be multiplicative and exponential; i.e.:

$$
Q = AP^{\beta}Y^{\gamma}
$$

and, to apply least squares regression validly, the error term also must be multiplicative, as in (15-11).

yield an estimate of the price elasticity of demand $|\hat{\beta}|$ and the income elasticity of demand $|\hat{\gamma}|$.

It is evident that (15-22) is fundamental, and accordingly it is proved in Section 15-3. Meanwhile, it may be illustrated with the following simple calculation.

Example 15-1. Suppose that the demand for a certain product is:

$$ln\ Q = 10 - 5\ ln\ P \qquad (15\text{-}24)$$

(a) Suppose that price rises from $P_0 = \$3.00$ to $P_1 = \$3.06$. Calculate the elasticity in 4 steps:
  (i) Calculate the initial and final quantities Q_0 and Q_1.
  (ii) Calculate the relative change in Q.
  (iii) Calculate the relative change in P.
  (iv) Calculate the elasticity, i.e., the ratio of (ii) to (iii).
(b) Does your answer in (a) agree with the coefficient $|-5|$, as given by (15-22)?

Solution. (a)
  (i) From (15-24):

$$\ln Q_0 = 10 - 5 \ln P_0 = 10 - 5 \ln 3.00$$

$$= 10 - 5(1.099) = 4.505 \qquad (15\text{-}25)$$

$$Q_0 = \text{antilog } 4.505 = 90.5$$

Similarly, we may calculate:

$$\ln Q_1 = 10 - 5 \ln(3.06) = 4.405 \qquad (15\text{-}26)$$

$$Q_1 = \text{antilog } 4.405 = 81.9$$

  (ii) Change in $Q = Q_1 - Q_0 = 81.9 - 90.5 = -8.6$. Relative change in Q

$$= \frac{Q_1 - Q_0}{Q_0} = \frac{-8.6}{90.5} = -.095 = -9.5\% \qquad (15\text{-}27)$$

  (iii) Relative change in P

$$= \frac{P_1 - P_0}{P_0} = \frac{3.06 - 3.00}{3.00} = .02 = 2\%$$

(iv) From (15-20):

$$\text{elasticity} = \left| \frac{-9.5\%}{2\%} \right| = 4.75$$

(b) The calculated value of elasticity does indeed approximately[10] agree with the coefficient $|-5| = 5.0$ given by (15-22).

---

[10]This agrees only approximately because, strictly speaking, elasticity is defined only at a single point (on a demand curve), rather than over a line segment; in other words, we should have considered only a very, very small (limiting) decrease in price, and calculated how many times greater the relative increase in Q would be—as we did in (15-21). But for our illustrative purposes, it is easier to consider a line segment (through the range P_0 to P_1), and this introduces the approximation error.

Proofs of Growth and Elasticity Coefficients

In this section, we give an important theorem about logs, and use it to establish results about growth rates and elasticities. The theorem[11] is this: for small changes in any variable x,

---

*[11]**Proof of (15-28).** We shall use a result from calculus: for any number u near 1,

$$ln\ u \simeq u - 1 \tag{15-29}$$

(In the special case when $u = 1$ exactly, this is recognized to be the familiar statement that $ln\ 1 = 0$).

In (15-28), we consider a small change in x, from its original value x_0 to a new value x_1. Then the left side of (15-28) is:

$$\text{Change in } ln\ x = ln\ x_1 - ln\ x_0$$

$$\text{Change in } ln\ x = ln\left(\frac{x_1}{x_0}\right) \tag{15-30}$$

Now substitute x_1/x_0 for u in (15-29) (and note that this substitution is justified because, by keeping the change in x small, $x_1/x_0 \simeq 1$):

$$ln\left(\frac{x_1}{x_0}\right) \simeq \left(\frac{x_1}{x_0}\right) - 1$$

Substituting this into (15-30):

(cont'd on page 456)

$$\boxed{\text{Change in } ln \ x \simeq \textit{relative} \text{ change in } x} \qquad (15\text{-}28)$$

For example, in (15-25) and (15-26), we saw that the change in $ln \ Q$ was from 4.505 to 4.405, a drop of $-.10$; and in (15-27), we saw that the relative change in Q was $-.095$, approximately[12] the same value.

Now let us prove (15-10), that β is the rate of growth (i.e., relative change in P) in the model $P = Ae^{\beta T}$. Expressing this equation in log form:

$$ln \ P = ln \ A + \beta T \qquad (15\text{-}31)$$

In one year, what is the change in $ln \ P$? Since A is a constant, $ln \ A$ does not change, and T changes by 1. Thus, in (15-31), the change in $ln \ P$ is β. Therefore, according to (15-28),

$$\text{the relative change in } P \text{ is } \beta \qquad (15\text{-}10) \text{ proved}$$

Finally let us prove (15-22), that β is the elasticity if Q is related to P by:

$$ln \ Q = \alpha + \beta ln \ P$$

Since α is constant, it contributes nothing when we calculate changes; that is:

$$(\text{change in } ln \ Q) = \beta \ (\text{change in } ln \ P)$$

Now substitute (15-28):

$$(\text{relative change in } Q) = \beta \ (\text{relative change in } P)$$

Thus:

$$\text{Change in } ln \ x \simeq \left(\frac{x_1}{x_0}\right) - 1$$

$$= \frac{x_1 - x_0}{x_0}$$

$$= \text{relative change in } x \qquad (15\text{-}28) \text{ proved}$$

[12]As we explained in footnote 10 on page 455, these two values are not exactly the same because of approximation error.

$$|\beta| = \left|\frac{\text{relative change in } Q}{\text{relative change in } P}\right|$$

By (15-20):

$$|\beta| = \text{elasticity} \qquad\qquad \text{(15-22) proved}$$

PROBLEMS

15-7 Suppose that a regression study of quantity demanded as a function of price yielded the following equation:

$$ln\ Q = 5.2 - 1.3\ ln\ P$$

(a) What is the elasticity?

(b) If price increased by 3%, by how much would the quantity demanded fall?

(c) What price decrease would be required to increase the quantity demanded by 10%?

REVIEW PROBLEMS

15-8 (a) Unrestrained by either food or space limitations, a bacteria colony will double within a certain period of time τ; and τ, of course, depends not on the size of the colony but on the reproductive metabolism of the individual bacteria. Write down an appropriate mathematical model to describe such growth.

(b) If the following data were observed for a particular species, fit your model by least squares, and hence estimate the doubling time τ.

time (minutes)	0	100	200
size of colony	2,000	6,300	18,000

15-9 Consider the model:

$$P = 49.5e^{.022\,t} \qquad\qquad \text{(15-18) repeated}$$

that is:

$$ln\ P = (\text{some constant}) + .022t \qquad\qquad \text{(15-32)}$$

Fill in the blanks:

(a) If t increases by 1 year, then P will increase by _____%.

(b) Thus, if P is 1,000 in some year, then P will be _____ a year later.

(c) And if P is 5,000 in some year, then P will be approximately _____ 2 years later.

15-10 Consider a model similar to (15-32) above, except for a different interpretation of the variables:

$$ln \ Y = (\text{some constant}) + .046E \tag{15-33}$$

where

$$Y = \text{the person's income (dollars per week)}$$

and

$$E = \text{the person's education (in years)}$$

Fill in the blanks:

(a) If E increases by 1 more year, then Y will increase by _____%.

(b) Thus if a person with a certain education earns \$200 per week, a person with one more year of education earns _____.

(c) And if a person with 12 years of education earns \$300 per week, a person with 16 years of education earns approximately _____.

15-11 In a 1964 study[13] of how personal income was related to education, IQ, color, and other variables, the following multiple regression was estimated from a sample of 1,400 U.S. veterans:

$$LINC \ \ = .046E + .0010AFQT + .17COLOR + \text{other variables}$$
$$\text{(st. error)} \ \ (.007) \quad (.0004) \quad \quad \quad (.05)$$

where

$$LINC = \text{natural log of the veteran's weekly income}$$

$$E = \text{number of years of additional education (during and after military service)}$$

---

[13]Zvi Griliches and W. M. Mason, "Education, Income, and Ability," in A. S. Goldberger and O. D. Duncan, *Structural Equation Models in the Social Sciences* (New York: Seminar Press, 1973), Chapter 13, p. 296, Equation 4.

$AFQT =$ the veteran's percentile rating on the Armed Forces Qualification Test Score, which roughly measures IQ

$COLOR =$ dummy variable for race, being 0 for blacks and 1 for whites

other variables $=$ age, amount of military service, amount of schooling before military service, father's education and occupational status, and degree of urbanization of his childhood home.

Other things being equal it is estimated that:

(a) A veteran with one more year's additional education earns _____% more income.

(b) A veteran who is white earns _____% more income than one who is black.

(c) A veteran who rates 1 point higher on the $AFQT$ (for example, by being in the 51st percentile instead of the 50th percentile) earns _____% more income.

(d) A black veteran who scores in the 80th percentile in the $AFQT$ would earn _____% more than a white veteran who scores in the 50th percentile.

15-12 Assuming that the model in Problem 15-11 is specified correctly, by how much should you hedge your estimates in (a) and (b) in order to be 95% certain of being correct?

REVIEW PROBLEMS (CHAPTERS 11–15)

15-13 An anthropologist sampled 4 women from a certain population, and recorded their weights and ages:

Weight W (Pounds)	Age A (Years)
130	30
150	50
140	60
100	20
$\overline{W} = 130$	$\overline{A} = 40$

(a) What is the equation of the line that you would use to predict W from A? Graph the line and the 4 points, to check that the line fits the data reasonably well.

(b) Predict the weight of a woman who is 40 years old. In order to give the prediction a 90% chance of being correct, by how much should the predicted value be hedged?

(c) If the regression line of W on A were computed for the whole population, in what range would you expect (with 95% confidence) to find the slope β?

15-14 (a) If a parabola were fitted to the whole population in Problem 15-13, what would you expect the sign of the coefficient of A^2 to be?

(b) Fit a parabola to the 4 data points. Does the sign of the coefficient of A^2 agree with part (a)?

(c) Do you think that the straight line or the parabola is a better model to fit the data? Why?

(d) Using the better model that you chose in (c), predict the weight of a woman who is:

  (i) 30 years old.
  (ii) 50 years old.

15-15 A sample of 6 men were scaled from 1 to 10 on their degree of authoritarianism (A-variable), and also had their years of education (E-variable) recorded, as follows:

A	E
9	5
1	12
5	12
7	9
1	13
4	9

Calculate:

(a) The regression line of A against E.

(b) The 95% confidence interval for the population slope β.

(c) The 2-sided prob-value for the null hypothesis that $\beta = 0$.

15-16 (a) When the data of Problem 15-15 were broken down by class, it was discovered that the first, third, and fourth men were middle-class, while the others were upper-class. Reanalyze the data to show how authoritarianism is related to both education and class.

(b) Graph your results to show clearly how this analysis is different from the simple regression calculated in Problem 15-15.

15-17 (a) Continuing Problem 15-16, suppose that the education variable had not been measured for these 6 men; consequently, authoritarianism might have been related naively only to class by simple regression (with class as a dummy variable). Carry out this analysis (including a confidence interval for the slope).

(b) Reanalyze part (a), using (8-32). Do you reach exactly the same conclusion?

(c) Graph your results in (a) to show clearly how this analysis is different from the multiple regression in Problem 15-16.

*15-18 To appreciate the bias that the multiple regression in Problem 15-16 avoids, verify (13-27) for the bias in:

(a) Problem 15-15.

(b) Problem 15-17.

15-19 An economist fitted quarterly unemployment figures from 1974 to 1976 with the following very simple model:

$$U = a + bt + cd$$

where

U = unemployment, in %

T = time (in quarter years, starting with $T = 1$ for the first quarter of 1974)

$D = 1$ for first quarter (winter) of each year

   = 0 otherwise

Select the most appropriate statement (and criticize the others):

The coefficient c is interpreted as:

(a) The 3-year trend in first-quarter unemployment—specifically, the slope of this trend, measured in percentage points per year.

(b) The number of months that serious unemployment lasts each year, due to the severe winter season.

(c) The average increase in unemployment (in percentage points) in the first quarter compared with the rest of the year, allowing for trend.

(d) The rise in unemployment that occurs every first quarter.

(e) The fall in unemployment that occurs every first quarter.

15-20 Answer true or false; if false, correct it.

(a) In simple regression, it is assumed that:

$$Y_i = \alpha + \beta x_i + e_i$$

where the e_i are independent errors, with positive means and decreasing variances.

(b) The least squares estimators $\hat{\alpha}$ and $\hat{\beta}$ are biased, consistent estimators of α and β.

(c) The more distant X_0 is from $\overline{X}$, the greater is the error in predicting Y, given X_0.

15-21 Answer true or false; if false, correct it.

(a) One severe limitation of multiple regression is that it cannot include factors that are categorical (nonnumerical, e.g., sex, region).

(b) $R^2 = \dfrac{\text{variation of } Y, \text{ explained by all regressors}}{\text{total variation of } Y}.$

(c) Multicollinearity of X and Z occurs when $r_{XZ} = 0$; then we can get a more reliable estimate of the regression of Y on X and Z.

(d) Multicollinearity often is a problem in the social sciences, when the regressors have high correlation. On the other hand, in the experimental sciences, the values of the regressors often can be designed to avoid multicollinearity.

15-22 In each part of Figure 14-2, suppose that the sample is split by the vertical line into two subsamples. Referring to Problem 14-14, but without going through the calculations again, guess how the correlations in the subsamples are related to the correlation in the whole sample.

15-23 Is the following a true summary of the preceding problem? If not, correct it.

(a) In the rather typical cases (a) and (c), in which X and Y are roughly linearly related, we find that the correlation has a smaller magnitude in each subsample than it has in the whole sample. To explain this, from (14-29) and (14-20) we have:

$$r^2 = \dfrac{\text{total variation} - \text{unexplained variation}}{\text{total variation}}$$

$$= 1 - \dfrac{\text{unexplained variation}}{\text{total variation}}$$

Upon dividing each variation by $n - 1$, we obtain variances:

$$r^2 = 1 - \frac{\text{unexplained variance}}{\text{total variance}}$$

$$= 1 - \frac{s^2}{s_Y^2} \qquad\qquad (15\text{-}34)$$

where

$s^2 = $ the unexplained (residual) variance about the fitted regression line,

and

$s_Y^2 = $ the variance of the Y values.

For a subsample taken from one half of the scattergram, s^2 will tend to remain the same, while s_Y^2 will tend to be much less. Thus from (15-34) r^2 tends to be smaller—i.e., r tends to be smaller in magnitude—in the subsample.

This idea is very similar to that shown in Figure 12-5: in panel (a), where the X values are less spread out, there is less "leverage" to estimate the true linear relation, and hence a smaller r^2.

(b) Cases (b) and (d) in Figure 14-2 are freak cases of perfect correlation, which is maintained in the subsamples. Case (e) also is peculiar, having zero correlation in the whole sample. In each half-sample, the correlation might be slightly nonzero, due to sampling fluctuation. But this fluke will disappear in a very large sample or population, such as shown in Figure 14-8.

(c) Finally, case (f) is interesting as the only case where, in splitting the data, the correlations in the two subsamples are essentially larger in magnitude than the correlation in the whole sample. This is because of the great curve (nonlinearity) in the data.

REVIEW PROBLEMS (CHAPTERS 1–15)

15-24 To investigate the effect of 3 rather similar drugs (drugs X, Y and Z), a scientist selected 24 rats of similar genetic background and size, and divided them at random into 4 groups of 6 each. One group was kept as a control, while the other 3 groups were injected with drugs X, Y, and Z, respectively. For two weeks, the 24 rats were then kept under identical conditions. At the end of this two-week period, the following gains were recorded:

$\begin{array}{c}\text{Drug} \rightarrow \\ \downarrow Rat\end{array}$	Control	X	Y	Z
1	56	60	61	65
2	59	65	59	61
3	50	61	59	60
4	62	66	63	63
5	52	60	58	59
6	63	60	54	64
	$\overline{C} = 57$	$\overline{X} = 62$	$\overline{Y} = 59$	$\overline{Z} = 62$

Select the most appropriate remark, and criticize the others:

(a) This is an improper design—there should have been a control group of 18 rats to match the 18 drugged rats.

(b) The appropriate analysis is to fit a least squares line to the data, as follows:

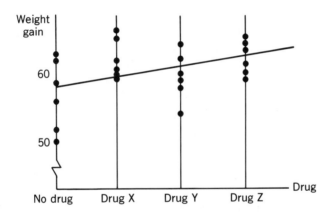

(c) This is a "1-factor experiment"—the factor being *drugs*. Thus, an analysis similar to Table 10-5 is appropriate.

(d) This is a "2-factor experiment"—the factors being *drugs* and *rats*. Thus, an analysis similar to Table 10-9 is appropriate.

(e) This is a "3-factor experiment"—the factors being *drugs* and *rats* and *weight gain*. Thus, an analysis extending Table 10-9 is appropriate.

15-25 (a) A physician investigated the number of hours of sleep required, comparing men and women. He collected a random sample of 5 men and an independent random sample of 5 women, with the following results:

Hours of Sleep
per Day

Men	Women
10	10
9	10
11	11
7	10
8	9

Analyze appropriately.

(b) A second physician suspected that age was relevant, and so reexamined the two samples, recording their ages as follows:

Men

Y = Hours of Sleep per Day	X = Age
10	30
9	35
11	20
7	55
8	45

Women

Y = Hours of Sleep per Day	X = Age
10	20
10	30
11	20
10	20
9	30

Outline how you would analyze this data. (Do not carry out the computations; merely state what method you would use, and refer to the relevant formulas or analogous examples in the text.)

Graph Y against X, and so roughly state how your conclusions would differ from (a).

(c) Which method is better, (a) or (b)? Why?

*15-26 (a) In a random sample of $n = 24$ workers drawn from an assembly line, performance scores Y had a mean $\overline{Y} = 70$ and standard deviation $s_Y = 10$. When another (25th) worker is drawn from the same population, we may predict his score with 95% certainty by means of a formula analogous to (12-43):

$$Y_0 = \overline{Y} \pm t_{.025}\, s_Y \sqrt{\frac{1}{n} + 1}$$

where t has $(n - 1)$ d.f. With this formula, calculate the actual 95% prediction interval.

(b) In order to predict better, it was decided to exploit a variable Z for marital status, defined as follows: $Z = -1$ for divorced or separated, $Z = 0$ for single or widowed, $Z = 1$ for married. In the sample of 24 workers, Z had a correlation of .80 with performance score Y; also a mean of .20 and a standard deviation of .60.

If the 25th worker was married, what would you predict his performance score Y to be, with 95% certainty?

(c) Briefly express in simple words how the interval in (b) is better than the one in (a).

15-27 Would it ever make sense to take logs of the response before applying ANOVA? Why, or why not?

15-28 Suppose that the regression model (12-3) is changed slightly, as follows: e_i has variance σ_i^2 (no longer a constant σ^2). The least-squares estimator:

$$\hat{\beta} = \frac{\sum x_i Y_i}{\sum x_i^2}$$

may still be calculated, although it no longer may have the optimal properties that it enjoyed formerly. What are the mean and variance of $\hat{\beta}$? (Incidentally, this sometimes is called the "heteroscedastic model.")

15-29 A certain drug (e.g., tobacco, alcohol, marijuana) is taken by a proportion of the American population. To investigate its effect on health (for example, mortality rate) suppose that certain people are to be studied for a five-year period, at the end of which each person's "mortality" will be recorded as follows:

$$M = 0 \text{ if he lives}$$

$$= 1 \text{ if he dies}$$

For each person, also let D represent his average monthly dose of the drug, and let subscripts 1 and 2 refer to drug users and nonusers, respectively.

Criticize the scientific merit of the following five proposals. (If you like, criticize their ethical and political aspects too.) Which proposal do you think is scientifically soundest? Can you think of a better proposal of your own?

(a) Draw a random sample of n persons. For each person, record the drug dose D that he chooses to take, and his mortality M after five years. From these n points, calculate the regression line of M against D, interpreting the coefficient of D as the effect of the drug.

(b) Again, draw a random sample of n persons. For each person, record such characteristics as age, sex, grandparents' longevity, etc., as well as drug dose D and mortality M after five years. Then calculate the multiple regression of M on all the other variables, interpreting the coefficient of D as the effect of the drug.

(c) Once again, draw a random sample of n persons. Then construct a 95% confidence interval for the difference in mortality rates between drug users and nonusers, using (8-32):

$$\text{drug effect, } (\mu_1 - \mu_2) = (\overline{M}_1 - \overline{M}_2) \pm t_{.025}\, s_p\, \sqrt{(1/n_1) + (1/n_2)}$$

where n_1 and n_2 are the numbers of drug users and nonusers, respectively (so that $n_1 + n_2 = n$, the size of the random sample), and s_p^2 is the pooled sample variance.

(d) Ask for volunteers who have never used the drug. Suppose that there are many volunteers, so that we may select from among them a random sample of n (where n is even). Divide the volunteer sample at random into two equal groups (control and treatment), each group being of the size $m = n/2$.

The control group of volunteers is allowed no drug, while the treatment group is given a standard dose, over the five-year period. Then a 95% confidence interval for the difference between drug users and nonusers would be, using (8-32) again:

$$\text{drug effect } (\mu_1 - \mu_2) = (\overline{M}_1 - \overline{M}_2) \pm t_{.025}\, s_p\, \sqrt{(1/m) + (1/m)}.$$

(e) Again, ask for volunteers who have never used the drug. Suppose that there are many, many volunteers, so that we now may select from among them a group of m matched pairs—each pair consisting of two volunteers of similar age, sex, grandparents' longevity, etc. From each pair, select at random (for example, by the flip of a coin) one of the two volunteers to go into the treatment group, while the other volunteer goes into the control group.

The control group is allowed no drug, while the treatment group is given a standard dose, over the five-year period. Then a 95% confidence interval for the difference between drug users and nonusers would be given in the equation above.

PART IV

Further Topics

CHAPTER 16

Nonparametric Statistics

Public agencies are very keen on amassing statistics—they collect them, add them, raise them to the nth power, take the cube root and prepare wonderful diagrams. But what you must never forget is that every one of those figures comes in the first instance from the village watchman, who just puts down what he damn pleases.

Sir Josiah Stamp

Strictly speaking, the classical statistics that we have considered so far (like t) require the assumption of population normality. If this assumption is seriously violated, we should look for techniques that are free of this distribution requirement. Such statistics are called *distribution free* or *nonparametric*. They are preferred for two reasons:

1. The corresponding classical statistic may be invalid. (That is, its confidence level may not actually be as high as 95%.)

2. But even in applications where the classical statistic is reasonably valid, a nonparametric statistic may be much more efficient (have smaller variance) and hence have a narrower confidence interval. In fact, this is the more important reason for using nonparametric statistics.

For nearly every parametric test (e.g., t test, ANOVA), there usually are several corresponding nonparametric tests. Since a description of all of these would fill several volumes, we begin by developing two: the sign test, which corresponds to the 1 sample t test (8-21), and the W test, which corresponds to the 2 sample t test (8-32).

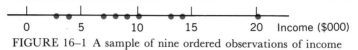

FIGURE 16–1 A sample of nine ordered observations of income

16-1   THE SIGN TEST

(a) Sign Test, Small Samples

Suppose that the median 1971 family income in the Southern U.S. was reported to be $3,800. But in a random sample[1] of 9 families shown in Figure 16-1, 8 have an income above $3,800, while only one has an income below. Does this evidence allow us to reject this report?

The null hypothesis may be stated formally as:

$$H_0: \text{population median} = \$3,800 \qquad (16\text{-}1)$$

That is, half the population incomes lie above $3,800; or, if an observation is drawn randomly, the probability that it lies above $3,800 is:

$$H_0: \pi = \tfrac{1}{2} \qquad (16\text{-}2)$$

We recognize (16-2) as being just like the hypothesis that a coin is fair. To state it more explicitly, we have two events that are equivalent mathematically:

> "random observation will fall above the median"
> is equivalent to
> "a coin will show heads"   (16-3)

If H_0 is true, the sample of $n = 9$ observations is just like tossing a coin 9 times. The total number of successes S (families above $3,800) will have the binomial distribution, and H_0 may be rejected if S is too far away from its expected value to be explained reasonably by chance.

The alternative hypothesis, of course, is that the population median is higher than this $3,800 report. To judge between these two competing hypotheses, we calculate as usual the prob-value for H_0; i.e., the probability (assuming that H_0 is true) that we would observe an S of 8 or more. Thus:

$$\text{prob-value} = \Pr(S \geq 8) \qquad (16\text{-}4)$$

from Appendix Table III(c):

$$= .0195 \qquad (16\text{-}5)$$

---

[1] From the 1974 *American Almanac*, Table 533.

Since this prob-value for H_0 is so low, we can reject H_0 in favor of H_1 at the 5% level (according to (9-19)). In other words, we conclude that the population median is higher than the $3,800 report.

(b) Two Matched Samples

With a little imagination, we can use the sign test for two matched samples (just as we used the t test in (8-37)).

Example 16-1.   Suppose that a small sample of 8 men had their motor capacity measured before and after a certain treatment. The results are shown in Table 16-1. Use the binomial distribution to calculate the prob-value for the null hypothesis that the treatment has no effect.

Solution.   The original matched pairs can be forgotten, once the differences (improvements) have been found in the third column. These differences D form a single sample, to which we now shall apply the sign test. The null hypothesis (treatment provides no improvement) is:

$$H_0: \text{population median} = 0$$

that is:

$$\pi = \text{Pr(observing positive } D) = \tfrac{1}{2}$$

The question is: are the observed 6 positive Ds (i.e., 6 "heads") in a sample of 8 observations ("tosses")[2] consistent with H_0? The probability of this is:

TABLE **16-1**
Motor Capacity of 8 Patients,
Before and After Treatment

X (Before)	Y (After)	Difference (D = Y − X)
750	850	+100
860	880	+20
950	930	−20
830	860	+30
250	300	+50
680	740	+60
720	760	+40
810	800	−10

$$\text{prob-value} = \Pr(S \geq 6)$$

$$= .1445$$

Since this prob-value is so high, we cannot reject H_0 (that the treatment is ineffective) at the 5% level.

---

[2]Any observation that might occur right on the hypothetical median (i.e., any D observed to be exactly zero) should not be counted as an observation at all, and should be discarded immediately.

Incidentally, you will recognize that Table 16-1 is remarkably similar to Problem 4-34.

The origin of the term "sign test" now should be clear. It is a test based on the binomial distribution, where the test statistic is the number of positive differences ("heads").

PROBLEMS

16-1 A random sample of annual incomes (thousands of dollars) of 10 brother–sister pairs was ordered according to the man's income, as follows:

Brother's Income (M)	Sister's Income (W)
9	14
14	10
16	8
16	14
18	13
19	16
22	12
23	40
25	13
78	24

Calculate the prob-value for the following null hypotheses about the population:
(a) That the men's median income is as low as 15.
(b) That the women's median income is as high as 15.
(c) That, on the whole, men earn no more than women.

16-2

Random Sample of Heights (Inches) of 8 Brother–Sister Pairs	
65	63
67	62
69	64
70	65
71	68
73	66
76	71
77	69

Calculate the prob-value for the following null hypotheses about the population:
(a) That the men's median height is as low as 66″.
(b) That the women's median height is as low as 63″.
(c) That, on the whole, men are no taller than women.

16-3 In a sample of 25 bolts from a production line, only 6 were above 10.0 cm. What is the prob-value for the hypothesis that the population median length is as high as 10.0 cm.?

16-2 CONFIDENCE INTERVALS FROM ORDER STATISTICS

We can give the sign test of the previous section a nice twist, to provide a confidence interval for the population median. To emphasize its similarity to the construction of a confidence interval for the population mean μ, we shall denote the population median by the Greek letter ν, pronounced "nu."

(a) Small Samples

Suppose that we suspected the reported median of $3,800 in Section 16-1 of simply being inaccurate, either on the high side or on the low side. It then would be appropriate to calculate the two-sided prob-value for H_0 $(\nu = \$3,800)$. From (16-5), therefore:

$$\text{Prob-value} = 2(.0195) = .039 \tag{16-6}$$

Now suppose that we wish to make a classical test of H_0 $(\nu = \$3,800)$ at the level of, say, $\alpha = .040$. According to (9-19), with the prob-value of

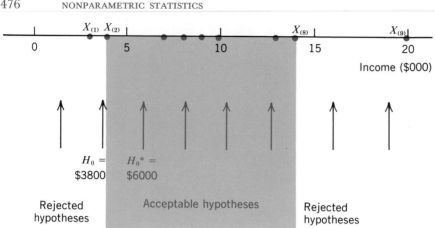

FIGURE 16–2 Nonparametric confidence interval for the population median ν

.039 given in (16-6), we could just barely reject H_0. Of course, Figure 16-2 shows us that any other hypothesis below $X_{(2)} = \$4,000$ would be rejected for the same reasons. Such rejected hypotheses are shown as black arrows.

On the other hand, consider an hypothesis above $X_{(2)} = \$4,000$, such as:

$$H_0^* : \nu = \$6,000 \tag{16-7}$$

Since there now are 7, rather than 8, "heads" (incomes above this median),

$$\text{prob-value} = 2\Pr(S \geq 7)$$
$$= 2(.0898) = .1796 \tag{16-8}$$

Since this now exceeds $\alpha = .04$, H_0^* is an acceptable hypothesis. By the same logic, even higher hypotheses such as $\nu = \$8,000$ or $\nu = \$11,000$ also are acceptable. These acceptable hypotheses are shown as colored arrows in Figure 16-2. Of course, at the very high end of the range we again encounter rejected hypotheses, in strong conflict with the observed data; (in a symmetric two-tailed test with $n = 9$ observations, if $X_{(2)}$ is one critical cutoff point between acceptance and rejection, then, by symmetry, $X_{(8)}$ must be the other). Finally, we note that the set of acceptable hypotheses form an interval.

Now we recall the crucial connection between hypothesis testing and confidence intervals:

A confidence interval is just the set of acceptable hypotheses	(16-9)

$$(9\text{-}5) \text{ repeated}$$

Thus, in Figure 16-2, we have constructed a confidence interval for the population median, the level of confidence being $1 - \alpha = 96\%$. In algebraic language, the 96% confidence interval for the population median, when $n = 9$, is:

$$X_{(2)} \le \nu \le X_{(8)} \tag{16-10}$$

that is:

$$4{,}000 \le \nu \le 14{,}000 \tag{16-11}$$

Note that, since the binomial distribution is discrete, confidence levels are discrete too. For example, if we considered that (16-11) had a higher confidence level (96%) than necessary, the next possible level would be 82%.[3] Of course, as sample size increases, the binomial becomes nearly continuous, and this problem tends to disappear.

(b) Generalization

To generalize, consider a sample of n ordered observations $X_{(1)}, X_{(2)}, \ldots, X_{(n)}$ from a population with unknown median ν. The confidence interval is defined by counting off q observations from each end:

$$\boxed{X_{(q)} \le \nu \le X_{(r)}} \tag{16-12}$$

where q are r are symmetrically chosen. Accordingly,

$$r = n - q + 1 \tag{16-13}$$

The problem, then, is to solve for the level α, which may be obtained from the binomial distribution, using $\pi = 1/2$ in Table IIIc. For S successes in n trials:

$$\boxed{\alpha = 2\,\Pr(S \ge r)} \tag{16-14}$$

Finally, this calculation allows us to specify the level of confidence $(1 - \alpha)$ for the interval (16-12).

---

[3]Using (16-8), it follows that $X_{(3)} \le \nu \le X_{(7)}$ has confidence level $1 - 2(.0898) \simeq 82\%$. Actually, a 95% confidence interval can be approximated by interpolating between $X_{(2)}$ and $X_{(3)}$, and between $X_{(7)}$ and $X_{(8)}$.

(c) Large Samples

Suppose that a large sample of n incomes is taken in order to find a confidence interval for the median. With this larger sample, it is evident that there is no longer any great problem of discreteness; thus it becomes convenient to use the standard technique of first setting the confidence level, and then solving for the interval. If the confidence level is set at 95% (i.e., $\alpha = .05$), then (16-14) becomes:

$$\Pr(S \geq r) = \frac{\alpha}{2} = .025 \qquad (16\text{-}15)$$

where S is binomial with $\pi = 1/2$. As in Example 6-10, the normal approximation to the binomial is easily obtained if we convert to proportions by dividing by n:

$$\Pr\left(\frac{S}{n} \geq \frac{r}{n}\right) = .025$$

The sample proportion $P = S/n$ has an approximately normal distribution with moments given by:

$$\mu_P = \pi \qquad \qquad \text{(6-18) repeated}$$
$$= \frac{1}{2}$$

$$\sigma_P = \sqrt{\pi(1 - \pi)/n} \qquad \text{(6-19) repeated}$$
$$= \frac{1}{2\sqrt{n}}$$

The critical value of S/n (which we call r/n) must be 1.96 standard deviations above the mean (to leave $2\frac{1}{2}\%$ in the upper tail, as specified by (16-15)). Thus:

$$\frac{r}{n} = \mu_P + 1.96\, \sigma_P$$

$$\frac{r}{n} = \frac{1}{2} + 1.96\, \frac{1}{2\sqrt{n}}$$

$$\boxed{r = \frac{n}{2} + .98\sqrt{n}} \qquad (16\text{-}16)$$

Once r is found, q is obtained from (16-13), and finally the confidence interval (16-12) is formed.

Example 16-2.   A random sample of 25 working students yielded the following summer incomes (in \$100s, arranged in order):

$$4,5,6,6,7, \quad 7,8,9,11,12, \quad 13,16,16,17,19,$$

$$20,20,22,24,24, \quad 26,29,33,34,38$$

Construct a 95%-confidence interval for the population median income.

Solution.   Substituting $n = 25$ into (16-16):

$$r = \frac{25}{2} + .98\sqrt{25}$$

$$= 12.5 + 4.9 = 17.4$$

To be safe, we should round up[4] (make the upper confidence limit higher), so that:

$$r \simeq 18$$

By (16-13):

$$q = n - r + 1 = 25 - 18 + 1 = 8$$

From the given data, $X_{(8)} = \$900$ and $X_{(18)} = \$2,200$. Thus, (16-12) yields the 95%-confidence interval:

$$\$900 \leq \nu \leq \$2,200$$

---

[4] In constructing confidence intervals, we should always round up to the *next highest* integer (not the *nearest* integer, as often is done). This is because we have ignored the continuity correction, which requires adding $1/2$ to r.

PROBLEMS

16-4 In Problem 16-1, construct a 98% nonparametric confidence interval for:
  (a) Median men's income.
  (b) Median women's income.
  (c) Median difference in income.

16-5 In Problem 16-2, construct a 93% nonparametric confidence interval for:
  (a) Median men's height.
  (b) Median women's height.
  (c) Median difference in heights.

16-6 (a) Referring to Problem 16-1, do you think that the population of men's incomes is distributed normally?
  (b) Do you think that the nonparametric confidence interval in Problem 16-4(a) would be narrower than the classical confidence interval based on t?
  (c) To check your conjecture in (b), go ahead and calculate the 98% confidence interval of the form (8-21).

16-7 (a) Referring to Problem 16-2, do you think that the population of men's heights is distributed normally?
  (b) Do you think that the nonparametric confidence interval in Problem 16-5(a) would be narrower than the classical confidence interval based on t?
  (c) To check your conjecture in (b), go ahead and calculate the 93% confidence interval of the form (8-21). (You will have to interpolate the t values in Table V.)

16-8 Write a summary of what you learned from Problems 16-6 and 16-7.

16-9

Random Sample of 9 Family Incomes, Southern U.S. 1971[a]
$X_{(1)} = 3,000$
$X_{(2)} = 4,000$
$X_{(3)} = 7,000$
$X_{(4)} = 8,000$
$X_{(5)} = 9,000$
$X_{(6)} = 10,000$
$X_{(7)} = 13,000$
$X_{(8)} = 14,000$
$X_{(9)} = 20,000$

[a]Same data as Figure 16-1, from the 1974 *American Almanac*, Table 533.

Suppose that a statistician used this data to construct the following confidence intervals for the median. What is her level of confidence for each?
(a) $3{,}000 \leq \nu \leq 20{,}000$.
(b) $4{,}000 \leq \nu \leq 14{,}000$.
(c) $7{,}000 \leq \nu \leq 13{,}000$.

16-3   THE W TEST FOR TWO SAMPLES

(a) Small Samples

The Wilcoxon–Mann–Whitney (W–M–W, or W) test is used for two independent samples (in contrast with the sign test, which used matched pairs). The objective again is to detect whether the two underlying populations are centered differently. For example, suppose that independent random samples of family income were taken from two different regions of the U.S. in 1971,[5] and then ordered as in Table 16-2.

TABLE **16-2**
2 Samples of income

South X	Northeast Y
4,000	5,000
8,000	10,000
9,000	11,000
14,000	12,000
	15,000
	22,000

size $m = 4$ $n = 6$

Let us test the null hypothesis that the two underlying populations are identical. Suppose that the alternative hypothesis is that the South is poorer than the Northeast, so that a one-sided test is appropriate.

We first rank the combined X and Y observations, as shown in Table 16-3.

The actual income levels now are discarded in favor of this ranking, providing a test that is not affected by skewness, or any other distributional peculiarity—in other words, a distribution-free test. Then W is defined as the sum of all the X ranks; in this case:

---

[5] From the 1974 *American Almanac*, Table 533.

TABLE **16-3**

Combined Ranking Yields the
W Statistic

Combined Ordered Observations		Combined Ranks	
X	Y	X	Y
4,000		1	
	5,000		2
8,000		3	
9,000		4	
	10,000		5
	11,000		6
	12,000		7
14,000		8	
	15,000		9
	22,000		10

$W = 16$

$$W = 1 + 3 + 4 + 8 = 16 \qquad (16\text{-}17)$$

Obviously, the lower this value, the stronger the evidence for rejecting H_0. For our data, Appendix Table VIII shows that the one-sided prob-value is .129 $\simeq$ 13%, which is weak evidence of a lower family income in the South.

(b) Generalization

In general, suppose that there are m observations in the smaller sample (call them $X_1, X_2, \ldots, X_m$) and $n \geq m$ observations in the larger sample (call them $Y_1, Y_2, \ldots, Y_n$). For a one-sided test, start counting from the end where the Xs predominate (this keeps the X ranks low).[6] Adding up all the X ranks yields the rank sum W.

Appendix Table VIII gives the corresponding one-sided prob-value for $n \leq 7$ and prob-value $\leq .25$. For samples larger than those covered by this table, W is approximately normal[7] (if H_0 is true), with:

---

[6] Or more precisely, start counting from the end where the Xs would tend to predominate if H_1 were true.

[7] It is remarkable that just as the binomial statistic is approximately normally distributed, so is W, and in fact all the other test statistics that we shall study in this chapter. The proof consists of generalizing the central limit theorem.

The approximately normal distribution for W must not be confused with the parent distribution of the Xs and Ys, which may be very nonnormal.

$$E(W) = \frac{1}{2} m(m + n + 1)$$

$$\text{var}(W) = \frac{1}{12} mn(m + n + 1)$$

(16-18)

Example 16-3. In Table 16-4, suppose that we have acquired a larger sample of incomes than in Table 16-3. Calculate the prob-value for H_0 in the light of this increased information.

TABLE **16-4**
W Statistic, Large-Sample

Combined Ordered Observations		Combined Ranks	
South X	Northeast Y	South X	Northeast Y
4,000		1	
	5,000		2
6,000		3	
	7,000		4
8,000		5	
9,000		6	
	10,000		7
	11,000		8
	12,000		9
14,000		10	
	15,000		11
	21,000		12
	22,000		13

size $m = 5$ $n = 8$ $W = 25$

Solution. Since $m = 5$ and $n = 8$ exceeds the capacity of Table VIII, we resort to the normal approximation. Substituting $n = 8$ and $m = 5$ into (16-18):

$$\left. \begin{array}{l} E(W) = 35 \\ \text{var}(W) = 46.7 \end{array} \right\}$$

(16-19)

To test the null hypothesis:

$$\text{prob-value} \triangleq \Pr(W \leq 25)$$

$$= \Pr\left(\frac{W - \mu_W}{\sigma_W} \leq \frac{25 - 35}{\sqrt{46.7}}\right)$$

$$= \Pr(Z \leq -1.46)$$

$$= .0722 \simeq 7\% \tag{16-20}$$

This provides stronger evidence than before of a lower family income in the South. In fact, at the level $\alpha = 10\%$, this result is statistically discernible.

(c) Ties

Observations that are tied with one another should be given the same rank—their average rank. Unless most of the observations are tied, we can then continue as usual. For example, suppose that the 13 observations of Table 16-4 were slightly different, with several ties occurring. Then we would assign ranks as shown in Table 16-5.

TABLE **16-5**
W Statistic, When Ties Occur

Combined Ordered Observations		Combined Ranks	
X	Y	X	Y
4,000		1	
6,000	6,000	2.5	2.5
7,000		4	
9,000	9,000 9,000	6	6,6
	11,000		8
	12,000		9
14,000		10	
	15,000		11
	21,000		12
	22,000		13

$$W = 23.5$$

$E(W)$ and var (W) may still be approximated by (16-18); the prob-value, taken from the normal tables, is:

$$\text{prob-value} \overset{\Delta}{=} \Pr(W \le 23.5) \tag{16-21}$$

$$= \Pr\left(\frac{W - \mu_W}{\sigma_W} \le \frac{23.5 - 35}{\sqrt{46.7}}\right)$$

$$= \Pr(Z \le -1.68)$$

$$= .0465 \tag{16-22}$$

PROBLEMS

16-10 Two makes of cars were sampled randomly, to determine the mileage (in thousands) until the brakes required relining. Calculate the prob-value for the null hypothesis that make A is no better than make B.

Make A	Make B
30	22
41	26
48	32
49	39
61	

16-11 Recalculate the prob-value in Problem 16-10:
(a) Using the large sample formula (16-18), just to see how well it works.
(b) Using the large sample formula (16-18), with continuity correction (w.c.c.). Do you think the c.c. is worthwhile?

*16-12 A random sample of 8 men's heights and an independent random sample of 8 women's heights were observed, and ordered as follows:

Men's Heights (M)	Women's Heights (W)
65	62
67	63
69	64
70	65
71	66
73	68
76	69
77	71

(a) Calculate the two-sided prob-value for the null hypothesis H_0 that the two population distributions are the same. (Although it may be unrealistic, let the alternative hypothesis be two-sided throughout this problem.)

(b) Calculate the two-sided prob-value for the hypothesis that men's heights are distributed 2″ higher than women's heights. (*Hint:* If this hypothesis were true, then by subtracting 2″ from each man's height, we would obtain a sample from the same distribution as the women's heights.)

(c) Construct a 93% nonparametric confidence interval for the population median difference in men's and women's heights. (*Hint:* Consider many possible shifts, as in part (b), and retain only those that are acceptable at the 7% level).

(d) Note that the tabled numbers in this problem are the same as in Problem 16-2, although the way in which they were obtained is different. How does the confidence interval in this problem compare with the one in Problem 16-5(c). Why?

(e) By interpolating the *t* table, construct a 93% parametric confidence interval for the mean difference in men's and women's heights. What assumption are you making? Does it seem reasonable? What benefit does it provide?

16-4   TESTS FOR RANDOMNESS

(a) Runs Test

One of the most crucial assumptions (even more important than the normality assumption) that we have relied upon in previous chapters is the assumption that our sampling is random.[8] Now we shall develop a test for even that assumption.

By definition, a random sample consists of observations that are drawn *independently* from a *common* population (identically distributed). Thus, if the observations are graphed in the order in which they were sampled, the graph should look somewhat like Figure 16-3(a). On the other hand, if the observations are correlated, they will display some "tracking," as in Figure 16-3(b). Or, if the observations come from 2 different populations, they may appear to be displaced, as in Figure 16-3(c). How can we quantify the differences that are obvious to the eye in this figure, and find some numerical measure to test the null hypothesis of randomness? We note that when H_0 is true, the path of the observations

---

[8]For example, independent observations were crucial in developing the variance of $\overline{X}$ in (6-7), which played a key role in estimating μ.

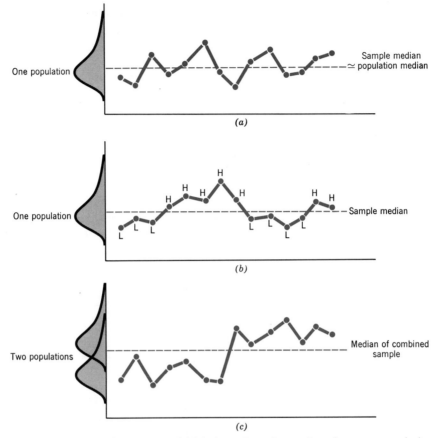

FIGURE 16–3 Sample sequence of (a) independent observations from one population. (b) Correlated observations from one population. (c) observations from two populations.

crosses the median line quite frequently; but when H_0 is not true, this happens much less frequently. This is the basis for the runs test.

For example, in Figure 16-3(b), we mark observations H (for high) or L (for low), depending on whether they fall above or below the sample median. With slashes indicating the crossovers, this sequence is

$$L\,L\,L/H\,H\,H\,H\,H/L\,L\,L\,L/H\,H \qquad (16\text{-}23)$$

The number of runs R is defined as the number of blocks separated by slashes; in this case $R = 4$. The more tracking, the fewer runs there are.

Let us suppose in general that there are n observations.[9] When H_0 is true, the distribution of R is approximately normal, with:

$$E(R) \simeq \frac{n}{2} + 1$$

$$\text{var } R \simeq \frac{n(n-2)}{4(n-1)} \simeq \frac{(n-1)}{4}$$

(16-24)

To calculate the prob-value for the null hypothesis of randomness, note that the usual alternative hypothesis results in too few runs (i.e., the observations are positively correlated, as in Figure 16-3(b), or drawn from different populations, as in Figure 16-3(c)). Thus, a one-sided prob-value usually is appropriate. For the example in Figure 16-3(b), $n = 14$, so that (16-24) yields:

$$E(R) \simeq \frac{14}{2} + 1 = 8$$

$$\text{var } (R) \simeq \frac{(14-1)}{4} = 3.25$$

Using the normal approximation:

$$\text{prob-value} = \Pr(R \leq 4)$$

$$= \Pr\left(\frac{R - \mu_R}{\sigma_R} \leq \frac{4-8}{\sqrt{3.25}}\right)$$

$$= \Pr(Z \leq -2.22) = .0132 \qquad (16\text{-}25)$$

which is strong evidence that the sample is not random.

(b) Mean Squared Successive Difference

Now we will develop an alternative test for randomness that exploits the difference between observations, $(X_i - X_{i-1})$. In fact, let us square to get rid of the sign, obtaining $(X_i - X_{i-1})^2$. Now average over the sample, thus defining the *mean squared successive difference*:

---

[9]If the sample size is odd, the median line will pass through the median observation, which should be counted neither L nor H. Instead it should be discarded, and n should be made to refer to the even number of observations remaining. When the sample size is even, this issue does not arise.

$$\text{MSSD} \triangleq \frac{1}{n-1} \sum_{i=2}^{n} (X_i - X_{i-1})^2 \tag{16-26}$$

If H_0 (sample is random) is true, the expected value[10] is:

$$E(\text{MSSD}) = 2\sigma^2 \tag{16-27}$$

If we knew σ^2, we could divide by it appropriately, obtaining:

$$E\left(\frac{\text{MSSD}}{\sigma^2}\right) = 2 \tag{16-28}$$

But since we do not know σ^2, it is reasonable instead to divide by the unbiased estimator s^2. Therefore, we finally define:

$$d \triangleq \frac{\text{MSSD}}{s^2}$$

that is:

$$\boxed{d = \frac{\sum\limits_{i=2}^{n} (X_i - X_{i-1})^2}{\sum\limits_{i=1}^{n} (X_i - \overline{X})^2}} \tag{16-29}$$

Small observed values of d call for rejection of the randomness hypothesis. For example, in Figure 16-3(b), the terms in the numerator of (16-29) will tend to be small, since X_i tends to follow X_{i-1} closely. And in Figure 16-3(c), the terms in the denominator will tend to be large, since each block of X_is is so far removed from the sample mean.

---

[10]**proof**   First, we derive the expected value of one individual squared successive difference:

$$E(X_i - X_{i-1})^2 = E[(X_i - \mu) - (X_{i-1} - \mu)]^2$$
$$= E(X_i - \mu)^2 + E(X_{i-1} - \mu)^2 - 2 \text{ cov } (X_i, X_{i-1})$$
$$= \sigma^2 + \sigma^2 - 0 = 2\sigma^2$$

Now the expected value of the average of all these is also $2\sigma^2$, as given in (16-27), since the sample mean is an unbiased estimator of the "population" mean $2\sigma^2$.

To illustrate, suppose that we drew the following sequence of 14 observations (similar to Figure 16-3(b), except that they have been drawn from a normal population):

$$18, 19, 18, 21, 22, 24, 23, 21, 21, 19, 17, 18, 18, 21 \quad (16\text{-}30)$$

Substituting into (16-29), and using $\overline{X} = 20$, we calculate:

$$d = \frac{(19 - 18)^2 + (18 - 19)^2 + (21 - 18)^2 + \cdots}{(18 - 20)^2 + (19 - 20)^2 + \cdots}$$

$$= \frac{39}{60} = .65 \quad (16\text{-}31)$$

Finally, we must calculate the prob-value—i.e., interpret $d = .65$ in terms of the distribution of all possible values of d. If the sample is taken from a normal population in which H_0 is true:

$$\boxed{E(d) \simeq 2} \quad (16\text{-}32)$$
$$\text{like } (16\text{-}28)$$

and:

$$\boxed{\text{var}\,(d) \simeq \frac{n - 2}{n^2}} \quad (16\text{-}33)$$

In our example (16-30), $n = 14$, so that:

$$\text{var}\,(d) = \frac{14 - 2}{14^2}$$

$$= .0612$$

$$\text{prob-value} = \Pr(d \le .65)$$

$$= \Pr\left(\frac{d - \mu_d}{\sigma_d} \le \frac{.65 - 2}{\sqrt{.0612}}\right) \quad (16\text{-}34)$$

$$= \Pr(Z \le -5.46)$$

$$\ll .001$$

which provides extraordinarily strong evidence of nonrandomness.

PROBLEMS

16-13

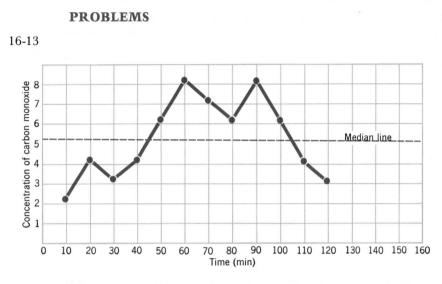

The above graph shows a sample of 12 air pollution readings, taken every 10 minutes over a period of 2 hours. To what extent can we claim that these are statistically independent observations from a fixed (rather than drifting) population? Answer by calculating the prob-value for the null hypothesis:

(a) Using the runs test.

(b) Using the mean square successive difference.

(c) What are the advantages and disadvantages of method (a) relative to method (b)?

16-5 THE ADVANTAGES OF NONPARAMETRIC TESTS

(a) Validity

(i) Introduction

A test is said to be valid if its prob-values and confidence levels are correct, as specified. Consider a very simple and familiar example: a valid 95% confidence interval for a normal population μ is:

$$\mu = \overline{X} \pm t_{.025} \frac{s}{\sqrt{n}}$$

(16-35)
(8-21) repeated

But if we carelessly use $z_{.025}$ instead of $t_{.025}$, obtaining:

$$\mu = \overline{X} \pm 1.96 \frac{s}{\sqrt{n}}$$

(16-36)

the test becomes invalid for small samples. For example, if $n = 6$ so that d.f. $= 5$, the interval (16-36) is not only narrower than the correct 95% confidence interval, it is even narrower than the correct 90% confidence interval:

$$\mu = \overline{X} \pm t_{.05} \frac{s}{\sqrt{n}} \qquad (16\text{-}37)$$

$$= \overline{X} \pm 2.015 \frac{s}{\sqrt{n}}$$

Thus, in a sample of 6, the true confidence level of (16-36) is less than 90%, rather than the specified 95%.

A 95% confidence interval may be invalid for either of two reasons:

1. Its true confidence level may be really lower than 95%, as in the example above. Such a case is called *unsafe*. Thus, we can view this problem as an overstatement of confidence level; alternatively, we might say that the interval is unjustifiably narrow.

2. Its true confidence level may be really higher than 95%. Such a case is called *too safe,* or *conservative.* Such confidence intervals are wider (vaguer) than necessary.

(ii) t Tests Versus Nonparametric Tests

The single-sample confidence interval for the population mean based on *t* may be invalid if the population is very nonnormal (this is especially true of the small-sample, one-tailed use of *t* in a skewed population). On the other hand, confidence intervals for the population median based on the sign test are perfectly valid for *any* population whatsoever.

In two-sample confidence intervals, *t* and *W* make one common assumption: that the two populations are identically shaped, their only possible difference being in location. This is illustrated in Figure 16-4;

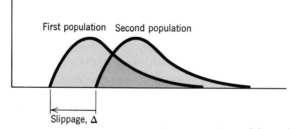

FIGURE 16–4 Slippage assumption of two-sample confidence intervals

the difference in location is called the "slippage" or "difference" Δ, and is given as:

$$\Delta = \nu_1 - \nu_2 = \mu_1 - \mu_2 \tag{16-38}$$

If this assumption does not hold, both t and W may be invalid. Thus, we see that even nonparametric tests require some assumptions, even though they are free of the normality assumption.

The two-sample t also nominally assumes that the populations are normal. If this assumption does not hold, however, t remains nearly valid, and is therefore called *robust*. To find the essential difference between t and W, therefore, we must look beyond their validity and consider their efficiency.

(b) Efficiency

(i) Introduction

We first considered efficiency in Section 7-2; recall that, in estimating the center of a normal distribution, the efficiency of the sample mean relative to the sample median was given by the ratio of their variances as:

$$\frac{(\pi/2)(\sigma^2/n)}{\sigma^2/n} = \frac{\pi}{2} = 157\% \tag{16-39}$$
$$(7\text{-}6) \text{ repeated}$$

It is evident that these two variances can be equalized by increasing the sample size for the median (i.e., increasing n in the numerator) by 57%. Thus the relative efficiency of two estimators may be viewed as the ratio of sample sizes that are required to make them equally accurate. In the same way we shall define:

> The relative efficiency of two *tests*[11] is the ratio of sample sizes that are required to give equally accurate tests; i.e., tests with the same probability of type I and type II errors.

(ii) Comparisons

In Table 16-6, efficiencies are compared; but first, a few words of explanation. Since the sign test is for the population median ν, while the t test is

---

[11]Similarly, we define the relative efficiency of two confidence intervals as the ratio of sample sizes that are required to give equally accurate intervals; i.e., intervals with the same confidence level and the same length.

TABLE **16-6**

Nonparametric Tests Compared with t Tests
(in symmetric populations where mean and median coincide)

Test		Population Parameter	Tested by Sample	Assumptions about Population Required for Validity[a]	Asymptotic[b] Relative[c] Efficiency When the Normal Assumption is True[d]	Asymptotic[b] Relative[c] Efficiency When the Normal Assumption is False, and Instead the Population Is		
						(a) Rectangular	(b) Two-tailed Exponential	(c) Cauchy
1 Sample	sign test	median, ν	median, $X_{\left(\frac{n+1}{2}\right)}$		64%	33%	200%	∞
	t test	mean, μ	mean, $\bar{X}$	normal[f]				
2 Sample	W test	slippage $\Delta = \nu_1 - \nu_2 = \mu_1 - \mu_2$	rank sum W	populations that are identical except for different locations	95%	100%	150%	∞
	t test	$\mu_1 - \mu_2$	$\bar{X}_1 - \bar{X}_2$	populations that are nominally[e] normal, and identical except for different locations				

[a]That is, to make confidence level $(1 - \alpha)$, as claimed. Of course, in all tests we make the standard assumption that the sample is *random*, as specified by (6-2).

[b]In small samples, relative efficiencies differ slightly from those given in this table.

[c]"Relative" means the efficiency of the nonparametric test relative to the efficiency of the t test.

[d]In the normal case for which it was developed, the t test is absolutely efficient, i.e., more efficient than any other test. (However, note that the W test is a close competitor.)

[e]As mentioned previously, the two-sample t test is robust, i.e., is very nearly valid even for nonnormal populations.

[f]As mentioned previously, the one-sample t test is fairly robust, i.e., is nearly valid even for a nonnormal population—except for one-tailed tests in a skewed population.

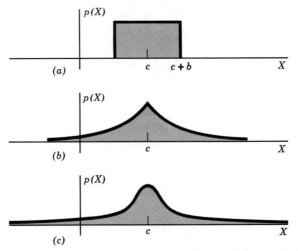

FIGURE 16–5 Three symmetric distributions. (a) Short-tailed rectangular. (b) Long-tailed two-tailed exponential. (c) Extremely long-tailed Cauchy.

for the population mean μ, comparisons make sense only for symmetric distributions where the median and the mean coincide:

$$\nu = \mu = c, \text{ say} \qquad (16\text{-}40)$$

Thus we chose 3 symmetric distributions as well as the normal, as a basis of comparison; their graphs are given in Figure 16-5, and their formulas are as follows.

1. The rectangular distribution has no tails. Its formula is:

$$p(X) = \frac{1}{2b} \qquad c - b < X < c + b$$

2. The two-tailed exponential has longer and thicker[12] tails than the normal. Its formula is:

$$p(X) = \frac{1}{2b} e^{\frac{-|X-c|}{b}}$$

---

[12]More precisely, "thicker at its extremities," since both distributions extend to infinity. Henceforth, we shall use the simple abbreviation "longer tails."

3. The Cauchy distribution has extremely long tails—so long, in fact, that its variance does not exist.[13] Its formula is:

$$p(X) = \left(\frac{b}{\pi}\right)\frac{1}{(X-c)^2 + b^2}$$

(iii) Conclusions

From the details of Table 16-6, we may draw two brief conclusions.

1. Although the t test is somewhat more efficient for the normal distribution, it is very much less efficient for the long-tailed distributions.

2. The W test shows up exceptionally well. This explains its growing popularity.

(c) Outliers

Outliers are defined as observations that are so far away from the rest of the sample that they should be discarded, or at least modified somehow. They may occur, for example, because of a clerical error such as misplacement of a decimal point, or because an observation was left blank and then counted as zero by a computer, or because of a wild measurement, such as a ridiculously false reply to a questionnaire ("What is your income?" "$10 million a year."). Outliers are particularly dangerous if data are analyzed impersonally, without a human eye to catch the outlier.

The problem is that we can never be certain whether or not an extreme observation is an outlier. (If we knew for sure *a priori*, we could just drop it from the sample, and there would be no problem.) For example, in Figure 16-6, we show three possible positions for the right-hand observation: in panel (a), it seems obviously an outlier; in panel (b), it seems obviously not; and in panel (c), we have the common case where it is not at all clear. Should this extreme observation be (1) discarded (trimming the sample); (2) kept as is; or (3) transformed to a more useful form? (Transformation is a compromise, and often the wisest choice.)

First, we briefly must build a mathematical model of the outliers. We suppose that the whole population of possible outliers forms a probability distribution that is very disperse; in a sense this "contaminates"

---

[13] Even its mean μ does not exist, in a strict mathematical sense. Thus it is no surprise that the sample mean $\overline{X}$ is a pretty useless estimator of the center c; note that the t test has zero relative efficiency in Table 16-6.

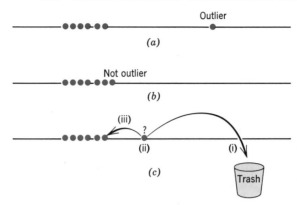

FIGURE 16–6 Is the extreme observation an outlier? (a) Probably. (b) Probably not. (c) Maybe. What should be done? (i) Discard (ii) keep as is, or (iii) transform?

the regular distribution, as shown in Figure 16-7. All the possible observations, both those from the regular distribution (to be estimated) and from the confusing or "noisy" outlier distribution, form the total distribution. Since outliers are slipped innocently into the sample with the rest of the observations, in practice the observer faces the undifferentiated total distribution, which has such long tails that it is wise, in view of the efficiency results in Table 16-6, to use a nonparametric test such as the W test, rather than the t test.

Ranking observations as in the W test may be viewed simply as a way of transforming the outliers. (It is true that we transform all the observations to ranks, but it is the outliers that are changed most.) For example, if the observations in Figure 16-6 are ranked, we see that in all three panels

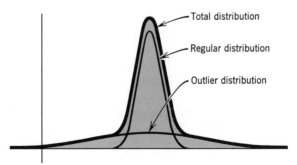

FIGURE 16–7 Outliers form a widespread distribution that contaminates the regular distribution

the right-hand observation counts the same—its rank is 7. Thus, we have taken the outlier in the first or third panel and "pulled it in" toward the rest of the distribution. On the other hand, the t test, which uses $\overline{X}$, would give the right-hand observation far more influence in the first panel—the very case in which it is most likely to be an outlier deserving less influence.

An alternative way to treat the outliers was described in Section 7-4—the best easy systematic estimator (BES), where the upper and lower quarters of the sample simply were trimmed off.[14] However, BES was only a point estimate. In this chapter, we have discussed the more crucial issue of interval estimates, and tests.

(d) Conclusions

The advantages of nonparametric statistics, in decreasing order of importance, are:

1. Greater efficiency in long-tailed distributions, including the case of outlier contamination.

2. Validity, in the sense that their confidence level (or type I error level) really is exactly as specified.

3. Easier computation in some cases.

REVIEW PROBLEMS

16-14 Two samples of children were selected randomly to test two art-education programs, A and B. At the end, each child's best painting was collected, to be judged by an independent artist. In terms of creativity, the artist ranked them as follows:

Rank of Child	1	2	3	4	5	6	7	8	9	10	11	12	13	14	15	16
Child's Art Program	B	A	B	B	B	A	B	B	A	B	A	B	A	A	A	A

Test whether program B (the intensive program) is better than program A.

16-15 Is the following true or false? Where false, correct it. If true, elaborate on it.

(a) If you have prior knowledge that the distribution is normal (or nearly normal), you gain by exploiting this in using the t test. However, if your prior knowledge is false, the t test that you have used may cost

---

[14]This is not quite the same as saying that these end observations are entirely useless. It is true that their specific values are ignored; but their presence tells us which are the more reliable central observations.

you dearly. A nonparametric test, especially a test like the W test, is a relatively risk-free alternative to the t test.

(b) Nonparametric tests are invaluable for data that are ordered, for example from best to worst (on an ordinal scale), rather than for data given a numerical value.

16-16 A production manager wishes to test whether soft background music will increase labor productivity on an assembly line. Suppose that you find the following effect on a sample of 8 workers (each observed with and without music).

Per Hour Output

Without Music	With Music
27	29
32	29
23	36
28	29
23	27
27	29
25	26
30	34

(a) Calculate the prob-value of H_0 (no favorable effect of music), using a nonparametric test, and then the t test.

(b) If the normality assumption is a reasonable one, what can you say about each of these tests in terms of its:

(i) Validity?

(ii) Relative efficiency?

(c) Repeat (b), if the normality assumption cannot be made (and if the parent population is like the two-tailed exponential or Cauchy distribution).

*16-17 Suppose that you have observed the following sequence of interest rates and bankruptcies in the construction industry.

(a) Calculate the regression of Y on X, and the correlation coefficient.

(b) If you believe that it is likely that the last observation is an outlier, explain why the analysis in (a) is unsatisfactory. Then recalculate (a), dropping the last observation.

(c) Rank all X values numerically, from smallest to largest. Do the same for Y. Then calculate a "rank correlation," using all the ranks. Compare with (a) and (b), and explain.

	Interest Rate (X)	Bankruptcies (Y)
First year	5%	3,300
Second year	6	3,000
Third year	5	2,600
Fourth year	7	4,200
Fifth year	7	3,800
Sixth year	6	4,000
Seventh year	6	3,600
Eighth year	10	4,300

CHAPTER 17

Chi-Square Tests

The age of chivalry is gone; that of sophisters, economists, and calculators has succeeded.

Edmund Burke

Chi square (χ^2) is a very popular form of hypothesis testing, and one that is subject to substantial abuse. Besides the type I and type II errors that we already encountered in Chapter 9, we now must face yet another kind of error—the error of asking the wrong question. Statisticians unaware of it may commit this error wholesale, and chi-square tests are likely to be their favorite vehicle. We shall, therefore, spend considerable effort in giving alternatives to chi square that are more appropriate under certain conditions.

17-1 χ^2 TESTS FOR GOODNESS OF FIT

(a) Example

Suppose that we wish to test the null hypothesis that births in Sweden occur equally often throughout the year. Suppose that the only available data is a random sample of 88 births distributed over the year, but grouped into seasons of differing length. These observed frequencies O_i are given in Table 17-1.

TABLE 17-1

Distribution of $n = 88$ Births Among 4 Cells

(Seasons of the Year)

	Given		χ^2 calculations			
			(1)	(2)	(3)	(4)
		Observed Frequency O_i	*Probability (if H_0 true)* π_i	*Expected Frequency* $E_i = n\pi_i$	*Deviation* $(O_i - E_i)$	*Deviation Squared and Weighted* $(O_i - E_i)^2/E_i$
i	*Season*					
1	Spring Apr–June	27	$91/365$ $= .25$	$.25(88)$ $= 22.0$	$27 - 22$ $= +5.0$	$25/22$ $= 1.14$
2	Summer July–Aug	20	.17	15.0	+5.0	1.67
3	Fall Sept–Oct	8	.167	14.7	−6.7	3.05
4	Winter Nov–Mar	33	.413	36.3	−3.3	.30
		$n = 88$	1.00 ✓	88 ✓	0 ✓	$\chi^2 = 6.16$

prob-value $\simeq .11$

How well does the data fit the hypothesis? The notion of goodness of fit may now be developed in several steps.

1. First consider the implications of H_0, the null hypothesis that every birth is likely to occur in any given season with a probability proportional to the length of that season. Spring is defined to have 3 months, or 91 days; thus π_1—the probability (given H_0) of a birth occurring in the first period, spring—is $91/365 = .25$. Similarly, all the other probabilities π_i are calculated in column (1) of Table 17-1.

2. Now calculate what the expected frequencies in each cell (season) would be if the null hypothesis were true. For example, consider the first cell; if H_0 were true, $\pi_1 = .25 = 25\%$, so that the expected frequency would be 25% of 88 = 22 births. Similarly, column (2) displays the calculation of all other expected frequencies[1] E_i:

$$\boxed{E_i = n\pi_i} \qquad (17\text{-}1)$$

---

[1] Formula (17-1) is just a restatement of formula (6-23) for the mean of a binomial distribution. In order to use Table VI(a) with assurance, it is necessary that $E_i \geq 5$ in each cell. If this condition is not met, then this table should be used only with considerable reservation. Better still, cells should be redefined more broadly until this condition is met.

3. The question now is: "By how much does the observed frequency deviate from the expected frequency?" For example, in the first cell, this deviation is $27 - 22 = 5$. Similarly, all the other deviations $(O_i - E_i)$ are set out in column (3).

4. The deviations always would sum algebraically to 0, so they are not a useful criterion. This problem is avoided by taking the square[2] of the deviation $(O_i - E_i)^2$. Then, to show its relative importance, each squared deviation somehow must be compared with the expected frequency in its cell, E_i. Therefore, we calculate the ratio $(O_i - E_i)^2/E_i$, as shown in column (4). Finally, we sum the contributions from all cells to obtain an overall measure of how the observations deviate from the null hypothesis. This is denoted as:

$$\text{chi-square goodness-of-fit statistic,}^3 \; \chi^2 = \sum_{i=1}^{k} \frac{(O_i - E_i)^2}{E_i} \qquad (17\text{-}2)$$

$$= 6.16 \qquad (17\text{-}3)$$

This whole argument obviously has strong similarities to the one that we used in developing ANOVA; specifically, the χ^2 statistic is very similar to the F statistic (10-7), which also measured discrepancy from a null hypothesis. Hence, the χ^2 statistic may be analyzed in a similar way.

First, χ^2 is a random variable that fluctuates from sample to sample; this fluctuation in χ^2 occurs because of the fluctuation in the observed cell frequencies O_i that comprise it. In passing, note that these frequencies are not independent: since $O_1 + O_2 + O_3 + O_4 = n$, any one of them may be expressed in terms of the other. For example, $O_4 = n - (O_1 + O_2 + O_3)$; thus, the last cell is determined by the previous three. Therefore, in this case χ^2 has only 3 degrees of freedom; and in general, with k cells, there are $k - 1$ degrees of freedom.

---

[2]This squaring of the deviations likewise occurred in the definition:

$$\text{MSD} = \sum (X_i - \overline{X})^2 \left(\frac{f_i}{n} \right) \qquad (2\text{-}5b) \text{ repeated}$$

Also note the weighting by relative frequency (f_i/n); a similar weighting will be used in defining χ^2.

[3]Since a large χ^2 value means a very bad fit to H_0, it would have been more appropriate to call it the "chi-square badness-of-fit statistic."

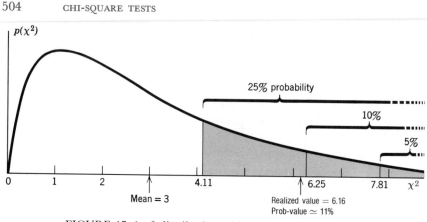

FIGURE 17–1 χ^2 distribution with 3 d.f., when H_0 is true

If H_0 is true, the χ^2 distribution is approximately[4] graphed in Figure 17-1, with the critical points (10%, 5%, etc.) given in Appendix Table VI(a). By interpolating Table VI(a) we find that the approximate prob-value for H_0 (births evenly distributed) is about 11%, as shown in Figure 17-1. Finally, several other χ^2 distributions with various other degrees of freedom are shown in Figure 17-2, with the critical points also given in Table VI(a).

Example 17-1.   A die was cast 1,000 times with the following results:

Face	Relative Frequency
1	.183
2	.161
3	.142
4	.174
5	.181
6	.159

---

[4]This approximation is like the normal approximation to the binomial—adequate, but not perfect. Perhaps this issue may be clarified by distinguishing between the χ^2 goodness-of-fit statistic (17-2), and the χ^2 *distribution* in Figure 17-1. The χ^2 *distribution* with r d.f. is defined rigorously as the distribution of the sum of r independent standard normal variables squared, and thus is easily tabulated. It is used for many purposes, including confidence intervals for σ^2 in (8-54), and in this chapter as an approximation to the distribution of the χ^2 goodness-of-fit statistic (17-2) and the χ^2 statistic to test independence, (17-10) below.

The definition of the χ^2 goodness-of-fit statistic is justified by intuitive arguments in Section 17-1(a); but this is not enough, since alternative measures—equally appealing on intuitive grounds—could have been devised. The χ^2 goodness-of-fit statistic really is justified because its distribution is very well approximated by the known χ^2 distribution.

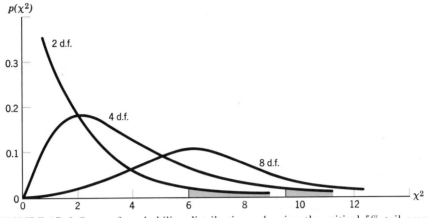

FIGURE 17–2 Some χ^2 probability distributions, showing the critical 5% tail areas

(a) Find the prob-value for H_0 (the die is fair).
(b) Can you reject H_0 at the 10% level?

Solution. (a) To use χ^2, we first must express the data as observed frequencies O_i by multiplying the relative frequencies by $n = 1,000$. Then we calculate the expected frequencies E_i by stating the null hypothesis:

$$H_0: \pi_1 = \pi_2 = \cdots = \pi_6 = \tfrac{1}{6}$$

Thus:

$$E_i = n\pi_i = 1,000(\tfrac{1}{6}) \simeq 167$$

Face i	O_i	E_i	$(O_i - E_i)$	$(O_i - E_i)^2$	$(O_i - E_i)^2/E_i$
1	183	167	$+16$	256	1.53
2	161	167	-6	36	.22
3	142	167	-25	625	3.74
4	174	167	$+7$	49	.29
5	181	167	$+14$	196	1.17
6	159	167	-8	64	.38

$$\chi^2 = 7.33$$

Referring to Table VI(a) with d.f. $= k - 1 = 6 - 1 = 5$, we find that the observed value of 7.33 is bracketed by $\chi^2_{.25} = 6.63$ and $\chi^2_{.10} = 9.24$. Thus:

$$.10 < \text{prob-value} < .25$$

$$\text{prob-value} \simeq .20$$

(b) According to (9-19), since prob-value $>.10$, we cannot reject H_0.

(b) Limitations of χ^2

Continuing the example in (a), suppose that we gather a very large sample; in fact, *all* the births in Sweden in 1935 are shown in Table 17-2. With a little stretch of the imagination, this may be considered a random sample from an infinite conceptual population. χ^2 is calculated to be 128.47, which exceeds our last tabular value by so much that we conclude that the prob-value $\ll .001$. At any reasonable level, H_0 is rejected.

At this point, the χ^2 test is seen to be quite inappropriate; it asks the wrong question. To see why, recall H_0: births occur with probability proportional to the length of the season. Even before *any* data were gathered, we knew that H_0 could not be *exactly* the truth—a good approximation, perhaps, but not exactly true. So when the data call for rejection

TABLE **17-2**

Distribution of $n = 88,273$ Births Among 4 Cells (Seasons of the Year), Sweden, 1935

i	Season	(1) Observed Frequency O_i	(2) Probability (if H_0 true) π_i	(3) Expected Frequency $E_i = n\pi_i$	(4) $(O_i - E_i)$	(5) $(O_i - E_i)^2/E_i$
1	Spring Apr–June	23,385	.24932	22,008	1,377	86.16
2	Summer July–Aug	14,978	.16986	14,994	-16	.02
3	Fall Sept–Oct	14,106	.16712	14,752	-646	28.29
4	Winter Nov–Mar	35,804	.41370	36,519	-715	14.00
		88,273	1.0000 ✓	88,273 ✓	0 ✓	$\chi^2 = 128.47$
						prob-value $\ll .001$

of H_0, the only sensible reaction is, "So what?—I could have told you so beforehand. All you have to do to reject *any* specific null hypothesis is collect a large enough sample."

(c) Alternative to χ^2: Confidence Intervals

With the relevance of χ^2 hypothesis tests in question, what then *is* relevant? The reason why this question is asked so seldom is that no single answer will work in all cases. However, we shall give one possible answer to illustrate an important principle: an imaginative and common sense use of confidence intervals may yield more information than a routine χ^2 hypothesis test.

Let us continue to regard the 88,273 observed births in Table 17-2 as a very large sample from an infinite conceptual population. Now consider P, the sample proportion of births observed in the first season, spring:

$$P = \frac{23,385}{88,273} = .265$$

Next, use this to construct a 95% confidence interval for the corresponding population proportion (probability of spring births). From (8-40):

$$\text{true } \pi = .265 \pm 1.96 \sqrt{\frac{(.265)(.735)}{88,273}} \qquad (17\text{-}4)$$

$$= .265 \pm .0029$$

Now compare this with the probability of births in the spring if H_0 were true: since spring is defined to have 91 of the 365 days of the year,

$$\text{null } \pi = \frac{91}{365} = .249 \qquad (17\text{-}5)$$

as shown in column (2) of Table 17-2. Consider the ratio:

$$\frac{\text{true } \pi}{\text{null } \pi} = \frac{(.265 \pm .0029)}{.249}$$

$$= 1.063 \pm .012 \qquad (17\text{-}6)$$

This is easy to interpret: in spring, births were about 6% above normal— more precisely, somewhere between 5.1% and 7.5% above normal, with 95% confidence. The darkly shaded bands in Figure 17-3 illustrate (17-6) and similar estimates for the other 3 seasons.

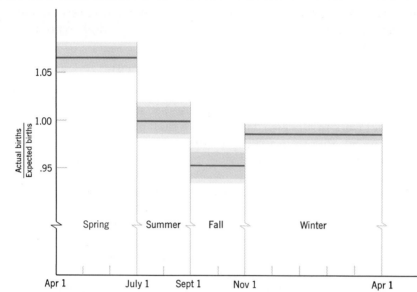

FIGURE 17–3 Ratio of actual births to expected births if H_0 were true. Darker shading shows individual 95% confidence intervals. Lighter shading shows *simultaneous* 95% confidence intervals.

Finally, we come to a problem that we already encountered in the ANOVA chapter: although we can be 95% confident of each individual interval in Figure 17-3, we can be far less confident that the whole system of intervals is true; there are 4 ways (intervals) where we could go wrong. How can we construct a system of confidence intervals that all are *simultaneously* true with 95% confidence, analogous to the multiple comparisons in (10-16)? One very simple solution is to cut the error rate of each interval from 5% to 5%/4 = 1.25%. Then the overall error rate will be at most (1.25%)4 = 5%, as required.[5] This solution, shown as the lightly shaded bands in Figure 17-3, is overly conservative but effective.[6]

---

[5] The proof, like the footnote on page 291, is as follows. Let E_i be the event that the ith confidence interval is in error. Then:

$$\Pr(E_1 \text{ or } E_2) = \Pr(E_1) + \Pr(E_2) - \Pr(E_1 \cap E_2) \le \Pr(E_1) + \Pr(E_2) \qquad \text{(3-17) repeated}$$

Similarly:

$$\Pr(E_1 \text{ or } E_2 \text{ or } E_3 \text{ or } E_4) \le \Pr(E_1) + \Pr(E_2) + \Pr(E_3) + \Pr(E_4) \le (1.25\%)4 = 5\%$$

that is: (cont'd)

PROBLEMS

17-1 According to the Mendelian genetic model, a certain garden pea plant should produce offspring that have white, pink, and red flowers, in the long-run proportion 25%, 50%, 25%. A sample of 1000 such offspring were colored as follows:

white, 21%; pink, 52%; red, 27%.

(a) Find the prob-value for the Mendelian hypothesis.

(b) Can you reject the Mendelian hypothesis at the 5% level?

$$\Pr(\text{any error at all}) \le 5\%$$

that is:

$$\Pr(\text{all correct}) \ge 95\%$$

*[6] A more refined and exact solution is to construct a 95% confidence region, using χ^2 in the following, unusual way. Noting (17-2) and (17-1), we can state with 95% probability that:

$$\sum \frac{(O_i - n\pi_i)^2}{n\pi_i} \le \chi^2_{.05} \tag{17-7}$$

With O_i, n, and $\chi^2_{.05}$ known, this defines a 95% confidence region for the unknown set of population probabilities $(\pi_1, \pi_2, \ldots, \pi_k)$. It must be emphasized that, in (17-7), the π_is are not fixed by a rigid null hypothesis; instead, they are unknown parameters to be solved for. The solution is a statement about all π_is simultaneously, so that the confidence region is of the same simultaneous type as the multiple comparisons in (10-16).

The major difficulty with (17-7) is a technical one: it is hard to graph, and impossible to express as a set of k individual inequalities (intervals) like (10-16). About the best that can be done is the following approximation.

In (17-7), as n increases, O_i approaches $n\pi_i$, so that for large enough samples this substitution in the denominator yields the following approximate 95% confidence region:

$$\sum \frac{(O_i - n\pi_i)^2}{O_i} \le \chi^2_{.05} \tag{17-8}$$

By appropriately dividing by n, we may reexpress this as:

$$\sum \frac{[(O_i/n) - \pi_i]^2}{(O_i/n)} \le \frac{\chi^2_{.05}}{n}$$

This form, called *relative* χ^2, is recognized as a quadratic function of π_i, and hence an ellipsoidal region in $(k-1)$ dimensions; (when $k-1$ of the π_i are specified, the last π_k, of course, is determined].

In our example, the 95% confidence ellipsoid is defined as all those combinations of π_1, π_2, and π_3 satisfying:

$$\frac{(.265 - \pi_1)^2}{.265} + \frac{(.169 - \pi_2)^2}{.169} + \frac{(.160 - \pi_3)^2}{.160} + \frac{(.405 - \pi_4)^2}{.405} \le .000089$$

where $\pi_4 = 1 - \pi_1 - \pi_2 - \pi_3$.

17-2

Period of Day	Number of Accidents
8–10 A.M.	17
10–12 A.M.	15
1–3 P.M.	20
3–5 P.M.	28

This table classifies the industrial accidents in an auto plant into 4 time periods.

(a) Find the prob-value for H_0 (that accidents are equally likely to occur at any time of day).

(b) Can you reject H_0 at the 5% level?

17-3 Throw a fair die 30 times (or simulate it with random digits in Table II(a)).

(a) Use χ^2 to test H_0 (that it is a fair die) at the 25% level.

(b) If each student in a large class correctly carries out the test in (a), what proportion will reject H_0 in the long run?

17-4 Repeat Problem 17-3 for an unfair die. (Since you do not have an unfair die available, use the table of random numbers to simulate a die that is biased towards aces; for example, let the digit 0 as well as 1 represent the ace, so that the ace has twice the probability of any other face.)

17-5 Is there a better test than χ^2 for the die in Problem 17-4 that is suspected of being biased toward aces? If so, use it to recalculate Problem 17-4.

17-6 The data of Table 17-2 were condensed from the following table of monthly births in Sweden, 1935:

TABLE 17-3
Monthly Detail on Births

Jan.	7,280	July	7,585
Feb.	6,957	Aug.	7,393
Mar.	7,883	Sept.	7,203
Apr.	7,884	Oct.	6,903
May	7,892	Nov.	6,552
June	7,609	Dec.	7,132
		Total	88,273

(a) Calculate a few cells of the χ^2 test of the null hypothesis (H_0: births are equally likely to occur on all days of the year).

(b) Analyze a few cells of the data in a more appropriate way, as in Figure 17-3.

(c) What are the advantages and disadvantages of using monthly data instead of seasonal data?

*17-7 Prove that the expectation of χ^2 in (17-2) is $(k - 1)$. (*Hint:* Use (5-32) and (6-23). This is another interpretation of d.f.)

17-2 CONTINGENCY TABLES

(a) Example

Contingency means dependence, so a contingency table is simply a table that displays how two or more characteristics depend on each other; for example, Table 17-4 shows the dependence of income on region in a sample of 200 U.S. families in 1971. To test the null hypothesis of no dependence in the population, χ^2 again may be used in a goodness-of-fit test. In these special circumstances, it customarily is renamed the χ^2 test of independence.

In Table 17-4 let π_{ij} denote the underlying bivariate probability distribution; for example, π_{14} is the probability that a family is in the South earning \$15,000 or more. Let π_i and π_j similarly denote the marginal probability distributions. Then the null hypothesis of statistical independence may be stated precisely:

$$\boxed{H_0: \pi_{ij} = \pi_i \pi_j}$$

(17-9)
like (5-13)

TABLE 17-4
Observed Frequencies O_{ij} of 400 U.S. Families Classified by
Region and Income, 1971[a]

i	j Region	Income (\$000)	1 0–5	2 5–10	3 10–15	4 15 and above	Total Frequency	$\hat{\pi}_i = $ Relative Frequency
1	South		28	42	30	24	124	.31
2	Other		44	78	78	76	276	.69
	Total Frequency		72	120	108	100	400	
$\hat{\pi}_j = $	Relative Frequency		.18	.30	.27	.25		

[a]From the 1974 *American Almanac*, Tables 533 and 14.

TABLE **17-5**

χ^2 Calculations

1. Assuming independence, estimated bivariate probabilities $\hat{\pi}_{ij} = \hat{\pi}_i \hat{\pi}_j$

i \ j	1	2	3	4	$\hat{\pi}_i$
1	.056	.093	.084	.077	.31
2	.124	.207	.186	.173	.69
$\hat{\pi}_j$	.18	.30	.27	.25	

2. Expected frequencies $E_{ij} = n\hat{\pi}_{ij}$

22.4	37.2	33.6	30.8
49.6	82.8	74.4	69.2

3. Deviations $(O_{ij} - E_{ij})$

5.6	4.8	−3.6	−6.8
−5.6	−4.8	3.6	6.8

4. $(O_{ij} - E_{ij})^2/E_{ij}$

1.40	.62	.39	1.50
.63	.28	.17	.67

sum, $\chi^2 = 5.66$

To test how well the data fit this hypothesis, in Table 17-5 we perform the sequence of χ^2 calculations analogous to Table 17-1:

1. First, work out the implications of H_0. The best estimates of π_i and π_j are the marginal relative frequencies shown in Table 17-4. Substituting them into (17-9) yields the estimated probabilities $\hat{\pi}_{ij}$ for each cell.[7]

2. Calculate the expected frequencies $E_{ij} = n\hat{\pi}_{ij}$.

3. Calculate the deviations, $(O_{ij} - E_{ij})$.

4. Square and weight by the expected frequencies; in other words, calculate $(O_{ij} - E_{ij})^2/E_{ij}$. Then sum to get an overall measure of discrepancy:

---

[7]This step is very much like the fitting of each cell in 2-way ANOVA, except that here a probability is fitted by multiplying two component probabilities, whereas in ANOVA a *numerical* response is fitted by the *addition* of two component effects, as in (10-31).

$$\boxed{\text{chi-square independence test, } \chi^2 \overset{\Delta}{=} \sum\sum \frac{(O_{ij} - E_{ij})^2}{E_{ij}}} \qquad \begin{array}{l} (17\text{-}10) \\ \text{like } (17\text{-}2) \end{array}$$

$$= 5.66$$

Letting r designate the number of rows in the table, and c the number of columns, the degrees of freedom of this test are:[8]

$$\boxed{\text{d.f.} = (r - 1)(c - 1)} \qquad (17\text{-}12)$$

$$= (2 - 1)(4 - 1) = 3$$

The observed $\chi^2 = 5.66$ is bracketed by $\chi^2_{.25} = 4.11$ and $\chi^2_{.10} = 6.25$ found in Table VI(a). Thus:

$$.10 < \text{prob-value} < .25 \qquad (17\text{-}13)$$

Thus the χ^2 test cannot reject H_0 at the customary 5% level (or even at the 10% level).

(b) Alternative to χ^2: a Confidence Interval Once Again

A serious drawback of the χ^2 test is that it does not exploit the numerical nature of the income factor. Thus the test completely misses the essential question: *How much* do incomes differ between regions? Even the secondary question of testing may be ineffectively answered by χ^2, since it may have relatively low power. To overcome these faults, the data in Table 17-4 will be reworked: since income is numerical, let us give it a numerical symbol X, and then estimate the difference in means for the two regions using the familiar approach set out in Section 8-3.

---

[8]The d.f. may be calculated from a general principle that we shall find useful for several applications:

$$\text{d.f.} = (\# \text{ cells}) - 1 - (\# \text{ estimated parameters}) \qquad (17\text{-}11)$$

To apply this, we have to know the number of estimated parameters, i.e., the number of estimated probabilities. Consider first the r estimated row probabilities $\hat{\pi}_i$. Once the first $(r - 1)$ are estimated, the last one is strictly determined, since $\Sigma\,\hat{\pi}_i = 1$. Thus, there are only $(r - 1)$ independently estimated row probabilities, and by the same argument only $(c - 1)$ column probabilities. Thus, from (17-11):

$$\text{d.f.} = rc - 1 - [(r - 1) + (c - 1)]$$

from which (17-12) follows.

We calculate $\overline{X}$ and s^2 for each of the two regions, approximating each observation by its cell midpoint; in the last cell, which is open-ended and has no midpoint, we arbitrarily use 17.5. Because of the large sample size, we may use s_i^2 as a proxy for σ_i^2 in (8-28), obtaining the 95% confidence interval:

$$(\mu_1 - \mu_2) = (\overline{X}_1 - \overline{X}_2) \pm 1.96 \sqrt{\frac{s_1^2}{n_1} + \frac{s_2^2}{n_2}}$$

$$= (14.52 - 15.87) \pm 1.96 \sqrt{\frac{26.3}{124} + \frac{28.0}{276}}$$

$$= -1.35 \pm 1.96(.56) \tag{17-14}$$

$$= -1.35 \pm 1.10$$

That is, for the South relative to the rest of the country, the mean income differential is:

$$\mu_1 - \mu_2 = \$-1,350 \pm \$1,100 \tag{17-15}$$

The secondary question of testing H_0 (no difference between regions) now may be answered immediately: At the 5% level, H_0 now can be rejected, since 0 does not lie in the confidence interval (17-15).

We can obtain the same conclusion using prob-value. From (17-14), $z = -1.35/.56 = -2.41$. From Table IV:

$$\text{two-sided prob-value} = 2(.008) = .016 \tag{17-16}$$

Thus, at the 5% level, we now can reject H_0. In conclusion, therefore, the test in this section is superior to the χ^2 test because it allows for rejection of H_0—or, more precisely, because the prob-value in (17-16) is much smaller and sharper than the χ^2 prob-value in (17-13).

Of course, if more than two regions were to be compared, we would use ANOVA and multiple comparisons instead of (17-14). But the conclusion would be generally the same: the χ^2 test is less powerful.

(c) General Alternatives to χ^2 Tests of Independence

The lesson to be drawn from the example above is clear:

> Whenever numerical variables appear, they should be analyzed with a tool (such as multiple comparisons or regression) that exploits their numerical nature. A χ^2 hypothesis test fails to do this. (17-17)

In fact, even if a variable is not naturally numerical but merely ordered (for example, a variable such as social class, or degree of success), it often is wise to code the various levels of the variable by 0, 1, 2, 3, . . . , and then proceed with this new numerical variable. Although this may seem arbitrary, it usually will yield a more powerful test of H_0 than χ^2, and also will give at least a rough-and-ready answer to the question, "*How much* do things differ?" As a special case of this, any variable with just 2 levels can be made numerical by coding it 0 and 1 (i.e., by making it a dummy variable). Problems 17-8 to 17-11 illustrate this.

PROBLEMS

In each of the following problems, a random sample of several hundred Americans was classified according to two characteristics. In each case:
  (a) Calculate χ^2 to test whether the two characteristics are independent.
  (b) Analyze in a better way, if possible.

17-8   Educational Attainment, by Color, 1964

Education / Color	Elementary School Attendance	High School Attendance	University Attendance
White	12,835	20,778	9,468
Nonwhite	2,321	1,956	477

17-9   Employment in Various Occupations, by Color, 1965

Occupation / Color	White Collar	Blue Collar	Household and Other Services	Farm Workers
White	95	88	74	85
Nonwhite	5	12	26	15

Note that this sample was designed to have exactly the same number of people (100) in each occupation. Therefore, the relative frequency in each occupation is 100/400, and cannot be an estimate of the population proportion. Thus, this sample differs from the simple random sample of Table 17-4.

Nevertheless, it turns out that the standard χ^2 test still remains valid— so go ahead and calculate it. When used in this way, it often is called the χ^2 test of homogeneity. [Are the various occupations homogenous, i.e., similar in terms of color?] Note that the next three problems also are tests of homogeneity.

17-10   Employment in Selected Industries, by Sex, 1965

Industry Sex	Durable Goods	Nondurable Goods	Mining	Trade
Male	162	124	190	118
Female	38	76	10	82

17-11   Income, by Sex, 1964

Income ($s) Sex	Male	Female
Less than 5,000	53	90
More than 5,000	47	10

17-12   Classification of 300 Newspaper
                            Readers, by Social Class

Social Class Newspaper	A	B	C
Poor	31	11	12
Lower middle class	49	59	51
Middle class	18	26	31
Rich	2	4	6

*17-3 χ^2 TESTS OF DISTRIBUTION SHAPE

(a) χ^2 with Estimated Parameters

χ^2 is often used to determine whether a sample is drawn from a normal or some other kind of population distribution. As a simple example, Table 17-6 displays a sample distribution of male offspring in families of 4 children. Suppose that we wish to test the null hypothesis:

$$H_0: \text{ the population distribution is binomial} \qquad (17\text{-}18)$$

To calculate χ^2, the first step is to derive the cell probabilities π_i, assuming that H_0 is true—where π_i represents the probability of i boys in 4 births. These probabilities will be binomial, with $n = 4$; but what value of π (i.e., the probability of a male in a single birth) should be used? If π is set at .50, we would be testing too narrow a null hypotheses, H_0': the

TABLE **17-6**

Distribution of Males in Completed Families of 4 Children

(1) Number of Males X_i	(2) Observed Frequency O_i	(3) Probability π_i (based on $\hat{\pi} = .55$)	(4) Expected Frequency $E_i = n\pi_i$	(5) Deviation $(O_i - E_i)$	(6) Deviation Squared and Weighted $(O_i - E_i)^2/E_i$
0	1	.041	4.1	-3.1	2.35
1	17	.200	20.0	-3.0	.45
2	49	.368	36.8	12.2	4.06
3	27	.300	30.0	-3.0	.30
4	6	.091	9.1	-3.1	1.06
		$1.000 \sqrt{}$	$100.0 \sqrt{}$	$0 \sqrt{}$	$\chi^2 = 8.2$

distribution is binomial, with $\pi = .50$. But (17-18) calls for no restrictions whatever on π. Therefore, π is estimated from the data: in 100 families of 4 births each, there are a total of 400 births, of which 220 are males (220 $= 1(17) + 2(49) + 3(27) + 4(6)$). Thus $\hat{\pi} = 220/400 = .55$. This estimate now can be used to calculate the binomial probabilities from the binomial formula (4-7). Then these probabilities π_i are entered into column (3) of Table 17-6, and the calculation of χ^2 proceeds as before, obtaining:

$$\chi^2 = 8.2$$

The only new twist is the d.f. Recall that in our earlier χ^2 test in (17-3),

$$\text{d.f.} = \text{\# cells} - 1$$

But in that case it was possible for the statistician to define the cell probabilities π_i (in column (3) of Table 17-6) directly from H_0; on the other hand, in this present example the cell probabilities π_i could not be defined directly from H_0, but only after π had been estimated. One degree of freedom is lost as a consequence,[9] and in general:

---

[9] The loss of a degree of freedom for every parameter that must be estimated beforehand (first encountered in the footnote to (8-17)) is by now a familiar principle. Because π_i are calculated from the *data* as well as from H_0, E_i fits O_i a little better. This makes χ^2 a little smaller. Therefore, the reduction of d.f. is justified. (Note that the critical points of χ^2 are lower for 3 d.f. than for 4.)

Finally, we confirm that π must be estimated, rather than prespecified. If we had arbitrarily set $\pi = .50$, then we would not know how much of the calculated χ^2 (i.e., how much of the difference between E_i and O_i) was because the observed distribution was not binomial, and how much was because it was not centered on .50.

$$\boxed{\text{d.f.} = (\# \text{ cells}) - 1 - (\# \text{ estimated parameters})}\ (17\text{-}11) \text{ repeated}$$

To conclude, our observed $\chi^2 = 8.2$ has $5 - 1 - 1 = 3$ d.f., and slightly exceeds the critical $\chi^2_{.05} = 7.8$ in Table VI(a). Thus the evidence contradicts the null hypothesis of a binomial distribution, at the 5% level.

(b) Alternative to χ^2: Dispersion Test and Confidence Interval

It is time to consider the alternate hypothesis H_1, against which H_0 is being tested. If H_1 is not considered carefully, the result may be a very poor test. In Figure 17-4, we show the H_0 distribution (from column (3) of Table 17-6) as a reference, and several possible alternative distributions. The H_0 binomial distribution in panel (a) is centered at the same place as the data, because π was estimated from the data. The alternative distributions H_1 also are shown centered on the data, because the issue between the competing hypotheses is not where the distributions are centered, but how they are distributed around their center (binomial, or not?). The most natural question is: *How much* spread is there around the center? In this regard, do the data conform to H_0 or H_1? A good statistic for measuring spread is the sample variance, which is calculated in Table 17-7:

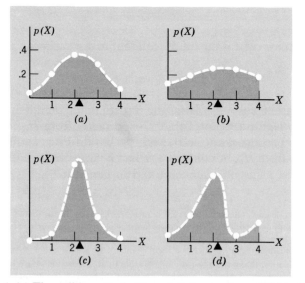

FIGURE 17–4 (a) The null hypothesis (binomial) for the data of Table 17–6. (a), (c), (d) Some alternative hypotheses

$$s^2 = .687 \qquad\qquad (17\text{-}19)$$

$$\text{d.f.} = n - 1 = 99$$

To see how well s^2 matches H_0, we calculate σ^2, the population variance if H_0 is true. From (6-23):

$$\sigma^2 = m\,\pi(1 - \pi)$$

Noting that m denotes the number of trials (the number of children, 4), and that π is estimated to be .55, it follows that:

$$\sigma^2 = 4(.55)(.45) \qquad\qquad (17\text{-}20)$$

$$= .990$$

Finally, compare (17-19) with (17-20) by forming the ratio:

$$C^2 = \frac{s^2}{\sigma^2} = \frac{.687}{.990} = .694 \qquad\qquad (17\text{-}21)$$

TABLE 17-7
Dispersion Test for the Data of Table 17-6

X_i	Observed Frequency O_i or f_i	$X_i f_i$	$X_i - \overline{X}$	$(X_i - \overline{X})^2 f_i$
0	1	0	−2.2	4.8
1	17	17	−1.2	24.5
2	49	98	−.2	2.0
3	27	81	.8	17.3
4	6	24	1.8	19.4

$$n = 100 \checkmark \qquad \overline{X} = \frac{220}{100} \qquad\qquad \sum (X_i - \overline{X})^2 f_i = 68.0$$

$$= 2.20 \qquad\qquad \text{from (2-6b), } s^2 = \frac{68.0}{99} = .687$$

If H_0 is true, the distribution of C^2 is approximately the modified χ^2 with $n - 1 = 99$ d.f., introduced in Section 8-6. From Table VI(b), we therefore find that the corresponding one-tailed probability is approximately .005; hence the two-sided prob-value is:

$$\text{prob-value} \simeq .01$$

(This must be a two-tailed test, since the competing hypotheses H_1 can involve a dispersion that is either larger or smaller than the binomial H_0). Since this prob-value is less than the corresponding value of .04 that was derived in the previous section, this is even stronger evidence against H_0; i.e., this dispersion test is a stronger statistical test of H_0 than the χ^2 test.

Of course, the dispersion test will have particularly high power against an alternative hypothesis with a variance that is very different from the null hypothesis σ^2. For example, Figure 17-4(b) shows an alternative hypothesis with a relatively high variance. As another example, Figure 17-4(c) shows a variance that is substantially lower than σ^2; in fact, the data probably came from a distribution like this.[10]

On the other hand, the dispersion test will have very little power against an alternative distribution such as Figure 17-4(d), which has about the same variance as the null distribution. However, since such an alternative is rare in practice, the dispersion test is generally a more powerful test than the classical χ^2 test.

As we have argued previously, more appropriate than an hypothesis test is the 95% confidence interval, calculated to be:

$$\frac{s^2}{C^2_{.025}} < \sigma^2 < \frac{s^2}{C^2_{.975}} \qquad \text{(8-54) repeated}$$

In our example, using interpolated values for $C^2_{.025}$ and $C^2_{.975}$ of approximately 1.3 and .74:

$$\frac{.687}{1.3} < \sigma^2 < \frac{.687}{.74}$$

$$.53 < \sigma^2 < .93$$

Since the null hypothesis value of σ^2 is .99 and therefore lies well outside this interval, our earlier rejection of H_0 at the 5% level is confirmed. At the same time, we see to what extent this variance is less than the binomial variance.

PROBLEMS

*17-13 From the data in Table 17-6, calculate the χ^2 test of the null hypothesis that the distribution of males is binomial, with $\pi = \frac{1}{2}$, (and $n = 4$, of course).

---

[10]Can you explain why this birth data might not be binomial? For a clue, read the title of Table 17-6 carefully.

*17-14 At an annual meeting of a large corporation, a sample of 80 stock-holders evaluated management performance on a scale 0 to 100, as follows:

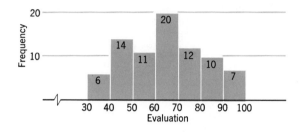

Suppose that you are considering a goodness-of-fit test for normality, i.e., a test of:

H_0: this sample is drawn from a normal population.

Before proceeding consider:
(a) If H_0 is rejected (at the 5% level say), what will you have shown?
(b) If H_0 is acceptable, what will you have shown? What further use might this information have?
(c) Before running through the calculations, explain how you would proceed to test H_0, indicating which parameters initially must be estimated.
(d) Calculate the χ^2 test of H_0.

*17-15 In Section 17-1(b), we argued that an hypothesis test about births in Sweden was inappropriate because it is known beforehand that H_0 cannot be exactly true. Is it equally inappropriate to test:

H_0: population binomial

when we are:
(a) Observing the number of males in completed families of 4 children (as in Table 17-6)?
(b) Observing the number of heads in 4 tosses of a coin?

REVIEW PROBLEMS

17-16 Suppose that a certain population is divided in the following proportions:

Region \ Income ($000s)	0–5	5–10	10–15	15 plus
North	.06	.12	.24	.18
South	.12	.16	.08	.04

To test whether both regions have the same distribution of income, use random digits to simulate drawing a random sample of 20 from each region.

(a) Then use χ^2 to calculate the prob-value for H_0 (both regions have the same mean income).

(b) Use a better test to calculate the prob-value.

(c) Use the test of part (b) to construct a 95% confidence interval.

17-17 A large corporation wishes to test whether each of its divisions has equally satisfactory quality control over its output. Suppose that the output of each division, along with the number of units rejected and returned by dealers is as follows:

	Division A	B	C
Output	1,200	800	2,000
Rejects	52	60	88

(a) Use χ^2 to test H_0 (no difference in divisions).

(b) Analyze in a better way, if possible.

*17-18 Consider the sample of 200 heights given in Table 2-2 and Figure 2-2. To see whether it may have come from the normal population with $\mu = 69$ and $\sigma = 3.2$ (as given in Figure 6-1), an anthropologist asks you to calculate the prob-values for each of the following hypotheses:

(a) Test whether $\mu = 69$ (using the t test).

(b) Test whether $\sigma = 3.2$ (using the C^2 test).

(c) Test whether the distribution shape is normal (χ^2 test).

(d) Combine parts (a), (b), and (c), by testing whether the population is the specific normal distribution with $\mu = 69$ and $\sigma = 3.2$.

CHAPTER 18

Maximum Likelihood
Estimation (MLE)

I have set my life upon a cast,
And I will stand the hazard of the die!

Shakespeare, *Richard III*

Maximum likelihood is a popular technique of estimation because it often has many of the attractive large-sample properties (such as asymptotic efficiency and consistency) that we discussed in Chapter 7. The basic idea is to find the population value that best matches the sample, i.e., the hypothetical population value that is more likely than any other to generate the observed sample. We shall prove in this chapter that many of the estimators already introduced, in particular the least squares regression estimators, also are maximum likelihood estimators, and so are justified further.

18-1   MLE FOR A PROPORTION π

(a) Example

Suppose that a shipment of radios is sampled for quality, and that 3 out of 5 are found defective. What should we estimate is the proportion π of defectives in the whole shipment (population)? Temporarily, try to forget the common sense method (estimate the population proportion π

with the sample proportion $P = 3/5 = .60$).[1] Instead, let us investigate an alternative method: consider a whole range of possible πs that we might choose, and then try to pick out the one that best explains the sample.

For example, is $\pi = .1$ a plausible value for the population? If

Table 18-1   Outline of Maximum Likelihood Estimation (MLE)

	0–1 Population, Special Case	0–1 Population, General Case	Normal Population
Given:	3 defectives in 5 observations	S successes in n observations	Observations X_1, X_2, X_3
Find:	MLE of π.	MLE of π.	MLE of μ.
As follows:	The probability of 3 defectives in 5 observations is: $p(3/\pi)$ $= \binom{5}{3}\pi^3(1-\pi)^2$ Since the number of defectives is fixed at its observed value of 3, we write this as the likelihood function of π only: $L(\pi)$ $\pi \quad = \binom{5}{3}\pi^3(1-\pi)^2$ $\begin{array}{ll} 0 & 0 \\ .1 & .008 \\ .2 & .051 \\ .3 & .132 \\ .4 & .230 \\ .5 & .312 \\ .6 & .346\Leftarrow \\ .7 & .309 \\ .8 & .205 \\ .9 & .073 \\ 1.0 & 0 \end{array}$	The probability of S successes in n observations is: $p(S/\pi) = \binom{n}{S}\pi^S(1-\pi)^{n-S}$ Since S is fixed at its observed value, we write this as the likelihood function of π only: $L(\pi) = \binom{n}{S}\pi^S(1-\pi)^{n-S}$ Try out all possible values of π, selecting the one that maximizes this likelihood function. Calculus shows that this is: $\pi = \dfrac{S}{n}$ $= P$, the sample proportion	The probability of our sample resulting from any given μ is: $p(X_1, X_2, X_3/\mu)$ $= \prod_{i=1}^{3}\left[\dfrac{1}{\sqrt{2\pi\sigma^2}}e^{-(1/2\sigma^2)(X_i-\mu)^2}\right]$ With (X_1, X_2, X_3) fixed at their observed values, we write this as likelihood function of μ: $L(\mu) = \prod_{i=1}^{3}\left[\dfrac{1}{\sqrt{2\pi\sigma^2}}e^{-(1/2\sigma^2)(X_i-\mu)^2}\right]$ Try out all possible values of μ, selecting that one that maximizes this likelihood function. Calculus shows that this is: $\mu = \frac{1}{3}(X_1 + X_2 + X_3)$ $= \overline{X}$, the sample mean.
Conclude:	The MLE of π is .6, the sample proportion P.	Hence, the MLE of π is P.	Thus the MLE of μ is $\overline{X}$.

---

[1]Actually, P is more than just a "common sense" estimator. It already has been justified in Chapter 6 as an estimator because it is a sample mean in disguise, while π is a disguised population mean. Hence, according to Section 12-3, P is BLUE (the Best Linear Unbiased Estimator). But

$\pi = .1$, then the probability of $S = 3$ defectives out of a sample of $n = 5$ observations would be given by the binomial formula (4-7):

$$\binom{n}{S}\pi^S(1 - \pi)^{n-S} = \binom{5}{3}.1^3 .9^2$$

by Table III(b):

$$\simeq .008 \qquad\qquad\qquad (18\text{-}1)$$

Regression Model $Y = \alpha + \beta X + \gamma Z + e$	Any Population $p(X/\theta)$
Observations $Y_1, Y_2, \ldots, Y_n$	Observations $X_1, X_2, \ldots, X_n$
MLE of α, β, γ.	MLE of θ.
The probability of our sample resulting from any combination of α, β, γ is: $p(Y_1, Y_2, \ldots, Y_n/\alpha, \beta, \gamma)$ $= \dfrac{1}{(2\pi\sigma^2)^{n/2}} e^{-(1/2\sigma^2)\Sigma[Y_i-(\alpha+\beta x_i+\gamma z_i)]^2}$ With the Y_i, X_i, and Z_i fixed at their observed values, this becomes the likelihood function of α, β, and γ, designated $L(\alpha, \beta, \gamma)$. Try out all combinations of α, β, and γ, selecting the combination that maximizes this likelihood function by minimizing the sum of squares in the exponent.	The probability of our sample resulting from any given θ is: $p(X_1, X_2, \ldots, X_n/\theta)$ $= p(X_1/\theta)\, p(X_2/\theta) \ldots p(X_n/\theta)$ $= \displaystyle\prod_{i=1}^{n} p(X_i/\theta)$ With $(X_1, X_2, \ldots, X_n)$ fixed at their observed values, we write this as the likelihood function of θ: $L(\theta) = \displaystyle\prod_{i=1}^{n} p(X_i/\theta)$
Thus the MLE is least squares.	Select the value of θ that maximizes this likelihood function.

here we disregard this, in order to concentrate on maximum likelihood, a completely different sort of justification.

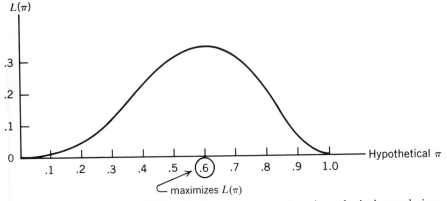

FIGURE 18-1 $L(\pi)$, the likelihood function that various hypothetical population proportions would yield the observed sample of 3 defectives in 5 observations

In other words, if $\pi = .1$, there are only about 8 chances in a thousand of getting the actual sample that we observed.

Similarly, if $\pi = .2$, we would find from Table III(b) that there are about 50 chances in a thousand of getting the sample that we observed. In fact, it seems quite natural to try out all values of π, in each case finding out how likely it is for such a population π to generate the sample that we actually observed. We simply read across Table III(b), and so obtain the first column of Table 18-1, or the graph in Figure 18-1. In this situation, where the sample value $S = 3$ is fixed, and the only variable is the hypothetical value of π, we call the result the *likelihood function* $L(\pi)$:

$$L(\pi) = \binom{5}{3}\pi^3(1 - \pi)^2 \qquad (18\text{-}2)$$

The *maximum likelihood estimate* (MLE) is simply the value of π that maximizes this likelihood function. From Figure 18-1 or Table 18-1, we see that this happens to be $\pi = .60$—which coincides nicely with the common sense estimator $P = 3/5$. In general, we similarly may put forth this definition:

> The MLE is the hypothetical population value that maximizes the likelihood of the observed sample.   (18-3)

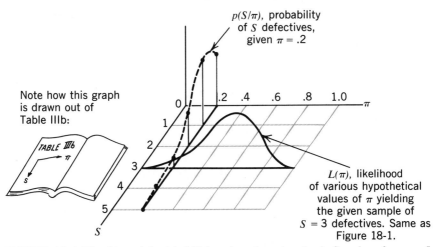

FIGURE 18–2 The binomial probabilities plotted against both S and π, for $n = 5$ observations

To show these same issues geometrically, in Figure 18-2 we graph the binomial probabilities as a function of both S and π. In Chapter 4, we thought of π fixed and S variable. For example, the dotted distribution in the S direction shows the probability of getting various numbers of defectives, if the population proportion were $\pi = .2$. But in this chapter, we regard S—the observed sample result—as fixed, while the population π is thought of as taking on a whole set of hypothetical values. For example, the solid curve in the π direction shows the likelihood that various possible population proportions would yield 3 defectives. Slices in the S direction are referred to as probability distributions, while slices in the π direction are called likelihood functions.

Next we shall apply the MLE principle (18-3) to the case of the general binomial, and then to some normal distributions as well. A summary of the results is given in Table 18-1 for reference.

(b) MLE for π in General

We shall show that our result in part (a) was no accident, and that the maximum likelihood estimate of the binomial π is *always* the sample proportion P.

Given an observed sample of S successes in n trials, the likelihood function is:

$$L(\pi) = \binom{n}{S} \pi^S (1 - \pi)^{n-S} \qquad (18\text{-}4)$$

With calculus, it easily can be shown[2] that the maximum value of this likelihood function occurs when $\pi = S/n = P$. Thus:

$$\boxed{\text{MLE of } \pi \text{ is } P, \text{ the sample proportion}} \qquad (18\text{-}5)$$

We argued in Chapter 1 that it is reasonable to use the sample proportion to estimate the population proportion; but in addition to its intuitive appeal, we now add the more rigorous justification of maximum likelihood: a population with $\pi = P$ has greatest likelihood of generating the observed sample.

PROBLEMS

18-1 Suppose that a sample of n observations is taken from a population, to estimate the proportion π of defectives. Graph the likelihood function, and show the maximum likelihood estimate of π for each of the following cases.

(a) $n = 8$, with 2 defectives.

(b) $n = 8$, with 4 defectives.

(c) $n = 2$, with 1 defective.

(d) $n = 2$, with 0 defective.

---

[2]To find where $L(\pi)$ is a maximum, set the derivative equal to zero.

$$\frac{dL(\pi)}{d\pi} = \binom{n}{S}[\pi^S(n - S)(1 - \pi)^{n-S-1}(-1) + S\pi^{S-1}(1 - \pi)^{n-S}] = 0 \qquad (18\text{-}6)$$

Divide by $\binom{n}{S} \pi^{S-1}(1 - \pi)^{n-S-1}$ to obtain:

$$-\pi(n - S) + S(1 - \pi) = 0$$

$$-n\pi + S = 0$$

$$\pi = \frac{S}{n}$$

You can easily confirm that this is a maximum (rather than a minimum or inflection point). Actually, this derivation assumed that $0 < S < n$. When $S = 0$, the proof of (18-5) does not even require calculus: from (18-4), when $S = 0$, then $L(\pi) = (1 - \pi)^n$, which is clearly a maximum at the end-point $\pi = 0$. Similarly, when $S = n$, then $L(\pi) = \pi^n$, which is clearly a maximum at the other end-point, $\pi = 1$. Thus we have proven (18-5) valid in all possible cases.

18-2 In Problem 18-1, for which cases is the likelihood function a parabola? For which cases is the MLE the sample proportion P?

18-2   MLE OF A NORMAL MEAN

Suppose that we have drawn a sample (X_1, X_2, X_3) from a parent population that is $N(\mu, \sigma^2)$; our problem is to find the MLE of the unknown μ. Because the population is normal, the probability[3] of observing any value X, given a population mean μ, is:

$$p(X/\mu) = \frac{1}{\sqrt{2\pi\sigma^2}} e^{-(1/2\sigma^2)(X-\mu)^2} \tag{18-7}$$

Specifically, the probability that we would get the value X_1 for the first observation is:

$$p(X_1/\mu) = \frac{1}{\sqrt{2\pi\sigma^2}} e^{-(1/2\sigma^2)(X_1-\mu)^2} \tag{18-8}$$

The probabilities of drawing the values X_2 and X_3 are, respectively:

$$p(X_2/\mu) = \frac{1}{\sqrt{2\pi\sigma^2}} e^{-(1/2\sigma^2)(X_2-\mu)^2} \tag{18-9}$$

and:

$$p(X_3/\mu) = \frac{1}{\sqrt{2\pi\sigma^2}} e^{-(1/2\sigma^2)(X_3-\mu)^2} \tag{18-10}$$

We assume, as usual, that X_1, X_2, and X_3 are independent, so that the joint probability is the product:

$$p(X_1, X_2, X_3/\mu) = \left[\frac{1}{\sqrt{2\pi\sigma^2}} e^{-(1/2\sigma^2)(X_1-\mu)^2} \right] \times \left[\frac{1}{\sqrt{2\pi\sigma^2}} e^{-(1/2\sigma^2)(X_2-\mu)^2} \right]$$

$$\times \left[\frac{1}{\sqrt{2\pi\sigma^2}} e^{-(1/2\sigma^2)(X_3-\mu)^2} \right] \tag{18-11}$$

---

[3] Strictly speaking, in the case of a continuous distribution such as the normal, we ought to use the phrase "probability density;" but hereafter we will abbreviate it to "probability."

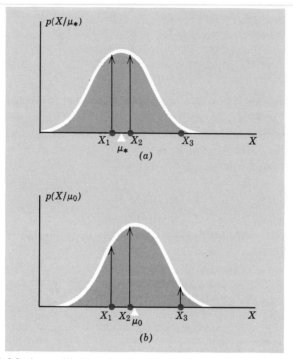

FIGURE 18-3 Maximum likelihood estimation of the mean μ of a normal population, based on three sample observations (X_1, X_2, X_3). (a) Small likelihood $L(\mu_*)$, the product of the three ordinates. (b) Larger likelihood $L(\mu^*)$.

Now the sample values X_1, X_2, X_3 are fixed while μ is thought of as varying over hypothetical values. Then (18-11) is called the likelihood function $L(\mu)$, and the MLE of μ is defined as the hypothetical value of μ that maximizes it. The MLE may be derived with calculus, but we consider only a geometric interpretation, in Figure 18-3.

We "try out" two hypothetical values of μ, say μ_* and μ^*. First, we note that a population with mean μ_* as in Figure 18-3(a) is not very likely to yield the sample we observed. Although the probabilities of X_1 and X_2 are large, the probability of X_3 (i.e., the ordinate above X_3) is very small because it is so far distant from μ_*. The product of all three probabilities (i.e., the likelihood of a population with mean μ_* generating the sample X_1, X_2, X_3) is therefore quite small.

On the other hand, a population with mean μ^* as in Figure 18-3(b) is more likely to generate the sample values. Since the X values are collectively closer to μ^*, they have a greater joint probability. Thus the likelihood is greater for μ^* than for μ_*; indeed, very little additional shift in μ^* is apparently required to maximize the likelihood of the sample. It seems that the MLE of μ might be centered right among the observations X_1, X_2, X_3—i.e., might be just the sample mean $\overline{X}$; in fact, this is proved in Problem 18-5(a):

$$\boxed{\text{In a normal population, MLE of } \mu \text{ is } \overline{X}} \qquad (18\text{-}12)$$

Finally, the reader who has carefully learned that μ is a fixed population parameter may wonder how it can appear in the likelihood function (18-11) as a variable. This is simply a mathematical convenience. The true value of μ is, in fact, fixed. But since it is unknown, in MLE we consider all of its possible or hypothetical values. The way to do this mathematically is to treat it as a variable.

18-3   MLE FOR NORMAL REGRESSION

(a) Simple Regression

We pointed out in Chapter 12 that estimating μ with $\overline{X}$ is just the simplest possible special case of least squares. Since we showed in the previous section that $\overline{X}$ is MLE, we now might ask if the least squares estimators are still MLE when they are applied in the full regression model? The answer is yes, provided that we make the assumption of normality of the error term (essentially, this was not required in Chapter 12).[4] Thus, to the standard assumptions of (12-3), we now add that the independent random variables e_i are normally distributed in the model:

$$Y_i = \alpha + \beta x_i + e_i \qquad (12\text{-}3) \text{ repeated}$$

Estimating α and β using MLE involves selecting those hypothetical population values of α and β that are more likely than any others to generate the sample values that we observed. Before addressing the algebraic derivation, it is best to clarify what is going on with a bit of geometry. To simplify, assume a sample of only three observations (Y_1, Y_2, Y_3).

---

[4]Except for small sample estimation—and this because of the general principle that small sample estimation requires a normally distributed parent population to strictly validate the t distribution.

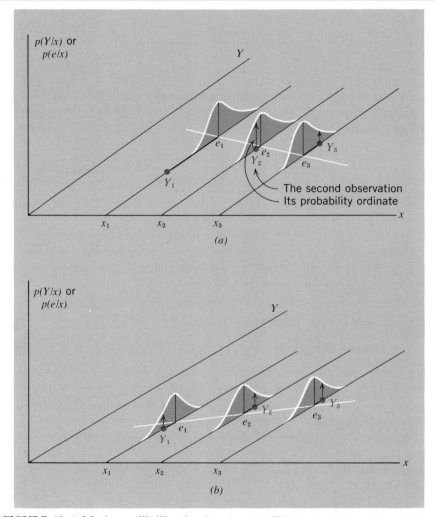

FIGURE 18–4 Maximum likelihood estimation. (a) This is *not* the true population; it is only a hypothetical population that the statistician is considering. But it is not very likely to generate the observed Y_1, Y_2, Y_3. (b) Another hypothetical population; this is more likely to generate Y_1, Y_2, Y_3.

First, let us try out the line shown in Figure 18-4(a). (Before examining it carefully, we note that it seems to be a pretty bad fit for our three observed points.) Temporarily, suppose that this were the true regression line; then the distribution of errors would be centered around it, as shown. The likelihood that such a population would give rise to the observations in our sample is the probability that we would get the particular set of the three e values that are shown in this diagram. The probability of each e value is shown rising vertically above it. Because our three observations are, by assumption, statistically independent, the likelihood of all three (i.e., the probability of getting the sample that we observed) is the product of these three ordinates. This likelihood seems relatively small, mostly because the miniscule ordinate above Y_1 reduces the product value. Our intuition that this is a bad estimate is confirmed; such a hypothetical population is not very likely to generate our sample values.

In Figure 18-4(b), it is evident that we can do much better. This hypothetical population is more likely to give rise to the sample that we observed. Since the disturbance terms are collectively smaller, their probability is consequently greater.

So the MLE technique speculates on various possible populations. How likely is each to give rise to the sample that we observed? Geometrically, our problem is to try them all out, by moving the population through all its possible values—i.e., *by moving the regression line and its surrounding e distribution through all possible positions.* Different positions correspond to different trial values for α and β. In each case, the likelihood of observing Y_1, Y_2, Y_3, would be evaluated. For our MLE, we choose that hypothetical population which maximizes this likelihood. It is evident that little further adjustment is required in Figure 18-4(b) to arrive at the MLE. This procedure intuitively seems to result in a good fit; moreover, since it seems similar to the least squares fit, it is no surprise that we shall be able to prove that the two coincide.

While geometry has clarified the method, it hasn't provided a precise means of arriving at the specific maximum likelihood estimate. This must be done algebraically. For generality, suppose that we have a sample of size n, rather than just 3. We wish to know the probability of the sample that we observed, expressed as a function of the possible values[5] of α and β:

$$p(Y_1, Y_2, \ldots, Y_n/\alpha,\beta) \tag{18-13}$$

Consider the probability of the first Y value:

$$p(Y_1) = \frac{1}{\sqrt{2\pi\sigma^2}} e^{-(1/2\sigma^2)[Y_1 - (\alpha + \beta x_1)]^2} \tag{18-14}$$

---

[5] And also of σ^2. But it turns out that, although the likelihood *function* depends on σ^2, the maximum likelihood *estimate* does not; hence σ^2 is ignored in this discussion.

This is simply the normal distribution, with the mean $(\alpha + \beta x_1)$ and variance (σ^2) substituted appropriately into (4-15). (In terms of the geometry of Figure 18-4, $p(Y_1)$ is the ordinate above Y_1.) The probability of the second Y value is similar to (18-14), except that the subscript 2 replaces 1 throughout, and so on, for all the other observed Y values.

The independence of the Y values justifies mutliplying all these probabilities together to find (18-13). Thus:

$$p(Y_1, Y_2, \ldots , Y_n/\alpha, \beta)$$

$$= \left[\frac{1}{\sqrt{2\pi\sigma^2}} e^{-(1/2\sigma^2)[Y_1-(\alpha+\beta x_1)]^2} \right] \left[\frac{1}{\sqrt{2\pi\sigma^2}} e^{-(1/2\sigma^2)[Y_2-(\alpha+\beta x_2)]^2} \right] \cdots$$

$$= \prod_{i=1}^{n} \left[\frac{1}{\sqrt{2\pi\sigma^2}} e^{-(1/2\sigma^2)[Y_i-(\alpha+\beta x_i)]^2} \right] \tag{18-15}$$

where $\prod\limits_{i=1}^{n}$ represents the product of n factors. Using the familiar rule for exponentials,[6] the product in (18-15) can be reexpressed by summing exponents:

$$p(Y_1, Y_2, \ldots , Y_n/\alpha, \beta) = \left(\frac{1}{\sqrt{2\pi\sigma^2}} \right)^n e^{\Sigma(-1/2\sigma^2)[Y_i-(\alpha+\beta x_i)]^2} \tag{18-16}$$

Recall that the observed Ys are given. We are speculating on various values of α and β. To emphasize this, we rename (18-16) the likelihood function:

$$L(\alpha, \beta) = \frac{1}{(2\pi\sigma^2)^{n/2}} e^{-(1/2\sigma^2)\Sigma[Y_i-\alpha-\beta x_i]^2} \tag{18-17}$$

Now we ask, what values of α and β produce the largest L? The only place α and β appear is in the exponent; moreover, maximizing a function with a negative exponent involves minimizing the magnitude of the exponent. Hence the MLE estimates are obtained by choosing α and β in order to:

$$\text{minimize} \sum [Y_i - \alpha - \beta x_i]^2 \tag{18-18}$$

Since the selection of maximum likelihood estimates of α and β to minimize (18-18) is identical to the selection of least squares estimates a and b to minimize (11-10), a very important conclusion follows:

---

[6] $e^a e^b = e^{a+b}$ for any a and b.

> In a normal regression model, MLE is identical to least squares.

This establishes a most important theoretical justification of least squares: it is the estimate that follows from applying maximum likelihood techniques to a regression model with normally distributed error.

(b) Multiple Regression

The multiple regression model is:

$$Y_i = \alpha + \beta x_i + \gamma z_i + e_i \tag{18-19}$$
$$\text{(13-2) repeated}$$

Maximum likelihood estimates of α, β, and γ are derived in the same way as in the simple regression case; again they coincide with least squares. Algebraically, the argument is similar to the analysis in part (a), and is left as an exercise.

18-4   MLE OF ANY PARAMETER FROM ANY POPULATION

We now state MLE in its full generality. A sample $(X_1, X_2, \ldots, X_n)$ is drawn from a population with probability distribution $p(X/\theta)$, where θ is any unknown population parameter that we wish to estimate. From our definition of random sampling (with replacement, or from an infinite population), the X_i are independent, each with the probability distribution $p(X_i/\theta)$; hence the joint probability distribution of the whole sample is obtained by multiplying:

$$p(X_1, X_2, \ldots, X_n/\theta) = p(X_1/\theta)\, p(X_2/\theta) \, \ldots \, p(X_n/\theta)$$

$$= \prod_{i=1}^{n} p(X_i/\theta) \tag{18-20}$$

But we regard the observed sample values as fixed, and ask, "Which of all the hypothetical values of θ maximizes this probability?" This is emphasized by renaming (18-20) the likelihood function:

$$L(\theta) = \prod_{i=1}^{n} p(X_i/\theta) \tag{18-21}$$

The MLE is that hypothetical value of θ that maximizes this likelihood function.

18-5   MLE AND MME (METHOD OF MOMENTS ESTIMATION)

The first principle of estimation that we used in this book was the most intuitive one: estimate a population mean μ with the sample mean $\overline{X}$, a population variance σ^2 with the sample variance s^2, etc. This principle of estimating a population moment with the corresponding sample moment can be generalized to cover more complicated cases, and is called *Method of Moments Estimation* (MME).

The principle of MLE that we gave in this chapter was more sophisticated. We found that MLE yields the same estimator as MME in many important cases (for example, the MLE of π is P, and for a normal population the MLE of μ is $\overline{X}$). Yet in other cases (such as Problem 18-6), MLE differs from MME; in these cases, MLE is usually superior. MLE has a strong intuitive appeal: since MLE is the population value that is most likely to generate the sample values observed, it is in some sense the population value that "best matches" the observed sample.

However, the real justification for MLE is that for large samples, it has many of the desirable characteristics of estimation described in Chapter 7: under broad conditions MLE has the following asymptotic properties.

1. *Efficiency,* with smaller variance than any other estimators.

2. *Consistency;* i.e., it is asymptotically unbiased, with variance tending to zero.

3. *Normal distribution,* with easily computed mean and variance; hence it may be readily used to make inferences.

For example, we already have seen that these three properties are true for $\overline{X}$, the MLE of μ in a normal population. (Property 3 follows from (6-10); Property 2 follows from (6-5) and (6-7); Property 1 is proved in advanced texts, and has been alluded to in (7-6).)

We emphasize that these properties are *asymptotic;* that is, true for large samples (as $n \to \infty$). But for small samples, MLE can be improved upon, as we shall see in the next chapter.

PROBLEMS

18-3 Suppose that a likelihood function has the following graph. What is the MLE of μ?

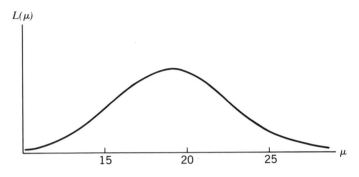

18-4 Consider the likelihood function (18-11), if $\sigma = 6$ and the sample turns out to be $X_1 = 17$, $X_2 = 22$, $X_3 = 18$.

*(a) (Computer exercise.) Compute $L(\mu)$ for $\mu = 10, 11, 12, \ldots, 30$, and so sketch the graph.

(b) Note that the graph is like Problem 18-3. What is the MLE of μ? What is $\overline{X}$? What principle does this illustrate?

*18-5 For a normal population, use calculus to derive:

(a) The MLE of μ. Is it biased?

(*Hint:* To simplify calculations, find the μ that maximizes $\ln L(\mu)$, the natural log of the likelihood function. This is legitimate, since large values of $\ln Y$ correspond to large values of Y. Use a similar technique in part (b)).

(b) The MLE of σ^2, assuming that μ is known. Is it biased?

18-6 As $N + 1$ delegates arrived at a convention, they were given successive tags numbered 0, 1, 2, 3, $\ldots$, N. In order to estimate the unknown number N, a brief walk in the corridor provided a sample of 5 tags, numbered 37, 16, 44, 43, 22.

(a) What is the mean of the population? (*Hint:* For a symmetric distribution, the mean is the midpoint.)

(b) Since $\mu = N/2$, we can get the MME estimator of N by solving $\overline{X} = N/2$, obtaining $\hat{N} = 2\overline{X}$. Is $\hat{N}$ biased?

*(c) What is the MLE of N? Is it biased?

*(d) If necessary, adjust the MME and MLE estimators to make them unbiased.

CHAPTER 19

Bayesian Inference

Life is the art of drawing sufficient conclusions from insufficient premises.

Samuel Butler

In Chapters 9 and 13, it became clear that prior belief about a parameter may play a key role in its estimation. Bayesian theory is the means of formally taking such prior information into account. Not only is it useful for its own sake, but it also sharpens our understanding of the limitations of classical statistics, especially MLE.

19-1   POSTERIOR PROBABILITIES

(a) Introduction

Again, we introduce this subject with an example that the student is urged to work out first before consulting our solution.

> *Example 19-1.*   In a certain country, suppose that it rains 40% of the days and shines 60% of the days. A barometer manufacturer, in testing his instrument, has found that although it is fairly reliable, it sometimes errs: on rainy days it

erroneously predicts "shine" 10% of the time, and on shiny days it erroneously predicts "rain" 30% of the time.

The best prediction of tomorrow's weather *before* looking at the barometer would be the *prior distribution* in Table 19-1. But *after* looking at the barometer and seeing it predict "rain," what is the *posterior distribution?* That is, with this new information in hand, can't we quote better odds on rain than Table 19-1? Intuitively, the answer should be yes; let's prove it.

TABLE **19-1**
Prior Probabilities,
$p(\theta)$

State θ	$p(\theta)$
θ_1 (Rain)	.40
θ_2 (Shine)	.60

Solution. We first set out the reliability of the barometer in Table 19-2.

TABLE **19-2**
Conditional Probabilities, $p(X/\theta)$

Prediction X \ State θ	X_1 ("Rain")	X_2 ("Shine")	Σ
θ_1 (Rain)	.90	.10	1.00
θ_2 (Shine)	.30	.70	1.00

This information is combined with the prior probabilities in Table 19-1 to define the sample space that is shown as the entire rectangle in Figure 19-1. If you like, you can think of this as an abstract representation of the millions of days that could occur. Now consider the proportion of those days in which the state of nature θ is rain and also in which the prediction X is "rain." 40% of the time, the state of nature is rain (Table 19-1), and when this occurs there is a 90% probability that the prediction will be "rain" (Table 19-2); accordingly, this combination occurs 90% of 40% = 36% of the time. This probability is shown in the upper-left corner of Figure 19-1. It may be written more formally on the basis of (3-27):

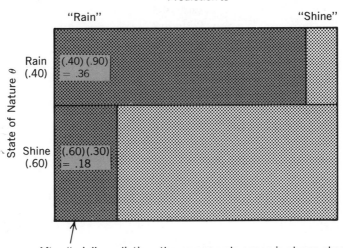

After "rain" prediction, the new sample space is shown shaded

FIGURE 19–1 How posterior probabilities are determined

$$p(\theta_1, X_1) = p(\theta_1)\, p(X_1/\theta_1) \qquad (19\text{-}1)$$

$$= (.4)(.9) = .36 \qquad (19\text{-}2)$$

Similarly, the probability of the state shine and the prediction "rain" is:

$$p(\theta_2, X_1) = p(\theta_2)\, p(X_1/\theta_2) \qquad (19\text{-}3)$$

$$= (.6)(.3) = .18 \qquad (19\text{-}4)$$

Of course, after "rain" has been predicted, the whole sample space in Figure 19-1 is no longer relevant; rather, only that shaded part covering the days in which there is a "rain" prediction is relevant. In that smaller sample space, we see that rain is twice as probable as shine (.36 versus .18). This produces the posterior distribution in Table 19-3.

Table 19-3 may be deduced more formally, as follows. From (19-2) and (19-4):

$$p(\text{prediction “rain”}) = p(X_1) = .36 + .18 = .54 \qquad (19\text{-}5)$$

TABLE **19-3**
Posterior
Probabilities,
$p(\theta/X_1)$

State θ	$p(\theta/X_1)$
θ_1 (Rain)	.67
θ_2 (Shine)	.33

According to (3-26):

$$p(\theta_1/X_1) = \frac{p(\theta_1, X_1)}{p(X_1)} = \frac{.36}{.54} = .67$$

Similarly: (19-6)

$$p(\theta_2/X_1) = \frac{p(\theta_2, X_1)}{p(X_1)} = \frac{.18}{.54} = .33$$

To keep the mathematical manipulations in perspective, we repeat the physical interpretation for emphasis. Before the evidence (barometer) is seen, the prior probabilities $p(\theta)$ give the proper betting odds on the weather. But after the evidence X_1 is available, we can do better; the posterior probabilities $p(\theta/X_1)$ now give the proper betting odds.

To generalize this example, we may write (19-6) in one equation, as follows:

$$p(\theta/X_1) = \frac{p(\theta, X_1)}{p(X_1)}$$

By (19-1) or (19-3):

$$\boxed{p(\theta/X_1) = \frac{p(\theta)\, p(X_1/\theta)}{p(X_1)}}$$ (19-7)

Now in this formula, as in Table 19-3, θ varies over all possible states, whereas X_1 is the fixed observation. Since X_1 is a constant, $p(X_1)$ also is a constant, and so we may write (19-7) as:

$$\boxed{p(\theta/X_1) = c\, p(\theta)\, p(X_1/\theta)} \tag{19-8}$$

where

$$c = \frac{1}{p(X_1)}$$

is a constant (that will make the total probability sum to 1 over all θ).

There is another interesting consequence of θ varying while X_1 is fixed. As we mentioned in Section 18-1, $p(X_1/\theta)$ is then called the *likelihood* function. Thus we may summarize (19-8) verbally, as follows:

$$\boxed{\text{(posterior distribution)} \propto \text{(prior distribution) (likelihood function)}}$$
$$\tag{19-9}$$

where $\propto$ means "equals, except for a constant," or "is proportional to." In conclusion, (19-9) states precisely how the final (posterior) distribution is calculated by combining the prior distribution with the sample information.

PROBLEMS

19-1 Suppose that another barometer is used: on shiny days it erroneously predicts "rain" 35% of the time, but on rainy days it always predicts "rain" correctly.

(a) With the prior probabilities in Table 19-1, calculate the posterior probability of rain, once this barometer has predicted "rain." What is the posterior probability of shine?

(b) True or false? If false, correct it. Since the barometer always predicts "rain" when it does rain, a "rain" prediction means that it is dead certain that it will rain.

(c) Explain why the posterior probability of rain is now *less* than in Table 19-3, even though this new barometer is a better predictor when it rains.

19-2 Suppose that you are in charge of the nationwide leasing of a specific car model. Your service agent in a certain city has not been perfectly reliable: he has shortcut his servicing in the past about 1/10 of the time. Whenever such shortcutting occurs, the probability that an individual will cancel his lease increases from .2 to .5.

(a) If an individual has canceled his lease, what is the probability that he received shortcut servicing?

(b) Suppose that the service agent is even more unreliable, shortcutting half the time. What is your answer to (a) in this case?

19-3 A factory has three machines (θ_1, θ_2, and θ_3) making bolts. The newer the machine, the larger and more accurate it is, according to the following table:

Machine	Proportion of Total Output Produced by this Machine	Rate of Defective Bolts
θ_1 (Oldest)	10%	5%
θ_2	40%	2%
θ_3 (Newest)	50%	1%

100% ✓

Thus, for example, θ_3 produces half of the factory's output, and of all the bolts it produces, 1% are defective.

(a) Suppose that a bolt is selected at random; *before* it is examined, what is the chance that it was produced by machine θ_1? by θ_2? by θ_3?

(b) Suppose that the bolt is examined and found to be defective; *after* this examination, what is the chance that it was produced by machine θ_1? by θ_2? by θ_3?

*19-4 (a) Suppose that, in an earlier quality control study, 100 steel bars had been selected at random from a firm's output, with each subjected to strain until it broke:

breaking strength, θ	200	210	220	230	240	250
relative frequency	.10	.30	.20	.20	.10	.10

Suppose that this is the only available prior distribution of θ. Graph it.

(b) Now suppose that a strain gauge becomes available that gives a crude measurement X of any bar's breaking strength θ (without breaking it). By "crude," we mean that the measurement error ranges from -20 to $+20$, so that X has the following distribution:

X	$\theta - 20$	$\theta - 10$	θ	$\theta + 10$	$\theta + 20$
$p(X/\theta)$	.2	.2	.3	.2	.1

Suppose that the measurement X of a bar purchased at random turns out to be $X_1 = 240$. What is the (posterior) distribution of θ, now that this estimate is available? Graph the likelihood, and posterior distributions. Where does the posterior distribution lie relative to the prior distribution and the likelihood?

19-2   POSTERIOR DISTRIBUTION FOR A POPULATION PROPORTION π

Now we shall apply the general principle (19-9) to an example in which there are many, many "states of nature," rather than just two. And each state of nature (formerly called θ) is now a population proportion π.

Example 19-2. Over the past year, the Radex Corporation has achieved a notorious reputation for bad quality control, having produced shipments of radios of widely varying quality. To be specific, let π represent the proportion of defective radios in a shipment. Then the past record of π for all shipments from Radex (i.e., the prior distribution of π) is given in Table 19-4(a).

TABLE **19-4**

Calculating the Posterior Distribution of a Population Proportion π

(a) Given Prior Distribution for π			(b) Calculations to Obtain Posterior Distribution		
(1) Proportion of Defectives π	(2) Number of Shipments	(3) Relative Number of Shipments	(4) Likelihood of π (from Table III(b)) Given Sample $n = 5, S = 3$	(5) Prior Times Likelihood (3) × (4)	(6) Divide by .160, Yields Posterior
0%	2	.01	0	0	0
10%	30	.15	.01	.002	.01
20%	40	.20	.05	.010	.06
30%	42	.21	.13	.027	.17
40%	34	.17	.23	.039	.24
50%	26	.13	.31	.040	.25
60%	16	.08	.35	.028	.18
70%	8	.04	.31	.012	.08
80%	2	.01	.20	.002	.01
90%		0	.07	0	0
100%		0	0	0	0
	200	1.00 ✓		.160	1.00 ✓

Now suppose that you have just been appointed purchaser for a large department store. Your first job is to make a decision on whether to return a shipment of radios from Radex that has been lying in your warehouse for 2 weeks. Your decision, of couse, will depend upon what you guess is the proportion of defectives π.

(a) Graph the prior distribution of π.

(b) Now suppose that you examine 5 radios at random to get sample evidence on π, and that 3 of the 5 turn out to be defective. Now what is the (posterior) distribution of π? Calculate it and graph it.

(c) Suppose that your department store will regard the shipment as satisfactory only if π is less than 25%. What would you say is the probability of this:

    (i) Before the sample?

    (ii) After the sample?

Solution. (a) The distribution of π prior to the sample is taken from Table 19-4(a) and graphed in Figure 19-2.

(b) To calculate the posterior distribution in Table 19-4(b),

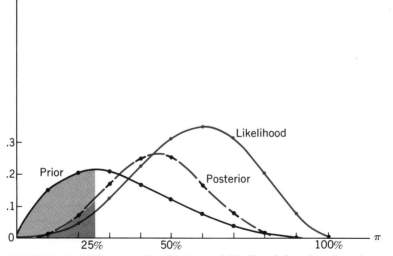

FIGURE 19–2 The prior distribution and likelihood function (based on sample information) are multiplied together to give the compromise posterior distribution of π

we shall use (19-9). We first need the likelihood function, i.e., the likelihood of getting the 3 defectives that we observed in our sample of 5. This, of course, is given by the binomial formula, for a fixed $S = 3$, and various values of π. That is, we read[1] Table III(b) horizontally along the row where $S = 3$, and record it in column (4) of Table 19-4(b).

Now, following (19-9), we multiply this likelihood function by the prior distribution. Finally, to "norm" the probabilities so they sum to 1.00, we must divide through by the constant[2] .160. This gives us the posterior distribution in the last column of Table 19-4(b), which we also graph in Figure 19-2.

(c) (i) From Table 19-4(a), the prior probability of less than 25% defectives is:

$$\Pr(\pi < 25\%) = .01 + .15 + .20 = .36$$

(ii) From the last column of Table 19-4(b), the posterior probability is:

$$\Pr(\pi < 25\%/S = 3) = 0 + .01 + .06 = .07$$

Thus the sample, with its large proportion of defectives, has lowered the probability of a satisfactory lot from .36 to .07.

---

[1]Alternatively, we could calculate the likelihood the hard way from the binomial formula:

$$p(S/\pi) = \binom{n}{S} \pi^{S}(1 - \pi)^{n-S} \tag{19-10}$$

where $S = 3$, $n = 5$, and π varies. We rename it $L(\pi)$ for simplicity:

$$L(\pi) = \binom{5}{3}\pi^{3}(1 - \pi)^{2} \tag{19-11}$$

Note that this is the same likelihood function that appeared in (18-2).

[2]Just as we divided both probabilities in (19-6) by .54.

This example displays several features that generally will be found to be true:

1. The posterior distribution is a compromise, peaking between the prior distribution and the likelihood function.

2. If we had multiplied either the prior or the likelihood by some convenient constant, it merely would have changed the second-last column of Table 19-4 by the same constant. But the adjusted or normed values in the last column[3] would be exactly as before. Accordingly:

> multiplying the prior or likelihood function by a convenient constant does not affect the posterior   (19-12)

3. The problem as stated had discrete values for $\pi (\pi = 0, .1, .2, \ldots)$. However, this was just a convenient way of tabulating a variable π that really is continuous. So we have sketched all the graphs as continuous.[4]

To generalize Example 19-2, let us consider first the possible prior distributions for π. There is a whole family of distributions, called the β distributions, that serve as a convenient approximation for almost any prior that we might encounter in practice; and the formula for the β-family is simple:

> β distribution for the prior, $p(\pi) = \pi^a(1 - \pi)^b$ (19-13)

(Note that this formula is remarkably similar to the likelihood function (19-15), below. In fact, this similarity will prove very convenient in the subsequent calculation of the posterior distribution.) In (19-13), the parameters (constants) a and b may be any numbers, although positive small integers are most common. For example, you may verify that the prior in column (3) of Table 19-4 is approximately[5] the β distribution with $a = 1$ and $b = 3$:

$$p(\pi) \simeq \pi(1 - \pi)^3 \qquad (19\text{-}14)$$

Next, to generalize the likelihood function, consider a sample of n observations that results in S "successes" and F "failures" (where

---

[3] Even if we do not norm the posterior distribution, we would get a graph that still is the right shape (it merely would be the wrong height). Therefore, we may omit the norming in our graphs. (However, we must remember to do the norming if probabilities are to be calculated as in part (c).)

[4] Is it really legitimate to shift the argument back and forth between discrete and continuous models? For our purposes, yes. The essential difference between a discrete probability function and the analogous continuous probability density is simply a constant multiplier (the constant being the cell width, as shown in (4-8)). And constant multipliers do not really matter, as stated in (19-12), above.

[5] In view of our remark (19-12) above, we have omitted in (19-13) and (19-14) an awkward multiplier (which is approximately 2).

$F = n - S$, of course). The likelihood function then is given by the general binomial formula:

$$p(S/\pi) = \binom{n}{S}\pi^S(1 - \pi)^F \qquad (19\text{-}15)$$

Since only π is varying, we may write this as:

$$\boxed{\text{likelihood, } L(\pi) \propto \pi^S(1 - \pi)^F} \qquad (19\text{-}16)$$

When the prior (19-13) is multiplied by the likelihood (19-16), we obtain the posterior:

$$p(\pi/S) \propto \pi^a(1 - \pi)^b\pi^S(1 - \pi)^F$$

$$\boxed{\text{posterior, } p(\pi/S) \propto \pi^{a+S}(1 - \pi)^{b+F}} \qquad (19\text{-}17)$$

Thus we see the real advantage of using a prior model of the β-function form: it has made the computation of the posterior distribution very simple.[6] In fact, all three distributions—prior, likelihood, and posterior—now have the same β function form: $\pi^i(1 - \pi)^j$ for some i and j. Furthermore, many such β functions are already tabulated, by reading across Table III(b) (ignoring the constant binomial coefficient, which does not really matter). This greatly simplifies computation, as the next example illustrates.

> *Example 19-3.*   Referring back to Example 19-2, suppose that your engineering expert tells you that Radex is now using a better technique; in fact, the prior distribution has so improved that it now may be approximated with a β function with $a = 0$, and $b = 4$:
>
> $$p(\pi) \propto \pi^0(1 - \pi)^4$$
> $$\propto (1 - \pi)^4$$
>
> But suppose that the sample turns out the same way (3 defectives out of 5). Without worrying about constant multipliers, graph the prior, the likelihood, and the posterior.

---

[6] Because the β prior distribution conveniently makes the posterior distribution turn out to be of the same β form, it is known technically as a *conjugate* prior.

Solution.   As already remarked, we may ignore constants such as binomial coefficients, which do not depend on π.

First we extract each distribution from the appropriate row of Table III(b). For example, the likelihood function is $L(\pi) \propto \pi^3(1 - \pi)^2$, which we find in Table III(b)—under $n = 5$ and $S = 3$. The prior distribution is $p(\pi) \propto \pi^0(1 - \pi)^4$, which we can also find in Table III(b)—under $n = 4$ and $S = 0$. Finally, the posterior distribution is $p(\pi/S) \propto \pi^3(1 - \pi)^6$, which we can again find in Table III(b)—under $n = 9$ and $S = 3$. Thus:

π	0	.1	.2	.3	.4	.5	.6	.7	.8	.9	1.0
$L(\pi)$	0	.01	.05	.13	.23	.31	.35	.31	.20	.07	0
$p(\pi)$	1.00	.66	.41	.24	.13	.06	.03	.01	.00	.00	0
$p(\pi/S)$	0	.04	.18	.27	.25	.16	.07	.02	.00	.00	0

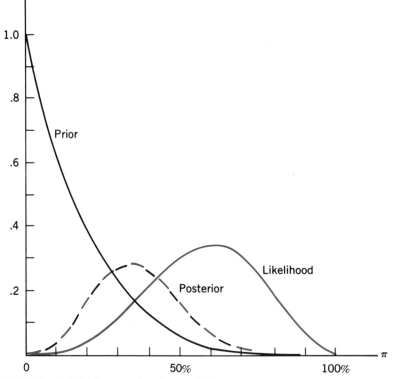

FIGURE 19–3 Different prior, hence different posterior, than in Figure 19–2

The graphs of all three distributions are given in Figure 19-3. Note again that the posterior is a compromise, peaking between the prior and the likelihood.

PROBLEMS

⇒19-5 (Continuation of Example 19-3.) In each of the following cases, graph the prior, likelihood, and posterior.

	Prior	Sample
(a)	$\pi^0(1 - \pi)^0$	3 defective, 6 good
(b)	$\pi^1(1 - \pi)^3$	2 defective, 3 good

⇒19-6 True or false? If false, correct it.

(a) In Problem 19-5(a), we started with an "informationless" prior[8] in the sense that no values of π were more probable than others; and we combined it with a relatively large sample of 9 observations. In Problem 19-5(b), on the other hand, we started with information in the prior, and a sample of 4 fewer observations. Yet we got the same posterior distribution, i.e., the same conclusion about the shipment in the warehouse. So this prior may be thought of as providing the same information as an extra 4 observations—1 defective and 3 good.

(b) We therefore may think of a prior distribution as a *quasi-sample* of extra observations. In general, the prior distribution of (19-13) may be considered a quasi-sample of a defective radios and b good radios.

(c) A person who insists on using only the sample, and ignores the prior, may be viewed just like a person who throws away part of a sample (insofar as statistical inference is concerned).

---

[8]There is some room for argument about what constitutes an "informationless" prior. For some purposes, mathematicians find it convenient to define the β function as:

$$p(\pi) = \pi^{a-1}(1 - \pi)^{b-1} \qquad \text{like (19-13)}$$

and then define an "informationless" prior as the case where $a = 0$ and $b = 0$. However, this raises issues that are too subtle for an introductory text.

19-3   POSTERIOR DISTRIBUTION FOR THE MEAN μ OF A NORMAL POPULATION

Once more, we shall apply the general principle (19-9), this time to an example where each "state of nature" is a mean μ of a normal population.

Example 19-4.   Suppose that Steelcorp sells steel beams; within each shipment, the breaking strengths of the beams are distributed normally around a mean μ, with variance $\sigma^2 = 300$. But μ changes from shipment to shipment because of poor quality control. In fact, suppose that when all shipment means μ are recorded in a bar graph, they turn out to be like Figure 19-4. That is, the distribution of μ is approximately normal, with mean $\mu_0 = 60$ and variance $\sigma_0^2 = 100$.

Now suppose that you have to make a decision on whether to return a specific shipment of beams that you bought from Steelcorp. Your decision, of course, will depend on what you guess is the mean strength μ. To guide you, you have the information contained in Figure 19-4 about what μ might be. Suppose that you also have available a sample of 12 beams from this particular shipment, and $\overline{X}$ turns out to be 70.
(a) Sketch the likelihood function.
(b) Sketch the posterior distribution obtained by multiplying the prior distribution of Figure 19-4 by the likelihood function.

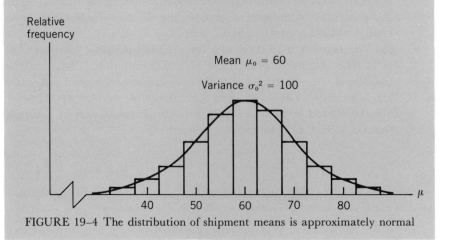

FIGURE 19-4 The distribution of shipment means is approximately normal

(c) Suppose that the shipment must be regarded as unsatisfactory if μ is less than 62.5. What would you say is the probability of this, as estimated from the sketched distributions,

(i) Before the sample was taken?

(ii) After the sample?

Solution. We shall use the techniques of Section 4-5 to sketch normal curves. And for simplicity, we shall ignore the constant multiplier in the normal distribution $(1/\sqrt{2\pi}\,\sigma)$; this will make each curve have a maximum value of 1.

(a) What is the likelihood of getting $\overline{X} = 70$? The distribution of $\overline{X}$ is normal, with a variance of $\sigma^2/n = 300/12 = 25$. Thus its equation is:

$$p(\overline{X}/\mu) \propto e^{-(1/2)(\overline{X}-\mu)^2/25} \qquad (19\text{-}18)$$

where $\overline{X} = 70$ and μ varies. We rename if $L(\mu)$ for simplicity:

$$L(\mu) \propto e^{-(1/2)(\mu-70)^2/25} \qquad (19\text{-}19)$$

We recognize this likelihood as a normal curve, centered at 70 with a standard deviation $\sqrt{25} = 5$. Its graph is sketched in Figure 19-5.

(b) For the posterior distribution, we must multiply the prior times the likelihood. We do this for several values[9] of μ, to obtain several points on the posterior distribution. From them, we sketch the graph of the posterior distribution shown in Figure 19-5.

(c) The probability that μ is below 62.5 can be estimated as the relative area under the distribution, shown shaded in Figure 19-5. This appears to be:

(i) For the prior, approximately .60.

(ii) For the posterior, approximately .10.

---

[9]For example, consider $\mu = 65$ as shown in Figure 19-5. We read off the prior probability (.9) and likelihood (.6) of $\mu = 65$ from the two appropriate graphs. Their product, $.9 \times .6 = .54$, is the desired point on the posterior distribution.

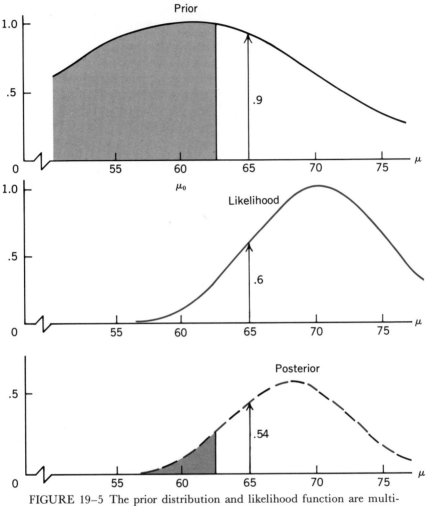

FIGURE 19–5 The prior distribution and likelihood function are multiplied to give the compromise posterior distribution of μ

In this example, the posterior seems to be distributed normally. This is no surprise, since it comes from a normal prior and normal likelihood. (This is just like our previous example: the posterior distribution for π was a β-function, just like the prior and likelihood.) Again, we note that the posterior peaks between the prior and the likelihood, and that it is closest to the curve with the least variance. That is, the more concentrated curve that accordingly provides the more precise and reliable information (in this case, the likelihood), has the greater influence in determining the posterior. These features generally are true, and may be stated more precisely, as follows.

We calculate the posterior distribution from two pieces of information:

1. The *actual* sample centered at $\overline{X}$, which is the basis of the likelihood function.

2. The prior distribution, which is equivalent to the information in *another* sample (often called a "hypothetical" or "quasi"-sample).[10]

This quasi-sample is, of course, centered on the prior mean μ_0, and is made up of n_0 hypothetical observations. The question is: how large is n_0? To answer this, note that the smaller the variance σ_0^2 of the prior distribution (relative to σ^2), the more influential the prior will be, and the larger n_0 consequently will be. So it is no surprise that the quasi-sample size n_0 is given by the formula:[11]

---

[10] Recall that this concept already has been encountered in Problem 19-6. Of course, the prior distribution is not *really* a sample. But its effect on the posterior distribution is exactly equivalent to the effect of a sample with n_0 observations; and such a "quasi"-sample provides a nice intuitive way to understand and remember the posterior formula (19-27).

[11] Here is the *rigorous* proof that definition (19-20) leads to the posterior distribution (19-27): We are given the prior distribution and likelihood function:

$$p(\mu) = K_1 e^{-(1/2\sigma_0^2)(\mu-\mu_0)^2} \tag{19-22}$$
$$\text{like (4-15)}$$

$$p(\overline{X}/\mu) = K_2 e^{-(n/2\sigma^2)(\overline{X}-\mu)^2} \tag{19-23}$$

$$\text{posterior} = (\text{prior})(\text{likelihood}) = K_1 K_2 e^{-(1/2\sigma_0^2)(\mu-\mu_0)^2 - (n/2\sigma^2)(\overline{X}-\mu)^2} \tag{19-9 repeated}$$

Let us consider in detail the exponent, which may be rearranged as a quadratic function of μ:

$$\text{exponent} = -\frac{1}{2}\left[\mu^2\left(\frac{1}{\sigma_0^2} + \frac{n}{\sigma^2}\right) - 2\mu\left(\frac{\mu_0}{\sigma_0^2} + \frac{n\overline{X}}{\sigma^2}\right) + K_3\right] \tag{19-24}$$

Now we exploit the definition of n_0; we substitute (19-21) into (19-24):

$$\text{exponent} = -\frac{1}{2}\left[\mu^2\left(\frac{n_0 + n}{\sigma^2}\right) - 2\mu\left(\frac{n_0\mu_0 + n\overline{X}}{\sigma^2}\right) + K_3\right]$$

$$= -\frac{1}{2}\left(\frac{n_0 + n}{\sigma^2}\right)\left[\mu^2 - 2\mu\left(\frac{n_0\mu_0 + n\overline{X}}{n_0 + n}\right) + K_4\right]$$

$$= -\frac{1}{2}\left(\frac{1}{\sigma^2/(n_0 + n)}\right)\left[\mu - \left(\frac{n_0\mu_0 + n\overline{X}}{n_0 + n}\right)\right]^2 + K_5$$

This exponent is recognized as that of a normal distribution, with:

(Cont'd on page 556)

$$n_0 = \frac{\sigma^2}{\sigma_0^2} \qquad (19\text{-}20)$$

That is:

$$\sigma_0^2 = \frac{\sigma^2}{n_0} \qquad (19\text{-}21)$$

When the n_0 "quasi-observations" centered at the prior mean μ_0 are combined with the n actual observations centered at $\overline{X}$, the overall mean is:

$$\text{posterior mean} = \frac{n_0\mu_0 + n\overline{X}}{n_0 + n} \qquad (19\text{-}25)$$

Since the total number of observations is $n_0 + n$, the variance of the posterior distribution is, by analogy with (6-7):

$$\text{posterior variance} = \frac{\sigma^2}{n_0 + n} \qquad (19\text{-}26)$$

In conclusion, therefore:

$$\boxed{\text{posterior distribution, } \mu \sim N\left(\frac{n_0\mu_0 + n\overline{X}}{n_0 + n}, \frac{\sigma^2}{n_0 + n}\right)} \qquad (19\text{-}27)$$

where:

$$n_0 = \frac{\sigma^2}{\sigma_0^2} \qquad (19\text{-}20) \text{ repeated}$$

and equation (19-27) is read, "μ has a normal distribution with mean $(n_0\mu_0 + n\overline{X})/(n_0 + n)$ and variance $\sigma^2/(n_0 + n)$."

$$\text{variance} = \sigma^2/(n_0 + n) \qquad \begin{array}{c}(19\text{-}26)\\ \text{proved}\end{array}$$

and:

$$\text{mean} = \left(\frac{n_0\mu_0 + n\overline{X}}{n_0 + n}\right) \qquad \begin{array}{c}(19\text{-}25)\\ \text{proved}\end{array}$$

Example 19-5. Let us rework Example 19-4 with formulas instead of graphs.

(a) Express the value of the prior distribution in terms of "quasi-observations."

(b) Calculate the posterior distribution.

(c) Calculate exactly the probability that μ is below 62.5,

   (i) Before the sample is taken.

   (ii) After the sample has been taken.

Solution. (a) From (19-20):

$$n_0 = \frac{\sigma^2}{\sigma_0^2} = \frac{300}{100} = 3$$

That is, the prior distribution is equivalent to a quasi-sample of 3 observations.

(b) From (19-27), the posterior distribution is normal, with:

$$\text{posterior mean} = \frac{n_0\mu_0 + n\overline{X}}{n_0 + n}$$

$$= \frac{3(60) + 12(70)}{3 + 12} = 68 \qquad (19\text{-}28)$$

$$\text{posterior variance} = \frac{\sigma^2}{n_0 + n} = \frac{300}{3 + 12} = 20 \qquad (19\text{-}29)$$

(c) (i) Before the sample, we use the prior distribution:

$$\Pr(\mu < 62.5) = \Pr\left(\frac{\mu - \mu_0}{\sigma_0} < \frac{62.5 - 60}{10}\right)$$

$$= \Pr(Z < .25) = 1 - .40 = .60$$

(ii) After the sample, we use the posterior distribution calculated in (b):

$$\Pr(\mu < 62.5) = \Pr\left(\frac{\mu - 68}{\sqrt{20}} < \frac{62.5 - 68}{\sqrt{20}}\right)$$

$$= \Pr(Z < -1.23) = .11$$

Remarks. Note that the precise answers calculated from formulas in (b) and (c) correspond very nicely to the rough graphical answers that we found before in Example 19-4. Since the formula answers are easier and more accurate, from now on we shall use formulas instead of sketching.

From the posterior distribution, we can calculate an interval that has a 95% chance of containing μ:

$$\mu = \text{posterior mean} \pm 1.96 \text{ posterior standard deviations} \quad (19\text{-}30)$$

Noting (19-25) and (19-26), we finally obtain:

95% Bayesian confidence interval[12]

$$\mu = \left(\frac{n_0\mu_0 + n\overline{X}}{n_0 + n}\right) \pm 1.96\sqrt{\frac{\sigma^2}{n_0 + n}} \quad (19\text{-}31)$$

---

where $n_0 = \sigma^2/\sigma_0^2$.

Compared with the classical confidence interval (8-14), the Bayesian confidence interval is centered at a compromise between μ_0 and $\overline{X}$, and is shorter because of the extra n_0 quasi-observations provided by prior information.

Example 19-6. For Example 19-5, calculate:
(a) The 95% Bayesian confidence interval.
(b) The 95% classical confidence interval.

Solution. (a) Substitute (19-28) and (19-29) into (19-30):

$$\mu = 68 \pm 1.96\sqrt{20} = 68 \pm 8.8$$

---

[12]The words "credibility interval" sometimes also are used, because it may be argued that it is an interval that can really be believed. By contrast, the 95% confidence interval is based only on the sample, and is not so credible if it conflicts with prior information.

(b) $$\mu = \overline{X} \pm 1.96\sqrt{\frac{\sigma^2}{n}} \qquad \text{(8-14) repeated}$$

$$= 70 \pm 1.96\sqrt{\frac{300}{12}} = 70 \pm 9.8$$

Note that this classical confidence interval is wider, since it takes no account of the information in the prior.

PROBLEMS

19-7 (Continuation of Example 19-4.) Suppose that the sample size is $n = 48$ instead of $n = 12$, but everything else is the same, namely: $\mu_0 = 60$, $\sigma_0^2 = 100$, $\sigma^2 = 300$, and the sample mean $\overline{X} = 70$.
(a) Calculate the posterior distribution, and graph it.
(b) Graph the prior and likelihood also. Does the posterior lie between them? Which is the posterior closer to? How much closer?
(c) On the basis of all the available evidence, find the probability that the shipment mean μ is below 62.5.
(d) Calculate the 95% Bayesian confidence interval for μ, and contrast it to the 95% classical confidence interval.

19-8 Repeat Problem 19-7, if the sample size is $n = 2$.

⇒19-9 Suppose that it is essential to estimate the length θ of a beetle that accidentally gets caught in a delicate piece of machinery. A measurement X is possible, using a crude device that is subject to considerable error: X is distributed normally about the true value θ, with a standard deviation $\sigma = 5$ mm. Suppose that a sample of 6 such observations yields an average $\overline{X} = 20$ mm.

Some information also is available from the local agricultural station. They tell us that this species of beetle has length that is distributed normally about a mean of 25 mm with a standard deviation of 2.5 mm.
(a) Find a 95% Bayesian confidence interval for the beetle's length θ.
(b) What is the probability that θ is at least 26 mm.?

19-10 The value of wheat falls with its moisture content, measured by the electrical conductivity of the wheat. Suppose that the measured moisture value is distributed normally about the true value, with an error that has a standard deviation of only one-fifth of a percentage point.

Suppose that the loads of wheat that are brought to a grain elevator during a certain week have moisture contents m varying from 14% to 16%, roughly; specifically, the values of m are distributed normally about a mean of 15%, with a standard deviation of .5%.

If a load has a measured value of 13.8%, what is the chance that its true value of m is less than 14%?

19-4   BAYESIAN REGRESSION

In the simple regression model, suppose that the slope β has a prior distribution:[13]

$$\text{prior distribution, } \beta \sim N(\beta_0, \sigma_0^2) \qquad (19\text{-}32)$$

The sample provides an estimator $\hat{\beta}$ that also is normal:[14]

$$\hat{\beta} \sim N\left(\beta, \frac{\sigma^2/\sigma_X^2}{n}\right) \qquad (19\text{-}33)$$

where σ^2 = the residual variance about the regression line, and

$$\sigma_X^2 = \frac{1}{n}\sum x^2 = \text{the variance of the } X\text{-values.}$$

When the prior information is combined with the sample information, we obtain a result[15] that is remarkably like the formula for μ:

$$\boxed{\text{posterior distribution, } \beta \sim N\left(\frac{n_0\beta_0 + n\hat{\beta}}{n_0 + n}, \frac{\sigma^2/\sigma_X^2}{n_0 + n}\right)} \qquad \begin{array}{l}(19\text{-}34)\\ \text{like } (19\text{-}27)\end{array}$$

---

[13] For simplicity, we assume that whatever prior information is available about α, it is independent of β, and so can be ignored in constructing inferences about β.

[14] Equation (19-33) comes from Figure 12-4, with the variance reexpressed according to (12-51). Incidentally, for $\hat{\beta}$ to be normal, we must make the same assumptions as in Figure 12-4: the Y observations are distributed normally, or the sample size n is large enough.

[15] To prove (19-34), we can appeal to the general normal case (19-27), with the following substitutions:

β for μ,

$\hat{\beta}$ for $\overline{X}$

σ^2/σ_X^2 for σ^2

where:

$$n_0 = \frac{\sigma^2/\sigma_X^2}{\sigma_0^2}.$$

From this, we can construct a confidence interval:

> 95% Bayesian confidence-interval
>
> $$\beta = \left(\frac{n_0\beta_0 + n\hat{\beta}}{n_0 + n}\right) \pm 1.96\sqrt{\frac{\sigma^2/\sigma_X^2}{n_0 + n}}$$

(19-35)
like (19-31)

Example 19-7. Suppose that the prior distribution for β is distributed normally about a central value of 5 with a variance of .25. A sample of 8 points provides the following statistics for estimating the slope β:

$$\sum xy = 2{,}400, \qquad \sum x^2 = 400, \qquad \text{residual } s^2 = 25.$$

Using s^2 to approximate σ^2, find:
(a) The quasi-sample size n_0 that is equivalent to the prior.
(b) The 95% Bayesian confidence interval (19-35).
(c) The 95% Bayesian confidence interval, correcting (19-35) by replacing 1.96 with $t_{.025}$ with d.f. $= n_0 + n - 2$.
(d) The 95% classical confidence interval, for comparison.

Solution. In standard notation, the prior information is:

$$\beta_0 = 5, \sigma_0^2 = .25 \qquad (19\text{-}36)$$

And from the sample we can calculate:

$$\hat{\beta} = \frac{\sum xy}{\sum x^2} = \frac{2{,}400}{400} = 6.0 \qquad (19\text{-}37)$$

$$\sigma_X^2 = \frac{1}{n}\sum x^2 = \frac{400}{8} = 50$$

(a)
$$n_0 = \frac{\sigma^2/\sigma_X^2}{\sigma_0^2} \simeq \frac{25/50}{.25} = 2$$

(b) $\beta = \left(\dfrac{n_0 \beta_0 + n\hat{\beta}}{n_0 + n}\right) \pm 1.96\sqrt{\dfrac{\sigma^2/\sigma_X^2}{n_0 + n}}$ (19-35) repeated

$\quad = \dfrac{2(5) + 8(6)}{2 + 8} \pm 1.96\sqrt{\dfrac{25/50}{2 + 8}}$

$\quad = 5.80 \pm 1.96(.224)$

$\quad = 5.80 \pm .44$

(c) From Table V, with d.f. $= n_0 + n - 2 = 8$ we find $t_{.025} = 2.306$, and hence:

$$\beta = 5.80 \pm 2.306(.224)$$

$$= 5.80 \pm .52$$

(d) Now we use d.f. $= n - 2 = 6$, and find $t_{.025} = 2.447$. Thus:

$$\beta = \hat{\beta} \pm t_{.025}\sqrt{\dfrac{s^2}{\sum x^2}} \qquad \text{(12-27) repeated}$$

$$= 6.00 \pm 2.447\sqrt{\dfrac{25}{400}}$$

$$= 6.00 \pm .61$$

Remarks. The Bayesian CI in (c) is shorter than the classical CI in (d), reflecting the value of the prior information. The Bayesian CI also is centered at a compromise between $\hat{\beta}$ and β_0—but 4 times closer to $\hat{\beta}$ because "the sample has 4 times as much information" ($n = 8$ while $n_0 = 2$).

Incidentally, replacing $z_{.025}$ with $t_{.025}$ in the Bayesian CI in part (c) is an approximation that is useful in applying (19-31) in any other case where s^2 has to be used as an estimator for σ^2.

Finally, Bayesian techniques can easily be applied to multiple regression, too. Since the calculations customarily are carried out by computer, however, we shall not attempt to do an example by hand.

PROBLEMS

19-11 A consulting economist was given a prior distribution for a regression slope β that was distributed normally about a mean of 2.0 with a standard deviation of .80. Calculate the Bayesian 95% confidence intervals for β as he successively gathers more data:

(a) $n = 5$, $\Sigma\,xy = 50$, $\Sigma\,x^2 = 25$, $s^2 = 12.8$.

(b) $n = 10$, $\Sigma\,xy = 80$, $\Sigma\,x^2 = 50$, $s^2 = 8.0$.

(c) $n = 20$, $\Sigma\,xy = 150$, $\Sigma\,x^2 = 100$, $s^2 = 9.6$.

REVIEW PROBLEMS

19-12 Suppose that your firm has just purchased a major piece of machinery, and that your engineers have judged it to be substandard. You know that the firm that produced it substitutes inferior domestic components for the standard imported components 1/4 of the time. You further know that such substitution of domestic components increases the probability that the machine will be substandard from .2 to .3. What is the probability that your machine has imported components?

19-13 A shipment of natural sponges is to be sampled to determine three characteristics:

(1) The proportion of defectives π.

(2) The mean weight μ.

(3) The slope β of the graph of absorbency Y as a linear function of weight X.

Suppose, on the basis of past shipments, that the following priors are deemed appropriate:

(1) $p(\pi) = \pi^3(1 - \pi)^{12}$.

(2) $\mu \sim N(150, 400)$.

(3) $\beta \sim N(4.0, .25)$.

A sample of $n = 20$ sponges yielded the following statistics:

(1) Sample proportion of defectives $P = 10\%$.

(2) For weight, $\overline{X} = 140$, $\Sigma(X_i - \overline{X})^2 = 22{,}800$.

(3) For regression, $\Sigma(X_i - \overline{X})^2 = 22{,}800$, $\Sigma(X_i - \overline{X})(Y_i - \overline{Y}) = 114{,}000$, and residual $s^2 = 2{,}850$.

On the basis of the prior and the data, find:

(a) The posterior distribution for π.

(b) The posterior distribution for μ, and its 95% Bayesian confidence interval.

(c) The posterior distribution for β, and its 95% Bayesian confidence interval.

CHAPTER 20

Bayesian Decision Theory

I find a coin in the street, flip it twice, and both times it comes up heads. The maximum likelihood estimate is that this coin has two heads. The Bayesian estimate is that it has one head. In this case, which is better?

20-1 OPTIMAL DECISIONS

In Example 19-4, we considered a shipment of beams. On the basis of the sample and prior information, we found a 10% probability of its being unsatisfactory ($\mu < 62.5$). Should it be returned or not? To make this decision, we also need to know the costs of a wrong decision. For example, if returning it involves costly delays, it might be wise to risk keeping the shipment, since it has only a 10% chance of being unsatisfactory. On the other hand, if an unsatisfactory shipment of steel would cause the collapse of a building or bridge, the shipment should be returned.

To probe further this question of how relative losses determine which action should be chosen, consider the following very simple example.

> *Example 20-1.* Suppose that a man runs the refreshment concession at a football stadium, selling drinks and umbrellas. He is paid a flat fee of $100 a game, from which he must deduct his losses; these, in turn, depend on how badly he matches his

merchandise with the weather. Suppose that he has just three possible options (actions, a_i):

a_1 = sell only drinks

a_2 = sell some drinks, some umbrellas

a_3 = sell only umbrellas

If he sells drinks (a_1) and it rains, his loss is $70. If it shines, however, he loses only $10.

Similarly, if he chooses action a_2 or a_3, there will also be certain losses. All this information may be assembled conveniently in the following loss table:

TABLE 20-1
Loss Function $l(\theta, a)$

State \ Action a	a_1	a_2	a_3
θ_1 (Rain)	70	40	20
θ_2 (Shine)	10	40	50

Suppose further that the probability distribution (long-run relative frequency) of the weather is as follows:

TABLE 20-2
Probability
Distribution of θ

State θ	$p(\theta)$
θ_1 (Rain)	.40
θ_2 (Shine)	.60

If he wants to minimize long-run losses, what is the best action for him to take? (As always, try to work this out before reading on; it will make the discussion much easier.)

Solution. If he chooses a_1, what would his loss be, on the average? Let us call it $L(a_1)$. To calculate it, we simply weight each possible loss with its relative frequency, and obtain

$$L(a_1) = 70(.40) + 10(.60) = 34 \qquad (20\text{-}1)$$

Similarly, we calculate his other possible average losses:

$$L(a_2) = 40(.40) + 40(.60) = 40 \qquad (20\text{-}2)$$

$$L(a_3) = 20(.40) + 50(.60) = 38 \qquad (20\text{-}3)$$

All these calculations are summarized in Table 20-3. We see that the average loss $L(a)$ is a minimum at a_1. Thus the optimum action is to sell drinks only.

TABLE **20-3**
Calculation of the
Optimal Action a, using
the prior distribution $p(\theta)$

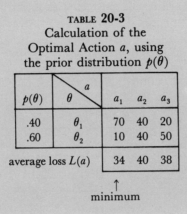

$p(\theta)$	θ	a_1	a_2	a_3
.40	θ_1	70	40	20
.60	θ_2	10	40	50
average loss $L(a)$		34	40	38

↑
minimum

Of course, this problem can be generalized to any number of states θ or actions a. For every action a, whenever a certain state θ occurs, there is a corresponding loss[1] $l(\theta, a)$. Our decision is called a *Bayesian* decision if we choose the action a that minimizes average (expected) loss, i.e.:

$$\boxed{\text{Choose } a \text{ to minimize } L(a) \overset{\Delta}{=} \sum_{\theta} l(\theta, a)\, p(\theta)} \qquad (20\text{-}4)$$

The probabilities $p(\theta)$, of course, should represent the best possible intelligence on the subject. For example, if the salesman cannot predict the weather, he will have to use the prior probabilities set out in Table 20-2; but if he can predict the weather, say by using a partially reliable barometer, then of course he should use the posterior probabilities that result from exploiting this sample information.

---

[1] The formulation of the problem in terms of losses is perfectly general, since gains may be represented simply as negative losses. Loss rather than gain is used in this chapter, since loss is more natural than gain in the context of estimation—and this is the major topic of this chapter.

Example 20-2. Suppose that the salesman in Example 20-1 has access to the partially reliable barometer in Example 19-1: if the barometer predicts "rain," then the posterior probability of rain is now as large as .67, while the posterior probability of shine is .33 (as shown in Table 19-3).

If the barometer predicts "rain," what should the salesman do?

Solution. The calculations of Table 20-3 are repeated in Table 20-4, now using the posterior probabilities.

TABLE **20-4**
Calculation of the Optimal Action
a, using the posterior
distribution, $p(\theta/X_1)$

$p(\theta/X_1)$	θ	a_1	a_2	a_3
.67	θ_1	70	40	20
.33	θ_2	10	40	50
average loss $L(a)$		50	40	30

↑
minimum

Thus, his optimal action now becomes a_3 (selling umbrellas only).

The logic of Bayesian decision making is summarized in Figure 20-1.

PROBLEMS

20-1 Suppose that the salesman running the refreshment concession finds it possible to make profits (negative losses) under some circumstances, so that the losses he faces are as follows:

State θ	a_1	a_2	a_3
θ_1 (Rain)	25	5	−20
θ_2 (Shine)	−10	5	10

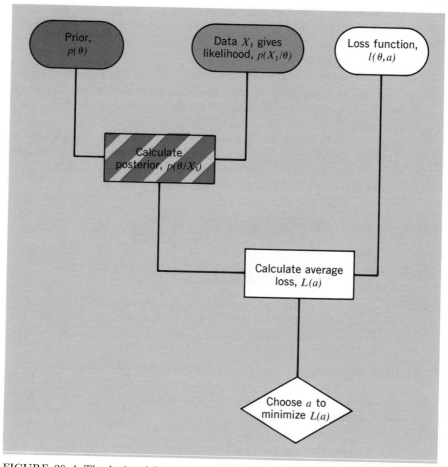

FIGURE 20–1 The logic of Bayesian decisions. The three given components—prior information, data, and the loss function—eventually lead to the optimum action a.

(a) If it rains 20% of the time and shines the remaining 80%, what is the optimal action?

(b) Suppose that the local weather prediction has the following record of accuracy: of all rainy days, 70% are predicted correctly to be "rain"; of all shiny days, 80% are predicted correctly to be "shine." What is the optimal action if the prediction is "rain"? If it is "shine"?

(c) Is this summary true or false? If false, correct it. If the salesman must decide before the weather report, he should choose the compromise

a_2. However, if he can wait for the weather report, then he should choose a_3 if the report is "rain," a_1 if the report is "shine." But this solution is obvious, without going to all the trouble of learning about Bayesian decisions.

*(d) How much is the weather report worth to the salesman? (*Hint:* What is the expected loss when "rain" is predicted? when "shine" is predicted? overall? Then compare with the loss without the prediction.)

20-2 A farmer has to decide whether to sell his corn as feed for animals (action A) or for human consumption (action H). His losses depend on its water content, (determined by the mill after the farmer's decision has been made) according to the following loss table:

State θ \ Action a	A	H
Dry	-10	20
Wet	30	10

(a) If his only additional information is that, through long past experience, his corn has been classified as dry 1/3 of the time, what should his decision be?

(b) Suppose that he has developed a rough-and-ready means of determining whether it is wet or dry—a method that is correct 3/4 of the time, regardless of the state of nature. If this indicates that his corn is "dry," what should his decision be?

⇒20-3 (a) Suppose that the steel bars in a large shipment have the following distribution of breaking strengths:

breaking strength θ	200	210	220	230	240	250
relative frequency	.10	.30	.20	.20	.10	.10

You have just purchased a bar at random, and want an estimate a of its breaking strength θ. You consult three statisticians, and receive three proposed estimates:

a_1 = the mode of the distribution

a_2 = the median

a_3 = the mean

Calculate each of these estimates.

(b) Suppose that the loss due to estimation error is proportional[2] to that error:

$$l(\theta, a) = |a - \theta| \qquad (20\text{-}5)$$

For the 6 possible values of θ and the 3 proposed values of a, tabulate $l(\theta, a)$. Then decide which estimate a is optimal (gives the minimum expected loss).

(c) Repeat (b) if the loss is heavier, now being proportional to the *square* of the error:

$$l(\theta, a) = |a - \theta|^2$$

(d) Repeat (b) if the loss is lighter, now being just the constant value 1, no matter how large the error—and being zero, of course, if there is no error at all.

⇒20-4 (Generalization of Problem 20-3.)

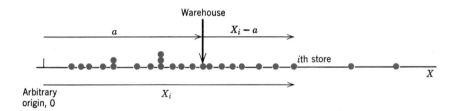

A warehouse is to be built to service many stores, all of the same size, and strung out along a single main road. In order to minimize total transportation cost, *guess* where the warehouse should be located (at the mean, median, mode, or midrange) in each of the following cases:

(a) If the transportation cost is zero for the stores right at the warehouse, and if it is a constant value (irrespective of distance) for the other stores.

(b) If the transportation cost for each store is strictly proportional to its distance from the warehouse. Thus, we wish to minimize:

$$\sum |X_i - a|$$

---

[2] In (20-5), we ought to write $l(\theta, a) = k|a - \theta|$. But for simplicity, we set $k = 1$. This does not in any way affect our decision.

(c) If there are not only transportation costs but also inventory costs that increase sharply with distance from the warehouse. Specifically, suppose that the cost is proportional to the square of the distance. Thus, we wish to minimize:

$$\sum (X_i - a)^2$$

(d) If the only cost involved is the time delay in reaching the farthest store. (This might occur, for example, if all stores had to receive a display-room sample of the new model before a crucial advertising campaign could start.) Thus we wish to minimize[3] the maximum distance.

20-5 Suppose that there were thousands of stores in Problem 20-4, and that the diagram showing their distribution now becomes:

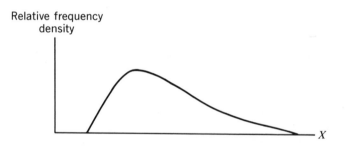

On this graph, roughly indicate the 4 solutions that you came up with for Problem 20-4, (a) through (d).

*20-6 Prove your answer in Problem 20-4.

---

[3]This criterion of putting all the emphasis on the maximum distance can be reexpressed: we wish to minimize:

$$\lim_{p \to \infty} \sum |X_i - a|^p$$

Similarly, the criterion in (a) could be reexpressed: we wish to minimize:

$$\lim_{p \to 0} \sum |X_i - a|^p$$

Thus, the criteria in (a) through (d) may be thought of as special cases of the general criterion:

$$\text{minimize} \sum |X_i - a|^p$$

for $p = 0, 1, 2, \infty$, respectively.

Note that these values of p are in the same order as the corresponding solutions in Problem 20-5 below (where mode < median < mean < midrange).

*20-2   POINT ESTIMATION AS A DECISION

In Problem 20-4, we saw an example in which θ had a distribution of possible values and we were required to make a "best guess" of what a should be. This optimal estimator turned out to be some form of central value (mean, median, or mode), depending on the loss function used. The principle that was illustrated in this example generally is true for any distribution of θ, as proved in Problem 20-6 and summarized in Table 20-5. Of course, the distribution of θ may be either a prior distribution, or a posterior distribution if data are available.

TABLE **20-5**

If θ has a Distribution of Possible Values, the Optimal
Estimator of θ depends on the Loss Function

Possible Loss Function $l(\theta, a)$	Corresponding Optimal Estimator a	
(a) $\begin{cases} 0 \text{ if } a = \theta \text{ exactly,} \\ 1 \text{ otherwise} \end{cases}$	Mode	(20-6)
(b) $\lvert a - \theta \rvert$	Median	(20-7)
(c) $(a - \theta)^2$ (quadratic)	Mean	(20-8)

Of the three loss functions suggested in Table 20-5, the one that usually is used is the *quadratic* loss function,

$$l(\theta, a) = (a - \theta)^2 \tag{20-9}$$

This is graphed in Figure 20-2; unless we specify otherwise, we will assume hereafter that every loss function is of this form.[4] If the *posterior*

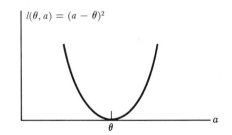

FIGURE 20-2 The quadratic loss function, for a given θ

---

[4]The quadratic loss function is the most tractable because it is easily differentiated. This same issue appeared in our discussion of the MAD estimator in Section 2-3. Note that the average quadratic loss is very similar to the mean squared error criterion in (7-7).

distribution is available, then the optimal estimator is the *posterior* mean, which we shall also call the *Bayesian estimator.* Now we shall derive the Bayesian estimator for the familiar case of a proportion π.

(a) Point Estimate of π

We already have found that the posterior distribution for a proportion π is:

$$\text{posterior, } p(\pi/S) \propto \pi^{a+S}(1 - \pi)^{b+F} \qquad \text{(19-17) repeated}$$

As we stated in Problem 19-6, we may regard the prior parameters a and b as a "quasi-sample" of a successes and b failures. Thus, the following definition makes sense:

equivalent number of successes

$$= \text{number of quasi successes (in the prior)}$$
$$+ \text{ actual successes (in the sample)}$$

that is:

$$S' = a + S \qquad (20\text{-}10)$$

Similarly, we could define:

$$F' = b + F \qquad (20\text{-}11)$$

Finally, the equivalent sample size then would be:

$$n' = S' + F' = a + S + b + F \qquad (20\text{-}12)$$

In this notation, the mean of the posterior distribution is proved in Appendix 20-A to be $(S' + 1)/(n' + 2)$. This is the Bayesian estimator of π (for the quadratic loss function)[5]; we shall denote it by $\tilde{\pi}$, to distinguish it from the MLE estimator $\hat{\pi}$. In conclusion, therefore,[6]

$$\boxed{\text{Bayesian estimator } \tilde{\pi} = \frac{S' + 1}{n' + 2}} \qquad (20\text{-}13)$$

---

[5] If we desired, we could work out the optimal estimator of π for some other loss function; for example, for the 0–1 loss function. This leads to the posterior mode, which is:

(cont'd)

Example 20-3. Calculate the Bayesian estimator $\tilde{\pi}$ for Example 19-3.

Solution. We are given $a = 0$, $S = 3$, $b = 4$, $F = 2$, so that:

$$\tilde{\pi} = \frac{S' + 1}{n' + 2} = \frac{(0 + 3) + 1}{(0 + 3 + 4 + 2) + 2}$$

$$= \frac{4}{11} = 36\% \tag{20-14}$$

(b) Point Estimate of the Mean μ of a Normal Distribution

The Bayesian estimator $\tilde{\mu}$ (for the quadratic loss function) is the posterior mean that we have already discussed thoroughly.[7] We repeat it here for completeness:

$$\boxed{\text{Bayesian estimator } \tilde{\mu} = \frac{n_0 \mu_0 + n\overline{X}}{n_0 + n}} \qquad \begin{array}{c} (20\text{-}15) \\ \text{like } (19\text{-}25) \end{array}$$

PROBLEMS

20-7 Assuming a quadratic loss function, and assuming that the prior is a β distribution $\pi^a(1 - \pi)^b$, calculate the optimal estimator $\tilde{\pi}$ in each of the following cases. Also calculate the MLE estimator $\hat{\pi}$ for comparison.

$$\tilde{\tilde{\pi}} = \frac{S'}{n'}$$

This result is just like the MLE estimator $\hat{\pi} = S/n$; and the proof also is just like the MLE proof in (18-6).

In Example 20-3, this estimator would be $\tilde{\tilde{\pi}} = S'/n' = 3/9 = 33\%$. We note that this is where the posterior distribution peaks in Figure 19-3—just as the MLE estimator $\hat{\pi} = 60\%$ is where the likelihood function peaked. (By contrast, the Bayesian estimator $\tilde{\pi} = 4/11 = 36\%$ is at the center of gravity of the posterior distribution.)

[6]If it helps, we may intuitively regard this estimator $\tilde{\pi}$ in (20-13) as being obtained by "throwing into the pot" yet one more success and one more failure, thus obtaining $S' + 1$ successes in $n' + 2$ trials, and then forming the usual ratio $(S' + 1)/(n' + 2)$. (One of the affects of adding that extra success and extra failure is to keep the estimator from ever getting too extreme—either as low as 0 or as high as 1—even if the prior has a or b as low as 0.)

[7]If we were to use the 0–1 loss function or the absolute value loss function $|a - \theta|$, we would obtain exactly the same estimator. This is because the posterior normal distribution is symmetric, and hence its mode and median coincide exactly with its mean.

		Given		Find	
	Prior Parameters		*Sample*		
			Successes	*Failures*	
	a	*b*	*S*	*F*	$\tilde{\pi}$	$\hat{\pi}$
(a)	0	0	1	0		
	0	0	0	1		
	0	0	5	0		
	0	0	0	5		
	0	0	2	8		
(b)	10	10	2	8		
	10	10	20	80		
	10	10	200	800		
(c)	−1	−1	2	8		
	−1	−1	1	0		

⇒20-8 Based on your solutions to Problem 20-7, answer true or false; if false, correct it.

(a) Consider the case of the prior in which $a = b = 0$. Because of the 1 and 2 appearing in the formula (20-13), $\tilde{\pi}$ is always "hedged" relative to $\hat{\pi}$. That is, $\tilde{\pi}$ is always closer than $\hat{\pi}$ to the compromise value of 50%. In particular, $\tilde{\pi}$ can never be too extreme—neither as low as 0 nor as high as 1.

(b) As the sample size increases, the Bayesian estimator $\tilde{\pi}$ draws further and further away from the MLE estimator $\hat{\pi}$. This is because the likelihood function overwhelms the prior distribution.

(c) For the peculiar prior,

$$p(\pi) = \frac{1}{\pi(1 - \pi)} \tag{20-16}$$

the Bayesian estimator coincides with the MLE.

20-9 Answer true or false; if false, correct it.

(a) A maximum likelihood estimator (MLE) is obtained by ignoring the prior, and just finding where the likelihood function peaks.

(b) If the sample is very large, however, MLE will be about the same as the Bayesian estimator.

20-10 Assume that the loss function is quadratic, that the prior distribution of μ is normal, and that the observations X_i are drawn from a normal population with mean μ and variance 150. Then, in each case, calculate the optimal estimator $\bar{\mu}$. Also calculate the MLE estimator $\hat{\mu}$ for comparison.

	Given				Find	
	Prior Parameters		Sample Data			
	μ_0	σ_0^2	X	n	$\bar{\mu}$	$\hat{\mu}$
(a)	100	25	200	6		
	100	250	200	6		
	100	2500	200	6		
(b)	100	25	200	6		
	100	25	200	60		
	100	25	200	600		

20-11 Based on your answers in Problem 20-10, answer true or false; if false, correct it.

As the prior information becomes more and more vague (as σ_0^2 decreases), the Bayesian estimator $\bar{\mu}$ gets closer and closer to the MLE estimator $\hat{\mu}$. This also occurs as the sample size gets bigger and bigger. In both cases, it is because the likelihood function overwhelms the prior distribution.

20-12 Suppose that the steel bars in a large shipment have breaking strengths that are distributed normally about a mean of 220 with a standard deviation of 15. A bar is purchased at random from this shipment, and crudely measured with strain gauges to estimate its breaking strength θ. Five such independent measurements turned out to be 230, 245, 235, 255, 235. Past experience shows that such measurements have a standard deviation of 10.
(a) What would you estimate the breaking strength to be?
(b) Calculate an interval that has a 95% chance of containing the true breaking strength.
(c) What assumptions did you make in (a) and (b)?

20-13 Over the past 20 quarters, the enrollment of students in Economics 100 has been fluctuating in a haphazard fashion:

enrollment	50	55	60	65	70
relative frequency	15%	15%	40%	20%	10%

Professor Smith, who must order texts for the next quarter, faces a dilemma: if he orders too many texts, the excess must be returned at a cost of $2 each for handling and postage; but if he orders too few, each student who is lacking a text must buy a temporary substitute costing $8 each.

How many texts should Smith order? (*Hint:* Consider the possible "estimators" or "orders" from 50 to 70. For each, tabulate the loss function and hence the average loss.)

20-14 Consider the following generalization of Problem 20-13, concerning an estimate a of a random variable θ.

Suppose that the loss function $l(a, \theta)$ is t times larger on the low side of θ as on the high side; that is:

$$l(\theta, a) = t\,|a - \theta| \quad \text{if } a < \theta$$
$$= |a - \theta| \quad \text{if } a > \theta$$

Then the optimal estimator a should tend to be high; specifically, it ought to be so high that t times as much of the distribution is below it as is above it.[8]
(a) Is this statement true for the special case where $t = 1$, whatever the distribution of θ?
(b) Would you guess it is true in general for any t and any distribution of θ?

20-15 In Problem 20-13, suppose instead that the enrollments over the past had been distributed normally about a mean of 60 with a standard deviation of 10. Now how many texts should he order? (Hint: Problem 20-14)

20-16 In Problem 20-10(b), compare the Bayesian and Classical 95% confidence intervals.

*20-17 The posterior distribution of π in (19-17) may be proved to have variance equal to:

---

[8]Technically, this is called the $[t/(t + 1) \times 100]$ percentile. In Problem 20-13, for example, where $t = 4$, this would be the 80th percentile. To get at least 80% of the distribution below (while 20% is above) requires going as high as 65 texts. (Incidentally, in a case like this in which the distribution is lumpy instead of continuous, we may imagine that each lump is slightly spread out so that we can divide it.)

$$\text{posterior variance} = \frac{\tilde{\pi}(1 - \tilde{\pi})}{n' + 3} \qquad (20\text{-}17)$$
$$\text{like } (6\text{-}19)$$

Furthermore, the posterior distribution is approximately normal, for reasonably large values of S' and F' (greater than 10, say). From this information, calculate the 95% Bayesian intervals for π in Problem 20-7(b). Also calculate the 95% classical confidence intervals for comparison.

*20-3   HYPOTHESIS TESTING AS A DECISION

(a) General

In the last section, we saw that in treating estimation as a form of decision, we were able to derive optimum estimators for many common situations. In this section, we likewise shall treat hypothesis testing as a form of decision, and so derive optimum tests that are free of the arbitrary features of classical hypothesis testing as it was introduced in Section 9-3.

Since hypothesis testing is just a form of decision, with its own special notation and calculations, we shall begin with an example that is remarkably similar in form to Example 20-1, our first illustration of decision making.

Example 20-4. Suppose that there are 2 species of beetle that might infest a young forest. The relatively harmless species, S_0, would do only $10,000 damage; the more harmful species, S_1, would do $100,000 damage. A perfectly harmless and effective insecticide spray is available for $30,000.

Now suppose that an infestation of beetles is sighted, but that unfortunately there is no information about which species it may be. Should the forest ranger decide to spray or not, if the only further information is that past records show:

(a) Species S_0 is three times as frequent as S_1?
(b) Species S_0 is nine times as frequent as S_1?

Solution. (a) It is convenient to summarize the given information and the solution in Table 20-6. The minimum average loss of $30,000 occurs for the action a_1. Thus the optimum action is a_1(spray).

TABLE **20-6**
Prior Probabilities, the Loss Function
(in thousands of dollars), and
Calculated Average Loss $L(a)$

$p(\theta)$	State θ / Action a	a_0 (Don't Spray)	a_1 (Spray)
.75	S_0 (Harmless)	10	30
.25	S_1 (Harmful)	100	30
Average Loss $L(a)$		32.5	30

$\uparrow$
minimum

(b) For the prior probabilities of .9 and .1, a similar calculation in the bottom row would give average losses of \$19,000 for a_0 and \$30,000 for a_1. Thus the optimum action in this case is a_0 (don't spray). It is now so unlikely that the beetle is harmful that it is worth taking the risk.

Remarks. The decision would have been trivial if it were possible to identify the species for sure: for example, if it were species S_0, we would merely scan the S_0 row, and come up with the smallest loss of 10 for the action a_0 (don't spray). Similarly, if it were species S_1, the optimum action would be a_1 (spray).

It is interesting to note that the optimal action remains a_0, even though species S_0 is not certain, but is only sufficiently likely (for example, 9 times as likely as S_1, as in part (b)). That is, if S_0 is sufficiently likely, we act just as if it were certain. We accept S_0 as a "working hypothesis"; even though we know it may not be true, it is better than the alternative hypothesis S_1. This exemplifies what hypothesis testing is all about—a search for the best working hypothesis, rather than for ultimate truth.

To make this example perfectly general, we need to make only a few changes, mostly notational:

1. We shall suppose that the data X has been observed so that the posterior probabilities $p(\theta/X)$ are available.

2. As well as θ_0 and θ_1, we shall call the hypothetical states H_0 and H_1 (null and alternative hypotheses), and then the possible actions are "accept H_0" and "accept H_1."

3. We shall abbreviate the loss function $l(\theta_i, a_j)$ to l_{ij}. Then Table 20-6 takes on the form[9] of Table 20-7.

TABLE 20-7
The Format for Bayesian Hypothesis Testing

$p(\theta/X)$	State θ	Action a / a_0 (Accept H_0)	a_1 (Accept H_1)
$p(\theta_0/X)$	H_0	l_{00}	l_{01}
$p(\theta_1/X)$	H_1	l_{10}	l_{11}

Now we must calculate the average losses. Recall in Example 20-4 that the first average loss was obtained by weighting each loss with its probability:

$$L(a_0) = (.75)10 + (.25)100 = 32.5 \qquad (20\text{-}18)$$

In the general case, we take the same sort of average:

$$L(a_0) = p(\theta_0/X)l_{00} + p(\theta_1/X)l_{10} \qquad (20\text{-}19)$$

The other average loss is calculated similarly:

$$L(a_1) = p(\theta_0/X)l_{01} + p(\theta_1/X)l_{11} \qquad (20\text{-}20)$$

Now we should choose a_0 iff:

$$L(a_0) < L(a_1) \qquad (20\text{-}21)$$

Substituting (20-19) and (20-20) into (20-21), and collecting like terms, we obtain the criterion: choose a_0 iff:

$$p(\theta_1/X)[l_{10} - l_{11}] < p(\theta_0/X)[l_{01} - l_{00}] \qquad (20\text{-}22)$$

The bracketed quantities are called "regrets," r_0 and r_1; i.e.:

$$r_0 \overset{\Delta}{=} l_{01} - l_{00} \qquad (20\text{-}23)$$

$$r_1 \overset{\Delta}{=} l_{10} - l_{11} \qquad (20\text{-}24)$$

---

[9]Note that Table 20-7 has the same headings as Table 9-1.

Thus, r_0 is the extent to which l_{01} exceeds l_{00}, i.e., the extra loss incurred by the wrong decision when H_0 is true. Similarly, r_1 is the extra loss (regret) incurred by the wrong decision when H_1 is true.

Returning to (20-22), we now may write it in terms of regrets:

$$p(\theta_1/X)r_1 < p(\theta_0/X)r_0 \qquad (20\text{-}25)$$

that is:

$$\frac{p(\theta_1/X)}{p(\theta_0/X)} < \frac{r_0}{r_1} \qquad (20\text{-}26)$$

The posterior probabilities in this equation now can be expressed using (19-7), and noting that $p(X)$ cancels:

$$\frac{p(\theta_1)\,p(X/\theta_1)}{p(\theta_0)\,p(X/\theta_0)} < \frac{r_0}{r_1}$$

Solve for the ratio of the likelihoods:

$$\frac{p(X/\theta_1)}{p(X/\theta_0)} < \frac{r_0\,p(\theta_0)}{r_1\,p(\theta_1)}$$

Recall that this is the criterion for action a_0 (accepting H_0). Thus we may summarize:

> Bayesian Likelihood Ratio Criterion:
> Accept H_0 iff:
> $$\frac{p(X/\theta_1)}{p(X/\theta_0)} < \frac{r_0\,p(\theta_0)}{r_1\,p(\theta_1)}$$

$$(20\text{-}27)$$

where r_i is the regret if θ_i is true, $p(\theta_i)$ is the prior distribution, and $p(X/\theta_i)$ is the likelihood function.

This criterion certainly is reasonable. If θ_1 is a sufficiently implausible explanation of the data (i.e., $p(X/\theta_1)$ is sufficiently less than $p(X/\theta_0)$), then the likelihood ratio will be small enough to satisfy this inequality. Thus H_0 will be accepted, as it should be.

To illustrate further, consider the very simple case in which the regrets r_i are equal, and the prior probabilities $p(\theta_i)$ also are equal. The right-hand side of (20-27) becomes 1; thus H_0 is accepted if the likelihood of θ_0 generating the sample ($p(X/\theta_0)$) is greater than the likelihood of θ_1 generating the sample ($p(X/\theta_1)$). Otherwise, the alternative H_1 is accepted. In simplest terms: we accept whichever hypothesis is more likely to generate the observed X, as shown in Figure 20-3(a). In Figure 20-3(b), we make the further assumption that the two likelihood functions (centered on θ_0 and θ_1, respectively) have the same symmetric and unimodal shape. Then (20-27) reduces to the very reasonable criterion:

$$\text{Accept } H_0 \text{ iff } X \text{ is observed closer to } \theta_0 \text{ than } \theta_1 \qquad (20\text{-}28)$$

(b) Using $\overline{X}$ to Test μ

The symbols X and θ in (20-27) may be interpreted very broadly, of course. For example, X may be a single observation as in Figure 20-3, or a

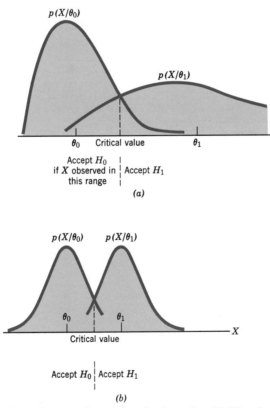

FIGURE 20–3 Hypothesis testing, using the Bayesian likelihood ratio (special case when $r_0 = r_1$ and $p(\theta_0) = p(\theta_1)$). (a) For any $p(X/\theta_i)$. (b) For $p(X/\theta_1)$ having the same symmetric shape as $p(X/\theta_0)$.

whole sample $X_1, X_2, \ldots, X_n$, or even a sample mean $\overline{X}$. We shall consider the case of $\overline{X}$ in some detail now, because it is very useful and amenable to algebraic simplification.

Let us assume that $\overline{X}$ is based on n observations drawn from a population with variance σ^2, and an unknown mean of either μ_0 or μ_1 (between which we must decide). Then $\overline{X}$ is approximately normal (by the Central Limit Theorem), with variance σ^2/n. So the criterion (20-27) for accepting H_0 becomes:

$$\frac{e^{-(n/2\sigma^2)(\overline{X}-\mu_1)^2}}{e^{-(n/2\sigma^2)(\overline{X}-\mu_0)^2}} < \frac{r_0\,p(\mu_0)}{r_1\,p(\mu_1)} \tag{20-29}$$

This may be reduced[10] to:

$$\boxed{\begin{array}{c} \text{Accept } H_0 \text{ iff:} \\[2mm] \overline{X} < \dfrac{\mu_1 + \mu_0}{2} + \dfrac{\sigma^2/n}{\mu_1 - \mu_0}\, ln\left[\dfrac{r_0\,p(\mu_0)}{r_1\,p(\mu_1)}\right] \end{array}} \tag{20-31}$$

----

[10] Both the numerator and denominator on the left hand side of (20-29) represent an evaluation of (4-15), with $\overline{X}$ being the normal variable with variance σ^2/n, and two possible means, μ_1 and μ_0. (Note how some of the terms of (4-15) cancel.)

Turning now to the proof of (20-31), we take the natural log of (20-29):

$$-\frac{n}{2\sigma^2}(\overline{X} - \mu_1)^2 + \frac{n}{2\sigma^2}(\overline{X} - \mu_0)^2 < K$$

where:

$$K = ln\left[\frac{r_0\,p(\mu_0)}{r_1\,p(\mu_1)}\right] \tag{20-30}$$

Rearrange:

$$\frac{n}{2\sigma^2}(2\,\mu_1\overline{X} - 2\,\mu_0\overline{X} - \mu_1^2 + \mu_0^2) < K$$

$$2\overline{X}(\mu_1 - \mu_0) - (\mu_1^2 - \mu_0^2) < \frac{2\sigma^2}{n}K$$

Solve for $\overline{X}$:

$$\overline{X} < \frac{(\mu_1^2 - \mu_0^2)}{2(\mu_1 - \mu_0)} + \frac{\sigma^2/n}{(\mu_1 - \mu_0)}K$$

By cancelling $\mu_1 - \mu_0$ in the middle term, and substituting for K in the last term, (20-31) follows.

where *ln* is the natural log to the base $e = 2.718$. (The common logs of Table I can be converted to natural logs by multiplying by 2.30.)

In the simplest case where the regrets r_i are equal, and the prior probabilities $p(\mu_i)$ also are equal, then the log in (20-31) is zero, and we have the simple result:

$$\overline{X} < \frac{\mu_1 + \mu_0}{2}$$

Since $(\mu_1 + \mu_0)/2$ is the half-way point between μ_0 and μ_1, this is similar to the criterion for accepting H_0 that we already found in (20-28); i.e., accept H_0 iff $\overline{X}$ is observed closer to μ_0 than μ_1.

Example 20-5. To continue Example 20-4(a), suppose that the two species of beetle are somewhat distinguishable by their length: beetle length X is distributed normally with a standard deviation $\sigma = 6$ mm., a mean $\mu_0 = 28$ mm. for species S_0 and a mean $\mu_1 = 30$ mm. for species S_1.

If a sample of $n = 4$ beetles is taken, how small must $\overline{X}$ be in order to accept H_0? In particular, if $\overline{X}$ turns out to be 28.2 mm., should we choose H_0 or H_1?

Solution. We first calculate the regrets from (20-23), (20-24) and Table 20-6:

$$r_0 = 30 - 10 = 20$$

$$r_1 = 100 - 30 = 70$$

In Table 20-6, we also find the prior probabilities, $p(\mu_0) = .75$ and $p(\mu_1) = .25$. When all these values are substituted into (20-31), we obtain:

$$\overline{X} < \frac{30 + 28}{2} + \frac{6^2/4}{30 - 28} \; ln \left[\frac{20}{70} \frac{.75}{.25} \right]$$

$$\overline{X} < 29 + 4.5 \, ln[.86] = 29 + 4.5(-.151)$$

$$\overline{X} < 28.32$$

Thus, the critical point below which we accept H_0 is 28.32. In particular, for $\overline{X} = 28.2$, we should accept H_0, i.e., not spray.

Remarks. Recall that following Example 20-4, we said, "If S_0 is sufficiently likely, we act just as if it were certain." This is a nice way to interpret our answer to this example: $\overline{X} = 28.2$ is very close to $\mu_0 = 28$; so the posterior probability of S_0 is sufficiently likely for us to accept it as the working hypothesis.

(c) Conclusion

Although Bayesian methods may be more complicated than classical methods, they are often more satisfactory. This is as true for testing as it was for estimation. A Bayesian test uses all the information in a classical test, and also exploits the prior distribution $p(\theta)$ and regrets (the loss function). A classical test sets the level of significance (probability of type I error) at 5% or 1%—sometimes arbitrarily, sometimes with implicit reference to vague considerations of loss and prior belief. Bayesians would argue that these considerations should be introduced explicitly—with all the assumptions exposed, and open to criticism and improvement.

PROBLEMS

20-18 As in Example 20-4, suppose that there were a threat from either a relatively harmless (S_0) or a harmful (S_1) species of beetle. Their relative probabilities and the losses associated with various actions are as follows:

Prob	Species	*Don't Spray* (Accept H_0)	*Spray* (Accept H_1)
.4	S_0	1	3
.6	S_1	10	2

Should H_0 be accepted or not?

20-19 Continuing Problem 20-18, suppose that one beetle in the threatening swarm has been captured, and its length X measured. Also suppose that a biologist provided the following distribution of lengths for each species:

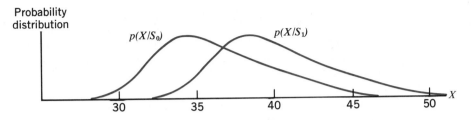

Now should H_0 be accepted or not, if X turns out to be:
(a) 30.
(b) 50.
(c) 33.
(d) 35.
(e) 38.
(f) 40.

20-20 Continuing Problem 20-18 in a different way, suppose that the distribution of beetle length $p(X/S_i)$ is normal with $\sigma = 2$, and with $\mu = 35$ for species S_0, or $\mu = 40$ for species S_1.
(a) Roughly graph these distributions.
(b) From the graph and (20-27), determine roughly the critical value for X (below which we should accept H_0).
(c) Using (20-31) with $n = 1$, determine exactly the critical point in part (b).
(d) If a sample of 5 beetles is available rather than just one, what is the critical value for $\overline{X}$ (below which we should accept H_0)?

(20-21) Repeat Problems 20-18 to 20-20, assuming that the table of Problem 20-18 is replaced with:

(i)

Prob	Action / Species	Accept H_0	Accept H_1
.25	S_0	5	10
.75	S_1	20	10

(ii)

Prob	Action / Species	Accept H_0	Accept H_1
.75	S_0	0	200
.25	S_1	700	100

20-22 (Acceptance sampling.) Suppose that 60% of the shipments of steel beams from Steelcorp come from their old plant, while the remaining 40% come from their new, better plant. Each shipment of beams has breaking strengths distributed normally with a standard deviation of 20, a mean of 250 from the old plant, and a mean of 275 from the new plant.

Suppose that you must make a decision on whether or not to return a shipment of 1,000 beams that has just arrived, and Steelcorp can't (or won't) tell you which plant they came from. So you take a sample of 4 beams, and find that their mean breaking strength is 265—unfortunately, not very conclusive. You are caught in a dilemma. If you accept a shipment of poor quality beams when you should have re-turned them, you will incur an extra cost of $10,000 required for additional bracing. On the other hand, if you return a shipment of good-quality beams when you should have kept them, the extra freight and delay costs are even worse—$12,000. Should you accept the shipment or not?

*20-23 A biologist collected many moths by random sampling in a certain region, and found that there were 2 species, beneficial (S_B) and harmful (S_H). He further found that these two species were easily (but im-perfectly) distinguishable on the basis of their antenna lengths, which were distributed as follows:

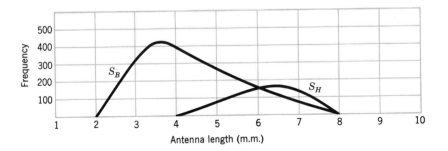

(These figures represent smoothed histograms, and you may take them as adequate representations of the populations; note from the areas enclosed by the two curves that species S_B is much more common than S_H.)

Consider the following classification rule for a randomly caught moth:
(1) Classify the moth as S_B if antenna length $< c$.
(2) Classify the moth as S_H if antenna length $\geq c$.

The biologist asks us to find the optimal value of the critical length c. Of course, we must take into account the relative regrets for the two misclassification errors, which we denote as:

$r_0 = $ regret for type I error (misclassifying an S_B moth as an S_H)

$r_1 = $ regret for type II error (misclassifying an S_H moth as an S_B)

Since much of your information is graphic, your solution will have to be graphic and approximate also. Find the optimum value for c if:

(a) $r_0 = r_1$.
(b) $r_0 = (\frac{1}{4})r_1$.
(c) $r_0 = 4r_1$.

*20-4   MLE AND BAYESIAN ESTIMATION

Under certain assumptions, a Bayesian estimator will coincide with MLE. Whenever these assumptions are questionable, MLE becomes questionable. This section, therefore, will provide a warning against using MLE too carelessly.

(a) Simplest Case: θ_0 Versus θ_1

Suppose that we wish to find the MLE of a parameter θ that is known to have only 2 possible values, θ_0 and θ_1. Then maximum likelihood estimation involves selecting θ_0 rather than θ_1 as the estimator, iff:

$$L(\theta_0) > L(\theta_1) \tag{20-32}$$

i.e., iff:

$$p(X/\theta_0) > p(X/\theta_1) \tag{20-33}$$

On the other hand, if we were to use Bayesian techniques to choose between θ_0 and θ_1, we would use the likelihood ratio criterion in (20-27). If the regrets are equal, and also the prior probabilities,[11] then this criterion reduces to:

---

[11] This set of assumptions need not be this restrictive; it is enough that:

$$\frac{r_0}{r_1} \cdot \frac{p(\theta_0)}{p(\theta_1)} = 1$$

accept θ_0 rather than θ_1 iff

$$\frac{p(X/\theta_1)}{p(X/\theta_0)} < 1$$

i.e.:

$$p(X/\theta_1) < p(X/\theta_0)$$

This is exactly the same as the MLE criterion (20-33), and therefore gives exactly the same answer. Thus a classical statistician who uses MLE is getting the same result as a Bayesian statistician who uses a constant loss function and a constant prior distribution. This is not a very flattering description of MLE as a routine procedure, since neither this prior nor this loss function may be easy to justify.

(b) Extension: Continuous θ

Suppose now that θ takes on a continuous range of values. If the statistician has absolutely no prior knowledge about θ, then in desperation he might use (just as in part (a)) the "equiprobable" prior:

$$p(\theta) = c, \text{ a constant} \tag{20-34}$$

Further suppose that, rather than using the familiar and attractive quadratic loss function, he opts for the 0–1 loss function. Consequently, according to (20-6), he will estimate θ with the mode of the posterior distribution. This posterior distribution is:

$$p(\theta/X) = \frac{p(\theta)\,p(X/\theta)}{p(X)} \tag{19-7 repeated}$$

which, according to (20-34), reduces to:

$$\left[\frac{c}{p(X)}\right] p(X/\theta) \tag{20-35}$$

To find the mode, he finds the value of θ that makes this largest. But since the bracketed term $[c/p(X)]$ does not depend on θ, he only needs to find:

the value of θ that makes $p(X/\theta)$ largest

However, this statement is recognized as just the definition of MLE. From this, we conclude that a classical statistician who uses MLE is getting the same result as a Bayesian statistician who uses a 0–1 loss function[12] and a constant prior distribution. This is not a very flattering description of MLE as a routine procedure, since neither this prior nor this loss function may be easy to justify.

(c) Special, Bizarre Case: a Proportion π

Because π can only take on values in the range 0 to 1, MLE estimation may yield even stranger results.

Suppose that, in a small poll of 12 students taken on a college campus, only 1 student is a Democrat. To estimate the population proportion π, the MLE estimator is the sample proportion:

$$P = \frac{S}{n} = \frac{1}{12} \simeq 8\% \tag{20-36}$$

We have shown in Problem 20-8(c) that a Bayesian would arrive at the same answer, if he used the quadratic loss function, and the prior distribution shown in Figure 20-4(a). This prior distribution is very peculiar indeed—it means that there is very likely a huge majority of Democrats ($\pi \simeq 1$) or Republicans ($\pi \simeq 0$). It would be more reasonable to guess "complete ignorance," as specified by the flat prior distribution in which $a = b = 0$, as shown in Figure 20-4(b). This leads to the Bayesian estimator:

$$\frac{S + 1}{n + 2} = \frac{2}{14} \simeq 14\% \tag{20-37}$$

But anyone with any knowledge of politics can do even better than this; for example, consider the prior distribution shown in Figure 20-4(c). This is one possible way of expressing the judgment that a huge majority is unlikely for either Democrats or Republicans. It leads to the Bayesian estimator:

$$\frac{S + 3}{n + 6} = \frac{4}{18} \simeq 22\% \tag{20-38}$$

---

[12]Or, if the posterior distribution is unimodal and symmetric, he would get the same result as a Bayesian using an equiprobable prior and *any* of the three loss functions in Table 20-5.

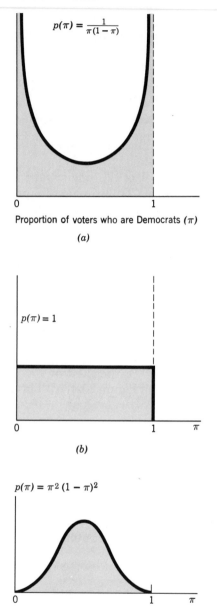

$$p(\pi) = \frac{1}{\pi(1 - \pi)}$$

Proportion of voters who are Democrats (π)

(a)

$p(\pi) = 1$

π

(b)

$p(\pi) = \pi^2 (1 - \pi)^2$

π

(c)

FIGURE 20–4 Various Bayesian estimators of party loyalty may be derived on the basis of a quadratic loss function and these priors: (a) The peculiar prior distribution that yields the MLE estimator $P = S/n$. (b) A less absurd prior distribution, yielding the estimator $(S + 1)/(n + 2)$. (c) A more plausible prior distribution, yielding the estimator $(S + 3)/(n + 6)$.

Now the same flawless logic is used in deriving all three estimators (20-36) to (20-38). They must be judged, therefore, by their differing prior distributions, with the classical MLE estimator (20-36) being the least satisfactory.

(d) Conclusions

When priors and losses are given objectively, nobody can quarrel with Bayesian techniques: they really *are* optimal in reducing average losses to a minimum. However, if priors and losses are not given objectively, and if subjective ones are used instead, Bayesian techniques often are criticized for being subjective. Yet to use classical MLE is a form of personal judgment, too. And it turns out that MLE often is a Bayesian estimator in disguise—an estimator with a very questionable prior and/or loss function.

20-5 BAYESIAN METHODS IN PRACTICE

(a) Introduction

In all the examples that we have discussed so far in this chapter, we have carefully avoided controversy. We have supposed that likelihood functions, prior distributions, and loss functions were given objectively. In practice, however, life is much more complicated. Personal judgment must be used at every stage.

For example, what is the prior distribution?[13] and the loss function? Often these must be guessed subjectively.[14] In other words, science (especially social science) cannot always be objective. So what should be done, in cases where there are no reasonably objective priors and losses?

> 1. The easiest suggestion to make is to collect more data. This, of course, is the all-purpose piece of advice that can be offered to a statistician facing *any* problem; unfortunately, it often is difficult or costly to implement. But if the sample size can be increased, then eventually the prior will be swamped by the likelihood function, and can be pretty well ignored. Or to put it another way: The prior distribution is equivalent to a "quasi-sample" of a few observations

---

[13]Sometimes a reasonable amount of empirical evidence does exist; for example, in acceptance sampling problems, there may be records of 10 or 15 previous shipments. These few values of θ constitute a prior "empirical distribution" that is a rough approximation to the prior probability distribution of θ. When this empirical distribution is used, the Bayesian procedure is called "Empirical Bayes."

[14]The form of the likelihood function (normal or otherwise?) often must also be guessed subjectively. But in these cases, MLE (which also is crucially based on the likelihood function) will be just as subjective as the Bayesian estimator.

that will become relatively insignificant as the number of *real* observations is increased. Rather than argue about a controversial prior, it may be easier instead to gather a few more real observations[15] and then use an "informationless" (usually flat) prior; or, equivalently, use a classical estimator like $\overline{X}$ that does not even mention the prior.

2. If, however, the only data you can get is a small, unreliable sample, then try to obtain as reliable a prior and loss as you can (some suggestions are given in sections (b) and (c) below). Be fair: be explicit about what evidence is objective, such as $\overline{X}$, and what is subjective, such as the prior and loss. Perhaps run an analysis with several reasonable alternative priors and losses, to show the range of reasonable answers.

Some people recommend avoiding the subjective aspects altogether, by using only data-based procedures such as MLE. But unfortunately, in small samples this does not eliminate the problem, but merely sweeps it under the rug. When we lift up the rug (as we did in Figure 20-4), we find that MLE is exactly equivalent to Bayesian procedures with priors that sometimes are more weird than personal priors ever could be.

(b) Subjective Priors

We have already discussed (in Problem 3-32) how personal probabilities may be roughly determined by getting the person to make bets. A simple extension of this idea will allow us to roughly determine a whole prior distribution.

(c) Subjective Losses

So far, we have assumed that expected monetary loss is the appropriate consideration. This may be valid enough if the stakes are small and the decision is made over and over again: whatever minimizes the expected monetary loss in each case will minimize total monetary loss in the long run.

Yet there are some decisions that are made for large stakes, in which expected monetary loss may not be the right criterion. For example, suppose that you were offered (tax-free) a choice between:

(a) \$100,000 for sure, or

(b) a 1/2 chance (lottery ticket) on a \$210,000 prize

---

[15] Physical scientists usually find it easy to gather more data by just conducting a bigger experiment. So they are often in a position where they can ignore Bayesian methods. Social scientists, on the other hand, are seldom so fortunate.

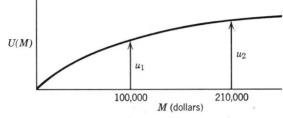

FIGURE 20–5 A subjective utility curve for money

Most people would prefer choice (a), even though its expected monetary value of \$100,000 is less than the \$105,000 expected value of choice (b). The reason is that most people value their first hundred thousand more than their second. (You easily can speculate on how you would spend the first hundred thousand. But once these purchases have been made, there would be less exciting opportunities for spending the second hundred thousand; the sports car already has been bought, and so on.) Thus we conclude that the above choice should be based not on money itself, but rather on a subjective valuation of money, or the "utility" of money.

As an illustration, Figure 20-5 shows a typical subjective utility[16] curve for money, $U(M)$. When expected utility, rather than expected money, is the basis, we reach a sensible decision: the expected utility of the gamble $(\frac{1}{2}u_2)$ is not as great as the expected utility of the \$100,000 for certain (u_1).

There are many decisions, such as those involving health or other non-monetary dimensions of human happiness, that are even more compli-cated than decisions involving large sums of money. In these cases as well, personal utility is the appropriate criterion for making a decision.

(d) Conclusions

In some problems of business or social science, estimation with large reliable samples is impossible or prohibitively expensive. Then Bayesian analysis provides a framework for making rational decisions.

On the other hand, when samples are abundant, simple classical MLE may be quite satisfactory. For example, while it would be foolish to ignore prior information in trying to guess an election on the basis of a sample of 3 voters, it would be unnecessary to bother with prior informa-tion if a sample of 3,000 voters is available.

---

[16]Utility is highly personal, and temporary. It is defined empirically for an individual by asking him a sequence of questions about which bets he prefers, in much the same way that personal probability was defined.

PROBLEMS

20-24 Before investigating the heights of a certain population of men, an anthropologist was asked what he thought the mean would be. His best guess was 70″, but he was reluctant to be so specific. Finally he agreed that his prior distribution was approximately as follows:

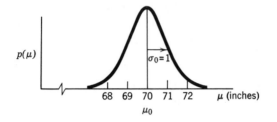

A random sample of 100 men then was taken, with a mean height of 71.3″, and variance 9 ($\sigma^2 \simeq s^2 = 9$).

(a) What is the Bayesian estimate of μ?

(b) What is the 95% Bayesian credibility interval for μ?

(c) How does (b) differ from the classical interval given in Problem 8-1(a)?

20-25 Suppose that a psychiatrist has to classify people as sick or well (hospitalized or not) on the basis of a series of tests, including interviews. The total test scores are distributed normally, with $\sigma = 8$, and mean $\theta_0 = 100$ if they are well or $\theta_1 = 120$ if they are sick. The losses (regrets) of a wrong classification are obvious: if a healthy person is hospitalized, human resources are wasted and the person himself may well be hurt by the treatment. Yet the other loss may be worse: if a sick person is not hospitalized, there may be a lot of damage. Suppose that this second loss is considered roughly five times as serious. From past records, it has been found that of the people taking the test, 60% are sick and 40% are healthy.

(a) (i) What should be the critical score above which the person is classified as sick?

   (ii) Then what is α? (Probability of type I error).

   (iii) What is β? (Probability of type II error).

(b) (i) If a classical test is used, arbitrarily setting $\alpha = 5\%$, what will be the critical score?

   (ii) Then what is β?

   *(iii) By how much has the average loss increased by using this less-than-optimal method?

(c) What would we have to assume the ratio of the two regrets to be in order to arrive at a Bayesian test having $\alpha = 5\%$? Do you think this is reasonable?

REVIEW PROBLEMS

20-26 After reading this chapter, a biologist wryly observed that a scientist was not a business man—his job was to discover the truth, not minimize costs. Do you agree?

20-27 Answer true or false; if false, correct it.
(a) Bayesian estimation combines data with a prior distribution and a loss function, whereas classical estimation uses only the data.
(b) As sample size increases, the difference between the classical estimate and the Bayesian estimate grows without limit.
(c) A Bayesian test uses the prior and loss functions to determine the critical point. The classical test instead uses an arbitrary level of α, usually $\alpha = 5\%$. The cost of this arbitrariness, however, is merely a loss of utility.

20-28 Suppose that an archaeologist has to classify skulls as "tribe A" or "tribe B," on the basis of their width. The populations of skull widths are normally distributed as follows:

Tribe	Mean	Standard Deviation
A	12 cm.	2 cm.
B	15 cm.	2 cm.

Tribe A is 5 times as numerous as tribe B. Misclassifying an A skull as a B is considered just as unfortunate as misclassifying a B skull as an A. How would you advise an archaeologist to classify each of the ten skulls whose widths are as follows:

$$11.6, \quad 15.9, \quad 16.8, \quad 13.5, \quad 13.9,$$

$$14.2, \quad 15.1, \quad 18.6, \quad 14.0, \quad 15.0$$

State clearly the basis for your decision.

20-29 When a manufacturing process is in control, the diameters of the bolts it produces are distributed normally, with mean 12.0 mm. and standard deviation .1 mm. When the process jumps out of control due to a sudden malfunctioning, the mean bolt diameter increases to 12.15 mm., while the standard deviation remains at .1 mm. Every 15 minutes a bolt is sampled, and if it is suspiciously large, the process is stopped in order to look for a possible malfunctioning. Past experience has shown that the process is in control 90% of the time; if it is stopped by mistake while it is in control, the cost in lost production time is $10. On the other hand, if the process is allowed to go on while it is out of control, the cost in wasted materials is $50.

How large should "suspiciously large" be defined in order to reduce the total costs to a minimum?

20-30 A TV manufacturer has a choice of 3 kinds of components to fill a crucial position in his TV sets: ordinary, at $2.00 each; reliable, at $2.55 each; or very reliable, at $3.00 each. The only difference the better kind makes is that it is less likely to break if the TV set is badly jarred in shipping; in fact, past shipping records show that the 3 kinds of components break 4%, 2%, and 1% of the time, respectively. If such a break costs $30 in labor and parts to repair, what kind of component should be used?

20-31 A poll is to be taken of your statistics class, to determine the proportion π who voted Democratic (or would have, if they had had the opportunity to vote) in the last presidential election.

(a) Our (the authors') personal prior distribution for π is approximately:

$$p(\pi) = \pi(1 - \pi)$$

Graph it.

(b) You (the reader) of course know your own class better than we do. How would your prior distribution differ from ours, roughly?

(c) The next time your class meets, get your instructor to take a random sample of 4 students. Suppose that S of these students turn out to vote Democratic. What will be the Bayesian estimate $\bar{\pi}$, and also the classical MLE estimate $\hat{\pi}$ (for comparison)? Now poll the whole class to find out what π actually is. Which estimator turns out to be better in this case, $\bar{\pi}$ or $\hat{\pi}$?

(d) Repeat step (c) several times, to try to average out the luck of the draw. What do you tentatively conclude?

*(e) If step (c) were repeated millions of times, tabulate what would be the relative frequency distribution of S. Then continue the table with the following columns:

		Bayesian				MLE			
S	$p(S)$	$\tilde{\pi}$	$(\tilde{\pi} - \pi)$	$(\tilde{\pi} - \pi)^2$	$(\tilde{\pi} - \pi)^2 p(S)$	$\hat{\pi}$	$(\hat{\pi} - \pi)$	$(\hat{\pi} - \pi)^2$	$(\hat{\pi} - \pi)^2 p(S)$
0									
1									
.									

*(f) What is $E(\tilde{\pi} - \pi)^2$? What is $E(\hat{\pi} - \pi)^2$? Which is smaller? Why?

APPENDIX 20-A

Proof of the Posterior Mean of π

To simplify notation, we shall denote $S' = a + S$ more briefly by s, and $F' = b + F$ more briefly by f. Using this new notation, we shall prove that for the posterior distribution:

$$p(\pi/s) \propto \pi^s(1 - \pi)^f \qquad (20\text{-}39)$$
$$\text{like } (19\text{-}17)$$

the mean is:

$$\tilde{\pi} = \frac{s + 1}{s + f + 2} \qquad (20\text{-}40)$$
$$\text{like } (20\text{-}13)$$

Proof. It may be proved[17] that:

$$\int_0^1 \pi^s(1 - \pi)^f \, d\pi = \frac{s! \, f!}{(s + f + 1)!} \qquad (20\text{-}41)$$

Thus, to make (20-39) a proper probability distribution with a total probability (area, integral) of 1, we should divide it by (20-41), obtaining:

$$p(\pi/s) = \frac{(s + f + 1)!}{s! \, f!} \pi^s(1 - \pi)^f \qquad (20\text{-}42)$$

---

[17] The proof of (20-41) is easy to carry out by induction, if s and f are positive integers. For other s and f, see a text in advanced calculus.

Then the mean is just:

$$\text{mean} = \int_0^1 \pi \, p(\pi/s) \, d\pi \qquad (20\text{-}43)$$
$$\text{like } (4\text{-}11)$$

Substitute (20-42) into (20-43), and take out the constant factorials:

$$\text{mean} = \frac{(s+f+1)!}{s! \, f!} \int_0^1 \pi \, \pi^s (1-\pi)^f \, d\pi \qquad (20\text{-}44)$$

To evaluate this, we return to (20-41), and restate it, with $s+1$ in the role of s:

$$\int_0^1 \pi^{s+1} (1-\pi)^f \, d\pi = \frac{(s+1)! \, f!}{(s+1+f+1)!} \qquad (20\text{-}45)$$

Substitute (20-45) into (20-44):

$$\text{mean} = \frac{(s+f+1)!}{s! \, f!} \, \frac{(s+1)! \, f!}{(s+f+2)!}$$

Now most of the factorials cancel out, leaving finally:

$$\text{mean} = \frac{s+1}{s+f+2} \qquad (20\text{-}40) \text{ proved}$$

PART V

Special Topics for Business and Economics

CHAPTER 21

Time Series Analysis

There's nothing constant in the universe
All ebb and flow, and every shape that's born
Bears in its womb the seeds of change.

Ovid

21-1   INTRODUCTION

There are two major categories of statistical information: cross section and time series. To illustrate, econometricians estimating how U.S. consumer expenditure is related to national income ("the consumption function") sometimes use a detailed breakdown of the consumption of individuals at various income levels at one point in time (cross-section); at other times, they examine how total consumption is related to national income over a number of time periods (time series); and sometimes they use a combination of the two. In this chapter, we shall use some familiar techniques (especially regression) and develop some new methods to analyze time series. Although our examples will use quarterly data, the techniques also are applicable to monthly data, weekly data, etc.

The main characteristic of a time series (which distinguishes it from a simple random sample) is that its observations have some form of dependence on time. The problem is that there are any number of patterns which this dependence may take. The major ones are illustrated in Figure 21-1. Panel (a) shows a time series with only a trend. Panel (b) shows a time series with only a quarterly pattern, repeated identically

603

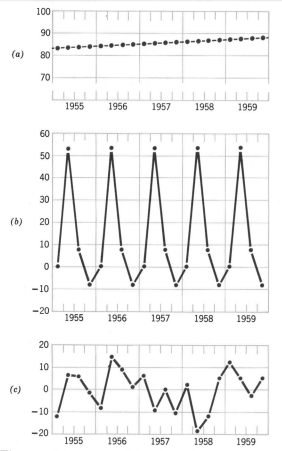

FIGURE 21–1 Three possible patterns of time dependence in a time series. (a) Trend. (b) Seasonal. (c) Random tracking.

every year; thus, for example, the fourth quarter of 1955 is the same as the fourth quarter of 1956 or any other year. Panel (c) displays a random-tracking time series of autocorrelated or serially correlated terms; that is, each value is related to the preceding values, with a random disturbance added. (There are many examples of "series that follow themselves" outside of business and the social sciences: for example, a garden hose leaves a random-tracking path of water along a wall.)

If a time series followed only one of these patterns, there would be no problem. In practice, however, it is typically a mixture of all three that is very difficult to unscramble. Consider, for example, the quarterly data on plant and equipment expenditures shown in Figure 21-2. What combination of these three patterns can be perceived? There appears to be some

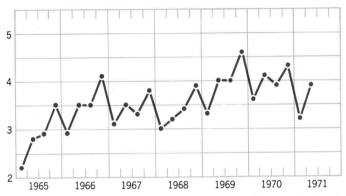

FIGURE 21–2 U.S. new plant and equipment expenditures, in durable manufacturing (in billions of dollars). Source: Survey of Current Business; U.S. Department of Commerce.

trend, some seasonal influence, and some random element. But how much of each is a mystery.

Time series analysis may be viewed as simply the attempt to break a time series up into these various components. We therefore shall consider each of the patterns in Figure 21-1 in turn.[1]

21-2   TREND

Trend is often the most important element in a time series. It may be linear, as shown in Figure 21-1(a), displaying a constant increase in each time period, or it may be exponential (a geometric series), displaying a constant *percentage* increase in each time period. Examples might be population growth in a developing country, or the growth of a trust fund over a long period. Then the logarithms of the observations will display a linear trend. The trend also may be a polynomial, or an even more complicated function.

The simplest way to deal with trend is with regression, simultaneously making seasonal adjustment as we discuss in the section below. We assume that the trend is linear, but if it is not, it may be possible to apply the regression techniques of Chapter 15.

21-3   SEASONAL

There may be seasonal fluctuation in a time series, as in Figure 21-1 (b), for several reasons. A religious holiday, in particular Christmas, results in

---

[1]To complicate matters even more, there often is a fourth component in a time series: a disturbance that substantially shifts its values onto another level. For example, a war will shift bond sales, as shown in Figure 13-8. Such a displacement often can be treated with a dummy variable.

completely different economic and purchasing patterns. Or the seasons may affect economic activity: in the summer, agricultural production is high, while the sale of ski equipment is low.

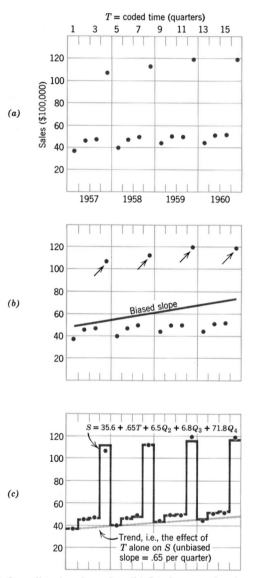

FIGURE 21–3 (a) Canadian jewelry sales. (b) Inadequate simple regression of S on T alone in which trend and seasonal influences are confounded. (c) Multiple regression of S on T and seasonal dummies, which explicitly distinguishes trend and seasonal effects.

In Figure 21-3(a) and in the first two columns of Table 21-1, we show a spectacular true-life example of seasonal fluctuation—jewelry sales. We note both the possibility of a very slight upward trend and an obvious seasonal pattern marked by the sharp rise in sales every fourth quarter because of Christmas. If we made the mistake of trying to estimate only trend, say by a simple linear regression of sales S on time T as shown in Figure 21-3(b), the result would be a substantial bias.[2] To avoid this, both trend *and* seasonal should be put into the regression model, in order to estimate their separate effects.[3]

The fourth-quarter observations may be treated in exactly the same way as the peculiar wartime observations of bond sales in Section 13-9—by the use of a dummy variable. Letting Q_4 be the fourth quarter dummy,[4] the model becomes:

$$S = \alpha + \beta_1 T + \beta_4 Q_4 + e \qquad (21\text{-}1)$$

TABLE 21-1
Canadian Department Store Jewelry
Sales and Seasonal Dummies

Time, T (Quarter Years)		Sales, S ($100,000's)	Q_4	Q_3	Q_2
1957	1	36	0	0	0
	2	44	0	0	1
	3	45	0	1	0
	4	106	1	0	0
1958	5	38	0	0	0
	6	46	0	0	1
	7	47	0	1	0
	8	112	1	0	0
1959	9	42	0	0	0
	10	49	0	0	1
	11	48	0	1	0
	12	118	1	0	0
1960	13	42	0	0	0
	14	50	0	0	1
	15	51	0	1	0
	16	118	1	0	0

Source: Statistics Canada, 63-002.

[2]The upward bias in slope largely is caused by the fact that the last observation is a high, fourth quarter one. (As an exercise, you can confirm that seasonal influences would exert a downward bias on slope if the first observation was for the 4th quarter of 1956 and the last was for the 3rd quarter of 1960.)

[3]Even with this much care, least squares regression still may involve problems, because of random-tracking residuals. This issue is discussed in detail in Section 21-8.

[4](See page 608.)

Even this model may not be adequate. If allowance also should be made for shifts in the other quarters, dummies Q_2 and Q_3 should be added. A dummy Q_1 is not needed for the first quarter, because Q_2, Q_3, and Q_4 measure the shift from a first-quarter base. (Whether or not to include the various regressors Q_4, Q_3, Q_2, can be decided on statistical grounds, by testing for statistical discernibility. It is common to include them all in such a test, and reject or accept them as a group. But such a statistical test on data as extreme as ours would be superfluous.) Our modified model is now:

$$S = \alpha + \beta_1 T + (\beta_4 Q_4 + \beta_3 Q_3 + \beta_2 Q_2) + e \qquad (21\text{-}2)$$

The least squares fit[5] is graphed in Figure 21-3(c). Notice that our seasonal adjustment is exactly the same every year; for example, each year there is the same upward shift $(\hat{\beta}_2)$ in our fit between the first and second quarters. The trend is shown as the slope of the bottom line in this diagram.

This is summarized on the left-hand (unshaded) side of Figure 21-4. Panels (a) and (b) show the fitted trend and seasonal components, in sum comprising the fitted regression in panel (d). The difference between this and the observed dots is the residual series, shown in magnified form in panel (c). Thus the total jewelry sales series has been broken down into its three components in panels (a), (b), and (c); except for the block-type artwork, this is the same sort of breakdown that we showed in Figure 21-1.

21-4   RANDOM TRACKING (AUTOCORRELATION)

We now return to Figure 21-4: with the trend and seasonal components of the time series removed by regression, this leaves the residual yet to be analyzed.

The simplest model for the random-tracking residual is to linearly

---

[4]There are three points in the analysis at which we might conclude that explicit account should be taken of seasonal swings. We may expect a strong seasonal influence from prior theoretical reasoning. Or, such an influence may be discovered after we plot the series. Finally, it may be discovered by examining residuals after fitting a regression; for example, after the simple regression in Figure 21-3(b) is fitted, we note that the observations indicated by arrows have consistently high residuals. To explain this, we look for something they have in common. Their common property is that they all occur in the fourth quarter. Hence, the fourth quarter is introduced as a dummy regressor. This technique of "squeezing the residuals till they talk" is important in every kind of regression, not just time series; used with discretion, it indicates which further regressors may be introduced in order to reduce bias and residual variance.

[5]The least squares fit to this model was calculated by a method similar to that of Table 13-2. Equation system (13-4) was extended to a system of five estimating equations for the five unknowns. However, since the calculations are carried out by computer, the details need not distract us.

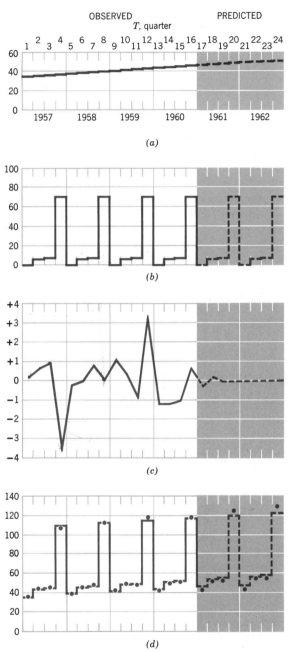

FIGURE 21–4 The three components of Canadian jewelry sales, projected with dashed lines. (a) Trend ($35.6 + .65T$). (b) Seasonal ($6.5Q_2 + 6.8Q_3 + 71.8Q_4$). (c) Serially correlated residual magnified 20 times ($e = -.21e_{t-1}$ for the projection). (d) Total (original) series as in Figure 21–3(c). Compare the dashed-line projection with the actual sales shown as dots.

relate each value e_t to its previous value e_{t-1}, with a random disturbance[6] v_t added:

$$e_t = \rho e_{t-1} + v_t \qquad (21\text{-}3)$$

with ρ representing the strength of the tracking influence.

To illustrate, let us assume that $\rho = 1$, and the random disturbance v_t is standard normal; accordingly, let us draw a sample of 20 independent values of v_t from Appendix Table II(b). Then, starting with an initial value of $e_0 = 5$, for example, we use (21-3) to generate $e_1, e_2, e_3, \ldots, e_{20}$, as shown in Figure 21-5. The positive nature of the autocorrelation is clear; e_t tends to be high whenever the previous value e_{t-1} is high.

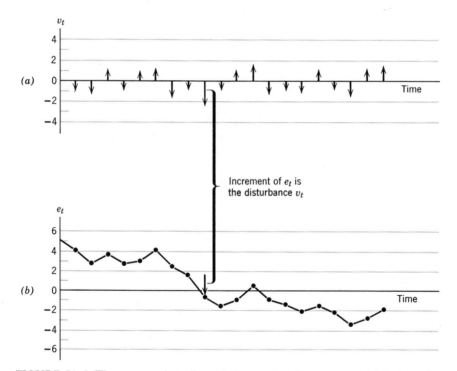

FIGURE 21–5 The construction of a serially correlated error term. (a) Independent disturbance v_t. (b) Generated error: $e_t = e_{t-1} + v_t$.

---

[6]In previous chapters, "error" and "disturbance" may have been used synonymously. However, in this chapter it is necessary to make a clear distinction between the original error e_t (which may be autocorrelated), and the further disturbance v_t (which is entirely random).

In practice, the parameter ρ typically is less than 1; moreover, it generally is not known. However, it may be estimated by applying simple regression to (21-3). As an example, the residuals of the jewelry sales data in Figure 21-4(c) were estimated[7] to follow the autoregressive scheme:

$$\hat{e}_t = -.21\hat{e}_{t-1} \qquad (21\text{-}4)$$

This negative estimate[8] of ρ shows that autocorrelation can be negative, as well as positive. In this case, values tend to change sign (a positive value is followed by a negative one), rather than track each other over time—as they did when positively correlated in Figure 21-5. (Inventory series often provide another example of negative serial correlation, if particularly large purchases for inventory result in overstocking, hence smaller purchases in the following period.)

In its more general form, (21-3) becomes:

$$\boxed{e_t = \rho_1 e_{t-1} + \rho_2 e_{t-2} + \cdots + \rho_k e_{t-k} + v_t} \qquad (21\text{-}5)$$

which sometimes is called a kth-order autoregression.

---

[7] The hats in (21-4) indicate that the residuals $\hat{e}_t$ are derived from the estimated regression line. That is, $\hat{e}_t$ are the estimated residuals, whereas e_t are the unobservable true residuals.

The estimate of ρ in (21-4) was calculated to be $-.21$ as follows: $\hat{e}_t$ is given in column (1) below; then column (2) follows directly from it; finally we regress the series in (1) on the series in (2).

Quarter T	Observed:		Calculation of $\hat{\rho}$	
	(1) $\hat{e}_t$	(2) $\hat{e}_{t-1}$	(3) $\hat{e}_t \cdot \hat{e}_{t-1}$	(4) $\hat{e}_{t-1}^2$
1	+ .2			
2	+ .7	+ .2	.14	.04
3	+1.0	+ .7	.70	.49
4	−3.6	+1.0	−3.60	1.00
5	− .1	−3.6	.	.
6	.	− .1	.	.
.	.	.	.	.
.	.	.	.	.
16	.5			

From (11-16):

$$\hat{\rho} = \frac{\sum \hat{e}_t \cdot \hat{e}_{t-1}}{\sum \hat{e}_{t-1}^2}$$

$$= -.21$$

[8] Later, in Section 21-8, we will show that (21-4) yields a biased estimate of ρ.

21-5   PROJECTION

One of the main reasons for decomposing a time series into its 3 components, as in Figure 21-4, is to predict its values into the gray future period in that diagram. A simple method suggested by that analysis is to project each of its components as in the extended dashed lines in panels (a), (b) and (c), and then sum them into the dashed-line projection[9] in panel (d).

Although the predicted values for the random tracking residual in panel (c) are included in the projection in panel (d), they quickly settle down to practically zero, since, from (21-4), each is only 1/5 the size of the previous one. This relatively small random-tracking element in the total projection makes good intuitive sense: in this time series, it is the least predictable of the three components, and hence makes almost zero projected contribution beyond the first few quarters. In fact, it sometimes is reasonable enough to set this equal to zero immediately (that is, forget about it), so that only the trend and seasonal are used. This is especially true if computational facilities are scarce, or if only long-run projections are desired.

To see how well the projection worked out, we compared it in panel (d) with the actual values that the time series took in 1961–1962, shown in dots. Although these values could not be used in the analysis (they were not yet known), they do provide a good test after the event of how well the projection turned out. Initially, the projection is not too bad, but it tends to get worse the farther it goes into the future—one more example of the dangers of extrapolation first mentioned in Section 12-7. Thus, the shorter the projection, the better; this suggests that if funds and computing facilities are available, the whole analysis should be recalculated and projections updated whenever new observations on the variables become available (i.e., delay your projections for 1962 until the 1961 figures be-

---

[9] For example, using the equations of Figure 21-4, the 17th and 18th quarters are predicted to be:

Projection of:	$T = 17$	$T = 18$
Trend $(35.6 + 65T)$	$35.6 + .65(17) = 46.7$	$35.6 + .65(18) = 47.3$
Seasonal $(6.5Q_2 + 6.8Q_3 + 71.8Q_4)$	Reference quarter $= 0$	Setting $Q_2 = 1, 6.5(1) = \ 6.5$
Residual $(-.21\hat{e}_{t-1})$	$-.21(.5) = -.1$	$-.21(-.1) = \ .02$
Total	46.6	53.82

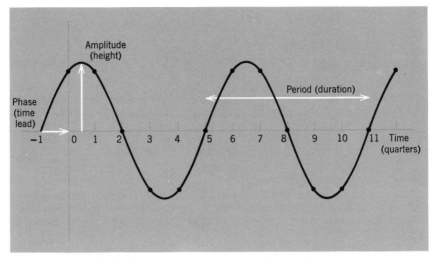

FIGURE 21–6 Defining characteristics of a cycle

come available; then each of the components in Figure 21-4 can be reestimated).

In our projections, we have limited ourselves to point estimates only, rather than to interval estimates. The reason is not only because a three-component time series is very complex, but also because the autocorrelation in the tracking residual invalidates prediction intervals like (12-43) that are based on the assumption of independent observations. This issue is discussed more fully in Section 21-8.

21-6 CYCLES

Figure 21-6 illustrates a cycle, showing its three defining characteristics— period, amplitude, and phase. We then define its frequency f as the number of cycles it completes every quarter (time period). For example, a cycle with period $\tau = 6$ (as in Figure 21-6) makes 1/6 complete cycles every quarter. In general:

$$f \stackrel{\Delta}{=} \frac{1}{\tau} \tag{21-6}$$

The decomposition of a series into cyclical components is called Fourier analysis or spectral analysis. Although Fourier analysis may be applied to any time series, it usually is restricted to residual series, after trend and seasonal components have been extracted. In other words, in

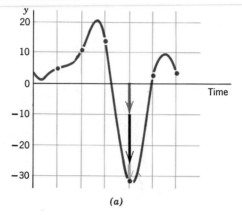

(a)

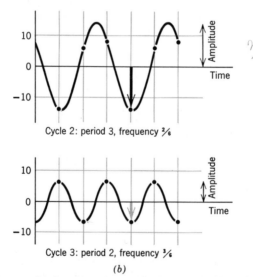

(b)

FIGURE 21–7 (a) y = the jewelry sales residual, reproduced from Figure 21–4(c) (The first six residuals, shifted slightly because a residual series requires a zero mean. Also rescaled by a factor of 10.) (b) Its decomposition into three cycles.

Figure 21-4 we split up a time series into its three components—one of which was a residual. In this section, we will be further decomposing a *residual* series into its cyclical *subcomponents.*

To keep things simple, in Figure 21-7(a) we illustrate by using only the first six of the residuals from the jewelry sales. With appropriate formulas, it is possible to calculate three cycles that, when added up, exactly generate the residual series y. At $T = 4$, for example, the three cyclical values indicated by arrows in panel (b) sum exactly[10] to the value y in panel (a). In general:

> A time series of length n may be decomposed into $\dfrac{n}{2}$ cycles.[11]   (21-7)

This can be used to decompose *any* series into a set of cycles. Hence, using this analysis to decompose a time series "into its cycles" does not establish that the series was generated by a "cyclical process." Therefore, before attaching much meaning to a specific cycle, we must not only find that its amplitude is relatively large but also establish its economic rationale. This often is difficult to do.

In many ways, the most plausible example of a cycle in business or economics is the replacement cycle. To illustrate, consider the invention of a new consumer good (e.g., TV) that initially is associated with abnormally heavy sales as the public stocks up. This is followed by more or less normal growth (as population and income grow, etc.). But when the original units begin to wear out, this normal growth usually is replaced by a time in which sales again are abnormally high, since they now include both normal growth *and* replacement of the original units. Theoretically, this cyclical process could continue, eventually fading out like an echo

---

[10]This is true not only at the integers like $T = 4$, but also at any time in between.

[11]We assume that the series has an even number of terms n, and a zero mean. In order to simplify both theory and computations, the frequencies of the cycles chosen to represent a given series y are spaced uniformly from 0 to 1/2. For example, in Figure 21-7, the frequencies are:

$$\frac{1}{6}, \frac{2}{6}, \frac{3}{6} \qquad (21\text{-}8)$$

and, in general, to decompose a time series of length n, we use frequencies:

$$f = \frac{1}{n}, \frac{2}{n}, \frac{3}{n}, \ldots, \frac{1}{2} \qquad (21\text{-}9)$$

With these frequencies prespecified, each cycle has two remaining parameters that may be fitted freely, the amplitude and phase. Since there are $n/2$ cycles, this makes n parameters altogether, which can be determined uniquely from the n values of y. This is why any cyclical decomposition (e.g., in Figure 21-7(b)) will always exactly fit the series (in Figure 21-7(a)) from which it was derived.

(since the lifetime of a product is not very clearly defined). But in practice, such echoes quickly become dominated by the appearance of new inventions and other disturbances. The same phenomena may apply in capital goods industries as well, initiated by the invention or development of a new kind of machine. (There may be an even more clearly distinct cycle in this case, if the depreciation rate allowed by the tax authorities tends to more clearly establish the lifetime of a machine.) Moreover, it is obvious that cycles are of widely varying duration (i.e., period). For example, a TV cycle will be of short duration compared to the long-term housing cycle that might be initiated by the substantial construction required following a war or a natural disaster.

Since other economic cycles are even harder to define than the replacement cycle, the exercise of decomposing economic series into cycles should be viewed with a healthy degree of skepticism; it's a bit like trying to sort out a set of echoes in a noisy sound chamber.

PROBLEMS

21-1 For the following car-sales data:

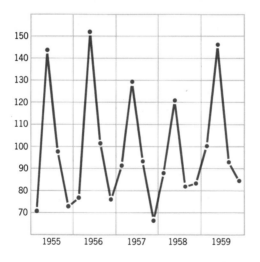

(a) Fit a linear trend by eye.

(b) Shift the linear trend so that it passes as close as possible to the first quarter (reference quarter) points. Then roughly estimate how much higher the second quarters rise, on average, above this reference line. Repeat for the third and fourth quarters.

(c) For this car sales data, the standard least squares (computerized) estimates of trend and seasonal are shown in Figure 21-1(a) and (b). Use them to judge the accuracy of your rough estimates in (a) and (b), above. (Incidentally, the residual in this series is shown in Figure 21-1(c); in other words, Figure 21-1 is the complete decomposition of the time series, above.)

21-2 (Seasonal Adjustment.)

Auto Sales, Y (from Problem 21-1)		Seasonal, S (from Figure 21-1(b))	(a) Seasonal Deviations, s	(b) Seasonally Adjusted Auto Sales, Y − s
1955	71	0		
	144	53		
	98	8		
	74	−9		
1956	77	0		
	153	53		
	102	8		
	76	−9		
1957	92	0		
	130	53		
	94	8		
	66	−9		
1958	89	0		
	122	53		
	82	8		
	84	−9		
1959	100	0		
	147	53		
	93	8		
	85	−9		

Auto sales Y and its seasonal component S are tabulated above. To fill in the missing columns:

(a) Calculate the mean of the seasonal series $\overline{S}$. Then calculate the seasonal series in deviation form, $s = S - \overline{S}$. This gives us a seasonal series that fluctuates up and down *around zero*.

(b) Subtract s from the original series of auto sales, Y. This new series estimates how auto sales would move if there were no seasonal component, and so is called the *seasonally adjusted* series.[12]

(c) Graph the seasonally adjusted series on the Figure in Problem 21-1 for comparison.

---

[12]In practice, the seasonal adjustment is *calculated* quite differently, but the *end result* (the seasonally adjusted series) is almost the same.

(d) Comparing the second quarter of 1957 with the first quarter, what was the increase in

(i) Auto sales?

(ii) Seasonally adjusted auto sales?

Which of these figures is more meaningful, in terms of indicating how well the auto industry fared in that quarter?

21-3 True or false? If false, correct it.

(a) A time series may be decomposed into three components:

(i) Trend.

(ii) Seasonal pattern.

(iii) Random-tracking residual.

(b) The fitting of trend and seasonal by eye, as in Problem 21-1, has the advantage of simplicity. However, if computer facilities are available, the least squares estimates have even more advantages, as follows:

(i) They are easy to calculate, since multiple regression is a standard library program.

(ii) They are more objective (and hence will not be biased, consciously or subconsciously, by the researcher).

(iii) They can be used for tests and confidence intervals.

(c) The tracking residual may be removed at the same time as the trend, using multiple regression with dummies. This then leaves the seasonal pattern, to be studied further with cyclical analysis.

(d) Only time series generated by some cyclical mechanism can be broken down by cyclical analysis.

21-4 Referring to Figure 21-5 and Equation (21-3) (and setting $\rho = 1$), construct a string of ten serially correlated disturbances e_t. (For v_t, use the table of standard normal numbers of Appendix Table II(b). Start with $e_0 = 0$ for simplicity.)

21-5 True or false? If false, correct it.

If the first observation in a residual time series is negative then:

(a) Positive serial correlation means that the odds favor (i.e., probability $> .50$) the next observation also being negative.

(b) Negative serial correlation means that the odds favor the next observation being positive.

(c) No serial correlation (i.e., a purely random residual) means that the next observation is equally likely to be positive or negative.

21-6 (a) In Figure 21-7, we showed the decomposition of a residual series into its cyclical components. In particular, we showed how the three components add up at $t = 4$. Show this again:

   (i) At $t = 2$.
   (ii) At $t = 3$.
   (iii) At even a fractional t, such as $t = 2.5$.

(b) Does Figure 21-7 prove that the series in panel (a) was "generated by" the three cyclical processes described in panel (b)?

*21-7   SPECTRAL ANALYSIS

Spectral analysis extends the Fourier analysis of Section 21-6 to determine which of the component cycles exerts most influence on the original series. In order to give an elementary presentation of this difficult subject, we defer many of the subtleties to the student's manual.

(a) The Variance Spectrum

First, we summarize the Fourier analysis of Figure 21-7 by recording the amplitude of each cycle. This is shown in Figure 21-8(a), and is called the amplitude spectrum.[13] The variance of each cycle[14] is:

$$V = \frac{A^2}{2} \qquad \text{except for the last cycle} \qquad (21\text{-}10)$$

$$= A^2 \qquad \text{for the last cycle} \qquad (21\text{-}11)$$

---

[13]The terminology of spectral analysis has some very interesting origins in the physical sciences. For example, the word "spectrum" is applied to the analysis of multicolored light. Each particular frequency of light (electromagnetic wave) strikes the eye as a particular color.

The word "harmonic" is applied in the theory of sound and music. When the vibration of a violin string is doubled in frequency (by halving the length of the string), the ear hears it raised an octave. Note that in the three cycles of Figure 21-7, the second is thus an "octave" higher than the first. Thus, the three cycles often are called three "harmonics," and Fourier analysis itself sometimes is called "harmonic" analysis.

[14]The variance of a cycle, as for any other sequence, can be found by squaring its deviations from its mean value (in this case, zero). Since these deviations usually are less than the amplitude, (see, for example, cycle 1 in Figure 21-7(b)), the variance will be less than A^2. In fact, it can be shown with trigonometry that:

$$V = \frac{A^2}{2} \qquad \qquad \text{(21-10) repeated}$$

There is one exception, however—the last cycle, with highest frequency, shown in our example as the last cycle in Figure 21-7(b). In this case, we see that the cycle hits its full amplitude at each integer, making:

$$V = A^2 \qquad \qquad \text{(21-11) confirmed}$$

These variances, graphed in Figure 21-8(b), are called the *variance spectrum, power spectrum,* or just plain *spectrum.*

The variance spectrum may be regarded as a decomposition of the variance of a time series y into its cyclical components—just as ANOVA was regarded as the decomposition of variation into factor components. Specifically, it has been proven that:

variance of a time series = sum of cycle variances

$$= \sum V \qquad\qquad \text{(21-12)}$$

like (10-27)

For example, from Figure 21-7(a), we may calculate the variance of the time series directly:

$$\text{variance}^{15} \text{ of } y = \frac{1}{n} \sum_1^n (y - \bar{y})^2 \qquad\qquad \text{(21-13)}$$

$$= \tfrac{1}{6}[4^2 + 10^2 + 13^2 + (-32)^2 + 2^2 + 3^2]$$

$$= 220 \qquad\qquad \text{(21-14)}$$

On the other hand, from Figure 21-8(b), we find that the frequency components approximately sum to:

$$\sum V = 81 + 99 + 40 \qquad\qquad \text{(21-15)}$$

$$= 220$$

which illustrates (21-12).

So far, our illustrations have been very short time series, for reasons of convenience. We shall see, however, that in practice, spectral analysis is worthwhile only for long series. We therefore illustrate, in Figure 21-9(a), the spectrum of a long annual series of European wheat prices.[16] This spectrum, having half as many components as the time series itself, is so detailed and irregular that it is hard to see the forest for the trees. There-

---

[15] Since the mean $\bar{y}$ is set equal to zero rather than estimated, the divisor n is used in calculating variance; otherwise, (21-13) is just like (2-6(a)), the standard calculation of variance for any set of observations.

[16] This spectrum is approximately the spectrum of wheat prices in Europe during the 17th and 18th centuries, with trend removed. Since the data are annual (not quarterly or monthly), there is no seasonal component to be removed.

The spectrum is graphed on a logarithmic scale, to keep the weak frequencies from being obscured entirely.

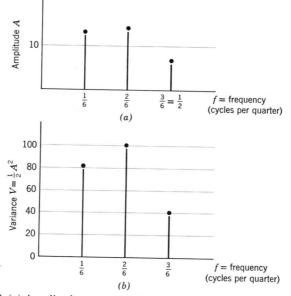

FIGURE 21–8 (a) Amplitude spectrum of the first jewelry sales residuals. (b) Variance spectrum, as given by (21–10) and (21–11).

fore, in Figures 21-9(b) and (c), we have averaged the spectrum lines into cells.

(b) Resolution versus Stability

By comparing Figures 21-9(b) and (c) it is evident that the broader we group, the less detail we can see and the less we can distinguish close frequencies; this is called loss of *resolution*. However, for this price we achieve substantial gains: note that the averaging process in panel (c) has given us a very *stable* bar graph, whereas in panel (a) we had an enormous amount of statistical fluctuation. This is why a long time-series is required: to allow the averaging process to reduce statistical fluctuation.

This stability issue may be expressed precisely. Imagine the time series extending indefinitely, without any change in the generating mechanism. Cut it into chunks of length n (where n is the size of the observed time series you have in hand), and put all these chunks into a bowl called the *population*. Every chunk has a spectrum like Figure 21-9(c) that can be calculated; the average of all these spectra is called the *population* spectrum. It is the target that we hope is reasonably well estimated by a single chunk (single time series of length n) sampled at random from the population bowl. Over any frequency band of width Δf, the sample spectrum V is an estimate of the population spectrum Φ. In fact:

$$\frac{V}{\Phi} \text{ has a } C^2 \text{ distribution}[17] \qquad (21\text{-}16)$$

$$\text{with d.f.}[18] \simeq 2n(\Delta f) \qquad (21\text{-}17)$$

where Δf represents the width of the frequency band. Thus for any frequency band we can construct a 95% confidence interval for Φ, just as we did for σ^2 in Section 8-6, obtaining:

$$\frac{V}{C^2_{.025}} < \Phi < \frac{V}{C^2_{.975}} \qquad \begin{array}{c} (21\text{-}18) \\ \text{like } (8\text{-}54) \end{array}$$

In Figure 21-10, we show the confidence intervals[19] for Φ based on V in Figure 21-9. Comparing Figures 21-10(b) and (c), we see explicitly how, according to (21-17), a wider frequency band Δf increases d.f., and hence narrows the confidence interval. Thus, greater stability is achieved at the price of less resolution.

(c) Cross-Spectral Analysis

Although a detailed examination of this technique would take us far beyond the scope of this book, let us offer a brief, intuitive discussion.

---

[17] For this to be valid, the spectrum should be fairly flat over the frequency band.

[18] Note that if $\Delta f = 1/2$, this represents maximum averaging; in fact, since there is only one frequency band, there is no decomposition of variance at all, so that $V = s^2$, which has $(n - 1)$ d.f. for a flat spectrum. From (21-17), we find about the same answer: d.f. $\simeq 2n(1/2) = n$.

In the first and last frequency bands, Δf (and consequently d.f.) are only half as large as usual.

[19] The calculations are as follows: it is easiest if we take logarithms of (21-18), obtaining:

$$\log V - \log C^2_{.025} < \log \Phi < \log V - \log C^2_{.975} \qquad (21\text{-}19)$$

For $n = 200$, and $\Delta f = 1/40$ as in Figure 21-9(b), we find from (21-17) that:

$$\text{d.f.} = 2(200)\left(\frac{1}{40}\right) = 10$$

and from Table VI(b):

$$C^2_{.025} = 2.05$$
$$C^2_{.975} = .325$$

Thus (21-19) becomes:

$$\log V - 0.31 < \log \Phi < \log V + .49$$

except for the first and last frequency bands, where

$$\text{d.f.} = 2(200)\left(\frac{1}{80}\right) = 5$$

$$C^2_{.025} = 2.57$$
$$C^2_{.975} = .166$$

$$\log V - .41 < \log \Phi < \log V + .78 \qquad (21\text{-}20)$$

This is illustrated in Figure 21-10(a). This shows yet another advantage in using a logarithmic axis: all the confidence intervals for Φ are the same length on the log scale, as shown in Figure 21-10(b).

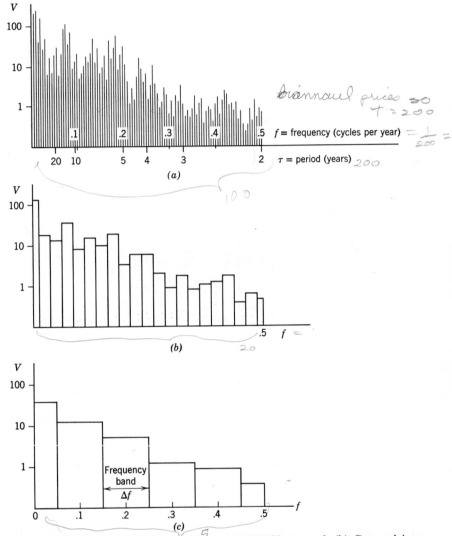

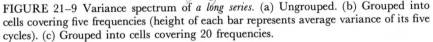

FIGURE 21–9 Variance spectrum of *a long series*. (a) Ungrouped. (b) Grouped into cells covering five frequencies (height of each bar represents average variance of its five cycles). (c) Grouped into cells covering 20 frequencies.

Suppose that we are interested not just in one series, but rather in the relation between two series; for example, consider the relation of Canadian GNP to U.S. GNP. Each series could be decomposed into its component cycles. Then, for every frequency (or, more precisely, over every frequency band), we could study the relation of the Canadian cycle to the U.S. cycle. One such comparison is shown in Figure 21-11; note that

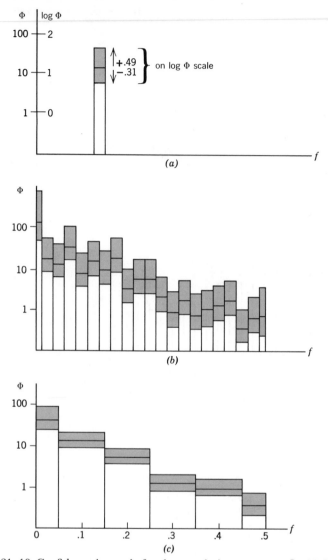

FIGURE 21–10 Confidence intervals for the population spectrum Φ. (a) Illustration for one frequency band in Figure 21–9(b). (b) For all frequency bands in Figure 21–9(b). (c) For all frequency bands in Figure 21–9(c).

the Canadian cycle is only 40% as high as the U.S. cycle, and lags the U.S. by about 1/3 of a cycle. These two values are known technically as the *gain* and *phase lag*. They may provide insight into the economic mechanisms linking the two series.

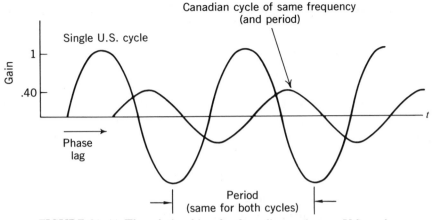

FIGURE 21–11 The relationship of a Canadian cycle to a U.S. cycle

(d) Conclusions

Spectral analysis is used to try to understand the nature of time series—which frequency bands dominate. It is particularly useful when two time series are analyzed together; then their cross-spectrum shows how their corresponding cycles are related.

On the other hand, spectral analysis is not helpful for prediction. For this, an autoregression of the form (21-5) is usually a simpler and more satisfactory procedure.[20]

PROBLEMS

*21-7 Can you guess at the result if you applied spectral analysis to the time series:

(a) In Figure 21-1(a)?

(b) In Figure 21-1(b)?

(c) In Figure 21-1(c) in the special case where the residual is purely random (serially uncorrelated)?

---

[20]Anyone who has worked through spectral analysis of a residual series may be tempted to use it for predictive purposes: the idea would be to pick the frequency band with the greatest amplitude, and then use it to predict the residual series. This is a temptation worth resisting, for several reasons.

In many cases, the estimated spectrum turns out to be jagged and ambiguous, with several peaks. Even if one frequency band does dominate, it is not clear which particular frequency within this band should be used. Moreover, the inevitable statistical error in the point estimate of the population parameter may be much more destructive in a time series than in, say, multiple regression; the reason is that in each successive time period, the estimated and true cycle become further and

(cont'd on page 626)

*21-8   HOW AUTOCORRELATION AFFECTS THE TREND AND SEASONAL ESTIMATES

(a) Introduction

In our initial analysis of a time series (Sections 21-2 and 21-3), we estimated the trend and seasonal components, leaving a residual series to be analyzed in Section 21-4 for autocorrelation. The problem is that the existence of a positively autocorrelated residual that is mixed up in the original series will reduce the reliability of our first-stage estimate of trend and seasonal (and indeed may reduce the reliability of our estimate of any other influence, for that matter). In this section we discuss this problem and suggest possible solutions. Before plunging into a lot of mathematics, it is important to understand the intuitive idea.

Autocorrelation means that the successive observations are dependent to some extent; thus, with positive autocorrelation, the second (or some later) observation tends to resemble, or repeat, the first observation, and hence gives little new information. Thus n autocorrelated observations give less information about trend (and other influences) than n independent observations would. Consequently, our estimates will be less reliable.

We turn now to the mathematical development. Our regression model is:

$$Y_t = \alpha + \beta X_t + e_t \tag{21-21}$$

where X_t represents time T, or seasonal dummies, or any other time series deemed influential in determining Y_t, or for that matter, any combination of these. In addition, the error e_t is assumed to be autocorrelated. Just how e_t is related to its previous values is, of course, unknown, just as α and β are unknown. It's true that we already have estimated it in a very simple way in our example in (21-4), but now we wish to look into its theoretical properties in more detail.

(b) A Simple Example

Just as we did in Figure 21-5, we again suppose that the error term e_t has the simple form of positive serial correlation given by $\rho = 1$ in (21-3). Then:

---

further out of phase (i.e., if the period of the cycle is slightly underestimated, then the first peak will be a little off target, the second peak a bit more, and so on).

Spectral prediction requires strong prior economic justification. We already have pointed out that many economic time series that appear to be cyclical may not have been generated by a cyclical process at all. Using apparent cycles to predict usually is much less satisfactory than using the relatively simple estimated autoregressive model (21-3).

$$e_t = e_{t-1} + v_t \qquad\qquad (21\text{-}22)$$

Now we must be even more explicit about the additional disturbance v_t, which is assumed to have the usual characteristics: zero mean, constant variance σ^2, and independence from the other disturbances v_{t-1}, v_{t-2}, etc. The initial value e_0 also is assumed to be random, with mean zero, and variance σ_0^2, say.

Suppose that the true regression line (defined by α and β) is as shown in Figure 21-12(a). Suppose further that the error terms e_1, e_2, . . . are those that we generated from the model shown in Figure 21-5: they are reproduced in Figure 21-12(b). Any observed Y (e.g., Y_2) is the sum of its expected value $(\alpha + \beta X_t)$ as given by the true regression, plus the corresponding error term (e_2). Since this error e_t is positively correlated, once our observations are above this true regression, they will tend to stay above it (i.e., once e_t is positive, it will be more likely to take on a positive value in the following time period). Similarly, whenever e_t becomes negative, it tends to stay negative.

In Figure 21-12(a), we immediately can see the difficulties that this serial correlation causes, by observing how badly an ordinary regression through this scatter would estimate the true regression.[21] Specifically, in this sample, β would be underestimated and α overestimated. But in another sample we might have observed precisely the opposite pattern of errors, with e_t initially taking on negative values, followed by positive ones. In this case, we would overestimate β and underestimate α. Either of these two types of estimation error seems equally likely; and there are other possibilities as well. Intuitively, it seems that the problem is not bias, since we are as likely to get an overestimate as an underestimate. Rather, the problem is that estimates may be badly wide of the target. Although unbiased, the estimates have a large variance. Nor is this primarily the fault of the ordinary least squares (OLS) regression procedure; any other estimating procedure (such as fitting by eye) would fit the "tilted" data about the same way. In fact, despite the large variance of its estimators, OLS may still be quite efficient; although there is another more sophisticated technique (Generalized Least Squares, GLS) that is more efficient, the improvement it gives over OLS may be quite small.

We are now in a position to draw our first conclusions.

1. Ordinary least squares estimates of α and β are unbiased (21-23)

---

[21]In the interests of clarity, this argument has been oversimplified, since it is assumed in Figure 21-12(a) that X_t is increasing over time regularly. Thus the argument holds best if X_t represents simply time T; but it still holds true even if X_t just tends to increase over time. On the other hand, if X_t alternates between low and high values, then our analysis would be complicated; in particular, our later conclusion about the efficiency of ordinary least squares would have to be modified.

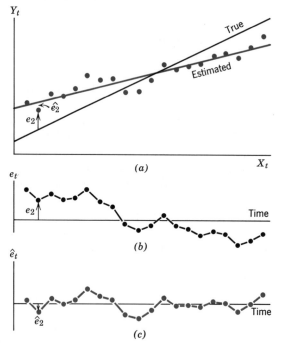

FIGURE 21–12 Regression with serially correlated error. (a) True and estimated regression lines. (b) True error terms (from Figure 21–5). (c) Observed error terms.

2. OLS may be relatively efficient   (21-24)

However, for an *interval* estimate of β (or α), OLS would be grossly deceptive. There are two reasons for this, which will be seen in comparing the serial correlation of Figure 21-12 to the case of no serial correlation in Figure 21-13. These two figures are alike in all respects other than the serial correlation; in particular, we emphasize that they have the same variance for the error e_t.

Since Figure 21-13 satisfies the assumptions of OLS, it will provide a valid confidence interval. But this is not so for the serially correlated data of Figure 21-12; in this case the standard confidence interval:

$$\beta = \hat{\beta} \pm t_{.025} \frac{s}{\sqrt{\sum x_i^2}} \qquad \begin{matrix} (21\text{-}25) \\ (12\text{-}27) \text{ repeated} \end{matrix}$$

will err for two reasons:

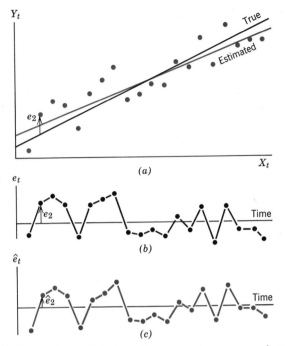

FIGURE 21–13 Regression with independent error (yet same variance as in Figure 21–12). (a) True and estimated regression lines. (b) True error terms. (c) Observed error terms.

1. It is likely to be centered at a more erroneous value of $\hat{\beta}$. We might hope that it would allow for this greater error by providing a wider confidence interval.

2. Yet it has a *narrower* confidence interval, because it has a smaller value of s. To establish this, we recall that the variance s^2 is calculated from the observed errors $\hat{e}_t$. In the case of serial correlation, the estimated regression line fits the tracking data rather well, leaving the small errors $\hat{e}_t$ shown in Figure 21-12(c).

To sum up: because the data is tracking, an OLS estimate of β is, in fact, less reliable. But it appears to the statistician to be *more* reliable because tracking data tends to yield smaller observed errors.

(c) Regression of First Differences

To estimate β in the linear regression model (21-21), the remedy for this special case of serial correlation is to transform the data so that it satisfies the assumptions of OLS. Since (21-21) holds true for any time t, it holds for time $(t - 1)$:

$$Y_{t-1} = \alpha + \beta X_{t-1} + e_{t-1} \tag{21-26}$$

We now can examine the change over time of our variables, by subtracting (21-26) from (21-21):

$$(Y_t - Y_{t-1}) = \beta(X_t - X_{t-1}) + (e_t - e_{t-1}) \tag{21-27}$$

Note from (21-22) that:

$$e_t - e_{t-1} = v_t$$

and define:

$$\left. \begin{aligned} Y_t - Y_{t-1} &\overset{\Delta}{=} \Delta Y_t \\ X_t - X_{t-1} &\overset{\Delta}{=} \Delta X_t \end{aligned} \right\} \tag{21-28}$$

which are called "differences." Then (21-27) can be written:

$$\boxed{\Delta Y_t = \beta \Delta X_t + v_t} \tag{21-29}$$

In this form, the error v_t has all the properties required by OLS. Thus, β may be estimated validly by OLS regression of ΔY on ΔX. However, we must emphasize that this works only for an error of the (usually unrealistic) form of (21-22).

(d) Generalized Differences–an Example of Generalized Least Squares (GLS)

Our original model for the error term (21-3) conforms more realistically to economic and business situations:

$$e_t = \rho e_{t-1} + v_t$$

$$(21-30)$$
$$(21-3) \text{ repeated}$$

where:[22]

---

[22]If $|\rho| \geq 1$ as in (21-22), a major problem arises because the error becomes "explosive"; that is, the variance of e_t increases infinitely with time.

$$|\rho| < 1$$

Again the error e_t is the sum of two components: its previous value *attenuated* by the factor ρ, plus the further disturbance v_t. (You can see that in the limiting case where $\rho = 0$, the error e_t reduces to the purely random v_t.)

We continue to assume the linear regression model:

$$Y_t = \alpha + \beta X_t + e_t \qquad (21\text{-}31)$$
$$(21\text{-}21)\ \text{repeated}$$

To estimate β, we transform the data in much the same way as before. Equation (21-31) for time $(t - 1)$ is multiplied by ρ:

$$\rho Y_{t-1} = \rho\alpha + \rho\beta X_{t-1} + \rho e_{t-1} \qquad (21\text{-}32)$$

Subtracting (21-32) from (21-31):

$$(Y_t - \rho Y_{t-1}) = \alpha(1 - \rho) + \beta(X_t - \rho X_{t-1}) + (e_t - \rho e_{t-1}) \quad (21\text{-}33)$$

Now define:

$$\left.\begin{array}{l} Y_t - \rho Y_{t-1} \stackrel{\Delta}{=} \Delta Y_t \\[2mm] X_t - \rho X_{t-1} \stackrel{\Delta}{=} \Delta X_t \end{array}\right\} \qquad (21\text{-}34)$$

These transformed values sometimes are called "generalized differences," with (21-28) being a special case. Then (21-33) can be written:

$$\boxed{\Delta Y_t = \alpha(1 - \rho) + \beta\, \Delta X_t + v_t} \qquad (21\text{-}35)$$

In this form, v_t is the error, and has all the properties required by OLS. Thus, after this transformation, we may regress ΔY_t on ΔX_t to estimate β. This technique is a form of generalized least squares (GLS).

However, prior to this regression, one additional adjustment must be made to the data. The *first* observed values Y_1 and X_1 cannot be transformed by (21-34) since their previous values are not available; instead, the appropriate transformation is:

$$\boxed{\begin{array}{l} \sqrt{1 - \rho^2}\, Y_1 \stackrel{\Delta}{=} \Delta Y_1 \\[2mm] \sqrt{1 - \rho^2}\, X_1 \stackrel{\Delta}{=} \Delta X_1 \end{array}} \qquad (21\text{-}36)$$

This adjustment to the data, along with a final adjustment to the regression procedure,[23] are very important; without it, the regression (21-35) may be no better than OLS on (21-21), and perhaps not even as good.[24]

To sum up, if ρ is known, the data is transformed by (21-34) and (21-36), with an adjusted regression of ΔY_t on ΔX_t in (21-35) providing the estimate of β. The problem with this approach, of course, is that ρ generally is not known and must be estimated.

(e) Statistical Estimation of ρ

To estimate ρ, equation (21-30) would be ideal if the true residuals e_t could be observed. Since they cannot, we are forced to use the observed residuals $\hat{e}_t$ that fall out of the OLS regression of Y_t on X_t, as in Figure 21-12(c). Thus ρ is estimated by applying OLS to:

$$\hat{e}_t = \rho \hat{e}_{t-1} + \text{error} \tag{21-37}$$

With this regression, we obtain an estimate r for ρ, which is consistent. Yet r has a bias, which can be seen from Figure 21-12(c). The estimated residuals $\hat{e}_t$ fluctuate around zero more (and hence have a smaller serial correlation) than do the true residuals[25] e_t. Thus r underestimates ρ on the average.

There are ways to allow for this underestimation; the most popular is *the Durbin–Watson test* for $\rho = 0$, against the alternative $\rho > 0$. Obviously, H_0 should be rejected when r is high. Durbin and Watson obtained an equivalent test of $\rho = 0$ using an alternative statistic:[26]

$$D = \frac{\displaystyle\sum_{t=2}^{n} (\hat{e}_t - \hat{e}_{t-1})^2}{\displaystyle\sum_{t=1}^{n} \hat{e}_t^2} \approx 2(1 - r) \tag{21-38}$$

---

[23]The constant regressor (whose coefficient is α) also must be transformed, just like the Y_t and X_t regressors. It becomes $(1 - \rho)$ for $t = 2, 3, \ldots, n$ in (21-35); it becomes $\sqrt{1 - \rho^2}$ for $t = 1$ by analogy with (21-36).

[24]The transformation (21-36) must be used with some care. It is appropriate if and only if the process (21-30) generating the error has been going on undisturbed for a long time previous to collecting the data. In practice, however, the first observation often is taken just after a war or some other catastrophe, which seriously disturbs the error.

[25]In Figure 21-12(b), the true residuals e_t cross the zero axis only three times. Yet in Figure 21-12(c), the observed residuals $\hat{e}_t$ bounce around the zero axis, crossing it eight times. Hence, they appear to have very little serial correlation; in fact, in this respect they behave like the uncorrelated residuals in Figure 21-13(b), which cross the zero axis ten times.

[26]We note that D is defined like d in (16-29). However, since D uses regression residuals $\hat{e}_t$ instead

Note how r and D are inversely related;[27] thus when r is as large as 1, D is as small as 0. (This easily is confirmed by examining the numerator in (21-38); the more the serial correlation, the more the $\hat{e}_t$ series tends to track itself, and the smaller become the differences $\hat{e}_t - \hat{e}_{t-1}$.) Therefore, there are two alternative ways in which to establish positive serial correlation:

1. If r is observed to be sufficiently high; or

2. if D is observed to be sufficiently low.

Durbin and Watson chose the latter, tabulating the critical values of D. This is reproduced in Appendix Table IX, and is accompanied by a graphical explanation. As well as allowing for the bias of r in estimating ρ, Durbin and Watson allowed for the dependence of D upon the configuration of X_t. Thus, there are two limiting values of D tabulated (D_L and D_U), corresponding to the two most extreme configurations of X_t; for any other configuration, the critical value of D will be somewhere between D_L and D_U.

Even when the existence of serial correlation has been established by the Durbin-Watson test in (21-38), there still remains the problem of estimating ρ. One possibility is to use the estimate that falls out of the regression (21-37); but we already noted the weaknesses of this estimate. A number of alternatives have been suggested, none of them foolproof; but one of the simplest has performed relatively well in one study.[28] It involves rearranging (21-33) into the following regression equation:

$$Y_t = \alpha(1 - \rho) + \rho Y_{t-1} + \beta X_t - \beta\rho X_{t-1} + (e_t - \rho e_{t-1}) \qquad (21\text{-}39)$$

where $e_t - \rho e_{t-1} = v_t$ has all the attractive properties of a residual. Accordingly, run a regression of Y_t on Y_{t-1}, X_t, and X_{t-1}, retaining only[29] $\hat{\rho}$, the estimated coefficient of Y_{t-1}; this provides a consistent estimate of ρ. This then can be used in Equations (21-34), (21-35) and (21-36) to run a generalized difference regression—or more precisely, a GLS regression—yielding estimates of α and β.

(f) Summary of Possible Procedures

When facing time series like (21-31) with an error with an unknown amount of serial correlation, one practical procedure would be to:

---

of original observations X_i, its distribution is more complicated. Nevertheless, many similarities remain; for example, when $r = 0$ in (21-38), we approximate (16-32).

[27] In our jewelery sales example, D was calculated to be 2.42. Since r was calculated in (21-4) to be $-.21$, the approximate relation between D and r given by (21-38) is confirmed.

[28] This was a Monte Carlo study comparing several alternative estimators.

[29] We call it $\hat{\rho}$ to distinguish it from the r calculated from regression (21-37); note that the two are alternative estimators of exactly the same population parameter ρ.

1. Apply OLS directly to (21-21), obtaining a time series of estimated residuals $\hat{e}_t$.

2. Use this series for the Durbin–Watson test (21-38). If serial correlation in the error is not established, use the OLS estimates of α and β that we already derived in step 1. The task of estimation now is complete. (It also usually is desirable to include the value of D in your results, as evidence of the indiscernible serial correlation.)

On the other hand, if the Durbin–Watson test establishes serial correlation, then proceed as follows:

3. Estimate ρ by running regression (21-39) and retaining $\hat{\rho}$ (the estimated coefficient of Y_{t-1}).

4. Finally use this $\hat{\rho}$ as the missing link to calculate and regress generalized differences according to (21-35); or, more precisely, use this $\hat{\rho}$ to apply GLS. This will provide the desired estimates of α and β. If steps 3 and 4 must be undertaken, will the result be much of an improvement over a simple and direct application of OLS to (21-21)? The answer seems to be that, while it is likely to provide an improvement, it may not be a very substantial one. Even when ρ is known, the improvement that can be achieved may be fairly modest; and when ρ is unknown and must be estimated, this introduces a source of error that tends to erode this advantage.

Estimation now is complete; but if we wish, in addition, to project, there are three additional steps:

5. The GLS estimates of α and β calculated in step 4 can now be used to estimate GLS residuals[30] $\hat{e}_t$, and in particular, the last residual $\hat{e}_n$.

6. The error term now may be projected. To estimate the true but unknown terms in:

$$e_{n+1} = \rho e_n + v_n \qquad (21\text{-}40)$$

$$\text{like (21-30)}$$

we may use:

$$\hat{e}_{n+1} = \hat{\rho}\hat{e}_n \qquad (21\text{-}41)$$

---

[30] In exactly the same way that we calculated OLS residuals in the second last column of Table 11-2.

Thus the projected residual $\hat{e}_{n+1}$ is calculated using the value of $\hat{\rho}$ from step 3 and the value of $\hat{e}_n$ from step 5 (and of course ignoring the random v_t).

7. We now can calculate the projected value of Y in the next period as:

$$\hat{Y}_{n+1} = \hat{\alpha} + \hat{\beta}\hat{X}_{n+1} + \hat{e}_{n+1} \tag{21-42}$$

where $\hat{X}_{n+1}$ is the projected value[31] of X_t, $\hat{e}_{n+1}$ has been derived in step 6, and $\hat{\alpha}$ and $\hat{\beta}$ are the GLS estimates calculated in step 4.

Finally, note that this more complicated procedure runs parallel to our simple-minded projection in section 21-5 above; the only differences are:

1. We hope to have a better projection of the residual ($\hat{e}_{n+1}$) because we now are estimating ρ from (21-39).

2. GLS has provided better estimates of α and β.

---

[31] It is even possible that X_{n+1} already may have been observed. In many institutions, like the U.S. Department of Commerce, that provide business, economic or social data to the public, there are frequent cases in which some data series (say X_t) become available more quickly than others (say Y_t). Thus, consider the problem that a firm may face in January 1978. It may be necessary for it to use an already available X_{1977} figure in order to "predict in the past" the Y_{1977} figure that it must have, but that is not yet available.

CHAPTER 22

Simultaneous Equations

*The chief difficulty Alice found at first was with her flamingo: . . .
generally, just as she had got its neck straightened out, and was going to
give the hedgehog a blow with its head, it would twist itself around and
look up in her face.*

Lewis Carroll

22-1   INTRODUCTION

Most business and social problems are considerably more complicated
than some of the techniques that we have introduced so far suggest.
Specifically, seldom is a variable determined by a single relationship
(equation). Instead, a variable normally is determined simultaneously
with many other variables in a whole *system* of simultaneous equations.
For example, the price of corn is determined simultaneously with the
price of rye, the price of hogs, etc.

An analogy may be useful. The economic or social system may be
similar in sense to the solar system, where the variables (positions
of the planets at any point in time) are determined simultaneously by a
whole system of equations or relationships (i.e., each indicates how the
position of a single planet depends on the position of all the others). Thus,
although the position of the moon is most substantially influenced by the
position of the earth, it is also influenced to some small degree by the position of the other planets as well. Similarly, the position (sales) of a business firm like Ford Motor Company will depend on the sales of its
"closest" competitors like GM and Chrysler and, to a lesser degree, its

637

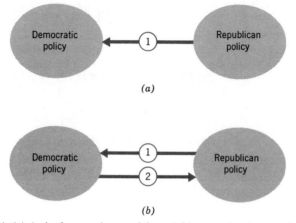

FIGURE 22–1 (a) A single equation model; variables not simultaneously determined. (b) A simultaneous model; variables jointly determined.

other competitors as well, like American Motors, Volkswagen, and Toyota.

Similarly, the position of the Democratic party on a national issue will depend on the position of the Republican party. If the Republican position is determined independently of the Democratic position, then the simple single equation model of previous chapters would suffice: the independent Republican variable can be "plugged into" the Democratic equation, thus determining the Democratic response. This is shown in Figure 22-1(a), where the arrow indicates the direction of influence. But the *simultaneous* determination of both the Republican and Democratic positions is more likely, as shown in Figure 22-1(b). Here, the Democratic position cannot be determined by plugging a Republican position into relationship (1), since the Republican position is not given; instead it is dependent on the Democrats, according to relationship (2). A single equation model will not work; instead the two-equation model, explaining the simultaneous determination of both variables, is required.

22-2   AN ECONOMIC EXAMPLE

Consider the following, very simple national income model of two equations:

$$Y = \alpha + \beta X + e \tag{22-1}$$

$$X = Y + I \tag{22-2}$$

Equation (22-1) is the standard form of the consumption function that relates consumer expenditure Y to income X. The parameters of this function that must be estimated are α (the intercept) and β (the slope, or marginal propensity to consume[1]). We assume that the error term is well-behaved: successive values of e are assumed to be independent and identically distributed, with mean 0 and variance σ^2. Equation (22-2) states that national income is defined as the sum of consumption and investment I. (Since both sides of this equation are equal by economic definition, no error term appears in this equation—but the analysis remains the same in any case).

An important distinction must be made between two kinds of variables in our system. By assumption, I is determined *outside* the system of equations; it often will be referred to as "exogenous" or "predetermined." The essential point is that its values are determined elsewhere, and are not influenced by Y, X, or e. In particular, we emphasize that:

$$I \text{ and } e \text{ are statistically independent} \qquad (22\text{-}3)$$

On the other hand, X and Y are jointly dependent, or endogenous variables; their values are determined *within* the model and thus are influenced by I and e. Since there are two equations to determine these two endogenous variables, the model is mathematically complete.[2]

Certain oversimplifications in this model immediately are evident. For example, it describes a closed economy with no government sector. The assumption that I is exogenous is also an oversimplification. In fact, I is likely to depend on Y or X. You are invited to experiment in setting up a three-equation model in X, Y, and I; such a system would involve an additional endogenous variable I and an additional equation relating I to another variable that might more reasonably be regarded as exogenous (e.g., the interest rate). Assumptions always can be made more realistic by increasing the size of the model—but then its mathematical complexity also increases. Statisticians constantly must make decisions that involve this sort of trade-off between realism and mathematical manageability. A larger model also would be more difficult to show geometrically. Since this is our next objective, we shall stick to our original two-equation model. (The assumption that only X and Y are jointly determined, and that I and other economic variables can be viewed as predetermined, is

---

[1]That is, the additional consumption that would result from a $1 increase in income. See (13-19).

[2]While there must be as many linear equations as unknowns to yield a unique solution, this is not always sufficient; these equations must also be linearly independent.

similar to the astronomer's assumption that the position of the moon may be determined by a simple earth/moon model; although other heavenly bodies theoretically should be included in the model also, this simplification yields a good approximation and keeps the mathematics manageable.)

A diagram will be useful in illustrating the statistical difficulties that are encountered. To highlight the problems the statistician will face, let us suppose that we have some sort of omniscient knowledge, so that we know the true consumption function $Y = \alpha + \beta X$, as shown in Figure 22-2. And let us watch what happens to the statistician—a mere mortal— who does not have this knowledge but must try to estimate this function by observing only Y and X. Specifically, let us show how badly things will turn out if he estimates α and β by fitting a line by ordinary least squares (OLS). To find the sort of scatter of Y and X that he will observe, we must remember that all observations must satisfy both equations (22-1) and (22-2).

Consider (22-1) first. Whenever e takes on a zero value, the observation of Y and X must fall somewhere along the true consumption function $Y = \alpha + \beta X$, shown in Figure 22-2. If e takes on a value greater than zero (say + \$50 billion), then consumption is greater as a consequence and the observation of Y and X must fall somewhere along $Y = \alpha + \beta X + 50$. Similarly, if e takes on a value of -50, the statistician will observe a point on the line $Y = \alpha + \beta X - 50$. According to the standard assumptions, e is distributed about a zero mean. To keep the geometry simple, we further assume that e is equally likely to take on any value between $+50$ and -50. Thus the statistician will observe Y and X falling within this band around the consumption function, shaded in Figure 22-2.

Any observed combination of Y and X also must satisfy (22-2). What does this imply? This condition can be rewritten as:

$$Y = X - I \qquad (22\text{-}4)$$

If I were zero, then Y and X would be equal, and any observation would fall on the 45° line where $Y = X$. Let us suppose that when I is determined by outside factors, it is distributed uniformly through a range of 100 to 250. If $I = 100$, then from (22-4) any observation of Y and X must fall along $Y = X - 100$, which is simply the line lying 100 units below the 45° line. Similarly, when $I = 250$, an observation of Y and X would fall along the line $Y = X - 250$. These two lines define the steeper shaded band within which observations must fall to satisfy (22-2).

Since any observed combination of Y and X must satisfy *both* conditions, all observations will fall within the parallelogram $P_1 P_2 P_3 P_4$. To clarify, this parallelogram of observations is reproduced in Figure 22-3.

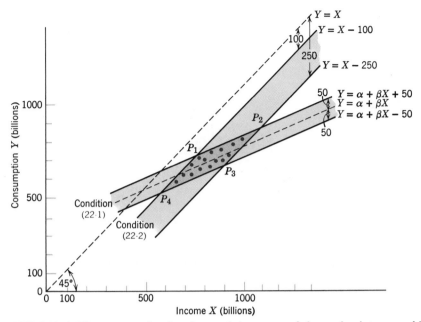

FIGURE 22–2 The consumption function, and the scatter of observed points around it

When the statistician regresses Y on X using OLS, the result is shown as $Y = \tilde{\alpha} + \tilde{\beta}X$. When this is compared with the true consumption function $(Y = \alpha + \beta X)$, it is clear that the statistician has come up with a bad fit; his estimate of the slope $(\tilde{\beta})$ has an upward bias. What has gone wrong?

The observations around P_2 have "pulled" the estimating line above the true regression; similarly, the observations around P_4 have pulled the estimating line below the true regression. It is the pull on both ends that has tilted this estimated line. Moreover, increasing sample size will not help to reduce this bias. If the number of observations in this parallelogram is doubled, this bias will remain.[3] Hence, OLS is inconsistent; this technique that worked so well in a single-equation model clearly is less satisfactory when the problem is to estimate one equation that is embedded within a system of simultaneous equations.

The reason is evident: for the single-equation model, it is assumed in (12-49) that the error e is independent of the regressor, and this is why

---

[3]With an increase in sample size, the reliability of $\tilde{\beta}$ as an estimator will be increased somewhat, because its variance will decrease towards zero. However, its bias will not be reduced.

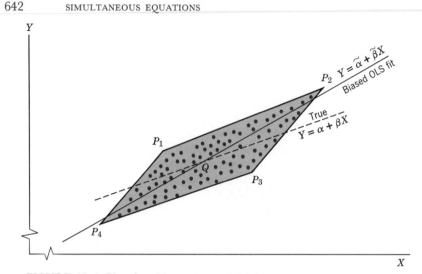

FIGURE 22–3 Biased and inconsistent OLS fit of the consumption function

OLS works so well. But in a multiple-equation model, this assumption no longer holds. Note, for example, in Figure 22-3 how e is correlated positively with X: e tends to be positive (i.e., the observed point lies above the true regression $Y = \alpha + \beta X$) when X is large, and e tends to be negative (with the observed point below the true regression) when X is small.[4] Consequently, the OLS fit has too large a slope. The reason is intuitively clear: in explaining Y, OLS gives as little credit as possible to the error, and as much credit as possible to the regressor X. When the error and regressor are correlated, then some of the effect of the error is attributed wrongly to the regressor.

This, then, is the problem: the single-equation technique of OLS is inconsistent in a simultaneous-equations context. A substantial body of econometric research has been devoted to developing more appropriate techniques; increasingly, business statisticians, sociologists, and other social scientists, all of whom must deal with situations of mutual dependence, are likely to turn their attention to these problems. The simplest solution is the use of the instrumental variable technique. Since this statistical method is highly useful in dealing with other problems as well, we now consider it in some detail.

---

[4]Note that the reason for this is precisely because of the second equation in the model: thus, in Figure 22-2, the observed points are not only determined by equation (22-1), but they are also determined, or "pulled," by equation (22-2). And it is this latter pull that results in the positive correlation of X and e.

22-3   INSTRUMENTAL VARIABLES (IV)

The instrumental variable technique, sometimes called the "covariance" method, has quite general application. In this section, we shall apply it to the simple single-equation model of Chapter 11, to show that IV yields the same result as OLS. Then, in Section 22-4, we shall apply it to the two-equation model ((22-1) and (22-2)) to show that IV yields a consistent result. Thus, in estimating simultaneous equations, it outperforms OLS.

(a) The Covariance Operator

First, recall from Chapter 14 how we can calculate the covariance of a sample of X and Y observations:

$$s_{XY} \triangleq \frac{\sum xy}{n-1}$$

(22-5)

(14-5) repeated

The instrumental variable method involves using this to estimate regression parameters.

To illustrate, consider again the model of Chapter 12:

$$Y = \alpha + \beta X + e$$

(22-6)

Taking covariances of X with each of the variables in this equation:

$$s_{XY} = \beta s_{XX} + s_{Xe}$$

(22-7)

To justify this, we recall that the n sample observations of X and Y are assumed in (22-6) to be generated in the following way:

$$Y_1 = \alpha + \beta X_1 + e_1$$
$$Y_2 = \alpha + \beta X_2 + e_2$$
$$\vdots$$
$$Y_n = \alpha + \beta X_n + e_n$$

(22-8)

We easily can show that with an appropriate translation of both the X and Y axes, the intercept term disappears[5] and we may write:

---

[5] **proof**

$$Y_i = \alpha + \beta X_i + e_i$$

Taking averages:

$$\overline{Y} = \alpha + \beta \overline{X} + \overline{e}$$

(cont'd on page 644)

$$y_1 = \beta x_1 + e'_1$$
$$y_2 = \beta x_2 + e'_2$$
$$\vdots$$
$$y_n = \beta x_n + e'_n \qquad\qquad (22\text{-}10)$$

where the y_i, x_i, and e'_i represent deviations from the sample mean. The first equation is multiplied by x_1, the second by x_2, and so on:

$$x_1 y_1 = \beta x_1 x_1 + x_1 e'_1$$
$$x_2 y_2 = \beta x_2 x_2 + x_2 e'_2$$
$$\vdots$$
$$x_n y_n = \beta x_n x_n + x_n e'_n \qquad\qquad (22\text{-}11)$$

When we sum all these equations and divide by $(n-1)$:

$$\frac{\sum xy}{n-1} = \frac{\beta \sum xx}{n-1} + \frac{\sum xe'}{n-1} \qquad\qquad (22\text{-}12)$$

Recalling from (22-9) that e' is just e expressed as deviations from the mean, (22-12) may be written:

$$s_{XY} = \beta s_{XX} + s_{Xe} \qquad\qquad (22\text{-}7)\text{ proved}$$

(b) The OLS Estimate

In order to estimate β, we divide (22-7) by s_{XX} (the variance of X):

$$\frac{s_{XY}}{s_{XX}} = \beta + \frac{s_{Xe}}{s_{XX}} \qquad\qquad (22\text{-}13)$$

From the observations of X and Y, s_{XY} and s_{XX} are easily calculated. But e is unobservable, so that s_{Xe} cannot be evaluated. However, if we can assume that s_{Xe} is small enough to neglect, we obtain the estimator:

Subtracting this equation from the previous one

$$y_i = \beta x_i + e'_i \qquad\qquad (22\text{-}10)\text{ proved}$$

where:

$$e'_i = e_i - \bar{e} \qquad\qquad (22\text{-}9)$$

$$\frac{s_{XY}}{s_{XX}} = \hat{\beta} \qquad (22\text{-}14)$$

This may be referred to as the IV estimator, using X as the instrumental variable. We also recognize this[6] to be the OLS estimator (11-16).

In conclusion, the IV or OLS estimator in (22-14) is seen to be appropriate if we can neglect the last term in (22-13). Provided that X and e can be assumed independent (as in the single-equation model of Chapter 12), this term can indeed be neglected; in fact, in a large enough sample, this last term will approach zero.[7] Hence the IV or OLS estimator in (22-14) is consistent.

22-4 SIMULTANEOUS EQUATION ESTIMATION

(a) The Inconsistency of OLS Confirmed

Now suppose that we try to apply this same technique (i.e., take covariances with respect to X) on the consumption function (22-1). The algebra in equations (22-6) to (22-14) remains the same. Now, however, X and e no longer are uncorrelated, since we are dealing with one equation in a two-equation model (recall how this correlation was established in Figures 22-2 and 22-3). Hence, s_{Xe} in (22-13) no longer approaches zero, and the OLS estimator in (22-14) no longer is consistent. Or, to put it another way: using X as an instrument simply does not work, because of its correlation with e.

(b) The Consistency of IV

Since X has been disqualified as an instrumental variable, why not use another variable, say Z, which *is* uncorrelated with e? Taking covariances of Z with respect to all the variables in (22-6) yields:

$$s_{ZY} = \beta s_{ZX} + s_{Ze}$$

---

[6]The OLS estimator may be written, according to (11-19):

$$\frac{\sum xy}{\sum xx} = \frac{\sum xy/(n-1)}{\sum xx/(n-1)} = \frac{s_{XY}}{s_{XX}}$$

[7]**proof** In large samples, the sample covariance approaches the population covariance, σ_{Xe}. According to (5-28), independence makes $\sigma_{Xe} = 0$. Of course, it also is necessary to assume that s_{XX} (the variance of X) is nonzero—that is, the Xs are spread out (multicollinearity avoided, as in Section 13-4).

To obtain an estimate of β, divide by s_{ZX}:

$$\frac{s_{ZY}}{s_{ZX}} = \beta + \frac{s_{Ze}}{s_{ZX}} \qquad (22\text{-}15)$$

Provided that s_{Ze} approaches zero (i.e., that Z and e are uncorrelated) while s_{ZX} does not approach zero (i.e., Z and X are correlated), then the last term approaches zero. Disregarding it, therefore, yields the consistent estimator:

$$\boxed{\frac{s_{ZY}}{s_{ZX}} = \hat{\beta}} \qquad (22\text{-}16)$$

OLS is recognized in (22-14) to be just the special case of IV estimation, when X itself is used as the instrumental variable.

(c) Requirements of an Instrumental Variable

From the above analysis, it is clear that:

> an instrumental variable Z must be
>
> (1) Uncorrelated with the error e, and
> (2) Correlated with the regressor X.   (22-17)

It now is appropriate to consider the second requirement in more detail. As is evident in (22-15), the higher the covariance s_{ZX}, the smaller will be the term s_{Ze}/s_{ZX}, which is the source of estimation error; hence, the better will be the estimator in (22-16).

As our instrumental variable, therefore, we should look for the most relevant variable (i.e., the instrumental variable that is most highly correlated with X) to get the most "statistical leverage." Thus, the exogenous variable I (investment) is a natural choice, since it is evident from (22-2) that I has a substantial effect on X. Rainfall in California or even the price of haircuts in Denver might satisfy the theoretical requirements of an instrumental variable; yet either would lack statistical leverage. To state this mathematically, if California rainfall R were used as an instrumental variable, its correlation with X would be so small that the error term s_{Re}/s_{RX} would be large. Only in extremely large samples would this ratio settle down to zero. The consistency of $\hat{\beta}$ in huge samples would be cold comfort for an economist who was faced with a small sample. Or, to look at it from another point of view, suppose that the price of haircuts in Denver alternatively were used as an instrumental variable; since this also would involve a large estimating error, the two small-sample estimates of β might vary widely—even though, in an infinitely large sample, they would coincide.

In view of this relevancy requirement for an instrument, it is desirable, if possible, to use as instrumental variables only those exogenous variables that explicitly appear in our model—that is, which appear in at least one of the equations in the system. It is evident that neither this approach, nor any other, entirely overcomes our difficulties; the decision on which variables may be used as instrumental variables merely is pushed onto the researcher who specifies the model and, in particular, the exogenous variables that are included. Specification of the model remains arbitrary to a degree, and this gives rise to some arbitrariness in statistical estimation. But this cannot be avoided. We can only conclude that the first task of specifying the original structure of the model is a very important one, since it involves a prior judgment of which variables are "close to" X and Y, and which variables are relatively "far away."

Finally, a note of caution. Estimating simultaneous equations is a very complex issue. Although IV has provided a consistent alternative to OLS, many further problems remain (for example, the "identification problem," introduced in Problem 22-6).

PROBLEMS

22-1 Draw a schematic diagram similar to Figure 22-1 to illustrate:
   (a) How the sales of Ford are determined. (For simplicity, assume that the only important variables are the "Big Three" auto manufacturers.)
   (b) How consumption and income are determined in (22-1) and (22-2).

22-2 Suppose that the true consumption function (22-1) is $Y = 10 + .6X$ and that the following combinations of Y, X, and I have been observed:

Y	X	I
46	60	14
31	45	14
61	75	14
58	80	22
43	65	22
73	95	22
70	100	30
55	85	30
85	115	30

(a) Graph the true consumption function and the scatter of (X, Y) observations.

(b) Regress Y on X using OLS, and graph the estimated consumption function. Is it consistent?

(c) Estimate the consumption function using I as an instrumental variable. Graph this estimated consumption function. Is it consistent?

(d) Explain why this small sample is a lucky one. What would you expect if your small sample is less lucky?

22-3 To estimate the consumption model (22-1), suppose that the covariance matrix (or table) of X, Y, I has been computed for a sample, as follows:

$$\begin{bmatrix} s_{XX} & s_{YX} & s_{IX} \\ s_{XY} & s_{YY} & s_{IY} \\ s_{XI} & s_{YI} & s_{II} \end{bmatrix} = \begin{bmatrix} 130 & 100 & 30 \\ 100 & 80 & 20 \\ 30 & 20 & 10 \end{bmatrix}$$

(a) Find a consistent estimate for the marginal propensity to consume.

(b) Estimate the marginal propensity to consume by applying OLS to the consumption function directly. Is this estimate consistent? Using a diagram, explain this to a student whose statistics is limited to understanding OLS.

22-4 Consider the single-equation model:

$$Y = \alpha + \gamma_1 X_1 + \gamma_2 X_2 + e$$

where both regressors (X_1 and X_2) are uncorrelated with the error term e. Show how γ_1 and γ_2 can be estimated:

(a) Using OLS.

(b) Using X_1 and X_2 in turn as instrumental variables. (Generate one estimating equation using X_1 as an instrumental variable, and a second using X_2). Do IV and OLS yield identical estimators? (This generalizes (22-14).)

22-5 (a) A sociologist observes that the number of children available for adoption Y depends on the number of births X. At the same time, births depend on income Z and on the number of children available for adoption Y. Draw a schematic figure similar to Figure 22-1 illustrating these relationships.

(b) Set up a system of two simultaneous equations similar to (22-1) and (22-2) showing how X and Y might be jointly determined.

(c) Assume that the equations in (b) are of the form:

$$Y = \alpha + \beta X + e \qquad (22\text{-}18)$$

$$X = \delta + \gamma Y + \psi Z + v \qquad (22\text{-}19)$$

where e and v are error terms and Z is exogeneous, of course. Show how β may be consistently estimated. Explain why this technique is consistent. Explain why OLS on (22-18) would not be.

(d) Can Equation (22-19), above, be consistently estimated? If so, how; if not, why?

*22-6 (This is a brief introduction to the "identification problem.") Consider the following system of two simultaneous equations[8] in the two jointly dependent variables Y_1 and Y_2:

$$Y_1 = \beta_2 Y_2 + e \qquad (22\text{-}20)$$

$$Y_2 = \beta_1 Y_1 + \gamma_1 Z_1 + v \qquad (22\text{-}21)$$

Note that there is only one appropriate instrumental variable, Z_1.

(a) Show that (22-20) can be consistently estimated, but (22-21) cannot. (We might say that "there are enough instrumental variables in this system (namely, one) to estimate the one parameter in (22-20), but not enough to estimate the two parameters in (22-21)." This is commonly restated as: "There are enough instrumental variables to *identify* the first question, but not the second.")

(b) Now suppose, from theory, that you have prior knowledge that the value of β_1 is unity. Can (22-21) now be identified?

(c) Suppose that, instead of prior knowledge of β_1, you know that Y_1 is dependent on Y_2 as before, and also on a new exogenous variable Z_2; that is, (22-20) is replaced by:

$$Y_1 = \beta_2 Y_2 + \gamma_2 Z_2 + e \qquad (22\text{-}20\text{b})$$

Is this equation still identified? Is the other equation in the system (22-21) still not identified?

(d) True or false? If false, correct it.

(i) Prior knowledge may help to identify an equation.

(ii) An equation is identified by instrumental (exogenous) variables that appear in *other* equations in the system—that is, by exogenous variables *excluded* from the equation that is to be estimated.

(iii) A requirement for identifying an equation is that the number of

---

[8] Except for dropping the constants and renaming the variables, this system of equations is like (22-18) and (22-19).

exogenous variables that are excluded from the equation must at least equal the number of endogenous variables that are included on the right-hand side of that equation.

(e) Now suppose that (22-20) is replaced by:

$$Y_1 = \beta_2 Y_2 + \gamma_2 Z_2 + \gamma_3 Z_3 + e \tag{22-20c}$$

where both Z_2 and Z_3 are exogenous variables. If the other equation (22-21) in the system is to be estimated, are there now too few, too many, or just the right number of instrumental variables? Is it true that (22-21) now can be estimated by selecting any two of the three available instrumental variables? (This surplus of instrumental variables is referred to as *overidentification*.)

Summary. If there are too few instrumental variables in the system, the equation is said to be "unidentified," and estimation is impossible.

If there are just barely enough instrumental variables, the equation is said to be "exactly identified," and IV estimation is straightforward.

Finally, if there is a surplus of instrumental variables, the equation is said to be "overidentified." Then estimation becomes complicated. The difficulty is that there is no longer a unique estimating procedure: any selection of just the right number of instrumental variables from the surplus available will yield a result, and these various results do not coincide (unless the sample is infinitely large). Consequently, a great deal of statistical and econometric research recently has been devoted to asking which selection of instrumental variables is optimal—or better still, how the large number of instrumental variables can be combined, thus reducing their number to just the right size. (For example, a very popular estimating technique called "Two-Stage Least Squares" may be viewed in just this way. The first stage involves combining the available instruments into a new set of just the right number, while the second stage may be interpreted as simply applying these new instruments to yield a unique result).

CHAPTER 23

Index Numbers

The time has come, the Walrus said,
To speak of many things.
Of shoes, and ships, and sealing wax,
Of cabbages—and kings.

Lewis Carroll

An index number is a single figure that shows how a whole set of related variables has changed over time or differs from place to place. For example, a price index shows the overall change in a set of prices. If we ask what has happened to prices over time, it is far more enlightening to reply that the price index has risen by 10%, rather than that the price of eggs is up 20%, the price of TVs down by 10%, etc.

Similarly, if we wish to compare the output of industrial goods in the United States and Canada, it is convenient to state that the industrial output index of Canada is 8% that of the United States, rather than that Canada produces, say, 9% of the U.S. output of autos, 2% of the U.S. output of aircraft parts, etc.

23-1 PRICE INDEXES

(a) Example

To construct the simplest possible price index, consider the hypothetical data in Table 23-1. The problem is to describe, with a single index number, how the price of this basket of goods has risen. Clearly, an erroneous way would be to sum all the 1970 prices and divide by the sum of the 1965 prices, thus:

$$\frac{\sum P_t}{\sum P_o}(100) = \left(\frac{1.15 + 2.20 + .27}{1.00 + 2.00 + .20}\right)(100) = 113$$

(Note that for convenience, an index value of 100 customarily is assigned to the base period; i.e., index numbers are expressed in percentage form, requiring the multiplication by 100.)

The conclusion is that, on average, prices have risen by about 13%. Why isn't this a good estimate? The answer is that this calculation is dependent on the arbitrary choice of units in which each good is measured. Thus, if bread were measured in ounces rather than pounds, bread would play a much less important role in this calculation; alternatively, if steak were measured in tons rather than pounds, the resulting index would be almost exactly 115, that is, completely dominated by the price of steak.

A simple way to avoid this problem would be to calculate the "price relative" for each item, as shown in the last column of Table 23-1. Each of these can be viewed as an index number for the single item involved. Then a simple average of these price relatives yields:

$$\frac{115 + 110 + 135}{3} = 120$$

TABLE 23-1
Hypothetical Prices of Selected Items,
1965 and 1970

Items Prices	1965 P_o	1970 P_t	Price Relative $P_t/P_o(100)$
Steak (per pound)	1.00	1.15	115
Pepper (per ounce)	2.00	2.20	110
Bread (per pound)	.20	.27	135

(b) The Laspeyres and Paasche Indexes

Although the average price relative is an improvement because it is independent of the unit of measurement of each item, one major flaw remains. Each item is given equal weight; thus, pepper is assumed to have as important a price as bread. In fact, it has an insignificant role in any normal expenditure pattern, and therefore it should be largely disregarded in favor of greater emphasis on the important consumption items of steak and bread. Thus the appropriate index should be a weighted average of the price changes, with the weights being the quantities consumed. Now consider the consumption pattern of the typical consumer, shown in the 1965 column of Table 23-2. When these weights are applied to the prices in Table 23-1, the resulting index is:

$$\text{Laspeyres price index} \triangleq \frac{\sum P_t Q_o}{\sum P_o Q_o}(100) \qquad (23\text{-}1)$$

$$= \frac{1.15(40) + 2.20(1) + .27(100)}{1.00(40) + 2.00(1) + .20(100)}(100) = 122$$

with this index depending heavily on prices of bread and steak, but little affected by the price of pepper.

The question then arises: why use the base period (i.e., the initial 1965) weights, rather than the weights in the comparison period 1970 (also shown in Table 23-2)? When the 1970 quantities are used as weights, they yield a different result:

$$\text{Paasche price index} \triangleq \frac{\sum P_t Q_t}{\sum P_o Q_t}(100) \qquad (23\text{-}2)$$

$$= \frac{1.15(50) + 2.20(1) + .27(90)}{1.00(50) + 2.00(1) + .20(90)}(100) = 120$$

TABLE **23-2**
Hypothetical Quantities
Consumed by the Typical
American, 1965 and 1970

Quantities / Items	1965 Q_o	1970 Q_t
Steak (pounds)	40	50
Pepper (ounces)	1	1
Bread (pounds)	100	90

Although there may be no theoretical reason for preferring one index to the other, there is a practical reason for using the Laspeyres index, if indexing is to be done for a number of years by hand rather than on a computer: the same base period is used in all calculations, whereas the Paasche index involves changing weights for each calculation.

Clearly, in the application of the Laspeyres index, the selection of a base year becomes a crucial issue. For example, the index would become almost meaningless if the initial base period were a wartime year in which steak was rationed, and very little was consumed. In this case, steak essentially would disappear from the calculation of price indexes for the later period when it might have been an important item in consumption. Hence, great care must be taken that the base year is a reasonably typical one, free of disasters or other unusual events that would distort consumption patterns.

(c) Fisher's Ideal Index: a Geometric Mean

First we digress to discuss the geometric mean, which is defined as the nth root of the product of n items:

$$\text{Geometric mean} \overset{\Delta}{=} \sqrt[n]{X_1 X_1 \cdots X_n} \qquad (23\text{-}3)$$

This mean has many attractive properties, especially for business and economics applications. It is the average that is appropriate in describing *rates* of increase. To illustrate, suppose that the population of a suburban town grew as follows:

	Population	Increase per Decade
1950	1,000	
1960	2,000	2 times
1970	16,000	8 times

Increase overall   16 times

The objective is to describe the average rate of increase per decade. First, note that population increased by a factor of 2 in the first decade, and by a factor of 8 in the second, so that the overall increase was 16 times. The simple arithmetic average of the two:

$$\frac{2 + 8}{2} = 5$$

clearly is seen to be inappropriate; if a fivefold increase occurred over two successive decades, the overall increase would be twenty-fivefold.

On the other hand, the geometric mean:

$$\sqrt{(2)(8)} = 4$$

is appropriate; if a fourfold increase occurred over two successive decades, the overall increase would be sixteenfold, which agrees with the actual value.

Now, consider the Laspeyres and Paasche indexes. Since they measure rates of change, the geometric mean is the appropriate way to obtain an average of the two. Thus (23-3) can be used to define:

Fisher's ideal price index $\overset{\Delta}{=} \sqrt{(\text{Laspeyres Index})(\text{Paasche Index})}$

$$= \sqrt{\left(\frac{\sum P_t Q_o}{\sum P_o Q_o}\right)\left(\frac{\sum P_t Q_t}{\sum P_o Q_t}\right)}(100) \quad (23\text{-}4)$$

$$= \sqrt{(122)(120)} = 121$$

23-2   FURTHER INDEXES

(a) Quantity Indexes

Suppose, in Table 23-2, that it is necessary to calculate an index of the change in the physical quantity of goods consumed. Steak purchases have gone up while bread purchases have gone down; which is more important? This depends on the relative importance of each in the consumer's budget, which in turn depends on the prices (in Table 23-1). Thus, prices should be used as weights in calculating a quantity index, just as quantities were used as weights in calculating a price index. We thus obtain:

$$\text{Laspeyres quantity index} \overset{\Delta}{=} \frac{\sum Q_t P_o}{\sum Q_o P_o}(100) \quad\quad (23\text{-}5)$$

$$= \frac{50(1.00) + 1(2.00) + 90(.20)}{40(1.00) + 1(2.00) + 100(.20)}(100) = 113$$

$$\text{Paasche quantity index} \overset{\Delta}{=} \frac{\sum Q_t P_t}{\sum Q_o P_t}(100) \quad\quad (23\text{-}6)$$

$$= \frac{50(1.15) + 1(2.20) + 90(.27)}{40(1.15) + 1(2.20) + 100(.27)}(100) = 111$$

Fisher's ideal quantity index

$$\overset{\Delta}{=} \sqrt{(\text{Laspeyres Index})(\text{Paasche Index})} \qquad (23\text{-}7)$$

$$= \sqrt{\frac{\sum Q_t P_o}{\sum Q_o P_o} \frac{\sum Q_t P_t}{\sum Q_o P_t}} \ (100)$$

$$= \sqrt{(113)(111)} = 112$$

(b) The Cost Index

So far, we have defined indexes of quantity as well as price. It finally would seem appropriate to measure *total cost* of the consumer's purchases. Since the cost of one item, such as steak, is:

$$\text{item cost} = PQ \qquad (23\text{-}8)$$

therefore, for all items:

$$\text{total cost} = \sum PQ \qquad (23\text{-}9)$$

Total costs in the two time periods may be compared by forming their ratio; thus we obtain the:

$$\boxed{\text{cost index} \overset{\Delta}{=} \frac{\sum P_t Q_t}{\sum P_o Q_o}(100)} \qquad (23\text{-}10)$$

$$= \frac{(1.15)(50) + (2.20)1 + (.27)(90)}{(1.00)(40) + (2.00)1 + (.20)(100)}(100) = 135.5$$

Now, if our indexes of price, quantity, and cost are good indexes, they should satisfy the same sort of relation (23-8) that individual items satisfy; that is, we want to check that:

$$(\text{price index})(\text{quantity index}) \overset{?}{=} \text{cost index} \qquad (23\text{-}11)$$

For example, if the price index increases twofold and the quantity index increases threefold, then the cost index should increase sixfold; if not, there must be something wrong with the price and quantity indexes.

Unfortunately, this so-called "factor-reversal test" (23-11) often is *not* satisfied, as simple calculation will show:[1]

$$
\begin{aligned}
\text{Laspeyres indexes} \quad & (122)(113) \overset{?}{=} 135.5 \quad \text{(a)} \\
& 138 > 135.5 \\
\text{Paasche indexes} \quad & (120)(111) \overset{?}{=} 135.5 \quad \text{(b)} \\
& 133 < 135.5 \\
\text{Fisher's ideal index} \quad & (121)(112) \overset{?}{=} 135.5 \quad \text{(c)} \\
& 135.5 \overset{\checkmark}{=} 135.5
\end{aligned}
\qquad (23\text{-}12)
$$

It easily is proved in general[2] that Fisher's ideal index always satisfies the test (23-11); this is one of the properties that justifies the name "ideal."

(c) Further Advantages of Fisher's Ideal Index

We note in (23-12a) that the Laspeyres indexes of price and quantity both are larger than the corresponding ideal indexes; thus their product is larger than it should be (i.e., larger than the cost index). This is not an isolated example; the Laspeyres indexes are too large in nearly all other cases as well, and it is interesting to see why. We shall give the reason in the case of the Laspeyres price index (leaving the quantity index as an exercise).

---

[1]To be mathematically correct in (23-11) and later, we ought to express indexes with *percent* signs; for example, in (23-12a), we actually should write:

$$(122\%)(113\%) \overset{?}{=} 135.5\%$$

that is:

$$(1.22)(1.13) \overset{?}{=} 1.355$$

In dropping the percent signs, we are deferring to customary usage. However, in mathematical proofs such as in (23-12d), we shall find it necessary to put the percent signs back—or even better, delete the factor $100\% = 1$.

[2]**proof**   (Fisher's price index) (Fisher's quantity index)

$$
= \sqrt{\left(\frac{\sum P_t Q_o}{\sum P_o Q_o}\right)\left(\frac{\sum P_t Q_t}{\sum P_o Q_t}\right)} \; \sqrt{\left(\frac{\sum P_o Q_t}{\sum P_o Q_o}\right)\left(\frac{\sum P_t Q_t}{\sum P_t Q_o}\right)}
\qquad (23\text{-}12\text{d})
$$

Many factors cancel, leaving:

$$\frac{\sum P_t Q_t}{\sum P_o Q_o} = \text{cost index}$$

The key to the argument is the numerator, given in (23-1) as $\Sigma P_t Q_o$. We see that current prices are weighted by the old quantities. Thus, whatever goods currently have high prices P_t and hence are not bought very much, are nonetheless weighted at the old relatively high quantities Q_o, making the product $P_t Q_o$ too high. Hence the Laspeyres index overstates.

In a similar way, we may argue that the Paasche index understates. Consequently the ideal index, being the geometric mean between the Laspeyres and Paasche indexes, is a very useful compromise.

PROBLEMS

23-1

Good	Prices		Quantities (per Person)	
	1960	1970	1960	1970
Bread (loaves)	$ .25	$ .30	100	80
Steak (pounds)	1.00	1.25	50	40
Milk (quarts)	.30	.35	100	100
Potatoes (pounds)	.05	.05	60	100

From the above hypothetical data, and using 1960 as the base year:

(a) Calculate the Laspeyres price index for 1970.

(b) Calculate the price relatives for each good, and pick out the smallest and largest. Is the Laspeyres index between them?

(c) Would you guess that the betweenness property of part (b) also is true of the Paasche and Fisher price indexes? Then calculate them to check your guess.

(d) Order the Laspeyres, Paasche, and Fisher price indexes according to size. Is it the same order as in equation (23-12)?

(e) Calculate the three quantity indexes. Order them also according to size. Is it the same order as in equation (23-12)?

(f) Calculate the cost index, and verify which of the three indexes passes the "factor-reversal" test (23-11).

(g) What difference to the various indexes would it make if all quantities were measured, not on a per person basis, but rather as total national consumption of the good?

(h) Try to generalize the result of (b). That is, either prove that the price index is *always* between the largest and smallest price relative, or else find an example for which it is not true ("counterexample").

23-2

Good	Prices				Quantities (per Person)			
	1940	1950	1960	1970	1940	1950	1960	1970
Steak (pounds)	$1.00	$1.20	$1.40	$1.20	40	50	70	50
Bread (pounds)	.10	.20	.40	.20	80	100	110	100

From the above hypothetical data, answer the following questions.
(a) Using 1940 as the base period, fill in the following table.

Type of Price Index	1940	1950	1960	1970
Laspeyres Paasche Ideal				

(b) Which of the three indexes is simplest to calculate? How important do you think simplicity is, as a practical issue today?
(c) In which decade was there the greatest increase in the cost of living?
(d) Fill in the following table.

Type of Price Index	1960 Index, Using 1950 Base	1970 Index, Using 1960 Base	Product of These 2 Indexes
Laspeyres Paasche Ideal			

Note that in 1970, both prices and quantities reverted to what they were in 1950; that is, during the 1960s, everything that happened in the 1950s was reversed. It therefore would be appropriate if the product of the two indexes, which measures the total change over the two decades, was 1 exactly. This so-called "time reversal test" is passed by which indexes?

23-3 Explain why:
(a) The Paasche price index usually is an understatement.
(b) The Paasche quantity index usually is an understatement.

23-3   PRACTICAL CONSIDERATIONS

(a) What's Really Happening to the Cost of Living?

A price index often is known popularly as a "cost of living" index, while a quantity index often is called a "standard of living" or "real income" index. We shall make several important observations about price indexes, which are equally true for quantity indexes.

First, it is necessary to select a representative sample of goods, obviously a much larger group than in our very simple example. Furthermore, another difficulty arises. The price of a television will appear in calculating today's cost of living. But suppose that this is to be compared with its price in the base year of 1940. Televisions were not part of consumer purchases at that time, hence there is no meaningful price that can be attached to them. Thus, since consumer purchasing patterns always are changing, the base period cannot be too far distant in the past; hence, any extended index series generally involves a number of points in time at which base weights have been changed in recognition of the introduction of new products.

Another problem that arises is that even when data does exist, it often is not comparable. Televisions again provide a good example. Suppose that we want to compare prices in 1977, with prices in 1947. Although TVs existed in 1947, they were not at all comparable with the 1977 product. To observe that their price has doubled may be of little consequence; if everyone judges that the increased size of screen, the addition of color, and so on have made a TV now "twice" as good, then isn't this twice the product for twice the price? Have prices changed at all? Unless quality adjustments are accounted for explicitly, such index calculations may not indicate the increase in the cost of living, but rather the increase in the cost of *better* living.

The difficulty with quality judgments is that it is almost impossible to make them. Is the present TV set twice as good, or three times as good as the first model? Autos are another example. At one extreme there are some who are skeptical that the auto is really any better today than it was 30 years ago. On the other hand, we might look to specific improvements (safety equipment, disc brakes, automatic choke, etc.) with a view to making a quality adjustment for each. The question is: how much? A specific estimate is feasible: the extra price of each item when it was first introduced as optional equipment. But even this is not satisfactory. From either the producer's or consumer's point of view, the value of a car does not necessarily increase by the sum of a set of options that become standard equipment. For the producer, the cost of these optional items may decrease when they become standard equipment since they no longer require the special care and attention of "custom items." For the consumer, there may be some items that he simply would not buy if they were options—hence he values them at less than their optional price. (Indeed there may even be items—such as automatic transmission—that some buyers would prefer not to have, even if they were *free* options.)

It may be concluded, therefore, that adjustments for changes in product quality should be made; yet in practice they become extraordinarily difficult. Our example of auto improvements (initially priced as options) is perhaps the most conducive to cold calculation; yet even here, any estimate becomes very difficult to defend. This sort of consideration has led many observers to conclude that, in a world of rapidly changing products, small changes in the cost of living index in the order of, say, 1% or 2% may not mean much.

Another important issue is the selection of the base year; that is, the year with which comparisons are to be made. Thus a Republican pointing to Truman's record (1945–1952) would conclude that this Democratic president had a poor record of controlling inflation. Implicitly, he is using 1945 as his base year. It is one that is quite inappropriate because of its special circumstances; it was at the end of a long period of wartime inflationary pressure that was held in check only by price controls that were difficult or impossible (for any administration) to maintain into the postwar period. Similarly, a Democrat who points to the rapid increase in output during the Roosevelt administration (1932–1945) is biasing his case by the selection of the atypical base year 1932, when employment and production levels were both abnormally low. Clearly, the selection of a base year can load an argument; hence it is crucial that a normal year be chosen. The problem is, of course, that there is no such thing as a perfectly "normal" year; yet some years are more normal than others. In this case, as in many others previously encountered, the good judgment of the statistician becomes crucial.

(b) Price Indexes as Deflators

The money value of U.S. Gross National Product (roughly all goods and services produced each year by Americans) is shown in Figure 23-1. This reflects not only an increase in the physical quantity of goods and services produced but also an increase in their price. Accordingly, to obtain the real (physical) GNP we must deflate money GNP with an appropriate price index.[3] Then we must deflate this series further by population

---

[3] Since this price index is to be used to adjust GNP, it should not be just a consumer price index; instead it is an index of prices of all goods that make up GNP (often called a GNP deflator). Hence it includes prices of investment machinery as well as consumer goods.

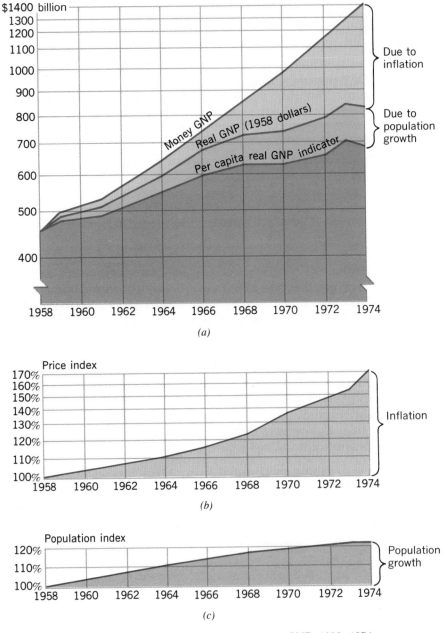

FIGURE 23–1 Deflation of U.S. money GNP, 1958–1974

growth if we wish to show the change in *per capita* real GNP (i.e., the improvement in real income[4] of the average American). Both these deflations are shown in Figure 23-1.

If the combination or mix of goods has changed (e.g., more autos, less bread), then any calculation of how an individual's real income has changed may err unless it is based on information about preferences between goods (e.g., between autos and bread.) This subtle problem is addressed in more detail in the next optional section.

PROBLEMS

23-4 (a) The U.S. Consumer Price Index (CPI) includes about 400 items of food, clothing, rent, medical charges, etc., which typically are purchased by the U.S. family. It is often used as a "wage deflator." Explain how.

(b) Money wages in a number of U.S. industries now are "tied" to the CPI; that is, escalator clauses in union–management bargains ensure that wage rates automatically adjust for any increase in the CPI. How are these justified? Do you see any problem arising if such escalator clauses become widespread in all (or most) industries?

23-5 (a) Would you prefer to spend an (after-tax) annual income of $10,000 by purchasing *entirely* out of the 1975 Sears catalogue, or *entirely* out of the 1940 Sears catalogue (with its lower prices, but more restricted selection of goods, which also are less up-to-date; and of course there will be many goods—like TVs—in the 1975 catalogue that just are not available in the 1940 catalogue.) Assume that not only all goods but also all services (such as medical care and transatlantic trips) are included. In other words, would you prefer to spend your entire income on 1975 goods and services at 1975 prices, or on 1940 goods and services at 1940 prices?

(b) Let us sharpen the answer to (a): how much money do you think you would need to spend in the 1975 catalogue in order to equal the satisfaction you would get by spending $10,000 in the 1940 catalogue? This ratio of expenditure (1975 to 1940) might be called the "subjective" price index.

(c) During this period, the U.S. Consumer Price Index actually rose from 42 to 161 (base = 1967). How does this compare with your answer in (b)? In your judgment, therefore, has the CPI overstated or understated price increases?

(d) Repeat (a) to (c) for an income of $10,000,000 instead of $10,000.

---

[4]Although GNP and national income are not identical concepts, they move together closely enough that changes in per capita GNP can be used to roughly approximate changes in average income.

23-6 "Indexes may be used for international or interspatial, as well as inter-temporal, comparisons." Suppose that the hypothetical data in Tables 23-1 and 23-2 represent comparisons for France and Germany, rather than for 1965 and 1970. With this substitution, interpret the price index (23-1) and the quantity index (23-5).

*23-4   INDEX COMPARISONS OF REAL INCOME

In this section, we will develop a more sophisticated view of index numbers. This is a starred section, since it becomes a bit more difficult toward the end and requires indifference curve analysis—which now will be very briefly reviewed.

(a) Indifference Curves

Our illustration will focus on changes in quantity (real income), rather than on prices. This will involve an examination for a typical consumer of the physical quantities of goods purchased in two time periods, using prices as weights.

Figure 23-2 illustrates a typical indifference curve, representing a consumer's preference for two goods X and Y. All points on this curve represent combinations of physical quantities of these two goods that leave the consumer equally satisfied, that is, indifferent—hence the name "indifference curve." Thus this individual is equally happy at Q_1 (with 100 units of X, and 300 of Y), or at Q_2 (with 200 of each); in other words, he is willing to trade 100 units of Y for 100 units of X. However, as he moves from Q_2 to Q_3, the scarcity of Y begins to pinch, and he is reluctant to keep trading Y away, unless he receives more X in return; in fact, for 100 units of Y, he now requires 200 units of X. This changing "rate of

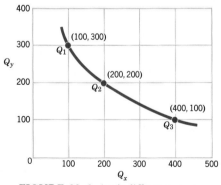

FIGURE 23–2 An indifference curve

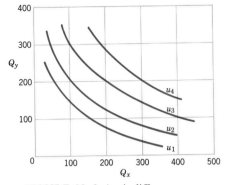

FIGURE 23–3 An indifference map

substitution" simply reflects the fact that the consumer's appetite for a specific good (Y) increases as the good becomes scarce; this in turn implies that the indifference curve will be concave.

Figure 23-3 illustrates a whole set of indifference curves, called an indifference system or map. The curve u_3 is reproduced from Figure 23-2. u_4 is a similar indifference curve, superior to u_3. In other words, any point on u_4 is preferred to any point on u_3.

This again illustrates the useful technique, often used by economists, of forcing three variables into a two-dimensional diagram. The three variables are Q_y, Q_x, and U, with U representing the consumer's satisfaction (also called "utility" or "real income"). Thus, this indifference mapping represents a "utility hill;" as we move from the origin to the northeast, we move up this hill onto higher and higher levels of satisfaction.

Figure 23-4 shows how the consumer will rationally select a combination of X and Y. He is limited by his budget, which we suppose is $6,000. If X costs $10 per unit and Y costs $30, then AC represents his budget limitation. For example, if he buys no Y, he can buy $6,000/10 = 600$ units of X, at point C. Or, if he buys no X, he can buy $6,000/30 = 200$ units of Y, at point A. Generally, we can confirm that this budget will just purchase any other combination of X and Y that lies on the straight line AC, by noting from (23-9) that:

$$P_xQ_x + P_yQ_y = \text{total cost of purchases} \qquad (23\text{-}13)$$

$$= \text{consumer's budget, } B \qquad (23\text{-}14)$$

and, in this example:

$$10Q_x + 30Q_y = 6,000 \qquad (23\text{-}15)$$

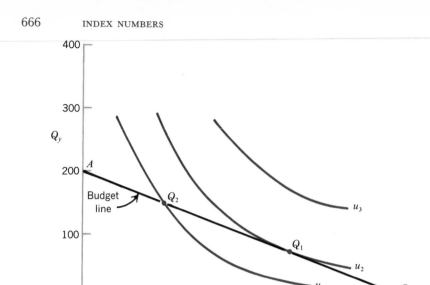

FIGURE 23–4 Consumer equilibrium

which is the equation of a straight line, called the *budget* line. As an exercise, you can confirm that the slope of this line is determined by relative prices; that is:

$$\text{slope} = \frac{-P_x}{P_y} = \frac{-10}{30} = -.33 \qquad (23\text{-}16)$$

while its distance from the origin is proportional to the size of the consumer's budget B.

The consumer wishes to use this budget to reach the highest level of utility (real income); that is, the highest u curve. It is easy to see that this will occur where his given budget line is tangent to an indifference curve. In our example, the highest indifference curve that he can reach is u_2: and he can reach this only by purchasing the quantity of X and Y that is represented by Q_1. Note how any other set of feasible purchases, say Q_2, necessarily would leave him on a lower indifference curve, u_1; at the same time, a higher indifference curve, say u_3, simply cannot be reached with his budget.

This very brief introduction to indifference curve analysis now can be used to illuminate the difference between the various index numbers.

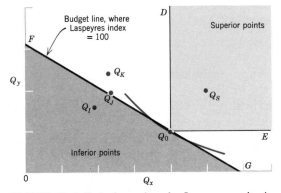

FIGURE 23-5 Indexing using the Laspeyres criterion

(b) Laspeyres Index and Revealed Preference

In these next three parts, we shall see how each of the three index numbers measures a change in real income (quantity of goods purchased). For example, in Figure 23-5, let Q_o represent the real income of the consumer in the base period. Do the other four points in this diagram, all representing possible consumption combinations in the later comparison period, represent an increase or decrease in his real income?

It is easy to see that Q_S represents an increase in real income, since the consumer receives more of both goods. Indeed, any point within the angle DQ_oE represents an increase in real income. However, a comparison of each of the other three points with Q_o is much more difficult, since each involves increased consumption of one good, but decreased consumption of the other. The only way that the consumer can be judged better off by moving from Q_o to Q_I, say, is to know what his indifference map looks like; specifically it is necessary to know whether his indifference curve running through Q_o passes above or below Q_I. But how can we possibly discover anything about the specific location of an indifference curve?

The answer comes from "revealed preference;" we can make inferences about an indifference system by examining how the consumer chooses between goods in the face of a specific set of prices. We note that the consumer decided to purchase Q_o with his base-year budget B_o when the prices of X and Y were P_{ox} and P_{oy}.

Thus the budget line, from (23-13) is:

$$P_{ox}Q_x + P_{oy}Q_y = B_o \qquad (23\text{-}17)$$

where the first subscript refers to the time, and the second to the good.

This is shown in Figure 23-5 as the line *FG*, which passes through Q_o with slope given by (23-16) as $-P_{ox}/P_{oy}$. The indifference curve must be tangent to this budget line, and therefore can be sketched roughly in Figure 23-5. Then point Q_o is seen to be superior to all other points on the budget line *FG*, and therefore superior to all points below the budget line. Thus we have shown that Q_o is superior to Q_I.

As well as delimiting the clearly inferior points, the budget line *FG* gives a very interesting interpretation of the Laspeyres quantity index: it is precisely the set of comparison points with a Laspeyres index[5] of 100. This is a very natural conclusion, of course, since both the budget line *FG* and the Laspeyres quantity index are based on the same *base* prices P_o. Yet these comparison points on the budget line have been shown to be inferior to Q_o, so that the Laspeyres index (which rates them equal to Q_o) is seen to overstate their utility. This overstatement is even more dramatic in the case of a comparison point that is slightly above the budget line, such as Q_J: given the indifference curve drawn into Figure 23-5, Q_J represents a reduction in utility from Q_o; but the Laspeyres index erroneously indicates that it is an increase.

Thus we have rigorously proved the point first made in Section 23-2(c): the Laspeyres quantity index *overstates* the utility (real income) of the comparison point.

(c) Paasche Index and Revealed Preference

Although the proof is more difficult,[6] the interpretation of Paasche indexing is similar to Laspeyres, as shown in Figure 23-6. The one difference is

---

[5]**proof**   The Laspeyres index (23-5) may be written in the notation of this section as:

$$\text{Laspeyres index} = \frac{P_{ox}Q_{tx} + P_{oy}Q_{ty}}{P_{ox}Q_{ox} + P_{oy}Q_{oy}}(100) \tag{23-18}$$

Since the point $Q_o = (Q_{ox}, Q_{oy})$ lies on the budget line, it satisfies the budget equation (23-17):

$$P_{ox}Q_{ox} + P_{oy}Q_{oy} = B_o \tag{23-19}$$

Any comparison point $Q_t = (Q_{tx}, Q_{ty})$ that also is on the budget line also satisfies equation (23-17):

$$P_{ox}Q_{tx} + P_{oy}Q_{ty} = B_0 \tag{23-20}$$

Substituting (23-19) and (23-20) into (23-18):

$$\text{Laspeyres index} = \frac{B_o}{B_o}(100) = 100$$

*[6]To rigorously interpret the Paasche index, we first address the question: how does a point such as Q_K in Figure 23-5 compare with Q_o? To answer this, it is necessary to have even more information (i.e., revealed preference) about the shape of the individual's indifference curves.

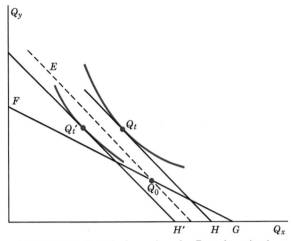

FIGURE 23–6 Indexing using the Paasche criterion

this: whereas all points with a Laspeyres index of 100 are on the line FQ_o (with slope determined by *old* prices), all points with a Paasche index of 100 are on the line EQ_o (with slope determined by the *new* prices). Finally, whereas the Laspeyres index overstates the utility of any comparison point such as Q'_t, the Paasche index understates it.

(d) Fisher's Ideal Index: a Compromise

In Figure 23-7, we summarize the conclusions of the previous sections. While we may say that Laspeyres indexing, using old prices P_o, is "in the

---

This information can be obtained from the new budget line, shown in Figure 23-6 as Q_tH, which is sloped according to the new prices P_t and passes through Q_t, the quantity of goods actually purchased in the comparison period; (note that Q_o may lie either above or below this line). Of course, all points below this line are inferior to Q_t. If Q_o is such a point, as shown in Figure 23-6, then the issue is decided.

However, if the new quantity consumed is at Q'_t, so that Q_o lies above the budget line Q'_tH', then it is not clear whether Q'_t or Q_o is superior. It therefore is appropriate to draw a line through Q_o, parallel to Q_tH, which we denote by Q_oE. This line divides the points like Q_t that are clearly superior to Q_o, from the points like Q'_t that are ambiguous.

The new budget line Q_tH gives a very interesting interpretation of the Paasche quantity index: Q_t has a Paasche index of 100 when compared to any base point that lies precisely on Q_tH. (This is entirely similar to the interpretation of the Laspeyres index, the only difference being that it is now the new prices P_t which give both the slope of the budget line Q_tH and the weights for the Paasche quantity index. The proof is just like the proof for the Laspeyres index in footnote 5.) Yet such a base point on the new budget line would in fact be inferior to Q_t; thus the Paasche index (which rates them equal) overstates the utility of the base point relative to the comparison point Q_t.

Turning this statement around, the Paasche index *understates* the utility of the comparison point Q_t relative to the base point Q_o.

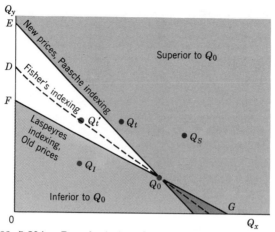

FIGURE 23–7 Using Paasche index, Q_t still superior to Q_0 but Q_t' is not

FQ_o direction," Paasche indexing, using new prices P_t, is in the EQ_o direction. Fisher's indexing is the geometric mean, and is therefore in some intermediate direction, say DQ_o, with DQ_o made up of all points with a Fisher index of 100.

Points like Q_I, that are below FQ_o and in fact are inferior to Q_o, correctly are given an index less than 100 by all three indexes—Laspeyres, Paasche, and Fisher's. Similarly, points like Q_t that are above EQ_o and are, in fact, superior to Q_o, are also correctly given an index greater than 100 by all three indexes. However, the wedge EQ_oF is ambiguous. Some points (near the lower line FQ_o) are, in fact, inferior, while other points (near the top line EQ_o) are, in fact, superior. The Laspeyres index erroneously overstates them, giving them *all* an index more than 100, while the Paasche index erroneously understates them, giving them *all* an index of less than 100. On the other hand, Fisher's ideal index splits them, and hence is likely to make fewer errors. To avoid making any errors at all, of course, we would need more complete knowledge of the consumer's indifference map.

(e) Further Difficulties

Several points should be emphasized in conclusion. First, the analysis holds only if the consumer's preferences do not change over time, specifically between the base and comparison periods. Even more important, this discussion has concentrated on evaluating real income changes for an individual, rather than for a group of individuals. The problem that arises in any evaluation of real income for a group, community, or nation, is that an indifference system cannot be defined for a collection of individuals, unless quite restrictive assumptions are made about the similarity of individual preferences.

PROBLEMS

*23-7 In each blank, insert the correct word: Laspeyres, Paasche, or Fisher's.
(a) If the comparison quantities Q_t have _____ index greater than 100, they necessarily have greater utility than the base period quantities.
(b) If the comparison quantities Q_t have _____ index less than 100, they necessarily have less utility than the base period quantities.
(c) The _____ index is always between the other two indexes.

*23-8 Suppose that you are omniscient, and you know that a certain consumer's indifference map is the one below.

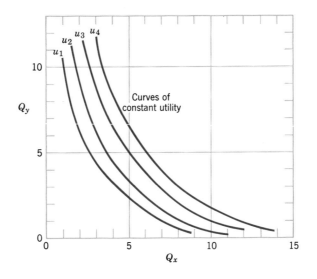

(a) Using the prices in Table A below, graph the budget line for each period, and hence verify the quantities purchased in Table B.
(b) Fill in Tables C, D, and E.

		Base year t_o	Comparison Year t_1	Comparison Year t_2
A.	P_x P_y Budget	\$ 300 600 3,000	\$ 500 250 3,000	\$ 400 200 2,000
B.	Q_x Q_y	8? 1?	3? 6?	2? 6?
C.	Quantity indexes: Laspeyres Paasche Fisher's	100 100 100	? ? ?	? ? ?
D.	What is the utility change since t_o?		?	?
E.	Which quantity index most accurately measures the utility change since t_o?		?	?

*23-9 Prove the statement made in footnote 6 that Q_t has Paasche quantity index 100 relative to any base point that lies precisely on the new budget line Q_tH.

CHAPTER 24

Sampling Designs

There are many in this old world of ours who hold that things break about even for all of us. I have observed for example that we all get the same amount of ice. The rich get it in the summertime and the poor get it in the winter.

Bat Masterson

The simple random sampling of Chapter 6 is conceptually and computationally the most straightforward method of sampling a population. However, alternative procedures do exist that may, in certain cases, be more efficient. Several of these are described in this chapter: each is seen to combine random sampling with some design features that are aimed at improving efficiency.

24-1   STRATIFIED SAMPLING

(a) Simplified Example

Let us consider first an exaggerated example, where the benefits of stratified sampling are clearest. Suppose that a large population is known to be 60% urban, 40% rural, and that the problem is to estimate average income. Suppose that the unknown income levels are $10,000 for each urban worker and $5,000 for each rural worker. Accordingly, the population earns a mean income:

$$\mu = (10,000)(.6) + (5,000)(.4) = 8000 \tag{24-1}$$

TABLE **24-1**
Random versus Stratified Sampling: an Oversimplified Example

Population			Various Samples				
			Typical Simple Random Samples			Stratified Samples	
		Mean Income (in $1,000 units)				(i) Proportional	(ii) Weighted
Stratum	Proportion (1)	(2)	(3)	(4)	(5)	(6)	(7)
Urban	.60	10	10	5	5	10	10
			5	5	10	10	
			10	5	10	10	
			10	10	5		
Rural	.40	5	10	5	10	5	5
						5	
Total	1.00	$\mu = 10(.6) + 5(.4) = 8$					
Sample Means			9	6	8	8 $\checkmark$	$10(.6) + 5(.4) = 8\checkmark$

How well would a simple random sample estimate μ? Several possible random samples of size 5 are displayed in columns (3), (4) and (5) of Table 24-1. Because the samples are completely random, the number of rural workers fluctuates; this in turn causes the sample mean $\overline{X}$ to fluctuate. Of course, occasionally we will be lucky, as in column (5), when the proportion of rural workers in the sample is exactly the same as in the population.

But if we have prior knowledge of the proportion of urban and rural workers, why depend on luck? Surely it would be wise to design the sampling proportions to be exactly equal to the population proportions. This proportional sampling is shown in column (6); note that the result in this greatly oversimplified case is an estimate that is dead on target. In fact, this same favorable result is guaranteed with a sample of just one observation from each stratum (column (7))—provided, of course, that instead of computing the ordinary mean, each observation is weighted according to the size of its stratum. This "weighted sampling" yields:

$$\overline{X} = 10,000(.6) + 5,000(.4) = 8,000 \qquad (24\text{-}2)$$

which, of course, exactly coincides with μ in (24-1). This is the perfect estimator, dead on, with a sample of only two (the number of strata in the population).

But we emphasize that the reason we have been able to do so well in this contrived example is because:

(i) The weights w_1 and w_2 (i.e., population proportions in each stratum) were known beforehand.

(ii) The income of each individual in each stratum was exactly the same (i.e., the variance within each stratum was zero; $\sigma_1^2 = \sigma_2^2 = 0$).

Normally, we would not be so lucky. But this does illustrate the essence of sample design: it is a means of increasing sampling efficiency (i.e., reducing the necessary size of the sample) by the imaginative application of prior information. Of the two components cited above, the more important gain in efficiency generally comes from prior knowledge of weights; this is fortunate because this is the sort of information that is more likely to be available—just as in our example, an independent source like the U.S. population census could be drawn upon for information on urban/rural proportions. Hence, in this analysis we will assume throughout that prior knowledge on weights is available. But we cannot assume, as in (ii) above, that all individuals in each stratum are identical. Therefore, the next step is to generalize the analysis to take this into account.

TABLE **24-2**
Generalization of Table 24-1

Stratum	Population			Stratified Sample	
	Proportion of Whole Population (weight)	Mean	Variance	Sample Size	Sample Mean
1	w_1	μ_1	σ_1^2	n_1	$\overline{X}_1 = \sum_j X_{1j}/n_1$
2	w_2	μ_2	σ_2^2	n_2	
$\vdots$					
i	w_i	μ_i	σ_i^2	n_i	$\overline{X}_i = \sum_j X_{ij}/n_i$
$\vdots$					
r	w_r	μ_r	σ_r^2	n_r	
	$\sum w_i = 1$	$\mu = \sum \mu_i w_i$		$n = \sum n_i$	$\hat{\mu} = \sum \overline{X}_i w_i$

(b) General Case

Table 24-2 is a generalization of Table 24-1; the variance within the first stratum is designated σ_1^2, and so on. The most general stratified sample consists of drawing a sample of size n_1 from the first stratum, and so on, so that the total sample size $n = \Sigma\, n_i$. Then calculate the sample mean for each stratum; for example $\overline{X}_1 = \Sigma_j\, X_{1j}/n_1$, where X_{1j} represents the jth observation in the first stratum.

Now the population mean is given by:

$$\mu = \mu_1 w_1 + \mu_2 w_2 + \cdots = \sum \mu_i w_i \qquad (24\text{-}3)$$
$$\text{like } (24\text{-}1)$$

where the weight w_i is the proportion of the population in the ith stratum. It is natural to estimate μ with:

$$\hat{\mu} = \overline{X}_1 w_1 + \overline{X}_2 w_2 + \cdots = \sum \overline{X}_i w_i \qquad (24\text{-}4)$$
$$\text{like } (24\text{-}2)$$

Using the rules for linear combinations, we may prove (Problem 24-4) that:

$$\boxed{E(\hat{\mu}) = \mu} \qquad (24\text{-}5)$$

that is, (24-4) is an unbiased estimator. Furthermore:

$$\boxed{\text{var}\,(\hat{\mu}) = \sum w_i^2 \frac{\sigma_i^2}{n_i}} \qquad (24\text{-}6)$$

This leaves the problem of how the n_i are specified; that is, how many observations should be drawn from each stratum? Three common methods are given below.

(c) Proportional Sampling

The most obvious way to choose n_i is to make n_i proportional to w_i; that is:

$$\boxed{\begin{array}{c} \text{proportional sampling requires} \\[4pt] n_i = w_i n \end{array}} \qquad (24\text{-}8)$$

as illustrated earlier in column (6) of Table 24-1. Then, substituting (24-8) into (24-6) yields:

$$\text{var}\,(\hat{\mu}) = \frac{1}{n} \sum w_i \sigma_i^2 \qquad (24\text{-}9)$$

(d) Stratification After Sampling

Sometimes it is not possible—or at least it is extremely inconvenient—to stratify before sampling. For example, suppose that it is known that 80% of the population of a city is White and 20% Black, and that income is being sampled in a house-to-house survey. There is no way of knowing before you knock on the door (i.e., before the sample is taken) whether a household is White or Black (i.e., from which stratum this sample value comes).[1] The solution is first to take a simple random sample; then to stratify it (i.e., allocate each observation to the appropriate stratum); and finally, to estimate the mean using the known weights w_i in (24-4).

In a very large sample, the sampling proportions that you observe (n_i/n) will be close to the population proportions (w_i), so that:

$$n_i \simeq w_i n \qquad (24\text{-}10)$$

That is, the n_i are approximately what you would have set them if you could have used the prespecified proportional sampling, as in (24-8). Thus the variance of $\hat{\mu}$ is approximately the same as in proportional sampling (24-9):

$$\text{var}\,\hat{\mu} \simeq \frac{1}{n} \sum w_i \sigma_i^2 \qquad (24\text{-}11)$$

However, in small samples, to the degree that the approximation (24-10) holds less well, the result is a slight loss in efficiency; that is, var $(\hat{\mu})$ exceeds that given in (24-11). This is the only cost of not being able to stratify in advance.

---

[1] Compare this with our earlier urban/rural example, in which any household clearly could be specified as urban or rural prior to sampling.

(e) Comparison of Proportional Sampling to Simple Random Sampling

To facilitate the comparison, we quote a formula for the whole population variance σ^2, in terms of the strata:[2]

$$\sigma^2 = \sum w_i(\mu_i - \mu)^2 + \sum w_i\sigma_i^2 \qquad (24\text{-}12)$$

If we take a simple random sample, the variance for $\overline{X}$ is σ^2/n, which may be expressed according to (24-12) as:

$$\text{var}\,(\overline{X}) = \frac{\sigma^2}{n} = \frac{1}{n}\sum w_i(\mu_i - \mu)^2 + \frac{1}{n}\sum w_i\sigma_i^2 \qquad (24\text{-}13)$$

If, on the other hand, we use the more efficient proportional stratified sampling technique, then the variance of the estimator is given by (24-9). This is recognized to be simply the last term in (24-13); thus the second-last term in this equation must represent the additional variance (extra cost) involved in simple random sampling. Therefore, (24-13) can be written:

$$\boxed{\begin{array}{c} \text{variance of } \overline{X} \\ \text{in simple random} \\ \text{sampling} \end{array} = \begin{array}{c} \text{variance that can be} \\ \text{eliminated by using} \\ \text{proportional stratified} \\ \text{sampling instead} \end{array} + \begin{array}{c} \text{variance that} \\ \text{must remain} \end{array}}$$

$$(24\text{-}14)$$

This leads us to another intuitively appealing conclusion: stratified sampling is seen in (24-13) to be particularly worthwhile when $(\mu_i - \mu)^2$ are large; that is, when differences between strata are large.

(f) Optimal Stratified Sampling

Finally, if the population can be stratified in advance, there is an alternative method with even greater efficiency than proportional sampling (i.e., with variance *less* than that given in (24-9)—in fact, with the smallest possible variance). Choose:

---

[2]Formula (24-12) may be stated verbally as: total variance = variance between strata + variance within strata. In this form, it is analogous to the analysis of variance (10-10), and its proof is similar.

$$\boxed{\text{optimum } n_i \text{ proportional to } w_i\sigma_i} \tag{24-15}$$

Then, as an exercise (Problem 24-4), we may prove that the variance is only:

$$\boxed{\operatorname{var} \hat{\mu} = \frac{1}{n}\left(\sum w_i\sigma_i\right)^2} \tag{24-16}$$

which is less than (24-9).

The interpretation of (24-15) corresponds to our intuition; a relatively large sample should be taken from a stratum:

(i) If w_i is large; that is, if the stratum is an important part of the population; and

(ii) If σ_i is large; that is, if the stratum is highly varied, and hence requires a large sample to estimate $\overline{X}_i$ accurately. (Or, to restate: the smaller the σ_i, the smaller the necessary sample; the limiting case in which a sample of only one is required in a stratum with zero variance already has been given in the example in column (7) of Table 24-1).

This method has restricted application because σ_i (as well as w_i) must be known beforehand. However, the fact that the sample s_i is a good enough estimator of σ_i—even in fairly small samples—may make it possible to proceed in two stages: as soon as a part of the sample has been taken on a purely random basis, the s_i observed up to that point may be used to estimate the σ_i; these then are used (along with prior information on the w_i) in (24-15) to estimate the optimal n_i, that is, the optimal design for the balance of the sampling. However, the gains from such interim estimation of σ_i tend to be small relative to the gains from prior knowledge of the w_i. Therefore, this procedure tends to be justified only if the sampling must be stopped to allow interim calculations for some other reason, as well.

24-2   OTHER SAMPLING DESIGNS

(a) Multistate (Cluster) Sampling

When a population is costly to sample, say because of transportation costs, multistage sampling may provide a way to cut costs. To illustrate, a simple random sample of the political views of 10,000 American men would be extremely costly in terms of time and transportation costs, as the interviewers would have to travel through thousands of counties spread over all 50 states. Therefore, it makes sense to just draw a random sample of, say, 10 states (first stage). Then within each state, draw a random sample of, say, 20 counties (second stage). Finally, within each county, draw a random sample of 50 men (last stage).

Clearly, this may provide great savings in cost; but it also is true that this procedure will tend to yield less information. To the extent that the men in a given county think alike, only the first man provides completely new information, while the next 49 tend increasingly to repeat. This loss of information is avoided in simple random sampling, where each observation (man) is drawn independently—thus providing completely new information.

The decision on whether or not multistage cluster sampling is advisable is thus seen to involve an evaluation of how the advantage of reduced sampling costs compares with the disadvantage of information loss.

(b) Systematic Sampling

As an example, we may wish to draw a sample of 20 from a deck of 2,000 IBM cards. The most straightforward way is simply to pull every 100th card. Unless strange periodicities occur in the data, this method has about the same precision as simple random sampling, so those formulas may be used as approximations.

The only drawback of systematic sampling is that its precision may badly deteriorate if systematic fluctuations do occur in the data. Although such a misfortune would be compounded by the fact that the sampler might receive no warning, this difficulty rarely arises in practice.

(c) Sequential Sampling

Sequential sampling is simple random sampling, with the sample size n no longer fixed, but rather determined by the information in the observations as they accumulate. For example, in testing the "true die" hypothesis, if the die keeps turning up ace each time, a sample of $n = 5$ (or perhaps 10) would suffice to make a decision (to reject the die); a predetermined sample size of 100 just is not necessary. This procedure already has been encountered in the footnote following Table 9-1: the idea is to take a small sample, and if the result is unambiguous (as in the case above), the job is done; but if the sample result is ambiguous (i.e., almost $\frac{1}{6}$ of the rolls are ace), a second, and perhaps even further stage of sampling is required until the ambiguity is resolved. This procedure is justified if the advantage of the smaller average sample size exceeds the disadvantage of having to sample several times.

24-3 CONCLUSIONS

In this chapter, we have discussed a number of imaginative ways of improving on simple random sampling; essentially these involve applying:

(i) Prior information (e.g., the w_i in stratified sampling); or

(ii) Information collected in the early stages of sampling (e.g., the initial sample results used in sequential sampling).

The general principle is that the sooner you have this information, the better use can be made of it. Thus in optimal sampling, prior knowledge of σ_i will be more effective than an estimate s_i derived after initial sampling. Of course, if neither (i) nor (ii) is available, then the only option is simple random sampling—where information is used only after all the sampling is complete.

The final observation is that, in practice, sampling often combines several of the above techniques. In stratified sampling, for example, cluster sampling often may be used within each stratum.

PROBLEMS

24-1 In a certain profession, suppose that men and women had the following annual incomes:

Income ($000)	Frequency Men	Frequency Women
10		500
15		500
20	100	500
25	200	
30	400	
35	200	
40	100	

(a) What is the overall mean income μ?
(b) For a sample of $n = 25$ incomes, what would be the variance of the estimator of μ for:
  (i) A simple random sample?
  (ii) A proportional stratified sample (stratified on sex)?
  (iii) An optimally stratified sample?

24-2 In order to investigate student political attitudes, suppose that you wish to take a sample of 100 students from your university; how would you take:

(a) A simple random sample?

(b) A stratified sample?

(c) A cluster (multistage) sample?

(d) Which sampling scheme do you think would be most economical (i.e., provide most accuracy for your dollar)?

24-3 Conduct a stratified sample of $n = 10$ students in your class, in order to estimate:

(a) Mean height

(b) Average number of novels read last year

(c) The proportion of car owners

(d) The proportion of cigar smokers

In each case, you must first think of an appropriate division of the student population to serve as the basis of stratification. In each case, do you think the stratification was justified, in the sense of providing a more accurate estimate?

24-4 (a) Prove (24-5) and (24-6).

(b) Prove (24-16).

24-5 In order to estimate the total inventory of its product that is being held, a tire company conducts a proportional stratified sample of its dealers, with these dealers being stratified according to their inventory held in the previous year. For a total sample size of $n = 1,000$, the following data were obtained:

Stratum (Last Year's Inventory)	Population		Proportional Sample (Current Inventory)		
	Number of Dealers	Proportion w_i	n_i	$\overline{Y}_i$	s_i^2
0–99	4,000	.20	200	105	1,600
100–199	10,000	.50	500	180	2,500
200–299	5,400	.27	270	270	2,500
300–	600	.03	30	390	5,600
	$N = 20,000$	$\sum w_i = 1$	$n = 1000$		

(a) From this stratified sample, estimate the mean inventory μ, and hence the total inventory.

(b) Estimate the optimum strata sizes n_i, using s_i^2 for σ_i^2 in (24-15).

(c) Again using s_i^2 for σ_i^2, and assuming a fixed sample size of $n = 1000$, estimate the variance of:

(i) $\hat{\mu}$, for the proportional stratified sample in (a).

(ii) $\hat{\mu}$, for the optimal stratified sample in (b).

(iii) $\overline{X}$, for a simple random sample.

(d) Express your answer to (c) in efficiency terms. That is, calculate the efficiency of:

(i) $\hat{\mu}$ (proportional stratified sampling) relative to $\overline{X}$ (simple random sampling);

(ii) $\hat{\mu}$ (optimal stratified sampling) relative to $\overline{X}$.

(e) On the basis of each of the estimators in (c), show how you would construct a 95% confidence interval for the mean inventory μ, and hence for the total inventory. Which of the three confidence intervals is best? Which is worst?

(f) Answer true or false; if false, correct it:

In sampling problems like this tire inventory problem, it often is wise to use proportional stratified sampling. It is true that optimal stratified sampling would be better, but the gains in efficiency that would result— even if full prior knowledge of σ_i^2 were available—often are relatively small.

24-6 Compare the regression analysis in Problem 12-15 with the stratified sampling in Problem 24-5. True or false? If false, correct it.

(a) Both methods involve the use of prior information (on last year's inventory X) to improve the estimate of this year's inventory Y.

(b) The two methods differ, however, in the way they use the prior information. In the case of regression, prior information on X was used *after* the sample of 6 was taken. It indicated that the sample of 6 dealers was a nontypical X group (i.e., their average inventory for last year $\overline{X}$ was smaller than the population average μ_X); hence, an upward correction was required. On the other hand, the proportional stratified sampling used prior information on last year's inventories *before* sampling to ensure that the sample taken was from a representative X group.

(c) Since stratification after sampling would involve application of knowledge about X after the sample is taken, it provides an even closer analogy to regression.

*(d) Proportional stratified sampling holds three advantages over regression.

(i) It provides a more accurate estimator because prior knowledge is used before the sample is taken, rather than after.

(ii) It can be applied to either a numerical (e.g., size of inventory) stratification, or a nonnumerical (e.g., male/female) one; but regression can only be used for a numerical stratification.

(iii) It does not require the assumption that the relation between X and Y is linear (which is required in regression analysis).

*(e) If the knowledge of last year's inventory had been used in the regression analysis of Problem 12-15 to fix X values at appropriate predetermined levels before the sample was taken, then the analogy to proportional stratified sampling no longer would hold.

CHAPTER 25

Game Theory

Gambling—the sure way of getting nothing for something.

Wilson Mizner

Anyone who has come this far deserves a little recreation. In this last chapter, therefore, we break away from the main theme of statistical inference in order to present game theory—a very simple and entertaining application of elementary probability.

"Games" are simply mathematical models of conflict situations, where the payoffs (outcomes) are determined by the strategies of the players. Because a player can choose his strategy, he has some control over the outcome of the game. But he is not in complete control, since the outcome also depends on the strategy of his opponents.

25-1   ZERO SUM GAMES

The simplest game occurs when one player's gain always equals the other player's loss, as in poker for example. Then the total amount won by the players equals zero, and the game is called "zero-sum." To further simplify, we shall just consider games involving two persons, A and B.

In Table 25-1, we give an example, in the form of a "payoff matrix" representing payments from B to A; for example, if A selects strategy 2 and B selects strategy 1, B pays A \$5. A will try to select a strategy to make the outcome as large as possible, while B will try to keep the outcome as small as possible.

Obviously, B will have no interest in playing this game as is, since he can do nothing but lose. So a payoff matrix normally involves some positive elements (where B pays A) and some negative ones (where A pays B). Alternatively, in order to induce B to play the game shown in Table 25-1, A might bribe B \$4 for each time he plays. This is the assumption we now make, in order to keep our payoff matrix all positive for easier geometric interpretation. The question is "With this \$4 side payment, is it in B's interest to play this game?"

Depending on who goes first, there are several ways in which the game may proceed, which we next consider one by one.

TABLE 25-1
An Example of a Payoff Matrix for A
(Loss Function for B)

A's Strategies \ B's Strategies	1	2	3
1	3	6	2
2	5	4	8

(a) A goes last

If B has to select his strategy first, what should he do? Suppose that he plays strategy 1. Then A would respond by picking the maximum in that first column (5), obtained by his playing strategy 2. Thus, \$5 would be A's gain, and B's loss. Similarly, B can easily see what it would cost him to play any other strategy—the maximum in that column. These maximums are listed along the bottom of Table 25-2.

In scanning all these possible losses, B will, of course, choose the minimum[1] value, 5. This *mini*mum of the *max*imums is called the *minimax value;* the corresponding strategy (strategy 1) is called the *minimax strategy*.

---

[1] Note how similar this is to the solution of Table 20-3, when we also chose the minimum loss listed along the bottom. The only difference is that Table 20-3 lists average losses (averaged over the states of nature), whereas Table 25-2 lists certain losses (chosen by an opponent).

TABLE **25-2**
Several Solutions to the
game of Table 25-1

B / A	1	2	3	Min
1	3	6	2	2
2	5	4	8	4 ←Maximin value
Max	5	6	8	

↑
Minimax value

In conclusion then, if B must go first, he will play strategy 1, A will respond with strategy 2, and the payoff from B to A will be $5.

(b) *B* goes last

Now if A must go first, what should he do? He will go through the same sort of agonizing as B did formerly. A can see what he could gain by playing any given strategy—the minimum in that row. These minimums are listed along the right of Table 25-2. As A looks at all these possible gains, obviously he will choose the maximum value, 4. This *maxi*mum of the *min*imums is called the *maximin value;* the corresponding strategy (strategy 2) is called the *maximin strategy.*

In conclusion, then, if A must go first, he will play strategy 2, B will respond with strategy 2, and the payoff from B to A will be $4. A notes that this payoff is not as great as the $5 he formerly got in section (a) when he went last, and was "sitting on top of B." In fact, it generally is true for games of any size that the:

$$\boxed{\text{maximin value} \leq \text{minimax value}} \tag{25-1}$$

(c) It doesn't matter who goes last

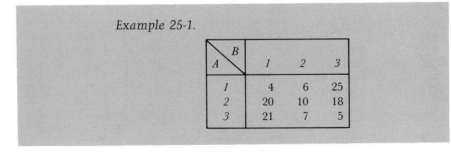

Example 25-1.

B / A	1	2	3
1	4	6	25
2	20	10	18
3	21	7	5

For the above matrix of payoffs from B to A, calculate the value of the game:
(a) If A goes last.
(b) If B goes last.

Solution.

A \ B	1	2	3	(b) Min
1	4	6	25	4
2	20	10	18	10 ←Maximin value
3	21	7	5	5
(a) Max	21	10	25	

Minimax value

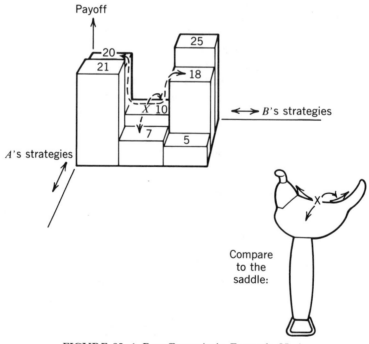

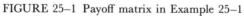

FIGURE 25–1 Payoff matrix in Example 25–1

In Example 25-1, we have a peculiar case in which the maximin value equals the minimax value, and so there is no advantage in going last. This happens because the entry 10 in the table enjoys a peculiar position: it is the highest value within its column, and at the same time the lowest value within its row. We show this feature graphically in Figure 25-1, where it is apparent that 10 can be called the *saddlepoint* of the matrix.

In conclusion, if a game has a saddlepoint, it will be both the minimax and maximin value.[2] Its row will be the optimal strategy for A, and its column the optimal strategy for B. In other words, it does not matter who goes first; for that matter, the solution would remain the same if each player chose simultaneously—the assumption frequently made in game theory.

PROBLEMS

25-1 For each of the following games, find the strategies (and consequent payoff from B to A) that will be chosen by rational players:
(i) If A goes first.
(ii) If B goes first.
Also state in which cases it makes no difference who goes first.

(a)

A \ B	1	2
1	5	10
2	8	6

(b)

A \ B	1	2	3
1	35	10	25
2	10	25	15

(c)

A \ B	1	2	3	4
1	3	−1	0	−2
2	2	2	1	3
3	−2	0	−1	−3

---

[2]And conversely, if the minimax and maximin values coincide, that common value will be a saddlepoint.

(d)

A \ B	1	2
1	15	2
2	10	4
3	8	6

25-2 In the game of "paper, scissors, rock," two players, A and B, each secretly write down either P, S, or R. Then they compare what they have written. Since paper can be cut by scissors, P loses a penny to S (more precisely, whoever wrote P would lose a penny to whoever wrote S); since scissors, in turn, can be broken by rock, S loses a penny to R; and finally, since rock can be covered by paper, R loses a penny to P. A tie such as P against P results in no money exchanged.

(a) Fill in the following matrix of payoffs from B to A:

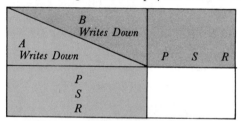

(b) If A can peek at B's strategy before writing down his own, what is going to happen?

(c) If B can peek at A's strategy before writing down his own, what is going to happen?

(d) If neither can peek, but instead they must play simultaneously, what may happen? Try playing this with a friend according to this "no peeking" rule. Repeat the game 5 or 10 times. Can you begin to detect the way in which you should play?

⇒25-3 To make the "paper, scissors, rock" game more exciting, let us modify it by having B win 2 pennies for a P–P tie. To compensate, have A win 2 pennies for an R–R tie. Otherwise, the game is the same. Intuitively, would you rather be A or B? Or does it matter?

Now answer the same questions as in Problem 25-2.

25-2   MIXED STRATEGIES FOR ZERO SUM GAMES

Let us change a basic assumption. Let us now assume that A and B play simultaneously, so that each has to guess what his opponent will do. Under these circumstances, it becomes a game of trying to outguess the opponent.

For example, the games of "paper, scissors, rock," in Problems 25-2 and 25-3 were played with these rules.

If the game is played over and over, each player should avoid a consistent stereotyped choice that could be exploited by his opponent. He should choose between his strategies as unpredictably as possible—at random. As a randomizing device, he might, in fact, make a spinner or roulette wheel, with various sectors indicating the various strategies that he should play.

Example 25-2.   The game in Problem 25-3 has the following matrix of payoffs from B to A:

A \ B	1	2	3
1	−2	−1	1
2	1	0	−1
3	−1	1	2

(a) Are there any circumstances under which A would play strategy 1 in preference to strategy 3?

(b) Suppose that A hires you to make him a spinner that will help him randomize between strategies 2 and 3 (strategy 1 being ruled out):

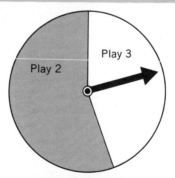

Suppose that A will be left with this spinner to play repeatedly against his shrewd opponent, B. If A has a spinner split .50–.50 for strategies 2 and 3, what will be B's response?

Solution.   (a) Whatever B may play, A can get a higher payoff from playing his strategy 3 instead of 1. (For example, suppose that B plays 2. Then A's strategy-3 payoff of 1 is higher than his strategy-1 payoff of -1.) Strategy 1 therefore should never be played by A, and is said to be *dominated* by strategy 3. (b) If A plays strategies 2 and 3 with .50–.50 probabilities, B can calculate his average loss. B does this for every possible strategy that he might choose, and minimizes it (just as in Table 20-3). B therefore chooses strategy 1, and on average A and B break even in this game.

Probability	B / A	1	2	3
.50	2	1	0	−1
.50	3	−1	1	2
B's average loss		0	.5	.5

↑
Minimum average loss

Example 25-3.   Following the preceding example, is there a better proportion than .50–.50 for A to choose for his spinner?

Solution.   If B is going to play his strategy 1, A should play his last strategy less often. (Of course, A still must play his last strategy sometimes, lest B switch to his strategy 3.) Exactly how much less should A play his last strategy? .60–.40? .80–.20? Just as A analyzed his .50–.50 spinner in example 25-2(b), so he might analyze a .60–.40 spinner, a .80–.20 spinner, or in general, a p, $1 - p$ spinner:

TABLE **25-3**
Determination of B's expected loss

Prob	B / A	1	2	3
p	2	1	0	−1
$1 - p$	3	−1	1	2
B's average loss, L		$1(p) - 1(1 - p)$ $= 2p - 1 = L_1$	$0(p) + 1(1 - p)$ $= 1 - p = L_2$	$-1(p) + 2(1 - p)$ $= -3p + 2 = L_3$

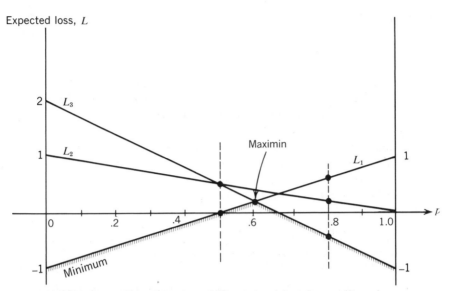

FIGURE 25–2 Determination of B's expected loss, hence A's optimum mixed strategy (based on Table 25–3)

In Figure 25-2, these average losses are graphed as a function[3] of p. When $p = .50$, for example, the calculations in Example 25-2(b) are shown: the dotted line shows the average losses $L_1 = 0$, $L_2 = .5$, $L_3 = .5$, of which B will choose the smallest. Or when $p = .80$, the dotted line shows the average losses $L_1 = .6$, $L_2 = .2$, $L_3 = -.4$, of which B again will choose the smallest. In fact, no matter what spinner A may use, B will respond by choosing his minimum loss, which is shaded in Figure 25-2.

When A sees these average payments that he might gain from various spinners, he chooses the maximum, of course. This occurs where L_1 and L_3 are the same height; i.e., where:

$$2p - 1 = -3p + 2$$

Solve for p:

$$p = \frac{3}{5} = .60$$

Then:

$$L_1 = L_2 = .2$$

Thus, by playing a .60–.40 mixture of his strategies, A can win .2 pennies on average. And there is nothing B can do about it, even if he knows what spinner A is using.[4]

This maximum of the minimums (.2 pennies) is again called the *maximin* value, and the corresponding strategy (using a .60–.40 spinner) is called A's *maximin strategy*. Often the terms *randomized maximin* or *mixed maximin* are used, in contrast to the term *pure maximin* (for the case discussed in Section 25-1).

---

[3]Each average loss function L is of first degree in p, and hence is a straight line. The easiest way to graph such a line is to plot the two end points. For example, consider the first average loss L_1; when $p = 1$, then $L_1 = 1$; when $p = 0$, then $L_1 = -1$. These values of L_1, incidentally, are the corresponding losses in the first column of Table 25-3.

[4]Of course, if B knew the result of a *particular spin*, he could exploit this information and get the better of A. But we are supposing that this is not possible.

PROBLEMS

25-4 In the game of "matching pennies," A and B each choose heads or tails, simultaneously and secretly. B pays A an amount, say \$20, if the faces differ. B receives from A the same amount, \$20, if the faces are the same.

(a) Guess whether A, or B, or neither has the advantage in this game.

(b) Prepare a spinner for A that he can use against B, that will maximize his profit even if B finds out what the spinner is. What, then, is A's maximum profit?

(c) Prepare a spinner for B that he can use against A, that will maximize his profit even if A finds out what the spinner is. What, then, is B's maximum profit?

(d) If A uses the spinner found in (b), does it make any difference what B does? If B uses the spinner found in (c), will it make any difference what A does? What happens if they both use their spinners?

25-5 To make "matching pennies" more exciting, let us modify it: Instead of paying A a straight \$20 for a mismatch, suppose that B pays sometimes \$10 and sometimes \$30; specifically, B pays A \$30 if A shows heads and B shows tails, and \$10 if A shows tails and B shows heads. (B continues to receive \$20 from A if the coins match.)

Answer the same questions as in Problem 25-4.

25-6 Suppose that you find yourself on a long sea voyage. You wish to match pennies, but your companion wants to play cards. He therefore suggests a compromise. You choose heads or tails while he selects an ace. If you select a head he pays you $15, $4, $-$5, and $1, respectively, depending on whether he's chosen the spade, heart, diamond or club ace. If you select a tail, he pays you $-$10, $-$2, $1, and $-$5, again depending on which ace he's chosen.

(a) Do you agree to play? Why? What strategies?

(b) If you were to play this game five times and found that you had won $5, what would you conclude?

(c) Are there any two lines in your diagram that do *not* intersect? From the diagram, show that, no matter what the circumstances, it always is preferable for him to select the club ace instead of the heart ace (i.e., the heart strategy is "dominated by" the club strategy). By initially examining the payoff matrix, couldn't he have dropped the heart strategy from all further consideration?

25-3   CONCLUSIONS

(a) Conclusions for Zero-Sum Games

In solving for the best game strategy, the first step is to test whether maximin and minimax coincide. If they do, this is a strictly determined game, and the single strategy to be used by each player is determined.

If minimax and maximin do not coincide, the game is not strictly determined. Randomized strategies are called for, and are found in simple cases geometrically or algebraically, as we have illustrated. In more complex cases, advanced techniques such as linear programming are required; but rather than extending the mathematical solution, it is more important to consider the fundamental philosophy and assumptions underlying game theory:

1. A player using his best randomized strategy can guarantee a certain expected value for the game, regardless of what his opponent may do. However, this is only the value towards which the average of many games will tend. If the game is only played a few times, luck may raise or lower this payoff.

2. In Problem 25-4, we noticed that the maximin and minimax randomized values of the game are equal. It has been proven that this is always true:

$$\boxed{\text{maximin randomized value} = \text{minimax randomized value}} \quad (25\text{-}2)$$

Another feature of Problem 25-5 also is generally true: if one player sticks to his minimax or maximin randomized strategy, it does not matter what the other player does (just so long as he avoids dominated strategies). And in view of (25-2), it further does not matter which of the players consistently uses the minimax or maximin randomized strategy.

3. In conclusion, then, the minimax or maximin theory of games that we have presented is conservative and defensive. It shows how to avoid being exploited by using a minimax or maximin strategy, but it does not give any advice on how to exploit the weakness of an opponent who fails to use a minimax or maximin strategy. It is appropriate if a game is being replayed many times against an intelligent opponent who knows the payoff matrix, and who can observe your strategy mix. If these conditions are not met, you usually can find a better strategy. To illustrate, consider an extreme example with payoff matrix given in Table 25-4.

TABLE 25-4
When Game Theory
May Be Inappropriate

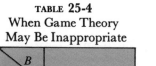

← minimax
= maximin.

Since maximin and minimax coincide, both A and B should play strategy 2 every time. But suppose that on the first play, B plays strategy 1! This may mean that he is unaware of the payoff matrix (and the $4,000 debacle that he faces). In these circumstances, A probably should drop maximin theory and play row 1, in the hopes that B again will mistakenly play his strategy 1.

4. In assuming an intelligent and antagonistic opponent, game theory of course becomes entirely inappropriate for "games against nature," as illustrated in Problem 25-7. Nor should it be used in any other situation in which the issues beyond your control ("states of nature") can be given a prior distribution. In these cases, Bayesian analysis is appropriate.

(b) More General Games

If a game is to be used as a model for any sort of realistic conflict situation (aside from parlor games), it obviously has to be much more complicated than the zero-sum, two-person game. For example, three-person games often involve collusion of two players against the third. Even if we confine ourselves to two-person games, the zero-sum restriction is highly unrealistic; a more interesting model would have a payoff matrix for A, and another separate payoff matrix for B, as shown in Table 25-5. (For example, we might think of A as management, and B as labor, in a wage dispute. Strategy 1 might represent a hard line, while strategy 2 might represent a more flexible bargaining attitude. When two hard lines come into conflict, they result in a long and costly strike that reduces each contestant's payoff to zero). In Table 25-5, A would play strategy 2, because it is better for him than strategy 1, whatever strategy B may choose. B, seeing that A will choose strategy 2, looks at the relevant second row of his payoff matrix. There he finds that his biggest payoff (100) occurs when he chooses strategy 1, and thus the game is solved.

TABLE **25-5**
A Nonzero-Sum Game

(a) Payoff Matrix for A

A \ B	1	2
1	0	100
2	50	110

(b) Payoff Matrix for B

A \ B	1	2
1	0	50
2	100	80

Most nonzero-sum games, however, do not have such an easy solution; in fact, many have no really satisfactory solution at all. This is one reason why they are less popular to game theorists than the less realistic two-person, zero-sum games, which always have a solution.

PROBLEMS

25-7 Suppose that you are trying to decide whether to hold a banquet indoors or outdoors, with your profit depending on both where the banquet is held and the weather, as follows:

You Hold Banquet \ Nature	Rain 1	Not Rain 2
Indoors 1	100	0
Outdoors 2	0	1,000

(a) According to game theory, what should your strategy, or mix of strategies, be? What would your expected profit be? Does this seem like a reasonable solution? If not, why not?

(b) Specifically, suppose that past weather records indicate that the probability of rain is $\frac{1}{5}$. What action would you take? What is the expected value of this action?

(c) True or false? If false, correct it.

If the other player's strategy is determined by some probabilistic mechanism (such as nature) that pays no attention to your losses, then maximin values are irrelevant. Instead, expected values are the proper criterion, resulting in a Bayesian analysis.

25-8 In the game of Table 25-5, suppose that A is thinking of hiring a bargaining agent. The agent announces loudly (so that both A and B can hear) that he will take the following inflexible schedule of fees from A's payoff matrix:

$$\begin{array}{|cc|} \hline 0 & 10 \\ 50 & 30 \\ \hline \end{array}$$

(a) What would be A's new (net) payoff matrix?

(b) Should A hire the agent?

(c) Suppose that a credible bargaining agent is not available to A. Can he achieve anything by announcing that he is determined to play strategy 1, no matter what happens?

(d) What happens if B counters A's announcement in (c) with an even louder announcement that he is determined to play strategy 1, no matter what happens? Is there any circumstance under which this would make B better off?

REVIEW PROBLEMS

25-9 Payoff matrix (money paid from B to A)?

A \ B	1	2
1	$6	$2
2	$4	$8

(a) Solve the above game (i.e., state what each player should do, and what the payoff will be, assuming that both players are rational):

(i) Assuming that A has to choose his strategy first.

(ii) Assuming that B has to choose his strategy first.

(b) Assume that A and B play over and over, each time choosing their strategies simultaneously. In order to be unpredictable, A prepares a spinner to randomize his choices.

(i) What probabilities should this spinner give to strategies 1 and 2?

(ii) How should B respond?

(c) Assume that A has been misinformed about the matrix, so that the spinner that he comes up with says, "Play 1 with probability .6, and play 2 with probability .4." How should B respond?

25-10 For each of the following matrixes showing payoffs from B to A, eliminate any dominated strategies. Then calculate what the game is worth to A if he uses his maximin mixed strategy. (*Hint:* If the maximin mixed value is too difficult to calculate, the minimax mixed value will be the same, according to (25-2)).

	(a)	
6	1	3
3	3	5
4	7	5

(b)	
3	2
1	6
2	5

	(c)	
2	5	9
5	3	5
3	8	4
6	3	8

TABLES

TABLE 1
Common Logarithms,[1] Log N
(To Obtain Natural Logarithms,[2] Multiply by 2.30)

$\overset{\rightarrow}{\downarrow N}$	.00	.01	.02	.03	.04	.05	.06	.07	.08	.09
				Second Decimal Place of N						
1.0	.0000	.0043	.0086	.0128	.0170	.0212	.0253	.0294	.0334	.0374
1.1	.0414	.0453	.0492	.0531	.0569	.0607	.0645	.0682	.0719	.0755
1.2	.0792	.0828	.0864	.0899	.0934	.0969	.1004	.1038	.1072	.1106
1.3	.1139	.1173	.1206	.1239	.1271	.1303	.1335	.1367	.1399	.1430
1.4	.1461	.1492	.1523	.1553	.1584	.1614	.1644	.1673	.1703	.1732
1.5	.1761	.1790	.1818	.1847	.1875	.1903	.1931	.1959	.1987	.2014
1.6	.2041	.2068	.2095	.2122	.2148	.2175	.2201	.2227	.2253	.2279
1.7	.2304	.2330	.2355	.2380	.2405	.2430	.2455	.2480	.2504	.2529
1.8	.2553	.2577	.2601	.2625	.2648	.2672	.2695	.2718	.2742	.2765
1.9	.2788	.2810	.2833	.2856	.2878	.2900	.2923	.2945	.2967	.2989
2.0	.3010	.3032	.3054	.3075	.3096	.3118	.3139	.3160	.3181	.3201
2.1	.3222	.3243	.3263	.3284	.3304	.3324	.3345	.3365	.3385	.3404
2.2	.3424	.3444	.3464	.3483	.3502	.3522	.3541	.3560	.3579	.3598
2.3	.3617	.3636	.3655	.3674	.3692	.3711	.3729	.3747	.3766	.3784
2.4	.3802	.3820	.3838	.3856	.3874	.3892	.3909	.3927	.3945	.3962
2.5	.3979	.3997	.4014	.4031	.4048	.4065	.4082	.4099	.4116	.4133
2.6	.4150	.4166	.4183	.4200	.4216	.4232	.4249	.4265	.4281	.4298
2.7	.4314	.4330	.4346	.4362	.4378	.4393	.4409	.4425	.4440	.4456
2.8	.4472	.4487	.4502	.4518	.4533	.4548	.4564	.4579	.4594	.4609
2.9	.4624	.4639	.4654	.4669	.4683	.4698	.4713	.4728	.4742	.4757
3.0	.4771	.4786	.4800	.4814	.4829	.4843	.4857	.4871	.4886	.4900
3.1	.4914	.4928	.4942	.4955	.4969	.4983	.4997	.5011	.5024	.5038
3.2	.5051	.5065	.5079	.5092	.5105	.5119	.5132	.5145	.5159	.5172
3.3	.5185	.5198	.5211	.5224	.5237	.5250	.5263	.5276	.5289	.5302
3.4	.5315	.5328	.5340	.5353	.5366	.5378	.5391	.5403	.5416	.5428
3.5	.5441	.5453	.5465	.5478	.5490	.5502	.5514	.5527	.5539	.5551
3.6	.5563	.5575	.5587	.5599	.5611	.5623	.5635	.5647	.5658	.5670
3.7	.5682	.5694	.5705	.5717	.5729	.5740	.5752	.5763	.5775	.5786
3.8	.5798	.5809	.5821	.5832	.5843	.5855	.5866	.5877	.5888	.5899
3.9	.5911	.5922	.5933	.5944	.5955	.5966	.5977	.5988	.5999	.6010

[1]To find the log of a number outside the range 1 to 10, just shift its decimal place until it falls within the range 1 to 10. Then look up the log, and add 1 for each place you shifted the decimal point left. For example, log 2310 = log 2.310 + 3 = .3636 + 3 = 3.3636. Similarly, log 0.0231 = log 2.31 − 2 = .3636 − 2 = −1.6364.

[2]Common logs use base 10, while natural logs (ln) use base $e \simeq 2.718$.

To find ln 4.7 for example, first find log 4.7 = .6721, and then multiply by 2.30, obtaining .6721 × 2.30 = 1.546. Thus ln 4.7 = 1.546.

TABLE **1** (Continued)

$\downarrow N \rightarrow$	.00	.01	.02	.03	.04	.05	.06	.07	.08	.09
4.0	.6021	.6031	.6042	.6053	.6064	.6075	.6085	.6096	.6107	.6117
4.1	.6128	.6138	.6149	.6160	.6170	.6180	.6191	.6201	.6212	.6222
4.2	.6232	.6243	.6253	.6263	.6274	.6284	.6294	.6304	.6314	.6325
4.3	.6335	.6345	.6355	.6365	.6375	.6385	.6395	.6405	.6415	.6425
4.4	.6435	.6444	.6454	.6464	.6474	.6484	.6493	.6503	.6513	.6522
4.5	.6532	.6542	.6551	.6561	.6571	.6580	.6590	.6599	.6609	.6618
4.6	.6628	.6637	.6646	.6656	.6665	.6675	.6684	.6693	.6702	.6712
4.7	.6721	.6730	.6739	.6749	.6758	.6767	.6776	.6785	.6794	.6803
4.8	.6812	.6821	.6830	.6839	.6848	.6857	.6866	.6875	.6884	.6893
4.9	.6902	.6911	.6920	.6928	.6937	.6946	.6955	.6964	.6972	.6981
5.0	.6990	.6998	.7007	.7016	.7024	.7033	.7042	.7050	.7059	.7067
5.1	.7076	.7084	.7093	.7101	.7110	.7118	.7126	.7135	.7143	.7152
5.2	.7160	.7168	.7177	.7185	.7193	.7202	.7210	.7218	.7226	.7235
5.3	.7243	.7251	.7259	.7267	.7275	.7284	.7292	.7300	.7308	.7316
5.4	.7324	.7332	.7340	.7348	.7356	.7364	.7372	.7380	.7388	.7396
5.5	.7404	.7412	.7419	.7427	.7435	.7443	.7451	.7459	.7466	.7474
5.6	.7482	.7490	.7497	.7505	.7513	.7520	.7528	.7536	.7543	.7551
5.7	.7559	.7566	.7574	.7582	.7589	.7597	.7604	.7612	.7619	.7627
5.8	.7634	.7642	.7649	.7657	.7664	.7672	.7679	.7686	.7694	.7701
5.9	.7709	.7716	.7723	.7731	.7738	.7745	.7752	.7760	.7767	.7774
6.0	.7782	.7789	.7796	.7803	.7810	.7818	.7825	.7832	.7839	.7846
6.1	.7853	.7860	.7868	.7875	.7882	.7889	.7896	.7903	.7910	.7917
6.2	.7924	.7931	.7938	.7945	.7952	.7959	.7966	.7973	.7980	.7987
6.3	.7993	.8000	.8007	.8014	.8021	.8028	.8035	.8041	.8048	.8055
6.4	.8062	.8069	.8075	.8082	.8089	.8096	.8102	.8109	.8116	.8122
6.5	.8129	.8136	.8142	.8149	.8156	.8162	.8169	.8176	.8182	.8189
6.6	.8195	.8202	.8209	.8215	.8222	.8228	.8235	.8241	.8248	.8254
6.7	.8261	.8267	.8274	.8280	.8287	.8293	.8299	.8306	.8312	.8319
6.8	.8325	.8331	.8338	.8344	.8351	.8357	.8363	.8370	.8376	.8382
6.9	.8388	.8395	.8401	.8407	.8414	.8420	.8426	.8432	.8439	.8445
7.0	.8451	.8457	.8463	.8470	.8476	.8482	.8488	.8494	.8500	.8506
7.1	.8513	.8519	.8525	.8531	.8537	.8543	.8549	.8555	.8561	.8567
7.2	.8573	.8579	.8585	.8591	.8597	.8603	.8609	.8615	.8621	.8627
7.3	.8633	.8639	.8645	.8651	.8657	.8663	.8669	.8675	.8681	.8686
7.4	.8692	.8698	.8704	.8710	.8716	.8722	.8727	.8733	.8739	.8745
7.5	.8751	.8756	.8762	.8768	.8774	.8779	.8785	.8791	.8797	.8802
7.6	.8808	.8814	.8820	.8825	.8831	.8837	.8842	.8848	.8854	.8859
7.7	.8865	.8871	.8876	.8882	.8887	.8893	.8899	.8904	.8910	.8915
7.8	.8921	.8927	.8932	.8938	.8943	.8949	.8954	.8960	.8965	.9971
7.9	.8976	.8982	.8987	.8993	.8998	.9004	.9009	.9015	.9020	.9025

TABLE 1 (Continued)

$\downarrow N$ $\rightarrow$	.00	.01	.02	.03	.04	.05	.06	.07	.08	.09
8.0	.9031	.9036	.9042	.9047	.9053	.9058	.9063	.9069	.9074	.9079
8.1	.9085	.9090	.9096	.9101	.9106	.9112	.9117	.9122	.9128	.9133
8.2	.9138	.9143	.9149	.9154	.9159	.9165	.9170	.9175	.9180	.9186
8.3	.9191	.9196	.9201	.9206	.9212	.9217	.9222	.9227	.9232	.9238
8.4	.9243	.9248	.9253	.9258	.9263	.9269	.9274	.9279	.9284	.9289
8.5	.9294	.9299	.9304	.9309	.9315	.9320	.9325	.9330	.9335	.9340
8.6	.9345	.9350	.9355	.9360	.9365	.9370	.9375	.9380	.9385	.9390
8.7	.9395	.9400	.9405	.9410	.9415	.9420	.9425	.9430	.9435	.9440
8.8	.9445	.9450	.9455	.9460	.9465	.9469	.9474	.9479	.9484	.9489
8.9	.9494	.9499	.9504	.9509	.9513	.9518	.9523	.9528	.9533	.9538
9.0	.9542	.9547	.9552	.9557	.9562	.9566	.9571	.9576	.9581	.9586
9.1	.9590	.9595	.9600	.9605	.9609	.9614	.9619	.9624	.9628	.9633
9.2	.9638	.9643	.9647	.9652	.9657	.9661	.9666	.9671	.9675	.9680
9.3	.9685	.9689	.9694	.9699	.9703	.9708	.9713	.9717	.9722	.9727
9.4	.9731	.9736	.9741	.9745	.9750	.9754	.9759	.9763	.9768	.9773
9.5	.9777	.9782	.9786	.9791	.9795	.9800	.9805	.9809	.9814	.9818
9.6	.9823	.9827	.9832	.9836	.9841	.9845	.9850	.9854	.9859	.9863
9.7	.9868	.9872	.9877	.9881	.9886	.9890	.9894	.9899	.9903	.9908
9.8	.9912	.9917	.9921	.9926	.9930	.9934	.9939	.9943	.9948	.9952
9.9	.9956	.9961	.9965	.9969	.9974	.9978	.9983	.9987	.9991	.9996

TABLE **II(a)**
Random Digits
(Blocked merely for convenience)

39 65 76 45 45	19 90 69 64 61	20 26 36 31 62	58 24 97 14 97	95 06 70 99 00
73 71 23 70 90	65 97 60 12 11	31 56 34 19 19	47 83 75 51 33	30 62 38 20 46
72 20 47 33 84	51 67 47 97 19	98 40 07 17 66	23 05 09 51 80	59 78 11 52 49
75 17 25 69 17	17 95 21 78 58	24 33 45 77 48	69 81 84 09 29	93 22 70 45 80
37 48 79 88 74	63 52 06 34 30	01 31 60 10 27	35 07 79 71 53	28 99 52 01 41
02 89 08 16 94	85 53 83 29 95	56 27 09 24 43	21 78 55 09 82	72 61 88 73 61
87 18 15 70 07	37 79 49 12 38	48 13 93 55 96	41 92 45 71 51	09 18 25 58 94
98 83 71 70 15	89 09 39 59 24	00 06 41 41 20	14 36 59 25 47	54 45 17 24 89
10 08 58 07 04	76 62 16 48 68	58 76 17 14 86	59 53 11 52 21	66 04 18 72 87
47 90 56 37 31	71 82 13 50 41	27 55 10 24 92	28 04 67 53 44	95 23 00 84 47
93 05 31 03 07	34 18 04 52 35	74 13 39 35 22	68 95 23 92 35	36 63 70 35 33
21 89 11 47 99	11 20 99 45 18	76 51 94 84 86	13 79 93 37 55	98 16 04 41 67
95 18 94 06 97	27 37 83 28 71	79 57 95 13 91	09 61 87 25 21	56 20 11 32 44
97 08 31 55 73	10 65 81 92 59	77 31 61 95 46	20 44 90 32 64	26 99 76 75 63
69 26 88 86 13	59 71 74 17 32	48 38 75 93 29	73 37 32 04 05	60 82 29 20 25
41 47 10 25 03	87 63 93 95 17	81 83 83 04 49	77 45 85 50 51	79 88 01 97 30
91 94 14 63 62	08 61 74 51 69	92 79 43 89 79	29 18 94 51 23	14 85 11 47 23
80 06 54 18 47	08 52 85 08 40	48 40 35 94 22	72 65 71 08 86	50 03 42 99 36
67 72 77 63 99	89 85 84 46 06	64 71 06 21 66	89 37 20 70 01	61 65 70 22 12
59 40 24 13 75	42 29 72 23 19	06 94 76 10 08	81 30 15 39 14	81 83 17 16 33
63 62 06 34 41	79 53 36 02 95	94 61 09 43 62	20 21 14 68 86	94 95 48 46 45
78 47 23 53 90	79 93 96 38 63	34 85 52 05 09	85 43 01 72 73	14 93 87 81 40
87 68 62 15 43	97 48 72 66 48	53 16 71 13 81	59 97 50 99 52	24 62 20 42 31
47 60 92 10 77	26 97 05 73 51	88 46 38 03 58	72 68 49 29 31	75 70 16 08 24
56 88 87 59 41	06 87 37 78 48	65 88 69 58 39	88 02 84 27 83	85 81 56 39 38
22 17 68 65 84	87 02 22 57 51	68 69 80 95 44	11 29 01 95 80	49 34 35 86 47
19 36 27 59 46	39 77 32 77 09	79 57 92 36 59	89 74 39 82 15	08 58 94 34 74
16 77 23 02 77	28 06 24 25 93	22 45 44 84 11	87 80 61 65 31	09 71 91 74 25
78 43 76 71 61	97 67 63 99 61	80 45 67 93 82	59 73 19 85 23	53 33 65 97 21
03 28 28 26 08	69 30 16 09 05	53 58 47 70 93	66 56 45 65 79	45 56 20 19 47
04 31 17 21 56	33 73 99 19 87	26 72 39 27 67	53 77 57 68 93	60 61 97 22 61
61 06 98 03 91	87 14 77 43 96	43 00 65 98 50	45 60 33 01 07	98 99 46 50 47
23 68 35 26 00	99 53 93 61 28	52 70 05 48 34	56 65 05 61 86	90 92 10 70 80
15 39 25 70 99	93 86 52 77 65	15 33 59 05 28	22 87 26 07 47	86 96 98 29 06
58 71 96 30 24	18 46 23 34 27	85 13 99 24 44	49 18 09 79 49	74 16 32 23 02
93 22 53 64 39	07 10 63 76 35	87 03 04 79 88	08 13 13 85 51	55 34 57 72 69
78 76 58 54 74	92 38 70 96 92	52 06 79 79 45	82 63 18 27 44	69 66 92 19 09
61 81 31 96 82	00 57 25 60 59	46 72 60 18 77	55 66 12 62 11	08 99 55 64 57
42 88 07 10 05	24 98 65 63 21	47 21 61 88 32	27 80 30 21 60	10 92 35 36 12
77 94 30 05 39	28 10 99 00 27	12 73 73 99 12	49 99 57 94 82	96 88 57 17 91

TABLE **II(b)**

Random Normal Numbers, $\mu = 0$, $\sigma = 1$
(Rounded to 1 Decimal Place)

.5	.1	2.5	-.3	-.1	.3	-.3	1.3	.2	-1.0
.1	-2.5	-.5	-.2	.5	-1.6	.2	-1.2	.0	.5
1.5	-.4	-.6	.7	.9	1.4	.8	-1.0	-.9	-1.9
1.0	-.5	1.3	3.5	.6	-1.9	.2	1.2	-.5	-.3
1.4	-.6	.0	.3	2.9	2.0	-.3	.4	.4	.0
.9	-.5	-.5	.6	.9	-.9	1.6	.2	-1.9	.4
1.2	-1.1	.0	.8	1.0	.7	1.1	-.6	-.3	-.7
-1.5	-.5	-.2	-.1	1.0	.2	.4	.7	-.4	-.4
-.7	.8	-1.6	-.3	-.5	-2.1	-.5	-.2	.9	-.5
1.4	.2	.4	.8	.2	-.7	1.0	-1.5	-.3	.1
-.5	1.7	-.1	-1.2	-.5	.9	-.5	-2.0	-2.8	-.2
-1.4	-.2	1.4	-.6	-.3	-.2	.2	.8	1.0	-.9
-1.0	.6	-.9	1.6	.1	.4	-.2	.3	-1.0	-1.0
.0	-.9	.0	-.7	1.1	-.1	1.1	.5	-1.7	.4
1.4	-1.2	-.9	1.2	-.2	-.2	1.2	-2.6	-.6	.1
-1.8	-.3	1.2	1.0	-.5	-1.6	-.1	-.4	-.6	.6
-.1	-.4	-1.4	.4	-1.0	-.1	-1.7	-2.8	-1.1	-2.4
-1.3	1.8	-1.0	.4	1.0	-1.1	-1.0	.4	-1.7	2.0
1.0	.5	.7	1.4	1.0	-1.3	1.6	-1.0	.5	-.3
.3	-2.1	.7	-.9	-1.1	-1.4	1.0	.1	-.6	.9
-1.8	-2.0	-1.6	.5	.2	-.2	.0	.0	.5	-1.0
-1.2	1.2	1.1	.9	1.3	-.2	.2	-.4	-.3	.5
.7	-1.1	1.2	-1.2	-.9	.4	.3	-.9	.6	1.7
-.4	.4	-1.9	.9	-.2	.6	.9	-.4	-.2	-.1
-1.4	-.2	.4	-.6	-.6	.2	-.3	.5	.7	-.3
.2	.2	-1.1	-.2	-.3	1.2	1.1	.0	-2.0	-.6
.2	.3	-.3	.1	-2.8	-.4	-.8	-1.3	-.6	-1.0
2.3	.6	.6	-.7	.2	1.3	.1	-1.8	-.7	-1.3
.0	-.3	.1	.8	-.6	.5	.5	-1.0	.5	1.0
-1.1	-2.1	.9	.1	.4	-1.7	1.0	-1.4	-.6	-1.0
.8	.1	-1.5	.0	-2.1	.7	.1	-.9	-.6	.6
.4	-1.7	-.9	.2	-.7	.3	-.1	-.2	-.1	.4
-.5	-.3	.2	-.7	1.0	.0	.4	-.8	.2	.1
.3	-.5	1.3	-1.2	-.9	.1	-.5	-.8	.0	.5
1.0	3.0	-.6	-.5	-1.1	1.3	-1.4	-1.3	-3.0	.5
-1.3	1.3	-.6	-.1	-.5	-.6	2.9	.5	.4	.3
-.3	-.1	-.3	.6	-.5	-1.2	-1.2	-.3	-.1	1.1
.2	-.9	-.9	-.5	1.4	-.5	.2	-.4	1.5	1.1
-1.3	.2	-1.2	.4	-1.0	.8	.9	1.0	.0	.8
-1.2	-.2	-.3	1.8	1.4	.6	1.2	.7	.4	.2
.6	-.5	.8	.1	.5	-.4	1.7	1.2	.9	-.3
.4	-1.9	.2	-.5	.7	-.1	-.1	-.5	.5	1.1
-1.4	.5	-1.7	-1.2	.8	-.7	-.1	1.0	-.8	.2
-.2	-.2	-.4	-.8	.3	1.0	1.8	2.9	-.8	-.1
-.3	.5	.4	-1.5	1.5	2.0	-.1	.2	.0	-1.2
.4	-.4	.6	1.0	-.1	.1	.5	-1.3	1.1	1.1
.6	.7	-1.1	-1.4	-1.6	-1.6	1.5	1.3	.7	-.9
.9	-.9	-.1	-.5	.5	1.4	.0	-.3	-.3	1.2
.2	-.6	.0	-.5	-.9	-.4	-.5	1.7	-.2	-1.2
-.9	.4	.8	.8	.4	-.3	-1.1	.6	1.4	1.3

TABLE III(a)

Binomial Coefficients $\binom{n}{s}$

n \ s	0	1	2	3	4	5	6	7	8	9	10
0	1										
1	1	1									
2	1	2	1								
3	1	3	3	1							
4	1	4	6	4	1						
5	1	5	10	10	5	1					
6	1	6	15	20	15	6	1				
7	1	7	21	35	35	21	7	1			
8	1	8	28	56	70	56	28	8	1		
9	1	9	36	84	126	126	84	36	9	1	
10	1	10	45	120	210	252	210	120	45	10	1
11	1	11	55	165	330	462	462	330	165	55	11
12	1	12	66	220	495	792	924	792	495	220	66
13	1	13	78	286	715	1287	1716	1716	1287	715	286
14	1	14	91	364	1001	2002	3003	3432	3003	2002	1001
15	1	15	105	455	1365	3003	5005	6435	6435	5005	3003
16	1	16	120	560	1820	4368	8008	11440	12870	11440	8008
17	1	17	136	680	2380	6188	12376	19448	24310	24310	19448
18	1	18	153	816	3060	8568	18564	31824	43758	48620	43758
19	1	19	171	969	3876	11628	27132	50388	75582	92378	92378
20	1	20	190	1140	4845	15504	38760	77520	125970	167960	184756

Note. $\binom{n}{s} = \dfrac{n(n-1)(n-2)\cdots(n-s+1)}{s(s-1)(s-2)\cdots 3\cdot 2\cdot 1}$; $\binom{n}{0} = 1$; $\binom{n}{1} = n$.

For coefficients missing from the above table, use the relation:

$$\binom{n}{s} = \binom{n}{n-s}, \quad \text{e.g.,} \quad \binom{20}{11} = \binom{20}{9} = 167960.$$

TABLE **III(b)**
Individual Binomial Probabilities $p(s)$

n	s	.10	.20	.30	.40	π .50	.60	.70	.80	.90
1	0	.9000	.8000	.7000	.6000	.5000	.4000	.3000	.2000	.1000
	1	.1000	.2000	.3000	.4000	.5000	.6000	.7000	.8000	.9000
2	0	.8100	.6400	.4900	.3600	.2500	.1600	.0900	.0400	.0100
	1	.1800	.3200	.4200	.4800	.5000	.4800	.4200	.3200	.1800
	2	.0100	.0400	.0900	.1600	.2500	.3600	.4900	.6400	.8100
3	0	.7290	.5120	.3430	.2160	.1250	.0640	.0270	.0080	.0010
	1	.2430	.3840	.4410	.4320	.3750	.2880	.1890	.0960	.0270
	2	.0270	.0960	.1890	.2880	.3750	.4320	.4410	.3840	.2430
	3	.0010	.0080	.0270	.0640	.1250	.2160	.3430	.5120	.7290
4	0	.6561	.4096	.2401	.1296	.0625	.0256	.0081	.0016	.0001
	1	.2916	.4096	.4116	.3456	.2500	.1536	.0756	.0256	.0036
	2	.0486	.1536	.2646	.3456	.3750	.3456	.2646	.1536	.0486
	3	.0036	.0256	.0756	.1536	.2500	.3456	.4116	.4096	.2916
	4	.0001	.0016	.0081	.0256	.0625	.1296	.2401	.4096	.6561
5	0	.5905	.3277	.1681	.0778	.0313	.0102	.0024	.0003	.0000
	1	.3280	.4096	.3602	.2592	.1563	.0768	.0283	.0064	.0004
	2	.0729	.2048	.3087	.3456	.3125	.2304	.1323	.0512	.0081
	3	.0081	.0512	.1323	.2304	.3125	.3456	.3087	.2048	.0729
	4	.0004	.0064	.0284	.0768	.1563	.2592	.3602	.4096	.3280
	5	.0000	.0003	.0024	.0102	.0313	.0778	.1681	.3277	.5905
6	0	.5314	.2621	.1176	.0467	.0156	.0041	.0007	.0001	.0000
	1	.3543	.3932	.3025	.1866	.0938	.0369	.0102	.0015	.0001
	2	.0984	.2458	.3241	.3110	.2344	.1382	.0595	.0154	.0012
	3	.0146	.0819	.1852	.2765	.3125	.2765	.1852	.0819	.0146
	4	.0012	.0154	.0595	.1382	.2344	.3110	.3241	.2458	.0984
	5	.0001	.0015	.0102	.0369	.0938	.1866	.3025	.3932	.3543
	6	.0000	.0001	.0007	.0041	.0156	.0467	.1176	.2621	.5314
7	0	.4783	.2097	.0824	.0280	.0078	.0016	.0002	.0000	.0000
	1	.3720	.3670	.2471	.1306	.0547	.0172	.0036	.0004	.0000
	2	.1240	.2753	.3177	.2613	.1641	.0774	.0250	.0043	.0002
	3	.0230	.1147	.2269	.2903	.2734	.1935	.0972	.0287	.0026
	4	.0026	.0287	.0972	.1935	.2734	.2903	.2269	.1147	.0230
	5	.0002	.0043	.0250	.0774	.1641	.2613	.3177	.2753	.1240
	6	.0000	.0004	.0036	.0172	.0547	.1306	.2471	.3670	.3720
	7	.0000	.0000	.0002	.0016	.0078	.0280	.0824	.2097	.4783

TABLE **III(b)** (continued)

n	s	.10	.20	.30	.40	π .50	.60	.70	.80	.90
8	0	.4305	.1678	.0576	.0168	.0039	.0007	.0001	.0000	.0000
	1	.3826	.3355	.1977	.0896	.0313	.0079	.0012	.0001	.0000
	2	.1488	.2936	.2965	.2090	.1094	.0413	.0100	.0011	.0000
	3	.0331	.1468	.2541	.2787	.2188	.1239	.0467	.0092	.0004
	4	.0046	.0459	.1361	.2322	.2734	.2322	.1361	.0459	.0046
	5	.0004	.0092	.0467	.1239	.2188	.2787	.2541	.1468	.0331
	6	.0000	.0011	.0100	.0413	.1094	.2090	.2965	.2936	.1488
	7	.0000	.0001	.0012	.0079	.0313	.0896	.1977	.3355	.3826
	8	.0000	.0000	.0001	.0007	.0039	.0168	.0576	.1678	.4305
9	0	.3874	.1342	.0404	.0101	.0020	.0003	.0000	.0000	.0000
	1	.3874	.3020	.1556	.0605	.0176	.0035	.0004	.0000	.0000
	2	.1722	.3020	.2668	.1612	.0703	.0212	.0039	.0003	.0000
	3	.0446	.1762	.2668	.2508	.1641	.0743	.0210	.0028	.0001
	4	.0074	.0661	.1715	.2508	.2461	.1672	.0735	.0165	.0008
	5	.0008	.0165	.0735	.1672	.2461	.2508	.1715	.0661	.0074
	6	.0001	.0028	.0210	.0743	.1641	.2508	.2668	.1762	.0446
	7	.0000	.0003	.0039	.0212	.0703	.1612	.2668	.3020	.1722
	8	.0000	.0000	.0004	.0035	.0176	.0605	.1556	.3020	.3874
	9	.0000	.0000	.0000	.0003	.0020	.0101	.0404	.1342	.3874
10	0	.3487	.1074	.0282	.0060	.0010	.0001	.0000	.0000	.0000
	1	.3874	.2684	.1211	.0403	.0098	.0016	.0001	.0000	.0000
	2	.1937	.3020	.2335	.1209	.0439	.0106	.0014	.0001	.0000
	3	.0574	.2013	.2668	.2150	.1172	.0425	.0090	.0008	.0000
	4	.0112	.0881	.2001	.2508	.2051	.1115	.0368	.0055	.0001
	5	.0015	.0264	.1029	.2007	.2461	.2007	.1029	.0264	.0015
	6	.0001	.0055	.0368	.1115	.2051	.2508	.2001	.0881	.0112
	7	.0000	.0008	.0090	.0425	.1172	.2150	.2668	.2013	.0574
	8	.0000	.0001	.0014	.0106	.0439	.1209	.2335	.3020	.1937
	9	.0000	.0000	.0001	.0016	.0098	.0403	.1211	.2684	.3874
	10	.0000	.0000	.0000	.0001	.0010	.0060	.0282	.1074	.3487

TABLE III(c)
Cumulative Binomial Probability in Right-Hand Tail

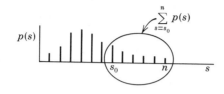

n	s_0	.10	.20	.30	.40	.50	.60	.70	.80	.90
2	1	.1900	.3600	.5100	.6400	.7500	.8400	.9100	.9600	.9900
	2	.0100	.0400	.0900	.1600	.2500	.3600	.4900	.6400	.8100
3	1	.2710	.4880	.6570	.7840	.8750	.9360	.9730	.9920	.9990
	2	.0280	.1040	.2160	.3520	.5000	.6480	.7840	.8960	.9720
	3	.0010	.0080	.0270	.0640	.1250	.2160	.3430	.5120	.7290
4	1	.3439	.5904	.7599	.8704	.9375	.9744	.9919	.9984	.9999
	2	.0523	.1808	.3483	.5248	.6875	.8208	.9163	.9728	.9963
	3	.0037	.0272	.0837	.1792	.3125	.4752	.6517	.8192	.9477
	4	.0001	.0016	.0081	.0256	.0625	.1296	.2401	.4096	.6561
5	1	.4095	.6723	.8319	.9222	.9688	.9898	.9976	.9997	1.0000
	2	.0815	.2627	.4718	.6630	.8125	.9130	.9692	.9933	.9995
	3	.0086	.0579	.1631	.3174	.5000	.6826	.8369	.9421	.9914
	4	.0005	.0067	.0308	.0870	.1875	.3370	.5282	.7373	.9185
	5	.0000	.0003	.0024	.0102	.0313	.0778	.1681	.3277	.5905
6	1	.4686	.7379	.8824	.9533	.9844	.9959	.9993	.9999	1.0000
	2	.1143	.3446	.5798	.7667	.8906	.9590	.9891	.9984	.9999
	3	.0159	.0989	.2557	.4557	.6562	.8208	.9295	.9830	.9987
	4	.0013	.0170	.0705	.1792	.3438	.5443	.7443	.9011	.9842
	5	.0001	.0016	.0109	.0410	.1094	.2333	.4202	.6554	.8857
	6	.0000	.0001	.0007	.0041	.0156	.0467	.1176	.2621	.5314
7	1	.5217	.7903	.9176	.9720	.9922	.9984	.9998	1.0000	1.0000
	2	.1497	.4233	.6706	.8414	.9375	.9812	.9962	.9996	1.0000
	3	.0257	.1480	.3529	.5801	.7734	.9037	.9712	.9953	.9998
	4	.0027	.0333	.1260	.2898	.5000	.7102	.8740	.9667	.9973
	5	.0002	.0047	.0288	.0963	.2266	.4199	.6471	.8520	.9743
	6	.0000	.0004	.0038	.0188	.0625	.1586	.3294	.5767	.8503
	7	.0000	.0000	.0002	.0016	.0078	.0280	.0824	.2097	.4783

TABLE III(c) (continued)

n	s_0	.10	.20	.30	.40	π .50	.60	.70	.80	.90
8	1	.5695	.8322	.9424	.9832	.9961	.9993	.9999	1.0000	1.0000
	2	.1869	.4967	.7447	.8936	.9648	.9915	.9987	.9999	1.0000
	3	.0381	.2031	.4482	.6846	.8555	.9502	.9887	.9988	1.0000
	4	.0050	.0563	.1941	.4059	.6367	.8263	.9420	.9896	.9996
	5	.0004	.0104	.0580	.1737	.3633	.5941	.8059	.9437	.9950
	6	.0000	.0012	.0113	.0498	.1445	.3154	.5518	.7969	.9619
	7	.0000	.0001	.0013	.0085	.0352	.1064	.2553	.5033	.8131
	8	.0000	.0000	.0001	.0007	.0039	.0168	.0576	.1678	.4305
9	1	.6126	.8658	.9596	.9899	.9980	.9997	1.0000	1.0000	1.0000
	2	.2252	.5638	.8040	.9295	.9805	.9962	.9996	1.0000	1.0000
	3	.0530	.2618	.5372	.7682	.9102	.9750	.9957	.9997	1.0000
	4	.0083	.0856	.2703	.5174	.7461	.9006	.9747	.9969	.9999
	5	.0009	.0196	.0988	.2666	.5000	.7334	.9012	.9804	.9991
	6	.0001	.0031	.0253	.0994	.2539	.4826	.7297	.9144	.9917
	7	.0000	.0003	.0043	.0250	.0898	.2318	.4628	.7382	.9470
	8	.0000	.0000	.0004	.0038	.0195	.0705	.1960	.4362	.7748
	9	.0000	.0000	.0000	.0003	.0020	.0101	.0404	.1342	.3874
10	1	.6513	.8926	.9718	.9940	.9990	.9999	1.0000	1.0000	1.0000
	2	.2639	.6242	.8507	.9536	.9893	.9983	.9999	1.0000	1.0000
	3	.0702	.3222	.6172	.8327	.9453	.9877	.9984	.9999	1.0000
	4	.0128	.1209	.3504	.6177	.8281	.9452	.9894	.9991	1.0000
	5	.0016	.0328	.1503	.3669	.6230	.8338	.9527	.9936	.9999
	6	.0001	.0064	.0473	.1662	.3770	.6331	.8497	.9672	.9984
	7	.0000	.0009	.0106	.0548	.1719	.3823	.6496	.8791	.9872
	8	.0000	.0001	.0016	.0123	.0547	.1673	.3828	.6778	.9298
	9	.0000	.0000	.0001	.0017	.0107	.0464	.1493	.3758	.7361
	10	.0000	.0000	.0000	.0001	.0010	.0060	.0282	.1074	.3487

TABLE **IV**
Standard Normal, Cumulative Probability in Right-Hand Tail
(For Negative Values of z, Areas are Found by Symmetry)

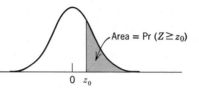

Area = Pr $(Z \geq z_0)$

0   z_0

$\downarrow z_0$ $\rightarrow$	.00	.01	.02	.03	.04	.05	.06	.07	.08	.09
				Second Decimal Place of z_0						
0.0	.5000	.4960	.4920	.4880	.4840	.4801	.4761	.4721	.4681	.4641
0.1	.4602	.4562	.4522	.4483	.4443	.4404	.4364	.4325	.4286	.4247
0.2	.4207	.4168	.4129	.4090	.4052	.4013	.3974	.3936	.3897	.3859
0.3	.3821	.3783	.3745	.3707	.3669	.3632	.3594	.3557	.3520	.3483
0.4	.3446	.3409	.3372	.3336	.3300	.3264	.3228	.3192	.3156	.3121
0.5	.3085	.3050	.3015	.2981	.2946	.2912	.2877	.2843	.2810	.2776
0.6	.2743	.2709	.2676	.2643	.2611	.2578	.2546	.2514	.2483	.2451
0.7	.2420	.2389	.2358	.2327	.2296	.2266	.2236	.2206	.2177	.2148
0.8	.2119	.2090	.2061	.2033	.2005	.1977	.1949	.1922	.1894	.1867
0.9	.1841	.1814	.1788	.1762	.1736	.1711	.1685	.1660	.1635	.1611
1.0	.1587	.1562	.1539	.1515	.1492	.1469	.1446	.1423	.1401	.1379
1.1	.1357	.1335	.1314	.1292	.1271	.1251	.1230	.1210	.1190	.1170
1.2	.1151	.1131	.1112	.1093	.1075	.1056	.1038	.1020	.1003	.0985
1.3	.0968	.0951	.0934	.0918	.0901	.0885	.0869	.0853	.0838	.0823
1.4	.0808	.0793	.0778	.0764	.0749	.0735	.0722	.0708	.0694	.0681
1.5	.0668	.0655	.0643	.0630	.0618	.0606	.0594	.0582	.0571	.0559
1.6	.0548	.0537	.0526	.0516	.0505	.0495	.0485	.0475	.0465	.0455
1.7	.0446	.0436	.0427	.0418	.0409	.0401	.0392	.0384	.0375	.0367
1.8	.0359	.0352	.0344	.0336	.0329	.0322	.0314	.0307	.0301	.0294
1.9	.0287	.0281	.0274	.0268	.0262	.0256	.0250	.0244	.0239	.0233
2.0	.0228	.0222	.0217	.0212	.0207	.0202	.0197	.0192	.0188	.0183
2.1	.0179	.0174	.0170	.0166	.0162	.0158	.0154	.0150	.0146	.0143
2.2	.0139	.0136	.0132	.0129	.0125	.0122	.0119	.0116	.0113	.0110
2.3	.0107	.0104	.0102	.0099	.0096	.0094	.0091	.0089	.0087	.0084
2.4	.0082	.0080	.0078	.0075	.0073	.0071	.0069	.0068	.0066	.0064
2.5	.0062	.0060	.0059	.0057	.0055	.0054	.0052	.0051	.0049	.0048
2.6	.0047	.0045	.0044	.0043	.0041	.0040	.0039	.0038	.0037	.0036
2.7	.0035	.0034	.0033	.0032	.0031	.0030	.0029	.0028	.0027	.0026
2.8	.0026	.0025	.0024	.0023	.0023	.0022	.0021	.0021	.0020	.0019
2.9	.0019	.0018	.0017	.0017	.0016	.0016	.0015	.0015	.0014	.0014
3.0	.00135									
3.5	.000 233									
4.0	.000 031 7		}	To interpolate carefully, see Table X.						
4.5	.000 003 40									
5.0	.000 000 287									

TABLE V
Student's t Critical Points

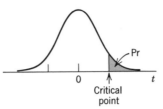

Pr d.f.	.25	.10	.05	.025	.010	.005	.0025	.0010	.0005
1	1.000	3.078	6.314	12.706	31.821	63.637	127.32	318.31	636.62
2	.816	1.886	2.920	4.303	6.965	9.925	14.089	22.326	31.598
3	.765	1.638	2.353	3.182	4.541	5.841	7.453	10.213	12.924
4	.741	1.533	2.132	2.776	3.747	4.604	5.598	7.173	8.610
5	.727	1.476	2.015	2.571	3.365	4.032	4.773	5.893	6.869
6	.718	1.440	1.943	2.447	3.143	3.707	4.317	5.208	5.959
7	.711	1.415	1.895	2.365	2.998	3.499	4.020	4.785	5.408
8	.706	1.397	1.860	2.306	2.896	3.355	3.833	4.501	5.041
9	.703	1.383	1.833	2.262	2.821	3.250	3.690	4.297	4.781
10	.700	1.372	1.812	2.228	2.764	3.169	3.581	4.144	4.537
11	.697	1.363	1.796	2.201	2.718	3.106	3.497	4.025	4.437
12	.695	1.356	1.782	2.179	2.681	3.055	3.428	3.930	4.318
13	.694	1.350	1.771	2.160	2.650	3.012	3.372	3.852	4.221
14	.692	1.345	1.761	2.145	2.624	2.977	3.326	3.787	4.140
15	.691	1.341	1.753	2.131	2.602	2.947	3.286	3.733	4.073
16	.690	1.337	1.746	2.120	2.583	2.921	3.252	3.686	4.015
17	.689	1.333	1.740	2.110	2.567	2.898	3.222	3.646	3.965
18	.688	1.330	1.734	2.101	2.552	2.878	3.197	3.610	3.922
19	.688	1.328	1.729	2.093	2.539	2.861	3.174	3.579	3.883
20	.687	1.325	1.725	2.086	2.528	2.845	3.153	3.552	3.850
21	.686	1.323	1.721	2.080	2.518	2.831	3.135	3.257	3.189
22	.686	1.321	1.717	2.074	2.508	2.819	3.119	3.505	3.792
23	.685	1.319	1.714	2.069	2.500	2.807	3.104	3.485	3.767
24	.685	1.318	1.711	2.064	2.492	2.797	3.091	3.467	3.745
25	.684	1.316	1.708	2.060	2.485	2.787	3.078	3.450	3.725
26	.684	1.315	1.706	2.056	2.479	2.779	3.067	3.435	3.707
27	.684	1.314	1.703	2.052	2.473	2.771	3.057	3.421	3.690
28	.683	1.313	1.701	2.048	2.467	2.763	3.047	3.408	3.674
29	.683	1.311	1.699	2.045	2.462	2.756	3.038	3.396	3.659
30	.683	1.310	1.697	2.042	2.457	2.750	3.030	3.385	3.646
40	.681	1.303	1.684	2.021	2.423	2.704	2.971	3.307	3.551
60	.679	1.296	1.671	2.000	2.390	2.660	2.915	3.232	3.460
120	.677	1.289	1.658	1.980	2.358	2.617	2.860	3.160	3.373
∞	.674	1.282	1.645	1.960	2.326	2.576	2.807	3.090	3.291

To interpolate carefully, see Table X.

TABLE **VI(a)**
χ^2 Critical Points

Pr d.f.	.250	.100	.050	.025	.010	.005	.001
1	1.32	2.71	3.84	5.02	6.63	7.88	10.8
2	2.77	4.61	5.99	7.38	9.21	10.6	13.8
3	4.11	6.25	7.81	9.35	11.3	12.8	16.3
4	5.39	7.78	9.49	11.1	13.3	14.9	18.5
5	6.63	9.24	11.1	12.8	15.1	16.7	20.5
6	7.84	10.6	12.6	14.4	16.8	18.5	22.5
7	9.04	12.0	14.1	16.0	18.5	20.3	24.3
8	10.2	13.4	15.5	17.5	20.1	22.0	26.1
9	11.4	14.7	16.9	19.0	21.7	23.6	27.9
10	12.5	16.0	18.3	20.5	23.2	25.2	29.6
11	13.7	17.3	19.7	21.9	24.7	26.8	31.3
12	14.8	18.5	21.0	23.3	26.2	28.3	32.9
13	16.0	19.8	22.4	24.7	27.7	29.8	34.5
14	17.1	21.1	23.7	26.1	29.1	31.3	36.1
15	18.2	22.3	25.0	27.5	30.6	32.8	37.7
16	19.4	23.5	26.3	28.8	32.0	34.3	39.3
17	20.5	24.8	27.6	30.2	33.4	35.7	40.8
18	21.6	26.0	28.9	31.5	34.8	37.2	42.3
19	22.7	27.2	30.1	32.9	36.2	38.6	32.8
20	23.8	28.4	31.4	34.2	37.6	40.0	45.3
21	24.9	29.6	32.7	35.5	38.9	41.4	46.8
22	26.0	30.8	33.9	36.8	40.3	42.8	48.3
23	27.1	32.0	35.2	38.1	41.6	44.2	49.7
24	28.2	33.2	36.4	39.4	32.0	45.6	51.2
25	29.3	34.4	37.7	40.6	44.3	46.9	52.6
26	30.4	35.6	38.9	41.9	45.6	48.3	54.1
27	31.5	36.7	40.1	43.2	47.0	49.6	55.5
28	32.6	37.9	41.3	44.5	48.3	51.0	56.9
29	33.7	39.1	42.6	45.7	49.6	52.3	58.3
30	34.8	40.3	43.8	47.0	50.9	53.7	59.7
40	45.6	51.8	55.8	59.3	63.7	66.8	73.4
50	56.3	63.2	67.5	71.4	76.2	79.5	86.7
60	67.0	74.4	79.1	83.3	88.4	92.0	99.6
70	77.6	85.5	90.5	95.0	100	104	112
80	88.1	96.6	102	107	112	116	125
90	98.6	108	113	118	124	128	137
100	109	118	124	130	136	140	149

To interpolate carefully, see Table *X*.

TABLE **VI(b)**
C^2 Critical Points ($C^2 = \chi^2/\text{d.f.}$)

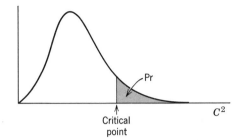

Pr d.f.	.995	.990	.975	.95	.90	.10	.05	.025	.010	.005
1	.000039	.00016	.00098	.0039	.0158	2.71	3.84	5.02	6.63	7.88
2	.00501	.0101	.0253	.0513	.1054	2.30	3.00	3.69	4.61	5.30
3	.0239	.0383	.0719	.117	.195	2.08	2.60	3.12	3.78	4.28
4	.0517	.0743	.121	.178	.266	1.94	2.37	2.79	3.32	3.72
5	.0823	.111	.166	.229	.322	1.85	2.21	2.57	3.02	3.35
6	.113	.145	.206	.273	.367	1.77	2.10	2.41	2.80	3.09
7	.141	.177	.241	.310	.405	1.72	2.01	2.29	2.64	2.90
8	.168	.206	.272	.342	.436	1.67	1.94	2.19	2.51	2.74
9	.193	.232	.300	.369	.463	1.63	1.88	2.11	2.41	2.62
10	.216	.256	.325	.394	.487	1.60	1.83	2.05	2.32	2.52
11	.237	.278	.347	.416	.507	1.57	1.79	1.99	2.25	2.43
12	.256	.298	.367	.435	.525	1.55	1.75	1.94	2.18	2.36
13	.274	.316	.385	.453	.542	1.52	1.72	1.90	2.13	2.29
14	.291	.333	.402	.469	.556	1.50	1.69	1.87	2.08	2.24
15	.307	.349	.417	.484	.570	1.49	1.67	1.83	2.04	2.19
16	.321	.363	.432	.489	.582	1.47	1.64	1.80	2.00	214
18	.348	.390	.457	.522	.604	1.44	1.60	1.75	1.93	2.06
20	.372	.413	.480	.543	.622	1.42	1.57	1.71	1.88	2.00
24	.412	.452	.517	.577	.652	1.38	1.52	1.64	1.79	1.90
30	.460	.498	.560	.616	.687	1.34	1.46	1.57	1.70	1.79
40	.518	.554	.611	.663	.726	1.30	1.39	1.48	1.59	1.67
60	.592	.625	.675	.720	.774	1.24	1.32	1.39	1.47	1.53
120	.699	.724	.763	.798	.839	1.17	1.22	1.27	1.32	1.36
∞	1.000	1.000	1.000	1.000	1.000	1.00	1.00	1.00	1.00	1.00

To interpolate carefully, see Table X.

TABLE VII
F Critical Points

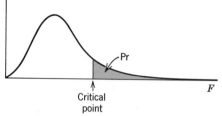

| | Pr | \multicolumn{11}{c}{Degrees of freedom for numerator} |
		1	2	3	4	5	6	8	10	20	40	∞
1	.25	5.83	7.50	8.20	8.58	8.82	8.98	9.19	9.32	9.58	.9.71	9.85
	.10	39.9	49.5	53.6	55.8	57.2	58.2	59.4	60.2	61.7	62.5	63.3
	.05	161	200	216	225	230	234	239	242	248	251	254
2	.25	2.57	3.00	3.15	3.23	3.28	3.31	3.35	3.38	3.43	3.45	3.48
	.10	8.53	9.00	9.16	9.24	9.29	9.33	9.37	9.39	9.44	9.47	9.49
	.05	18.5	19.0	19.2	19.2	19.3	19.3	19.4	19.4	19.4	19.5	19.5
	.01	98.5	99.0	99.2	99.2	99.3	99.3	99.4	99.4	99.4	99.5	99.5
	.001	998	999	999	999	999	999	999	999	999	999	999
3	.25	2.02	2.28	2.36	2.39	2.41	2.42	2.44	2.44	2.46	2.47	2.47
	.10	5.54	5.46	5.39	5.34	5.31	5.28	5.25	5.23	5.18	5.16	5.13
	.05	10.1	9.55	9.28	9.12	9.10	8.94	8.85	8.79	8.66	8.59	8.53
	.01	34.1	30.8	29.5	28.7	28.2	27.9	27.5	27.2	26.7	26.4	26.1
	.001	167	149	141	137	135	133	131	129	126	125	124
4	.25	1.81	2.00	2.05	2.06	2.07	2.08	2.08	2.08	2.08	2.08	2.08
	.10	4.54	4.32	4.19	4.11	4.05	4.01	3.95	3.92	3.84	3.80	3.76
	.05	7.71	6.94	6.59	6.39	6.26	6.16	6.04	5.96	5.80	5.72	5.63
	.01	21.2	18.0	16.7	16.0	15.5	15.2	14.8	14.5	14.0	13.7	13.5
	.001	74.1	61.3	56.2	53.4	51.7	50.5	49.0	48.1	46.1	45.1	44.1
5	.25	1.69	1.85	1.88	1.89	1.89	1.89	1.89	1.89	1.88	1.88	1.87
	.10	4.06	3.78	3.62	3.52	3.45	3.40	3.34	3.30	3.21	3.16	3.10
	.05	6.61	5.79	5.41	5.19	5.05	4.95	4.82	4.74	4.56	4.46	4.36
	.01	16.3	13.3	12.1	11.4	11.0	10.7	10.3	10.1	9.55	9.29	9.02
	.001	47.2	37.1	33.2	31.1	29.8	28.8	27.6	26.9	25.4	24.6	23.8
6	.25	1.62	1.76	1.78	1.79	1.79	1.78	1.77	1.77	1.76	1.75	1.74
	.10	3.78	3.46	3.29	3.18	3.11	3.05	2.98	2.94	2.84	2.78	2.72
	.05	5.99	5.14	4.76	4.53	4.39	4.28	4.15	4.06	3.87	3.77	3.67
	.01	13.7	10.9	9.78	9.15	8.75	8.47	8.10	7.87	7.40	7.14	6.88
	.001	35.5	27.0	23.7	21.9	20.8	20.0	19.0	18.4	17.1	16.4	15.8
7	.25	1.57	1.70	1.72	1.72	1.71	1.71	1.70	1.69	1.67	1.66	1.65
	.10	3.59	3.26	3.07	2.96	2.88	2.83	2.75	2.70	2.59	2.54	2.47
	.05	5.59	4.74	4.35	4.12	3.97	3.87	3.73	3.64	3.44	3.34	3.23
	.01	12.2	9.55	8.45	7.85	7.46	7.19	6.84	6.62	6.16	5.91	5.65
	.001	29.3	21.7	18.8	17.2	16.2	15.5	14.6	14.1	12.9	12.3	11.7
8	.25	1.54	1.66	1.67	1.66	1.66	1.65	1.64	1.63	1.61	1.59	1.58
	.10	3.46	3.11	2.92	2.81	2.73	2.67	2.59	2.54	2.42	2.36	2.29
	.05	5.32	4.46	4.07	3.84	3.69	3.58	3.44	3.35	3.15	3.04	2.93
	.01	11.3	8.65	7.59	7.01	6.63	6.37	6.03	5.81	5.36	5.12	4.86
	.001	25.4	18.5	15.8	14.4	13.5	12.9	12.0	11.5	10.5	9.92	9.33
9	.25	1.51	1.62	1.63	1.63	1.62	1.61	1.60	1.59	1.56	1.55	1.53
	.10	3.36	3.01	2.81	2.69	2.61	2.55	2.47	2.42	2.30	2.23	2.16
	.05	5.12	4.26	3.86	3.63	3.48	3.37	3.23	3.14	2.94	2.83	2.71
	.01	10.6	8.02	6.99	6.42	6.06	5.80	5.47	5.26	4.81	4.57	4.31
	.001	22.9	16.4	13.9	12.6	11.7	11.1	10.4	9.89	8.90	8.37	7.81

To interpolate carefully, see Table X.

TABLE VII (Continued)

		Degrees of freedom for numerator										
	Pr	1	2	3	4	5	6	8	10	20	40	∞
10	.25	1.49	1.60	1.60	1.59	1.59	1.58	1.56	1.55	1.52	1.51	1.48
	.10	3.28	2.92	2.73	2.61	2.52	2.46	2.38	2.32	2.20	2.13	2.06
	.05	4.96	4.10	3.71	3.48	3.33	3.22	3.07	2.98	2.77	2.66	2.54
	.01	10.0	7.56	6.55	5.99	5.64	5.39	5.06	4.85	4.41	4.17	3.91
	.001	21.0	14.9	12.6	11.3	10.5	9.92	9.20	8.75	7.80	7.30	6.76
12	.25	1.56	1.56	1.56	1.55	1.54	1.53	1.51	1.50	1.47	1.45	1.42
	.10	3.18	2.81	2.61	2.48	2.39	2.33	2.24	2.19	2.06	1.99	1.90
	.05	4.75	3.89	3.49	3.26	3.11	3.00	2.85	2.75	2.54	2.43	2.30
	.01	9.33	6.93	5.95	5.41	5.06	4.82	4.50	4.30	3.86	3.62	3.36
	.001	18.6	13.0	10.8	9.63	8.89	8.38	7.71	7.29	6.40	5.93	5.42
14	.25	1.44	1.53	1.53	1.52	1.51	1.50	1.48	1.46	1.43	1.41	1.38
	.10	3.10	2.73	2.52	2.39	2.31	2.24	2.15	2.10	1.96	1.89	1.80
	.05	4.60	3.74	3.34	3.11	2.96	2.85	2.70	2.60	2.39	2.27	2.13
	.01	8.86	5.51	5.56	5.04	4.69	4.46	4.14	3.94	3.51	3.27	3.00
	.001	17.1	11.8	9.73	8.62	7.92	7.43	6.80	6.40	5.56	5.10	4.60
16	.25	1.42	1.51	1.51	1.50	1.48	1.48	1.46	1.45	1.40	1.37	1.34
	.10	3.05	2.67	2.46	2.33	2.24	2.18	2.09	2.03	1.89	1.81	1.72
	.05	4.49	3.63	3.24	3.01	2.85	2.74	2.59	2.49	2.28	2.15	2.01
	.01	8.53	6.23	5.29	4.77	4.44	4.20	3.89	3.69	3.26	3.02	2.75
	.001	16.1	11.0	9.00	7.94	7.27	6.81	6.19	5.81	4.99	4.54	4.06
18	.25	1.41	1.50	1.49	1.48	1.46	1.45	1.43	1.42	1.38	1.35	1.32
	.10	3.01	2.62	2.42	2.29	2.20	2.13	2.04	1.98	1.84	1.75	1.66
	.05	4.41	3.55	3.16	2.93	2.77	2.66	2.51	2.41	2.19	2.06	1.92
	.01	8.29	6.01	5.09	4.58	4.25	4.01	3.71	3.51	3.08	2.84	2.57
	.001	15.4	10.4	8.49	7.46	6.81	6.35	5.76	5.39	4.59	4.15	3.67
20	.25	1.40	1.49	1.48	1.46	1.45	1.44	1.42	1.40	1.36	1.33	1.29
	.10	2.97	2.59	2.38	2.25	2.16	2.09	2.00	1.94	1.79	1.71	1.61
	.05	4.35	3.49	3.10	2.87	2.71	2.60	2.45	2.35	2.12	1.99	1.84
	.01	8.10	5.85	4.94	4.43	4.10	3.87	3.56	3.37	2.94	2.69	2.42
	.001	14.8	9.95	8.10	7.10	6.46	6.02	5.44	5.08	4.29	3.86	3.38
30	.25	1.38	1.45	1.44	1.42	1.41	1.39	1.37	1.35	1.30	1.27	1.23
	.10	2.88	2.49	2.28	2.14	2.05	1.98	1.88	1.82	1.67	1.57	1.46
	.05	4.17	3.32	2.92	2.69	2.53	2.42	2.27	2.16	1.93	1.79	1.62
	.01	7.56	5.39	4.51	4.02	3.70	3.47	3.17	2.98	2.55	2.30	2.01
	.001	13.3	8.77	7.05	6.12	5.53	5.12	4.58	4.24	3.49	3.07	2.59
40	.25	1.36	1.44	1.42	1.40	1.39	1.37	1.35	1.33	1.28	1.24	1.19
	.10	2.84	2.44	2.23	2.09	2.00	1.93	1.83	1.76	1.61	1.51	1.38
	.05	4.08	3.23	2.84	2.61	2.45	2.34	2.18	2.08	1.84	1.69	1.51
	.01	7.31	5.18	4.31	3.83	3.51	3.29	2.99	2.80	2.37	2.11	1.80
	.001	12.6	8.25	6.60	5.70	5.13	4.73	4.21	3.87	3.15	2.73	2.23
60	.25	1.35	1.42	1.41	1.38	1.37	1.35	1.32	1.30	1.25	1.21	1.15
	.10	2.79	2.39	2.18	2.04	1.95	1.87	1.77	1.71	1.54	1.44	1.29
	.05	4.00	3.15	2.76	2.53	2.37	2.25	2.10	1.99	1.75	1.59	1.39
	.01	7.08	4.98	4.13	3.65	3.34	3.12	2.82	2.63	2.20	1.94	1.60
	.001	12.0	7.76	6.17	5.31	4.76	4.37	3.87	3.54	2.83	2.41	1.89
120	.25	1.34	1.40	1.39	1.37	1.35	1.33	1.30	1.28	1.22	1.18	1.10
	.10	2.75	2.35	2.13	1.99	1.90	1.82	1.72	1.65	1.48	1.37	1.19
	.05	3.92	3.07	2.68	2.45	2.29	2.17	2.02	1.91	1.66	1.50	1.25
	.01	6.85	4.79	3.95	3.48	3.17	2.96	2.66	2.47	2.03	1.76	1.38
	.001	11.4	7.32	5.79	4.95	4.42	4.04	3.55	3.24	2.53	2.11	1.54
∞	.25	1.32	1.39	1.37	1.35	1.33	1.31	1.28	1.25	1.19	1.14	1.00
	.10	2.71	2.30	2.08	1.94	1.85	1.77	1.67	1.60	1.42	1.30	1.00
	.05	3.84	3.00	2.60	2.37	2.21	2.10	1.94	1.83	1.57	1.39	1.00
	.01	6.63	4.61	3.78	3.32	3.02	2.80	2.51	2.32	1.88	1.59	1.00
	.001	10.8	6.91	5.42	4.62	4.10	3.74	3.27	2.96	2.27	1.84	1.00

Degrees of freedom for denominator

717

TABLE **VIII**
Wilcoxon–Mann–Whitney (W) Test

The one-sided prob-value (Pr) corresponding to the rank sum W of the smaller sample, ranking from the end where this smaller sample is concentrated. For $n > 7$ or prob-value $> .25$, see equation (16-18).

n = 2		Larger Sample Size, n = 3						Larger Sample Size, n = 4							
m		Smaller Sample Size, m						Smaller Sample Size, m							
1	2	1		2		3		1		2		3		4	
W Pr	W Pr	W	Pr	W	Pr	W	Pr	W	Pr	W	Pr	W	Pr	W	Pr
1 .333	3 .167	1	.250	3	.100	6	.050	1	.200	3	.067	6	.029	10	.014
	4 .333	2	.500	4	.200	7	.100	2	.400	4	.133	7	.057	11	.029
				5	.400	8	.200			5	.267	8	.114	12	.057
						9	.350			6	.400	9	.200	13	.100
						10	.500					10	.314	14	.171
												11	.429	15	.243
														16	.343

Larger Sample Size, n = 5										Larger Sample Size, n = 6											
Smaller Sample Size, m										Smaller Sample Size, m											
1		2		3		4		5		1		2		3		4		5		6	
W	Pr	W	Pr	W	Pr	W	Pr	W	Pr	W	Pr	W	Pr	W	Pr	W	Pr	W	Pr	W	Pr
1	.167	3	.048	6	.018	10	.008	15	.004	1	.143	3	.036	6	.012	10	.005	15	.002	21	.001
2	.333	4	.095	7	.036	11	.016	16	.008	2	.286	4	.071	7	.024	11	.010	16	.004	22	.002
3	.500	5	.190	8	.071	12	.032	17	.016	3	.429	5	.143	8	.048	12	.019	17	.009	23	.004
		6	.286	9	.125	13	.056	18	.028			6	.214	9	.083	13	.033	18	.015	24	.008
		7	.429	10	.196	14	.095	19	.048			7	.321	10	.131	14	.057	19	.026	25	.013
				11	.286	15	.143	20	.075			8	.429	11	.190	15	.086	20	.041	26	.021
				12	.393	16	.206	21	.111					12	.274	16	.129	21	.063	27	.032
				13	.500	17	.278	22	.155					13	.357	17	.176	22	.089	28	.047
						18	.365	23	.210					14	.452	18	.238	23	.123	29	.066
						19	.452	24	.274							19	.305	24	.165	30	.090
								25	.345							20	.381	25	.214	31	.120
								26	.421							21	.457	26	.268	32	.155
								27	.500									27	.331	33	.197
																		28	.396	34	.242
																		29	.465	35	.294

TABLE **VIII** (Continued)

Larger Sample Size, n = 7

Smaller Sample Size, m													
1		2		3		4		5		6		7	
W	Pr	W	Pr	W	Pr	W	Pr	W	Pr	W	Pr	W	Pr
1	.125	3	.028	6	.008	10	.003	15	.001	21	.001	28	.000
2	.250	4	.056	7	.017	11	.006	16	.003	22	.001	29	.001
3	.375	5	.111	8	.033	12	.012	17	.005	23	.002	30	.001
4	.500	6	.167	9	.058	13	.021	18	.009	24	.004	31	.002
		7	.250	10	.092	14	.036	19	.015	25	.007	32	.003
		8	.333	11	.133	15	.055	20	.024	26	.011	33	.006
		9	.444	12	.192	16	.082	21	.037	27	.017	34	.009
				13	.258	17	.115	22	.053	28	.026	35	.013
				14	.333	18	.158	23	.074	29	.037	36	.019
				15	.417	19	.206	24	.101	30	.051	37	.027
				16	.500	20	.264	25	.134	31	.069	38	.036
						21	.324	26	.172	32	.090	39	.049
						22	.394	27	.216	33	.117	40	.064
						23	.464	28	.265	34	.147	41	.082
								29	.319	35	.183	42	.104
								30	.378	36	.223	43	.130
								31	.438	37	.267	44	.159
								32	.500	38	.314	45	.191
										39	.365	46	.228
										40	.418	47	.267

TABLE IX

Critical Points of the Durbin-Watson Test for Autocorrelation

This table gives two limiting values of critical D (D_L and D_U), corresponding to the two most extreme configurations of the regressors; thus, for every possible configuration, the critical value of D will be somewhere between D_L and D_U:

$P(D)$, if H_0 true

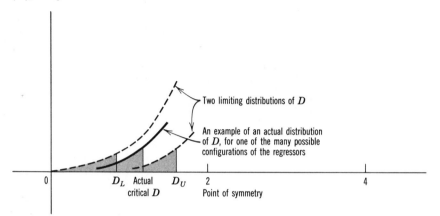

As an example of a test for positive serial correlation, suppose that there are $n = 15$ observations and $k = 3$ regressors (excluding the constant) and we wish to test $\rho = 0$ versus $\rho > 0$ at the level $\alpha = .05$. Then if D falls below $D_L = .82$, reject H_0. If D falls above $D_U = 1.75$, do not reject H_0. If D falls between D_L and D_U, this test is indecisive.

To test for negative serial correlation ($\rho = 0$ versus $\rho < 0$), the right-hand tail of the distribution defines the critical region. The symmetry of the distribution permits us to calculate these values very easily. With the same sample size, number of regressors, and level α as before, our new critical values would be $4 - D_L = 4 - .82 = 3.18$, and $4 - D_U = 4 - 1.75 = 2.25$. Accordingly, if D falls beyond 3.18, reject H_0. If D falls short of 2.25, do not reject H_0. If D falls between 2.25 and 3.18, this test is indecisive.

TABLE **IX** (continued)

Sample size = n	Pr = Probability in Lower Tail (Level, α)	\multicolumn{10}{c}{k = Number of Regressors (Excluding the Constant)}
		\multicolumn{2}{c}{1}	\multicolumn{2}{c}{2}	\multicolumn{2}{c}{3}	\multicolumn{2}{c}{4}	\multicolumn{2}{c}{5}
		D_L	D_U	D_L	D_U	D_L	D_U	D_L	D_U	D_L	D_U
15	.01	.81	1.07	70	1.25	.59	1.46	.49	1.70	.39	1.96
	.025	.95	1.23	.83	1.40	.71	1.61	.59	1.84	.48	2.09
	.05	1.08	1.36	.95	1.54	.82	1.75	.69	1.97	.56	2.21
20	.01	.95	1.15	.86	1.27	.77	1.41	.68	1.57	.60	1.74
	.025	1.08	1.28	.99	1.41	.89	1.55	.79	1.70	.70	1.87
	.05	1.20	1.41	1.10	1.54	1.00	1.68	.90	1.83	.79	1.99
25	.01	1.05	1.21	.98	1.30	.90	1.41	.83	1.52	.75	1.65
	.025	1.18	1.34	1.10	1.43	1.02	1.54	.94	1.65	.86	1.77
	.05	1.29	1.45	1.21	1.55	1.12	1.66	1.04	1.77	.95	1.89
30	.01	1.13	1.26	1.07	1.34	1.01	1.42	.94	1.51	.88	1.61
	.025	1.25	1.38	1.18	1.46	1.12	1.54	1.05	1.63	.98	1.73
	.05	1.35	1.49	1.28	1.57	1.21	1.65	1.14	1.74	1.07	1.83
40	.01	1.25	1.34	1.20	1.40	1.15	1.46	1.10	1.52	1.05	1.58
	.025	1.35	1.45	1.30	1.51	1.25	1.57	1.20	1.63	1.15	1.69
	.05	1.44	1.54	1.39	1.60	1.34	1.66	1.29	1.72	1.23	1.79
50	.01	1.32	1.40	1.28	1.45	1.24	1.49	1.20	1.54	1.16	1.59
	.025	1.42	1.50	1.38	1.54	1.34	1.59	1.30	1.64	1.26	1.69
	.05	1.50	1.59	1.46	1.63	1.42	1.67	1.38	1.72	1.34	1.77
60	.01	1.38	1.45	1.35	1.48	1.32	1.52	1.28	1.56	1.25	1.60
	.025	1.47	1.54	1.44	1.57	1.40	1.61	1.37	1.65	1.33	1.69
	.05	1.55	1.62	1.51	1.65	1.48	1.69	1.44	1.73	1.41	1.77
80	.01	1.47	1.52	1.44	1.54	1.42	1.57	1.39	1.60	1.36	1.62
	.025	1.54	1.59	1.52	1.62	1.49	1.65	1.47	1.67	1.44	1.70
	.05	1.61	1.66	1.59	1.69	1.56	1.72	1.53	1.74	1.51	1.77
00	.01	1.52	1.56	1.50	1.58	1.48	1.60	1.46	1.63	1.44	1.65
	.025	1.59	1.63	1.57	1.65	1.55	1.67	1.53	1.70	1.51	1.72
	.05	1.65	1.69	1.63	1.72	1.61	1.74	1.59	1.76	1.57	1.78

TABLE X
Interpolating Tables V to VII

For all the exercises in the text, simple linear interpolation gives a good enough approximation. However, important research data may deserve more careful interpolation, as follows.

Table V (Student's t) may require interpolation in either of two directions:

(a) Down (interpolating d.f.); or

(b) Across (interplating Pr).

We give examples of both kinds of interpolation, showing the tabled values in black, and interpolation calculations in color.

(a) Interpolation of d.f.

First, calculate $r = 1/\text{d.f.}$ Then interpolate linearly with r. For example, let us find $t_{.025}$ for d.f. $= 600$:

d.f.	$r = \dfrac{1}{\text{d.f.}}$	$t_{.025}$
120	$\dfrac{1}{120} = .00833$	1.980
600	$\dfrac{1}{600} = .00167$	$1.960 + (1.980 - 1.960)\left(\dfrac{.00167 - 0}{.00833 - 0}\right) = 1.964$
∞	$\dfrac{1}{\infty} = 0$	1.960

(b) Interpolation of Pr

First, calculate $L = \log \text{Pr}$. Then interpolate linearly with L. For example, when d.f. $= 4$, let us find Pr corresponding to $t = 1.800$:

Pr	.10	.0734	.05
$L = \log \text{Pr}$	-1.00	$-1.00 + (-1.301 + 1.00)\left(\dfrac{1.800 - 1.533}{2.132 - 1.533}\right) = -1.134$	-1.301
t	1.533	1.800	2.132

Finally, we note that Tables VI(b)[1] and VII can be interpolated similarly, while Table VI(a) can be calculated from Table VI(b).

---

[1] The right half of Table VI(b) can be interpolated similarly; the left half requires first calculating the tail probability $(1 - \text{Pr})$ in place of Pr.

I. (a) From H. A. Simmons, *Wiley Trigonometric Tables,* Section Edition. Copyright © 1945 by John Wiley and Sons, Inc. Reprinted by permission.

(b) Reprinted from John E. Freund, *Modern Elementary Statistics,* 3rd Edition, © 1967, by permission of Prentice-Hall Inc., Englewood Cliffs, New Jersey.

II. (a) Reprinted, by permission, from Clelland et al., *Basic Statistics with Business Applications,* copyright © 1966 by John Wiley and Sons, Inc.

(b) Reprinted, by permission, from the RAND Corporation.

III. Reprinted, by permission, from the *Chemical Rubber Company Standard Mathematical Tables,* 19th edition, 1971, courtesy of The Chemical Rubber Co., Cleveland, Ohio.

IV. Reprinted with permission of The Macmillan Company from *Introduction to Statistics* by R. E. Walpole. Copyright © by Ronald E. Walpole, 1968.

V. Reprinted, by permission, from E. S. Pearson and H. O. Hartley, *Biometrika Tables for Statisticians,* vol. 1, 2nd edition, Cambridge, 1962.

VI. (a) Reprinted with rounding, by permission, from E. S. Pearson and H. O. Hartley, *Biometrika Tables for Statisticians,* vol. 1, 2nd edition, Cambridge, 1962.

(b) From *Introduction to Statistical Analysis* by Dixon and Massey. 2nd Ed. Copyright © 1957 by McGraw-Hill, Inc. Used with permission of McGraw-Hill Book Company.

VII. Abridgment, by permission, from *Chemical Rubber Company Standard Mathematical Tables,* 19th edition, 1971, courtesy of The Chemical Rubber Co., Cleveland, Ohio. And from *Statistical Principles in Experimental Design* by B. J. Winer, 1971, McGraw-Hill Book Company; (this originally appeared in *Biometrica Tables for Statisticians,* vol. 1).

Answers, Odd-numbered Problems

1-1 (a) $\pi = .51 \pm .0253$
    (b) $\pi = .64 \pm .0243$
    (c) $\pi = .50 \pm .0253$
    (d) $\pi = .38 \pm .0246$

2-1 (b) $\overline{X} = 11.0$

2-3 (b) $\overline{X} = 15.8$

2-5 (a) $\overline{X} = 78.4$, mode $= 80$

2-9 (b) mean $= 15$, median $= 8$,
      mode $= 6$
    (c) 78

2-11 (a) 107, 105, 102, 105
     (b) True

2-13   range $= 20$, MAD $= 3.16$,
       MSD $= 15.4$, $s^2 = 16.0$,
       $s = 4.0$

2-15 $\overline{X} = 77.4$, $s_X = 11.0$

2-17 $\overline{X} = 51.8$, $s_X = 9.73$

2-19 (a)

x	0	1	2	3	4
f/n	.38	.40	.14	.04	.04

    (b) $\overline{X} = .96$
    (c) $s_X^2 = 1.06$, $s_X = 1.03$

2-21   27.6% (*NOT* 23.8%)

3-1 (c) relative frequency,
     when n $= 50$

3-3 (c) relative frequency,
     when n $= 50$ (or better yet,
     figure it out to be
     $8/36 = .222$)

3-1 (c) relative frequency,
     when n $= 50$

3-3 (c) relative frequency,
     when n $= 50$ (or better yet,
     figure it out to be
     $8/36 = .222$)

3-5 (a) $3/36 = .083$
    (b) $6/36 = .167$
    (c) $8/36 = .222$
    (d) $6/36 = .167$
    (e) $15/36 = .417$
    (f) $2/36 = .056$
    (g) $1/36 = .028$
    (h) Yes, of course.

3-7 (a) $\dfrac{17}{16}$, hence not a partition

    (b) a partition, hence $\sum \mathrm{Pr} = 1$

3-9 (a) $1 - (1/2)^{10} = .999$
    (b) $1 - 2(1/2)^{10} = .998$

3-11 (a) .34
     (b) .66
     (c) .04

3-13 $\mathrm{Pr}(E_1 \cup E_2 \cup E_3) =$
       $\mathrm{Pr}(E_1) + \mathrm{Pr}(E_2) + \mathrm{Pr}(E_3)$
       $- \mathrm{Pr}(E_1 \cap E_2)$
       $- \mathrm{Pr}(E_1 \cap E_3)$
       $- \mathrm{Pr}(E_2 \cap E_3)$
       $+ \mathrm{Pr}(E_1 \cap E_2 \cap E_3)$

3-17 (a) $1/3 = .33$
     (b) $1/14 = .071$
     (c) $1/42 = .024$

725

3-19 (a) $\dfrac{4 \times 3 \times 2 \times 1 \times 48}{52 \times 51 \times 50 \times 49 \times 48}$

$= 3.69 \times 10^{-6}$

(b) same as (a)

(c) 5 times as large, $= 18.5 \times 10^{-6}$

(d) 13 times as large again,

$= .00024$

3-21 (a) .476

(b) .49

3-23 (a) .67

(b) .33

3-25 (a) $.167 = .167$

(b) $.167 \neq .083$

(c) $.167 = .167$

(d) true

3-27 (a) independent

(b) not independent

3-29 (a) independent

(b) independent

(c) not independent

3-31 (a) .3

(b) misinformation

(c) 0

(d) $0 \le \Pr(e_4) \le .2$

3-35   valid

3-37 (a) .60

(b) .12

(c) .68

3-39 (a) True

(b) True

(c) False: A and B are *statistically independent.*

3-41 (a) .33

(b) .50

(c) 0

(d) .17

3-43 (a) $(1/2)^{10} = .001$

(b) 1/2

3-45   For n tosses,

$\Pr = \dfrac{.001}{.001 + (1/2)^n(.999)}$

Thus

(a) .0079

(b) .506

(c) .999

4-1 (a)

x	0	1	2	3	4
$p(x)$	$\dfrac{1}{16}$	$\dfrac{4}{16}$	$\dfrac{6}{16}$	$\dfrac{4}{16}$	$\dfrac{1}{16}$

(b)

y	0	1	2	3
$p(y)$	$\dfrac{2}{16}$	$\dfrac{6}{16}$	$\dfrac{6}{16}$	$\dfrac{2}{16}$

4-5 (a) $\mu = 2$, $\sigma^2 = 1$

(b) $\mu = 1.5$, $\sigma^2 = .75$

4-7 $\mu_Y = 11$, $\sigma_Y = 3.42$

4-9 (a) $2\mu - 8$, 2σ

(b) we cannot say, when transformation is nonlinear.

(c) $\dfrac{1}{10}\mu - \dfrac{5}{10}$, $\dfrac{1}{10}\sigma$

(d) 0, 1

4-11 (a) $n = 100$, $\pi = .5$, mean $= 50$

(b) $n = 100$, $\pi = 1/6$,

mean $= 16.7$

(c) $n = 25$, $\pi = .5$,

mean $= 12.5$

4-13 (c) $\mu = 2$, $\sigma^2 = 1$

4-17 (a) .46

(b) .87

(c) .57

(d) True

4-19 (a) .0548

(b) .0441

(c) .9495

(d) .1034

(e) .3174

(f) .9973

4-21 (a) .2590
    (b) .9332
    (c) .3830

4-23 (a), (b) 8.0

4-25 (a) 4.6
    (b) 1.5
    (c) 23.2
    (d) 2.04

4-27 (a) 3.5
    (b) 15
    (c) True

4-29 (a) True, by Table 4-2
    (b) True, by Table 4-2
    (c) False
    (d) True, by Table 4-2

4-31 (a) .0038
    (b) .4962

4-33 (a) .60
    (b) .648
    (c) .733

4-35 (a)

x	0	1	2
$p(x)$	$\dfrac{7}{12}$	$\dfrac{3}{12}$	$\dfrac{2}{12}$

$\mu = 7/12$

    (b) 1/4

4-37 (a) True, assuming independent
       questions
    (b) False. And there is only about
       one chance in 1000 that an
       innocent person will show
       guilty knowledge in 7 or more
       questions.

4-39 (i) (b) mean = median = mode
            = .50
      (c) $\sigma^2 = .050$
    (ii) (b) mean = .60, median
          = .614, mode = .667
      (c) $\sigma^2 = .040$

5-1 (a)

3			2/16		
2		2/16	2/16	2/16	
1		2/16	2/16	2/16	
0	1/16				1/16

$y \diagdown x$	0	1	2	3	4
$p(x)$	1/16	4/16	6/16	4/16	1/16

    (b) not independent

5-3 (b) not independent

5-5 (a),(b) 1.6
    (c) 8.0

5-7 (a)

s	2	3	4	5	6
$p(s)$	.20	.31	.31	.15	.03

$\mu_S = 3.5$, $\sigma_S^2 = 1.13$
    (b) $\mu_X = 1.9$, $\sigma_X^2 = .69$
       $\mu_Y = 1.6$, $\sigma_Y^2 = .44$
    (c) $E(X + Y) = E(X) + E(Y)$
       $var(X + Y) = var(X) + var(Y)$

5-9 (a),(b) $-.60$
    (c) $-.90$

5-11 Yes

5-15 (a)

15		.1	.1
10	.2	.2	.2
5	.1	.1	

$y \diagdown x$	10	15	20

    (b) $\mu_X = 15$, $\sigma_X^2 = 15$
    (c) $\mu_Y = 10$, $\sigma_Y^2 = 10$
    (d) $\sigma_{XY} = 5$
    (e) 8.33, 10.00, 11.67
    (f) $\mu_S = 25$, $\sigma_S^2 = 35$
    (g) $\mu_W = 17$, $\sigma_W^2 = 16.6$
    (h) $\mu_D = 5$, $\sigma_D^2 = 15$
    (i) Not necessarily

5-17

			400
			400
(a)	65	15	225
(b)	70	15.6	2200/9

5-19 (a)

s	2	3	4	5	6
$p(s)$	.1	.2	.3	.3	.1

$\mu_S = 4.1, \quad \sigma_S^2 = 1.29$

(b) $\mu_1 = 2.0, \quad \sigma_1^2 = .6$

$\mu_2 = 2.1, \quad \sigma_2^2 = .69$

$\sigma_{12} = 0$ by symmetry

5-21 (a) 2, 1, 9, 4

(b) True, $9 = 9$

(c) False, $4 \neq 3$

5-23 (a) independent, hence $\sigma_{XY} = 0$

(b)

x	0	1	2
$p(x)$	1/4	1/2	1/4

y	0	1
$p(y)$	1/2	1/2

s	0	1	2	3
$p(s)$	1/8	3/8	3/8	1/8

$\mu_X = 1.0 \quad \sigma_X^2 = .50$

$\mu_Y = .5 \quad \sigma_Y^2 = .25$

$\mu_S = 1.5 \quad \sigma_S^2 = .75$

5-25 (a)

7	1/9	1/9	1/9
6	1/9	1/9	1/9
2	1/9	1/9	1/9
x_2 $\diagup$ x_1	2	6	7

(b) X_1 and X_2 both have the same distribution, as follows:

x	2	6	7
$p(x)$	1/3	1/3	1/3

(c) independent, hence $\sigma_{12} = 0$

(d) $\mu_1 = \mu_2 = 5$

$\sigma_1^2 = \sigma_2^2 = 4.67$

(e) $\mu_{\overline{X}} = 5, \quad \sigma_{\overline{X}}^2 = 2.33$

$\overline{x}$	2	4	4.5	6	6.5	7
$p(\overline{x})$	$\dfrac{1}{9}$	$\dfrac{2}{9}$	$\dfrac{2}{9}$	$\dfrac{1}{9}$	$\dfrac{2}{9}$	$\dfrac{1}{9}$

hence $\mu_{\overline{X}} = 5, \quad \sigma_{\overline{X}}^2 = 2.33$

(f) no different

(g) correct

5-27 (a)

7	1/6	1/6	0
6	1/6	0	1/6
2	0	1/6	1/6
x_2 $\diagup$ x_1	2	6	7

(b) X_1 and X_2 have the same distribution, as follows:

x	2	6	7
$p(x)$	1/3	1/3	1/3

(c) dependent, $\sigma_{12} = -2.33$

(d) $\mu_1 = \mu_2 = 5$

$\sigma_1^2 = \sigma_2^2 = 4.67$

(e) $\mu_{\overline{X}} = 5, \quad \sigma_{\overline{X}}^2 = 1.17$

$\overline{x}$	4	4.5	6.5
$p(\overline{x})$	1/3	1/3	1/3

hence $\mu_{\overline{X}} = 5, \quad \sigma_{\overline{X}}^2 = 1.17$

(f) The answers would be about the same as in Problem 5-25. That is, $\mu_{\overline{X}} = 5$, $\sigma_{\overline{X}}^2 = 2.33$. That is, sampling without replacement is about the same as sampling with replacement, if the population is large.

(g) The conclusion should be changed to: "In sampling without replacement, the absolute frequency of the chips matters for the *variance* of $\overline{X}$."

5-29 (a) 68.33

(b) life expectancy would be 63.62, a reduction of 4.7 years

5-31 (a)

x	0	1	2	3
$p(x)$	.512	.384	.096	.008

(b) $\mu = .6, \quad \sigma^2 = .48$

(c) $13.88

5-33   .32

5-35 (a) 1

(b) 1.67

(c) 1

(d) $(n - k + k^2)/n$

6-1 (a) False. . . . its standard deviation would be $\sigma/\sqrt{n} = 1.01$ inches.

(b) True

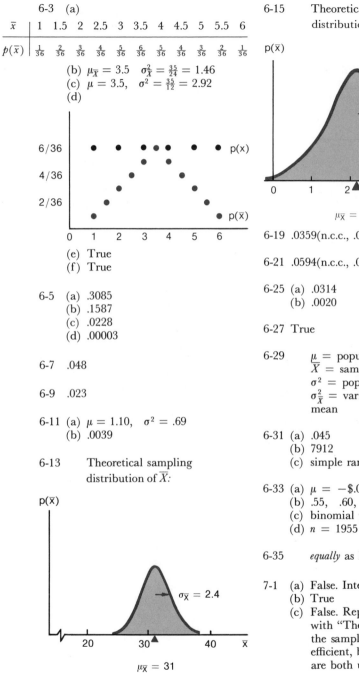

6-3 (a)

$\bar{x}$	1	1.5	2	2.5	3	3.5	4	4.5	5	5.5	6
$p(\bar{x})$	$\frac{1}{36}$	$\frac{2}{36}$	$\frac{3}{36}$	$\frac{4}{36}$	$\frac{5}{36}$	$\frac{6}{36}$	$\frac{5}{36}$	$\frac{4}{36}$	$\frac{3}{36}$	$\frac{2}{36}$	$\frac{1}{36}$

(b) $\mu_{\bar{X}} = 3.5$ $\sigma_{\bar{X}}^2 = \frac{35}{24} = 1.46$
(c) $\mu = 3.5,$ $\sigma^2 = \frac{35}{12} = 2.92$
(d)

(e) True
(f) True

6-5 (a) .3085
(b) .1587
(c) .0228
(d) .00003

6-7 .048

6-9 .023

6-11 (a) $\mu = 1.10,$ $\sigma^2 = .69$
(b) .0039

6-13   Theoretical sampling
distribution of $\bar{X}$:

6-15   Theoretical sampling
distribution of $\bar{X}$:

$\mu_{\bar{X}} = 2.00$

6-19 .0359(n.c.c., .0183)

6-21 .0594(n.c.c., .0793)

6-25 (a) .0314
(b) .0020

6-27 True

6-29 μ = population mean
$\bar{X}$ = sample mean
σ^2 = population variance
$\sigma_{\bar{X}}^2$ = variance of the sample
mean

6-31 (a) .045
(b) 7912
(c) simple random sample

6-33 (a) $\mu = -\$.0526$
(b) .55,   .60,   .72
(c) binomial yields .549
(d) $n = 1955$

6-35   *equally* as large

7-1 (a) False. Interchange μ and $\bar{X}$
(b) True
(c) False. Replace last sentence
with "The difference is that
the sample mean is more
efficient, by about 57%. (They
are both unbiased)"

7-3 (a) both are unbiased
     (b) 90%

7-5 (a) True
     (b) Perhaps it would be wiser to spend $99,000 collecting data, and $1,100 analyzing it.

7-7 (a) $\mu = \pi + \dfrac{1 - 2\pi}{n + 2}$

$\sigma^2 = \dfrac{n\pi(1 - \pi)}{(n + 2)^2}$

MSE $= \dfrac{1 + (n - 4)\pi(1 - \pi)}{(n + 2)^2}$

MSE = var $= \pi(1 - \pi)/n$
     (c) Smaller (better) value is circled:

π	$MSE(P^*)$	$MSE(P)$
0, 1.0	.007	⓪
.1, .9	.011	⟨.009⟩
.2, .8	⟨.014⟩	.016
.3, .7	⟨.016⟩	.021
.4, .6	⟨.017⟩	.024
.5	⟨.017⟩	.025

     (d) P^* is preferred to P when the sample size is small, and π is not extreme (not too close to 0 or 1).

7-9 (b) $E(\overline{X}) = \mu = 4$
     $E(s^2) = \sigma^2 = 8/3$
     $E(MSD) = \dfrac{1}{2}\sigma^2 = 4/3$

7-11 (b) On the average, $\overline{X}$ will have least error (least mean squared error) because it is most efficient for a *normal* population.

7-13 (a) True
     (b) True
     (c) False. If we *quadruple* the sample size, . . .

(d) False. The expected value of $\overline{X}$ is *always* exactly equal to the population mean, for random sampling.

7-15 (a) $\overline{X}, W_1, W_3$ are unbiased
     (b) The coefficients must sum to 1.
     (c) $\dfrac{1}{3}\sigma^2, \ \dfrac{3}{8}\sigma^2, \ \dfrac{3}{16}\sigma^2, \ .34\sigma^2$
     (d) $\overline{X}$
     (e) Here are the efficiencies relative to $\overline{X}$:
     for W_1, 8/9 = 89%
     for W_3, .333/.34 = 98%

7-17 (a) They really are different.
     (b),(c) The first estimator has bias σ^2/n. The second has bias σ^2.
     (d) no bias at all.

8-1 (a) $71.3 \pm .588 \simeq 71.3 \pm .6$
     NOTE: In these problems, we should round the confidence allowance (.6) to the same number of decimal places as the estimate (71.3). We give the unrounded value (.588) merely as a check on your arithmetic.
     (b) $71.3 \pm .774$

8-3 (a) $\$27.60 \pm \$7.44$
     (b) There is a 95% chance that the CI correctly brackets the true mean that you would get by taking a complete audit of all accounts.
     (c) Since $29.10 lies within $27.60 \pm \$7.44$, this CI turned out to be one of the usual cases that is correct.

8-5 $11.6 < \theta < 11.6/.05^{1/25}$
     i.e., $11.6 < \theta < 13.1$

8-7 (a) 54 ± 4.97
     (b) 54 ± 8.24

8-9 66.1 ± 3.75

8-11 (a) $.4 \pm 1.02$
(b) $.4 \pm .86$

8-13 (a) $\mu_W = 12.0 \pm 4.97$
(b) $\mu_B = 10.0 \pm 4.21$
(c) $\mu_W - \mu_B = 2.0 \pm 5.41$

8-15 (a) $\mu_M - \mu_W = 5.0 \pm 5.79$
(b) $\mu_M = 16.0 \pm 5.76$
(c) $\mu_W = 11.0 \pm 3.93$
(d) $\mu_M - \mu_W = 5.0 \pm 3.90$
    $\mu_M = 16.0 \pm 1.65$
    $\mu_W = 11.0 \pm 3.93$

8-17   With 95% confidence, B yields from 1.5 to 10.5 bushels more than A.

8-19 (a) $.01 < \pi < .57$ approximately
(b) $.06 < \pi < .41$ approximately
(c) $\pi = .20 \pm .0157$

8-21 (a) $\pi_1 - \pi_2 = -.19 \pm .101$
(b) $\pi_1 - \pi_2 = -.19 \pm .132$

8-23 (a) More people are making up their minds. (Among these people, there is also a trend for more to be opposed, but this trend is not well established: Between weeks 1 and 2, the increase may be explained by the rewording of the question; between weeks 2 and 3, the increase is overwhelmed by the confidence allowance.)
(b) We agree that "more and more people are making up their minds on the issue." (We cannot judge the rest of the claim on the basis of the data.) This may be supported by confidence intervals that show a real decrease in "no opinion" in week 2 and again in week 3.

8-25   The approximate CI is 25% too wide. As long as

$.2 \leq \pi \leq .8$, therefore, (8-42) is no more than 25% too wide.

8-27 (a) $\pi_W - \pi_E = .070 \pm .081$ where π_W (or π_E) represents the population proportion who would pass the West (or East) Coast Board.
(b) To ensure that (a) measures only the difference in standards of the two boards, extraneous factors should be balanced out. Consider, for example, the practice effect: half the students should take their first interview with board E, the other half with board W.

8-29 (a) $\mu > 1216$
(b) Yes, since the old mean was only 1200.

8-31 $73 < \sigma^2 < 232$

8-33 (a) False. *Less than 95% . . .*
(b) True. Best interpretation.
(c) False. *Probably, more than 95% . . .*
(d) False. *Almost none of . . .*

8-35 (b) (i) 95%
(ii) More than 5%, of course. It cannot be exactly calculated from the techniques in this text. However, from the class relative frequency P a 95% confidence can be constructed: $\pi = P \pm 1.96\sqrt{P(1 - P)/n}$. Very large sample sizes (say 300 instead of 3) would bring this probability near its ideal value of 1.00. It happens to be called the *power* of the confidence interval or test, as described in Table 9-1.

8-37 (a) $.04 \pm .019$
(b) $.10 \pm .024$
(c) $-.06 \pm .031$

8-39   Of those with an opinion in the population, let π be the proportion in favour of N. Then:
$$\pi = .595 \pm 1.96(.01084)$$
$$= .595 \pm .0212$$
Since π definitely exceeds 50%, we conclude that the new box is superior.

8-41   Let π_A and π_B represent the population pregnancy rates after and before. Then
$$\Delta\pi = \pi_A - \pi_B$$
$$= -.028 \pm 1.96(.00944)$$
$$= -.028 \pm .019$$
Thus $\dfrac{\Delta\pi}{\pi} = \dfrac{-.028 \pm .019}{.142}$
$$= -.20 \pm .13$$
That is, with 95% confidence there is a relative decline of $20\% \pm 13\%$ in the pregnancy rate.

8-43 $\Delta\pi = .16 \pm 1.96(.00737)$
$$= .16 \pm .0144$$
That is, the conviction rate of the presiding judge is higher than that of the jury, by 16 ± 1 percentage points.

9-1

Prob-lem	H_0	CI	Discernible?
8-39	$\pi = .50$	.60 $\pm .02$	yes
8-40	$\mu_A - \mu_B = 0$	$-.10$ $\pm .15$	no
8-41	$\pi_A - \pi_B = 0$	$-.028$ $\pm .019$	yes
8-42	$\pi_{68} - \pi_{63} = 0$	$-.45$ $\pm .09$	yes
8-43	$\pi_1 - \pi_2 = 0$	.16 $\pm .01$	yes
8-44	$\mu_A - \mu_B = 0$	.15 $\pm .07$	yes

9-3   Figure for Problem 8-39:

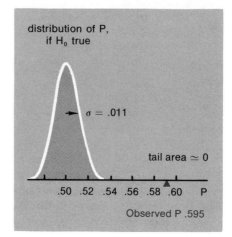

distribution of P, if H_0 true

$\sigma = .011$

tail area $\simeq 0$

.50 .52 .54 .56 .58 .60   P

Observed P .595

Prob-lem	H_0	H_1	z or t	Approx. prob-value
8-39	$\pi = .50$	$\pi > .50$	8.6	.000
8-40	$\mu_A - \mu_B = 0$	$\mu_A - \mu_B < 0$	-1.3	.09
8-41	$\pi_A - \pi_B = 0$	$\pi_A - \pi_B < 0$	-3.0	.002
8-42	$\pi_{68} - \pi_{63} = 0$	$\pi_{68} - \pi_{63} < 0$	-10.2	.000
8-43	$\pi_1 - \pi_2 = 0$	$\pi_1 - \pi_2 > 0$	21.7	.000
8-44	$\pi_A - \pi_B = 0$	$\pi_A - \pi_B > 0$	4.2	.000

9-5   (a) $\Pr(\overline{X}_T = \overline{X}_C) = 0$, *because* $\overline{X}_T$ *and* $\overline{X}_C$ *are continuous.*

(b) The conditional probability of the *data, given* H_0, is .013 — roughly speaking. (A more precise statement is given in (d)).

(c) The probability of H_0 is *impossible to determine from the data alone.*

(d) correct

(e) If we repeated the experiment and *if* H_0 *were true*, the probability is .987 of getting a t value *smaller than* the one we observed. (This is now the correct complement of (d)).

9-7 $\alpha = .264$ (But using normal approximation, $\alpha = .30$, or $\alpha = .15$ n.c.c.)

9-9   I, α. II, β. α, β.

9-11 (b) $\overline{X}_c = 338.6$
(c) No, do not reject H_0
(e) Do not reject H_0

9-13 (a) $\pi_M - \pi_W \geq .0487 \simeq 5\%$. That is, men outsmoke women by at least 5 percentage points.
(b) $.0002$ $(z = 3.57)$
(c) Is discernible.

9-15 (a) True
(b) True

9-17 $.1493 \simeq 15\%$

9-19 (a) $P_C = .228$
(b) $\alpha = 5\%$, $\beta \simeq 30\%$
(c) Accept H_0. This decision is unfortunate because it ignores extra-statistical information.
(d) β decreases, to about 20%.

9-23 (a) Measuring incomes in thousands of dollars,

H_0	t	prob-value	reject H_0?
30	5	$p \ll .001$	yes
35	1.88	$.05 < p < .10$	no
37.5	.31	$p > .50$	no

(Reject H_0 if $p < .05$, or $|t| > 2.064$)

(b) $34.70 < \mu < 41.30$
(c)

H_0	t	prob-value	reject H_0?
30	5	$p \ll .0005$	yes
35	1.88	$.025 < p < .05$	yes
37.5	.31	$p > .25$	no

(Reject H_0 if $p < .05$, or $t > 1.711$)

(d) $35.26 < \mu$

9-25 (a) Since $t = 1.99$, $.025 < p < .05$. Discernible.
(b) Since $|t| = 1.99$, $.05 < p < .10$. Not discernible.
(c) Answer (b) is consistent with 2-sided CI.
(d) Answer (a) is consistent with one-sided H_1, which we feel is appropriate.

9-27 (a) .625
(b) The prob-value for H_0 is practically zero $(z = 46)$.
(c) No, because H_0 is close enough to the truth.
(d) Births are independent, and the parents "stopping rule" doesn't depend on the sex split.

9-29 (a) The one-sided CI would shave closer on one side, at the cost of being completely vague on the other side.
(b) $\mu_G - \mu_P > \$570$
(c) Claims 1 and 2.
(d) $H_0: \mu_G - \mu_P = 0$ (or 500, etc.) $H_1: \mu_G - \mu_P > 0$ (or 500, etc.)

9-31   Assuming a one-sided test, we found the prob-value nearly 5%. Thus
(a) False . . . statistically *discernible*
(b) False. If we imagine that many polls *(of size n = 1500)* were taken of a population *where* H_0 *was true* $(\pi_{75} = \pi_{72} = 42\%)$, in about 5% of such polls we would find the percentage of fearful people *increasing* by 3 percentage points or more.
(c) False. The probability that the population opinion is unchanged is *not obtainable* from the data alone.

10-1 (b) To the extent that the class is large, the frequency distribution of F approximates the probability distribution in Figure 10-2, and approximately 5% of the F values exceed 3.89.

(c) Since the values of F now tend to be much larger, the frequency distribution of F is further to the right than in Figure 10-2. Consequently, many more than 5% of the values will exceed 3.89 in the long run.

10-3

source	variation	df
between sexes	62.5	1
residual	126	8
total	188.5 √	9 √

variance	F ratio	prob-value
62.5 15.75	3.97	$.05 < p < .10$

10-5

source	variation	df
between regions	8	3
residual	114	7
total	122 √	10 √

variance	F ratio	prob-value
2.67 16.3	.16	$p \gg .25$

10-7 (b) If the class were large, less than 95% (but more than $1.00 - 3(.05) = 85\%$) would succeed in having all three CI correct.

(c) If the class were large, at least 95% would succeed in having all three CI correct. (There is 95% confidence in *all contrasts;* so for the simple differences, there would be more than 95% confidence).

10-9 (a) -5.5 ± 3.43
(b) -5.5 ± 4.39

10-11 (a) -5.0 ± 8.2
(b) -5.0 ± 12.1

10-13 (a) Yes.
(b) True

10-15 (a)

source	variation	df
between hours	18	2
between men	78	2
residual	10	4
total	106 √	8 √

variance	F ratio	prob-value
9	3.6	$.10 < p < .25$
39 2.5	15.6	$.01 < p < .05$*

(b) differences in men, allow ± 4.81 for 95% confidence

i \ I	1	2	3
1		5*	7*
2	-5*		2
3	-7*	-2	

10-17 (a) Any hypothesis in the interval $150 < \mu < 190$ is called *acceptable at the 5% level,* while any hypothesis outside this interval is called *rejected.*

(b) The *sample* mean is a random variable with expectation μ (unknown) and standard deviation $\sigma/\sqrt{n}$ (20 represents $t_{.025}\, s/\sqrt{n}$).

(c) . . . would cover *the true but unknown mean* μ.

(d) See (b) for a correction.

(e) True.

10-19 $\Delta\pi = -.110 \pm .074$ i.e., approval dropped 11 ± 7 percentage points.

10-21 (a) differences in classes, allow ± 18.0 for 95% confidence

i \ I	1	2	3	4
1		1	21*	−10
2	−1		20*	−11
3	−21*	−20*		−31*
4	10	11	31*	

(b) 15.0 ± 8.68

(c) No. For statistical proof, it would require the students to be initially assigned to the 4 sections at random, then have a common exam, etc.

10-23 (a) $\pi_M - \pi_W = .097 \pm .018$ We therefore conclude that men have an overall admission rate that is higher than women's. But this cannot be called bias, in view of (c).

(b) *Humanities*

$$\pi_M - \pi_W = .023 \pm .024$$

Science

$$\pi_M - \pi_W = -.045 \pm .032$$

We therefore conclude that within faculty, the admission rates are only slightly different. The only discernible difference is in Science, where men have an admission rate slightly *lower* than women's. Is the Science faculty biased against men? We really cannot say, without a lot more information.

(c) True.

10-25 (a) bias of 18 percentage points towards R.

(b) Yes

11-1 (a) $Y = 80 + .070x$ or $= 59 + .070X$

(c) $a = 80$ is the estimated yield when the average amount (300 lb) of fertilizer is applied, while $a_o = 59$ is the estimated yield when no fertilizer is applied.

(d) MPP $= .070$ bushels per lb. of fertilizer.

(e) MRP $= \$.14$ per lb. of fertilizer. So it is economic to apply.

11-3 $C = .396 + .856X$ Slope is .856, which equals $1 - .144$, that is, the consumption coefficient is the complement of the savings coefficient.

11-5 (c) yes, the new equation is indeed $y = bx$

12-1 (f) The mean of $\hat{\beta}$ is about .10, and the standard error is about .03.

12-3 (a) var $\hat{\beta} = .0000125\ \sigma^2$ var $\tilde{\beta} = .0000056\ \sigma^2$ is less

12-5 (a)

Problem	*95% CI*
11-1	$.070 \pm .101$
11-2	$.144 \pm .148$
11-3	$.856 \pm .148$
11-4	$.50 \pm 2.64$

(b) $\alpha = .76 \pm .28$

12-7   only $\beta = .50$ can be rejected

12-9   (a) $\beta > .035$
   (b) $t = 3.11$, $.025 < p < .05$
   (c) Yes, reject H_0

12-11 The intervals are narrower:
   (a) $.47 \pm .41$
   (b) $.76 \pm .28$
   (c) $1.05 \pm .41$
   (d) $1.34 \pm .66$
   (e) Again, (b) is most precise,
      (d) is least precise because it
      is furthest from the center of
      the data.
   (f) Exactly the same as the
      confidence interval for α.

12-13 (b)   (i) 50% as large
      (ii) 25% as large
      (iii) 71% as large

12-15 (a) $Y = 140 + 1.27x$
      $= 0 + 1.27X$
   (c) $\mu_Y = 229 \pm 44$
   (d) $\mu_Y = 140 \pm 117$
   (e) centered better, and narrower

12-17 (a) Yes, model is linear
   (b) var $(Y/X) = 5.56$, 12.5,
      5.56
      Thus variance is not constant

13-1   (a) $S = .76 + .115x - .0294w$
      $= .105 + .115X - .0294W$
   (b) multiple regression (.115) is
      better, because it shows the
      relation of S to X if W *were
      constant.*
   (c) $880
   (d) $230
   (e) $30

13-3   (a) $\hat{\alpha} = \bar{S}$

$$\sum Sx = \hat{\beta}\sum x^2 + \hat{\gamma}\sum wx$$

$$+ \hat{\psi}\sum nx$$

$$\sum Sw = \hat{\beta}\sum xw + \hat{\gamma}\sum w^2$$

$$+ \hat{\psi}\sum nw$$

$$\sum Sn = \hat{\beta}\sum xn + \hat{\gamma}\sum wn$$

$$+ \hat{\psi}\sum n^2$$

   (b) $S = .76 + .105x - .0242w$
      $- .0381n$
      $= .252 + .105X - .0242W$
      $- .0381N$

13-5   (a) Rounding appropriately,
      2.7   2.2   3.6   8.2   11.2   .14
   ± 170 ± 16 ± 50 ± 120 ± 30 ± 730
   (b)   (i) False. ". . . would be
      centered around the true *but
      unknown* value.
      Therefore . . ."
      (ii) Controversial. We would
      only say, "There is practically
      no evidence that T is
      (positively) related to S."
      (iii) Controversial and false.
      We would say "Our prior
      expectation that Y affects S is
      strongly confirmed by the
      data. So we must reject the
      null hypothesis."
   (c) The most important
      determinants of a professor's
      salary, with the clearest
      evidence, are first of all, years
      of experience Y and number
      of Ph.D.'s supervised D, then
      books B, and excellent articles
      published E; way down on
      the list are ordinary articles A
      and teaching evaluation T.

13-7   Spring temperature, soil fertility,
   etc.

13-9   (a) Suppose US prices were
      completely independent of
      every one of the 100 variables.
      Since $\alpha = 5\%$, one could

expect "false alarms" 5 times in a 100 — and these may be the very 5 variables that he found "statistically discernible."

(b) His conclusions are still too bold. Several of the 20 discernible variables may be false alarms. To keep the overall error rate down to 5%, he could cut the individual error rate down to 5%/100 = .05%. But it is better to use *economic analysis* to narrow down the number of regressors, and better to use *multiple* regression.

13-11 (a) (i) False. Other things being equal, we estimate that a professor earns $230 more annually for *each* book he writes, on average.
(ii) False. . . . of writing *each book.*
(iii) False. The average salary increase would average *much more* than $190. There would be further increases because of increased performance (writing articles and books, etc.). There would also be important increases because of a generally rising wage level (which would increase starting salaries, too).

(b) Other things being equal, we *estimate* a professor annually earns on average:
(i) $18 more for each ordinary article he has written
(ii) $100 more for each excellent article he has written
(iii) $490 more for each Ph.D. he has supervised
(iv) $50 more for having a teaching evaluation in the top half (but $50 is a *very* unreliable estimate)

13-13 There likely is bias. For example, to the extent that smokers are careless of their health in other ways, the 5-year figure is biased to be too high.

13-15 (a) Yes, a positive bias.
(b) No bias.
(c) False. . . . will *not* introduce bias . . . . *But* s^2 *will be unnecessarily large.*

13-17 (a) 10.1
(b) larger bias, because $\hat{\beta}_{wy}$ is larger.
(c) negative bias, because $\hat{\beta}_{wy}$

13-19 (a), (b)

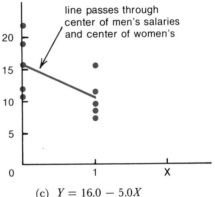

line passes through center of men's salaries and center of women's

(c) $Y = 16.0 - 5.0X$
(d) $\beta = -5.0 \pm 5.8$. Women earn on average $5,000 ($\pm$$5,800) less than men.
(e) Same
(f) No. To measure discrimination, men and women *with the same qualifications* should be compared.

13-21 We disagree. The $2400 figure may be due to discrimination, but it also may be partly due to some more subtle difference between men and women not measured in this study.

13-23 (a) $E = 13.625 + .000583(F\text{-}8000)$
$- .917(D\text{-}.5)$
where $D = 0$ for urban,
$D = 1$ for rural.

13-25 (a) False. . . . income that the
average *woman* earns more
than the average *man of the
same age and education.*
(b) True.

13-27 (a) For the AIRCAP regression,
all variables are discernible
except PASMOK.
For the BRONC regression,
the discernible variables are
AGE, PRSMOK, PASMOK,
CHEMW.
(b) 3605cc, 10.7%
(c) age, height, smoking habits,
350cc, 6.5 percentage points
(d) age, height, smoking habits,
170cc, 9.7 percentage points
(e) 39cc, .21 percentage points
(f) 180cc, 9.4 percentage points
(g) 4.6 years, important variables
being omitted from the
regression

14-1 (a) $r = .62$
(b) $-.47 < \rho < .95$
(c) H_0 *cannot* be rejected.

14-3 (a) $Y = 9 + .35x$
(b) $t = 2.18$, $.05 < p < .10$
(c) $t = 2.18$, $.10 < p < .20$
(d)

source	variation	d.f.
explained	12.25	1
unexplained	7.75	3
total	20.0	4

variance	F	prob-value
12.25	4.75	$.10 < p < .25$
2.58		

61% explained,
39% unexplained
$.10 < p < .25$

(e) Since $r = .78$,
$-.25 < \rho < .97$
Hence $\rho = 0$ is acceptable.
(f) Yes. In every case, the null
hypothesis cannot be rejected.
(g) $Y = 7.95$
$Y = 10.75$
$X = 16.75$

14-5 (a) True
(b) True
(c) True
(d) False. $\hat{\beta}_* = \dfrac{r^2}{\hat{\beta}}$

14-7 (a) $R = .63$
(b) Yes, $R = r$
(c) To be precise, $R = |r|$

14-9 $F = 28$,
$.01 < $ 2-sided $p < .05$ or,
$.005 < $ 1-sided $p < .025$

14-11 (a) 490
(b) requires n
(c) 77.8
(d) 67.2

14-13 (a) .404
(b) .262
(c) less. Usually true.

14-15 (a) Yes
(b) No, $\hat{\beta}' = 100\hat{\beta}$
(c) Yes
(d) Yes

15-1 (a) $Y = 55 - 28.2Q + 5.43Q^2$
(c) 18.3

15-3 (b) $P(T) = 968e^{.00549(T-1800)}$
$P(2000) = 2900$, which is
ridiculous, because growth
rate has *not* been constant. It
is rapidly increasing recently.
(c) 0.549% per year, very small

15-5 (a), (b), (e) Use multiple
regression
(c), (d) Take logs, then use
regression

15-7 (a) 1.30
      (b) 3.9%
      (c) 7.7%

15-9 (a) 2.2%
      (b) 1,022
      (c) 5,220

15-11 (a) 4.6%
      (b) 17%
      (c) 0.1%
      (d) -14% (i.e., 14% *less*)

15-13 (a) $\hat{W} = 130 + 1.00(A\text{-}40)$
      (b) $W = 130 \pm 46.2$
      (c) $\beta = 1.00 \pm 1.92$

15-15 (a) $\hat{A} = 4.50 - .93(E\text{-}10)$
      (b) $\beta = -.93 \pm .77$
      (c) $t = -3.38,\quad .02 < p < .05$

15-17 (a) $E = 7.0 - 5.0C$
          where $C = 0$ (or 1) for
          middle (or upper) class
          $\beta = -5.0 \pm 4.24$
      (b) $\mu_U - \mu_M = -5.0 \pm 4.24$

15-19 Only (c) is true

15-21 (a) False. One *advantage* of
          multiple regression is that, *by
          using dummy variables,* it *can*
          include factors that are
          categorical.
      (b) True
      (c) False. Multicollinearity of X
          and Y is *completely avoided*
          when $r_{XZ} = 0$ . . .
      (d) True

15-23 (a) True, except for one small
          correction: Upon dividing
          each variation by $n - 2$ *and*
          $n - 1$ *respectively,* we obtain
          variances:
          $$r^2 = 1 - \left(\frac{n-2}{n-1}\right)\frac{s^2}{s_Y^2} \simeq 1 - \frac{s^2}{s_Y^2}$$
      (b) True
      (c) True

15-25 (a) 95% CI for men's excess sleep:
          $\mu_M - \mu_W = -1.00 \pm 1.78$
      (b) A multiple regression of sleep
          on sex *and* age can roughly be
          carried out graphically,
          obtaining $Y \simeq$ constant
          $- .11$ age $+ 1\ S$, where
          $S = 0$ (or 1) for women (or
          men). Thus, men sleep about
          one hour *more* than women of
          the same age (not one hour
          less, as in (a)).
      (c) Multiple regression in (b) is
          better, since it gives the
          difference between men and
          women *of the same age.*

15-27 Yes, for the same reasons as in
      regression.

15-29 Plan (b) is better than (a).
      Plan (c) is cruder than (a).
      Plan (e) is better than (d).
      But it should use (8-37) instead
      of (8-32).
      Although (d) and (e) are free of
      bias, they are unnecessarily rigid.
      Let the "treated" volunteers take
      whatever dose they want.

16-1 (a) .0547
      (b) .1719
      (c) .0547 (For H_0: median
          difference $= 0$)

16-3 .0082(n.c.c., .0047)

16-5 (a) $67 \leqslant \nu \leqslant 76$
      (b) $63 \leqslant \nu \leqslant 69$
      (c) $\ 3 \leqslant \nu \leqslant 7$

16-7 (a) yes, approximately
      (b) no, wider
      (c) $67.8 \leqslant \mu \leqslant 74.2$

16-9 (a) 99.6%
      (b) 96.1%
      (c) 82.0%

16-11 (a) .025

(b) .033   Since this is much closer to the correct value of .032, the c.c. seems worthwhile.

16-13 (a) $z = -2.41$, $p = .0080$

(b) $z = -5.16$, $p \ll 10^{-6}$

(c) Method (a) is easier to calculate and is valid even for non-normal populations. Method (b) is more efficient for normal populations.

16-15 (a) True. In fact, Table 16-6 shows that the W test has 95% efficiency relative to t in the normal case (where t works best).

(b) True. Problem 16-14 illustrates this.

16-17 (a) $Y = 3600 + 272(X - 6.5)$
$r = .73$

(b) $Y = 3500 + 525(X - 6)$
$r = .75$

(c) rank correlation $= .84$

17-1 (a) $\chi^2 = 8.8$, $.010 < p < .025$

(b) yes, we can reject H_0

17-3 (b) 25%

17-5 With no continuity correction:

(a) Reject when proportion of aces $P \geq .212$

(b) Power $= .816$

17-9 (a) $\chi^2 = 18.4$, $p < .001$

(b) In each occupation, construct a 95% CI for the proportion of nonwhites, using Figure 8-5.

17-11 (a) $\chi^2 = 33.6$, $p \ll .001$

(b) The percentage of males earning over $5000 exceeds the corresponding percentage of females by 37 ± 11 percentage points with 95% confidence.

17-13 $\chi^2 = 10.7$, $.025 < p < .05$

17-15 (a) A test of H_0 is inappropriate because births are bound to be dependent to some slight degree, etc.

(b) A test of H_0 may be appropriate. Though even here, it is not likely that the tosses will be *exactly* independent.

17-17 (a) $\chi^2 = 13.2$ with 2 d.f. Hence $.001 < p < .005$

(b) 95% CI for proportion rejected:
$\pi_A = .043 \pm .0115$
$\pi_B = .075 \pm .0183$
$\pi_C = .044 \pm .0090$

18-1 (a) $\hat{\pi} = .25$

(b) $\hat{\pi} = .50$

(c) $\hat{\pi} = .50$

(d) $\hat{\pi} = 0$

18-3 $\hat{\mu} = 19$

18-5 (a) $\hat{\mu} = \overline{X}$, unbiased

(b) $\hat{\sigma}^2 = \dfrac{\sum (X_i - \mu)^2}{n}$, unbiased.

19-1 (a) .66, .34

(b) False. Since the barometer sometimes predicts "rain" when it shines, a "rain" prediction is uncertain.

(c) It is a worse predictor when it shines.

19-3 (a) .10, .40, .50

(b) .28, .44, .28

19-7   (a) $N(69.4, 5.88)$
     (b) Posterior is between, but 16 times closer to the likelihood function.
     (c) .0022
     (d) $\mu = 69.41 \pm 4.75$ (Bayesian)
        $\mu = 70.00 \pm 4.90$ (Classical)

19-9   (a) $\mu = 22.0 \pm 3.10$
     (b) .0057

19-11 (a) 2.00 ± 1.26 (if $z_{.025}$ used, ± 1.05)
     (b) $1.68 \pm .79$ (if $z_{.025}$ used, $\pm .70$)
     (c) $1.57 \pm .60$ (if $z_{.025}$ used, $\pm .57$)

19-13 (a) $p(\pi) \propto \pi^5 (1 - \pi)^{30}$
     (b) Approximately $\sim N(141.3, 52.2)$
        $\mu = 141.3 \pm 15.0$
        (if $z_{.025}$ used, ± 14.2)
     (c) Approximately $\sim N(4.67, .0833)$
        $\beta = 4.67 \pm .59$
        (if $z_{.025}$ used, $\pm .57$)

20-1   (a) a_1 yields $L(a_1) = -3$
     (b) if "rain," a_3 yields
        $L(a_3) = -4$
        if "shine," a_1 yields
        $L(a_1) = -7$
     (c) False. Action a_1 is best when "rain" is predicted, and also when no prediction is possible. Action a_3 is best when "shine" is predicted. Action a_2 is never best.
     (d) 3.1

20-3   (a) $a_1 = 210$
        $a_2 = 220$
        $a_3 = 222$
     (b) a_2
     (c) a_3
     (d) a_1

20-5

Relative frequency density

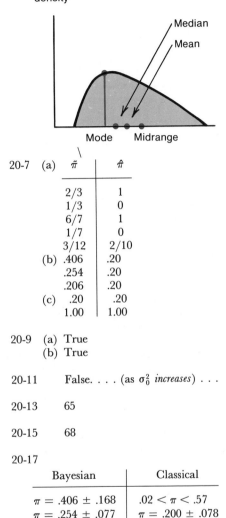

20-7   (a)

$\hat{\pi}$	$\hat{\pi}$
2/3	1
1/3	0
6/7	1
1/7	0
3/12	2/10

     (b)

.406	.20
.254	.20
.206	.20

     (c)

.20	.20
1.00	1.00

20-9   (a) True
     (b) True

20-11   False. . . . (as σ_0^2 *increases*) . . .

20-13   65

20-15   68

20-17

Bayesian	Classical
$\pi = .406 \pm .168$	$.02 < \pi < .57$
$\pi = .254 \pm .077$	$\pi = .200 \pm .078$
$\pi = .206 \pm .025$	$\pi = .200 \pm .025$

20-19 (a) accept H_0
     (b) accept H_1
     (c) accept H_0
     (d) accept H_1
     (e) accept H_1

(f) accept H_1
(accept H_0 iff $p(X/S_0)$
$> 6p(X/S_1)$, i.e., iff $X < 33.5$)

20-21

	20-18	20-19
(i)	accept H_1	accept H_0 iff $X < 33.5$
(ii)	doesn't matter	accept H_0 iff $X < 37$

	20-20(c)	20-20(d)
(i)	36.07	37.21
(ii)	37.5	37.5

20-23 (a) $c \simeq 6$
(b) $c \simeq 5$
(c) always accept H_0

20-25 (a) 103.54, $\alpha = .33$, $\beta = .020$
(b) 113.12, $\alpha = .05$, $\beta = .195$
Average loss increases by a factor of 3.16
(c) $r_0/r_1 = 4/1$, which is unreasonable.

20-27 (a) True
(b) False. As sample size increases, the difference between the classical estimate and the Bayesian estimate *approaches zero.*
(c) False. . . . The cost of this arbitrariness *may be a substantial loss of utility.*

20-29 12.114

20-31 (e) $p(S)$ is binomial, with $n = 4$ and $\pi = $ population (class) proportion.
(f) $E(\bar{\pi} - \pi)^2 = \text{MSE}(\bar{\pi}) = $ sum of last Bayesian column.
$E(\hat{\pi} - \pi)^2 = \text{MSE}(\hat{\pi}) = $ sum of last MLE column. The smaller MSE determines the better estimator.

21-3 (a) True
(b) True
(c) False. Interchange "tracking residual" and "seasonal pattern."
(d) False. *Any time series* can be broken down by cyclical analysis.

21-5 (a)(b)(c) True, if the residual is well-behaved.

21-7 (a) spectrum peaks near $f = 0$
(b) spectrum peaks at $f = 1/4$ and $f = 1/2$ cycle/quarter (i.e., 1 and 2 cycles/year), also near $f = 0$
(c) spectrum has no peaks—all frequencies are present to the same degree.

22-1 (a)
$$\begin{array}{ccc} & F & \\ \swarrow & & \searrow \\ C & \leftrightharpoons & GM \end{array}$$
(b)
$$\begin{array}{cc} & I \\ & \swarrow \\ X & \leftrightharpoons Y \end{array}$$

22-3 (a) .67
(b) .77, inconsistent.

22-5 (a) $Z \rightarrow X \leftrightharpoons Y$
(b) One possibility is (c)
(c) Z as IV on (22-18) yields $\hat{\beta} = s_{YZ}/s_{XZ}$
(d) No, (22-19) cannot be consistently estimated, because there is only one IV(Z), but two unknowns (γ and ψ)

23-1 (a) 120.8
(b) 120, 125, 117, 100. Yes, $100 \leq \text{LPI} \leq 125$
(c) Yes, PPI = 120, FPI = 120.4
(d) Yes, same order as (23-12): LPI $\geq$ FPI $\geq$ PPI
(e) Yes, same order as (23-12): LQI $\geq$ FQI $\geq$ PQI i.e., $88 \geq 87.7 \geq 87.4$
(f) Cost index = 105.6. Only Fisher's index satisfies (23-11): $(120.4)(87.7) = 105.6$

(g) Price indexes would be unchanged while quantity indexes would be inflated by population growth.

(h) always (b) is true.

23-3 (a) The key is in the denominator of (23-2), $\sum P_0 Q_t$. Whatever goods used to have high prices P_0 and hence were bought in small quantities, are nevertheless weighted with the new relatively high quantities Q_t, making the product $P_0 Q_t$ too high.

(b) The key is in the denominator of (23-6), $\sum Q_0 P_t$. Whatever goods used to be bought in large quantities Q_0 because prices were low, are nevertheless weighted with the new relatively high prices P_t, making the product $Q_0 P_t$ too high.

23-7 (a) Paasche
(b) Laspeyres
(c) Fisher's ideal

24-1 (a) $\mu = 21$
(b) (i)   3.04
(ii)   .88
(iii) .86

24-5 (a) $\hat{\mu} = 195.6$, $\hat{T} = 3,912,000$
(b) if $n = 1000$ still, then $n_i = 164,513,277,46$
(c) 2.41, 2.37, 6.80
(d) 282%, 287%
(e) $\mu = 195.6 \pm 3.1$
$\mu = \hat{\mu} \pm 3.0$
$\mu = \bar{X} \pm 5.1$
$T = 3,912,000 \pm 62,000$
$T = 20,000 \, \hat{\mu} \pm 60,000$
$T = 20,000 \, \hat{\mu} \pm 102,000$
(f) True

25-1 Denote by a_{ij} the payoff corresponding to A choosing strategy i and B choosing strategy j. Then:

	(ii) minimax value	(i) maximin value
(a)	$a_{21} = 8$	$a_{22} = 6$
(b)	$a_{22} = a_{13} = 25$	$a_{12} = a_{21} = 10$
(c)	$a_{23} = 1$	$a_{23} = 1$
(d)	$a_{32} = 6$	$a_{32} = 6$

In parts (c) and (d), it doesn't matter who goes first—because the minimax and maximin values and strategies are the same.

25-3 (a)
$$\begin{array}{rrr} -2 & -1 & 1 \\ 1 & 0 & -1 \\ -1 & 1 & 2 \end{array}$$

(b) A achieves minimax value of $+1$
(c) B achieves maximin value of -1
(d) Try it and see

25-5 (a) Intuitively, it is hard to see who has any advantage.
(b) A's spinner has probabilities $(.375, .625)$ for (H, T). It yields a maximin value of -1.25 (a loss for A).
(c) B's spinner has probabilities $(.625, .375)$ for (H, T). It yields a minimax value of -1.25 (a gain for B).
(d) No difference; no difference; again no difference.

25-8 (a)
$$\begin{array}{rr} 0 & 90 \\ 0 & 80 \end{array}$$

(b) Yes
(c) Yes. If B believes him, A does even better than by paying the agent as in (b).

(d) If B and A both insist on strategy one, it becomes the game of "chicken" with nobody chickening out. B will be better off only if he can make A back off to strategy 2.

25-10 (a) Row 2 is dominated.
Expected gain for A is 4.5
(b) Expected gain for A is 2.75
(c) Row 2 is dominated, then column 3, finally row 1.
Expected gain for A is 4.875

INDEX